DIFFERENTIAL EVOLUTION

Fundamentals and Applications in Electrical Engineering

DIFFERENTIAL EVOLUTION

Fundamentals and Applications in Electrical Engineering

Anyong Qing

IEEE
IEEE PRESS

John Wiley & Sons (Asia) Pte Ltd

Visit our Home Page on www.wiley.com

Other Wiley Editorial Offices

John Wiley & Sons, Ltd, The Atrium, Southern Gate, Chichester, West Sussex, PO19 8SQ, UK

John Wiley & Sons Inc., 111 River Street, Hoboken, NJ 07030, USA

Jossey-Bass, 989 Market Street, San Francisco, CA 94103-1741, USA

Wiley-VCH Verlag GmbH, Boschstrasse 12, D-69469 Weinheim, Germany

John Wiley & Sons Australia Ltd, 42 McDougall Street, Milton, Queensland 4064, Australia

John Wiley & Sons Canada Ltd, 5353 Dundas Street West, Suite 400, Toronto, ONT, M9B 6H8, Canada

Wiley also publishes its books in a variety of electronic formats. Some content that appears in print may not be available in electronic books.

Library of Congress Cataloging-in-Publication Data

Qing, Anyong.
Differential evolution / Anyong Qing.
p. cm.
Includes bibliographical references and index.
ISBN 978-0-470-82392-7 (cloth)
1. Evolution equations. I. Title.
QA377.3.Q22 2009
621.301'5196–dc22

2009015177

ISBN 978-0-470-82392-7 (HB)

Typeset in 10/12pt Times by Thomson Digital, Noida, India.
Printed and bound in Singapore by Markono Print Media Pte Ltd, Singapore.
This book is printed on acid-free paper responsibly manufactured from sustainable forestry in which at least two trees are planted for each one used for paper production.

Contents

Preface

1 Where Did the Idea of a Parametric Study on Differential Evolution Come From?

I got to know the subject of differential evolution in 2000 through the introduction of Prof. K.A. Michalski of Texas A & M University. At that time, I was still painfully debating with a Taiwanese professor over his dubious paper, which has been widely regarded as one of the pioneering works on the application of genetic algorithms in electromagnetic inverse scattering. Differential evolution was so amazing that I almost immediately implemented it in my study of electromagnetic inverse scattering, a topic I have been working on since 1994. The more I use it, the more I love it.

It is embarrassing to have to admit that before 2004 I was simply implementing the strategy and the intrinsic control parameter values described in Prof. Michalski's publications. However, in 2004, from limited collected publications, I began to notice inconsistencies in reported intrinsic control parameter values and the sensitivity the efficiency of differential evolution to intrinsic control parameters.

At the same time, solving application problems at my hand using differential evolution was a time-consuming, unreliable, and frustrating business. I was concerned with application problems such as electromagnetic inverse scattering, electromagnetic composite materials, and electromagnetic structures. I dreamt that I could have optimal intrinsic control parameter values for differential evolution so as to find the solutions for those application problems more robustly and efficiently.

Confused by the inconsistent intrinsic control parameter values reported, I decided to conduct my own search for optimal intrinsic control parameter values by means of sweeping. The sweep ranges were either natural, or chosen so as not to miss any representative scenarios. Thus arose the first idea for a parametric study on differential evolution.

It was computationally impractical to employ trial and error to look for optimal intrinsic control parameter values for the aforementioned application problems directly. I assumed that differential evolution would behave at least similarly if two or more problems had identical mathematical features. Fortunately, this hypothesis has been justified by the parametric study results presented in this book.

Hence, I decided to carry out the numerical experiment on toy functions. The non-uniqueness of solutions gave cause for concern since the target application problems are non-unique in essence. Accordingly, I proposed two toy functions, the translated sphere function and what has come to be known as the Qing function.

In 2006, I discovered dynamic differential evolution, or, more precisely, DDE/best/1/bin. I was very pleased to find that DDE/best/1/bin significantly outperforms CDE/best/1/bin. I submitted a paper to a reputable journal to report this finding. It was rejected. Two of the most critical comments were the small test bed and the disadvantage of using the best differential mutation base rather than a random differential mutation base.

Inspired by these comments, I did two literature surveys simultaneously, one on differential evolution, and the other on test beds for single-objective optimization. I also decided to expand the initial parametric study on differential evolution to cover the evolution mechanism and differential mutation base.

The results of the literature survey came as a great shock. It is hard to believe that there are so many false claims and dubious practices in the evolutionary computation community.

Through the literature surveys, I learnt that too much value was placed on crossover for genetic algorithms and that people believe that crossover is insignificant for differential evolution. However, I decided to challenge this consensus by further expanding the parametric study to cover crossover because the findings of the literature surveys and the preliminary parametric study results filled me with confidence that the study of crossover for differential evolution is equally valuable. Today, this study has been fully justified by the fruitful findings of the comprehensive parametric study.

2 Why Did I Write this Book?

As mentioned above, my journal paper reporting my findings on DDE/best/1/bin was rejected at first submission. I agree with the reviewers' comments that the implemented test bed is small, while disagreeing that best differential mutation base is at a disadvantage compared to random differential mutation base. I spent almost a year conducting the aforementioned literature surveys and the expanded parametric study on differential evolution. The findings were very promising. Therefore, I rewrote the paper and resubmitted it to the same journal. To my great surprise it was rejected again, even though the reviewers' comments were in general quite positive. Ridiculously, I was told that my paper implemented too big a test bed and advised to commercialize my work.

I well know that I am not a good businessman. I am more interested in academic study. Therefore, with the help of Prof. C.K. Lee of Nanyang Technological University, Singapore, I approached academic publishers with a view to publishing the parametric study results in book form. John Wiley & Sons and CRC promptly responded positively to my book proposal. On account of the coverage of the book and my electrical engineering background, I chose John Wiley & Sons as my publisher.

I was invited to submit a book proposal to another publisher. Being conservative, I accepted their invitation. Having submitted a book proposal, I am greatly surprised that I have not received any response. I do not mind rejection based on negative reviewers' comments, be they reasonable or unreasonable. But I am very angry at the lack of respect for commonly accepted rules and behavior.

3 What is Academic Novelty?

I learnt in June 2008 that it is now the policy of the journal which rejected my submissions not to publish any paper which does not present a new algorithm. What a stupid policy. Unfortunately, it is not rare. A similar policy is operated in many other fields, such as antennas. To pursue novelty, some researchers frequently change their research topic. I know of one professor who boasts that he changes his area of research every five years.

New algorithms are not and should not be the main aim of academic study. Inventing a new algorithm is not difficult. What is more difficult is to come up with a new algorithm that outperforms existing algorithms. I tried many new algorithms between 2004 and 2006 before I finally discovered DDE/best/1/bin. Such algorithms have been claimed by other researchers as new. However, I know from my simulation experience that they are not as good as their inventors have claimed and that evolutionary crimes have been committed in making such false claims. Most of them are worse, and any claims of improvements are based on unfair comparisons.

Many researchers today are busy writing papers for the sake of reputation or material return, or both. They are always rushing to publish their new ideas and they usually have plenty of them, never mind that none of these so-called new ideas really works. I know of one researcher who aims to publish three papers every two weeks, and of another whose aim is to notch up 1000 papers during his career. I really cannot imagine how these papers are prepared and how much I can learn from them.

A direct consequence of the obsession with novelty is the ignorance of fundamentals and lack of serious research, leading to unforgivable false claims. Even though the above literature surveys are not exhaustive, the flood of publications on new differential evolution algorithms is obvious while very few of the reported new differential evolution algorithms are useful. At the same time, inappropriate practices or even mistakes in preceding publications are perpetuated by even established researchers.

4 What is in this Book?

This book covers the fundamentals of differential evolution, a parametric study on differential evolution, and applications of differential evolution in electrical and electronic engineering.

The first part of the book presents fundamentals of differential evolution. Chapter 1 gives a brief introduction to optimization, including its formulation, applications, essential ingredients, and optimization algorithms. Characteristic features of different objective functions are discussed in detail. Attention is focused on optimization algorithms, including deterministic optimization algorithms and stochastic algorithms. Evolutionary algorithms are introduced as a major category of stochastic optimization algorithms. Evolutionary crimes widespread in the evolutionary computation community are identified for the first time.

Fundamentals of differential evolution are introduced in Chapter 2. After a brief description of the critical historical milestones in the history of differential evolution, a detailed overview of applications of differential evolution is given based on a comprehensive literature survey. The basics of differential evolution are presented. Unique features and corresponding advantages over other evolutionary algorithms are highlighted.

Advances in differential evolution are reviewed in Chapter 3, organized according to their optimization parameters, evolution mechanism, number of populations, initialization mechanism, differential mutation strategy, crossover scheme, selection scheme, objective and constraint function evaluation. Multi-objective optimization using differential evolution, parametric study on intrinsic control parameters of differential evolution, and adaptation of intrinsic control parameters of differential evolution are also discussed.

Because of the evolutionary crimes committed by many differential evolution practitioners and the desperate need for optimal intrinsic control parameters, or empirical rules for choosing intrinsic control parameters, a comprehensive parametric study on differential evolution has been conducted. The second part of this book is focused on the parametric study. Findings from the parametric study and recommendations on differential evolution strategies, intrinsic control parameters, and non-intrinsic control parameters are presented.

A framework for a comprehensive parametric study on differential evolution is constructed in Chapter 4. Full consideration is given to differential evolution strategies, intrinsic control parameters, non-intrinsic control parameters, problem features, and interactions between them. Defects of previous approaches are highlighted. A limited comparative study involving the standard binary genetic algorithm, real-coded genetic algorithm, and particle swarm optimization is also presented to demonstrate the advantages of differential evolution.

Benchmarking a single-objective optimization test bed for the parametric study on differential evolution is discussed in Chapter 5. Through a massive literature survey, numerous test problems over different existing test beds have been collected. Approaches to generating new test problems have also been summarized. A tentative benchmark test bed containing 13 toy functions has been constructed. All member toy functions satisfy relevant admission criteria.

Numerical results of the parametric study on differential evolution are presented in Chapter 6. Solid evidence against the notorious misconceptions on differential evolution has been observed. Recommendations on choosing appropriate differential evolution strategies for future applications are accordingly presented.

The optimal values of intrinsic control parameters by which differential evolution performs best are summarized in Chapter 7. The relationship between optimal intrinsic control parameters, evolutionary operators, and problem features is also numerically disclosed here.

The relationship between problem features, differential evolution strategies, intrinsic control parameters, and non-intrinsic control parameters is disclosed in Chapter 8 by conducting further parametric study over alternative search spaces. It has been observed that differential evolution enjoys less success in wider search spaces. However, it is encouraging to note that there is only a marginal difference between the optimal intrinsic control parameter values, apart from the optimal population size for the Ackley function.

Nowadays electrical and electronic products are an indispensable part of daily life. However, although differential evolution has been applied to many engineering fields, its application in electrical and electronic engineering is still in its infancy. Its potential here has not been evident to most application engineers involved and accordingly has yet to be exploited. To boost the awareness of differential evolution in electrical and electronic engineering, differential evolution is applied to solve some representative electrical and electronic engineering problems. The third part of this book is dedicated to presenting a consistent introduction to applications of differential evolution to various electrical and electronic engineering problems.

Chapter 9 presents an introductory survey on applications of differential evolution in electrical and electronic engineering. Communication, computer engineering, control theory and engineering, electrical engineering, electromagnetics, electronics, magnetics, power engineering, signal and information processing are covered.

Finally, Chapters 10–16 are given over to invited contributions from researchers with unique expertise in specific fields. Topics covered include the next generation internet, grid computing, antennas, integrated circuits, power engineering, license plate tracking, and color map generation. These further demonstrate the versatility and potential of differential evolution.

5 What is Available from the Companion Website

As already mentioned, two literature surveys were conducted, one on differential evolution, the other on test beds for single-objective optimization. Numerous publications have been gathered together into two separate bibliographies. Due to limitations of space, it is impossible to present a hard copy of these two bibliographies. More importantly, it might be more convenient and beneficial for fellow researchers in these areas to have an electronic and online copy. Therefore, these two bibliographies are posted online as part of the companion website.

Through the literature survey on test beds for single-objective optimization, more than 300 toy functions and tens of application problems for single-objective optimization have been collected. The formulation, mathematical features, and historical citation mistakes of each toy function and application problem have been documented. In addition, all toy functions and some of the application problems have been coded in Fortran 90/95. Such a collection will be extremely helpful to fellow researchers working on single-objective optimization as well as in others relevant fields.

If I find a book on algorithms, I will always check whether there are any source codes available. Given sufficient computational resources, I will try out those algorithms myself. This invariably gives me a better understanding of those algorithms. In this regard, I would like to share my Fortran 90/95 source codes on differential evolution with fellow researchers in the hope of improving understanding.

Acknowledgement

I owe a great deal to Prof. C.K. Lee of Nanyang Technological University, Singapore. He is the best man I have ever met in my life.

I would take this opportunity to thank Prof. H. Lim, director of Temasek Laboratories, National University of Singapore, for his support and encouragement of my study on differential evolution. I would also like to thank my colleagues and former colleagues of Temasek Laboratories, National University of Singapore, especially Mr Y.B. Gan, Dr X. Xu, and Mr C. Lin, for fruitful discussions and considerable assistance.

Special thanks go to Prof. S.W. Yang, Prof. Z.P. Nie, Ms M. Meng, and Mr Y.K. Chen of University of Electronic Science and Technology of China, Prof. Q.Y. Feng and Prof. F. Zhu of Southwest Jiaotong University, China, and the Ministry of Education of China.

My appreciation also goes to Mr J. Murphy and R. Bullen, editors at John Wiley & Sons. Their professionalism and patience have facilitated improvements to the book.

Writing a book is a long and arduous process. I could not have done it without the support of my family. Thank you very much, Jiaoli, Chen, and Tian, for your support.

I am greatly indebted to my brother, Anbing, and sister-in-law, Juhua Liu, for voluntarily taking on the responsibility of my upbringing when I suddenly lost my parents in 1984. Without their loving care, I would not have had the opportunities I have had and my life would doubtless have taken a less satisfying course.

List of Figures

List of Tables

1

An Introduction to Optimization

1.1 A General Optimization Problem

1.1.1 Definition

1.1.1.1 Qualitative Description

Throughout human being history, mankind has faced optimization problems and made great efforts to solve them. Loosely speaking, optimization is the process of finding the best way to use available resources, while at the same time not violating any of the constraints that are imposed. More accurately, we may say that we wish to define a system mathematically, identify its variables and the conditions they must satisfy, define properties of the system, and then seek the state of the system (values of the variables) that gives the most desirable (largest or smallest) properties. This general process is referred to as optimization.

It is not our purpose here to define a system. This is the central problem of various disciplines which are sciences or are struggling to become sciences. Our concern here is, given a meaningful system, what variables will make the system have the desirable properties.

1.1.1.2 Mathematical Formulation

The general optimization problem is mathematically defined as
Find $\mathbf{x}^* = [x_1^* \quad x_2^* \quad \cdots \quad x_N^*] \in D^N = D_1 \cap D_2 \cap \cdots \cap D_N$
where

$$f_i^{\min}(\mathbf{x}^*) \le f_i^{\min}(\mathbf{x})\ \forall \mathbf{x} = [x_1 \quad x_2 \quad \cdots \quad x_N] \in D^N, \quad 1 \le i \le N_{f^{\min}},$$

$$f_i^{\max}(\mathbf{x}^*) \ge f_i^{\max}(\mathbf{x})\ \forall \mathbf{x} \in D^N, \quad 1 \le i \le N_{f^{\max}},$$

$$c_i^{=}(\mathbf{x}^*) = 0, \quad 1 \le i \le N_{c^=},$$

$$c_i^{+}(\mathbf{x}^*) > 0, \quad 1 \le i \le N_{c^+},$$

$$c_i^{-}(\mathbf{x}^*) < 0, \quad 1 \le i \le N_{c^-}.$$

Differential Evolution Anyong Qing

$\mathbf{x}^*$ is the optimal solution in the N-dimensional search space D^N. N is the number of optimization parameters, or the dimension of the optimization problem. D_i, either continuous or discrete, is the search space of x_i, the ith optimization parameter. D_i and D_j are not necessarily identical, from the point of view of either type or size. $\mathbf{x}$ is an N-dimensional vector of optimization parameters. $f_i^{\min}(\mathbf{x})$ is the ith objective function to be minimized, $N_{f^{\min}}$ is the number of objective functions to be minimized. $f_i^{\max}(\mathbf{x})$ is the ith objective function to be maximized, $N_{f^{\max}}$ is the number of objective functions to be maximized. $c_i^=(\mathbf{x})$ is the ith equality constraint function, $N_{c^=}$ is the number of equality constraint functions. $c_i^+(\mathbf{x})$ is the ith positive constraint function, N_{c^+} is the number of positive constraint functions. $c_i^-(\mathbf{x})$ is the ith negative constraint function, N_{c^-} is the number of negative constraint functions.

An optimization problem is thus made up of three essential ingredients: optimization parameters $\mathbf{x}$; objective functions, $f_i^{\min}(\mathbf{x})$ and $f_i^{\max}(\mathbf{x})$; and constraint functions, $c_i^=(\mathbf{x})$, $c_i^+(\mathbf{x})$, and $c_i^-(\mathbf{x})$.

Objective and constraint functions in many real-world application problems can be formulated in many ways. There might be better formulation of objective and constraint functions to describe a particular optimization problem. Any knowledge about the optimization problem should be worked into the objective and constraint functions. Good objective and constraint functions can make all the difference.

1.1.1.3 Some Examples

1.1.1.3.1 Investment Fund Management An investment fund manager is authorized to invest a sum of money A in N investment funds which claim to offer return rates r_i, $1 \le i \le N$. The rate of administration charge for fund i is c_i. Suppose he invests an amount x_i in the ith fund. How can he plan his investment so that he will get maximum return?

1.1.1.3.2 Experimental Data Fitting Suppose a mathematical model $h(\mathbf{x})$ with unknown physical parameters $\mathbf{x}$ has been built for a physical phenomenon. A total of m experiments have been done during which experimental data $\mathbf{y}$ have been collected. It is then required to fit the mathematical model $h(\mathbf{x})$ to the collected experimental data $\mathbf{y}$.

1.1.1.3.3 Radome Design Radomes are used to protect antennas. Materials for a radome can only be chosen from the available materials database. A radome has to meet certain electromagnetic and mechanic demands. It is also subject to economic constraints and fabrication limitations.

1.1.2 Optimization Parameters

Optimization parameters $\mathbf{x}$ are critical for an optimization problem. They affect the value of objective and constraint functions. If there are no optimization parameters, we cannot define the objective and constraint functions. In the investment fund management problem, the optimization parameters are the amounts of money invested in each fund. In experimental data fitting problems, the optimization parameters are the parameters that define the model. In the radome design problem, the optimization parameters might include the material index in the materials database, material thickness, and some other parameters.

An optimization parameter can be continuous, discrete, or even symbolic.

1.1.3 Objective Functions

An objective function $f(\mathbf{x})$ is what we want to optimize. It is either $f_i^{\min}(\mathbf{x})$ or $f_i^{\max}(\mathbf{x})$, depending on the desirable properties of the optimization problem. For instance, in the investment fund management problem, the fund manager wants to maximize the return. In fitting experimental data to a user-defined model, we might minimize the total deviation of observed data from predictions based on the model. In the radome design problem, we have to maximize the strength and minimize the distortion and cost.

Almost all optimization problems have objective functions. However, in some cases, such as the design of integrated circuit layouts, the goal is to find optimization parameters that satisfy the constraints of the model. The user does not particularly want to optimize anything, so there is no reason to define an objective function. This type of problem is usually called a feasibility problem. On the other hand, in some optimization problems, there is more than one objective function. For instance, in the radome design problem, it would be nice to minimize weight and maximize strength simultaneously.

An objective function has at least one global optimum, and may have multiple local optima as shown in Figure 1.1. In the remainder of this book, optimum stands for the global optimum of an objective function, $f(\mathbf{x}^*)$, the objective function value at the optimal solution point $\mathbf{x}^*$. For an optimization problem with multiple objective functions, optimal solution points corresponding to different objective functions may be inconsistent.

Figure 1.1 A one-dimensional continuous multimodal objective function

An objective function has some characteristic features. These features are very important for choosing optimization algorithms to solve the optimization problem of interest.

1.1.3.1 Continuity

An objective function can be continuous as shown in Figure 1.1, or discontinuous, as shown in Figure 1.2.

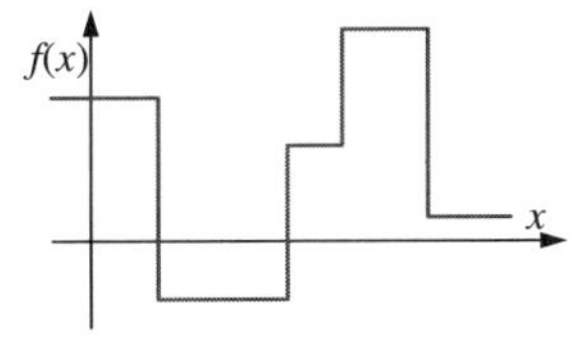

Figure 1.2 A one-dimensional discontinuous objective function

1.1.3.2 Decomposability

Decomposability is sometimes also referred to as nonlinearity, separability, or epistasis. If each optimization parameter x_i is independent of the other parameters x_j ($j \neq i$) in $f(\mathbf{x})$, it is decomposable and is relatively easy to optimize, since the landscape can be decomposed into simpler components. The optimization process of a decomposable objective function can be performed in a sequence of N independent optimization processes, where each parameter is optimized independently.

As an example, consider minimizing the two-dimensional sphere function

$$f(\mathbf{x}) = x_1^2 + x_2^2. \tag{1.1}$$

Regardless of the value of x_2, we can find the solution $x_1 = 0$ which minimizes $f(\mathbf{x})$, and vice versa for x_1.

However, many objective functions are not decomposable in this manner. The two-dimensional Chung–Reynolds function

$$f(\mathbf{x}) = (x_1^2 + x_2^2)^4 \tag{1.2}$$

is not decomposable but is very easy to optimize, since its first-order derivative is a product

$$\frac{\partial f(\mathbf{x})}{\partial x_i} = 8x_i(x_1^2 + x_2^2)^3. \tag{1.3}$$

Such a product yields a solution for $x_1 = 0$ that is independent of x_2, and vise versa.

From this observation, the following general condition was developed by Salomon [1] in order to determine whether an objective function $f(\mathbf{x})$ is easy to optimize or not:

$$\frac{\partial f(\mathbf{x})}{\partial x_i} = g(x_i)h(\mathbf{x}), \tag{1.4}$$

where $g(x_i)$ means any function of only x_i and $h(\mathbf{x})$ means any function of $\mathbf{x}$. If this condition is satisfied, $f(\mathbf{x})$ is partially decomposable and easy to optimize, because solutions for each x_i can still be obtained independently of all other parameters x_j, $j \neq i$. The Chung–Reynolds function is partially decomposable and is therefore easy to optimize.

On the other hand, the Rosenbrock saddle function

$$f(\mathbf{x}) = 100(x_2 - x_1^2)^2 + (x_1 - 1)^2 \tag{1.5}$$

is not decomposable since the equations below do not satisfy the above condition:

$$\frac{\partial f(\mathbf{x})}{\partial x_1} = 400x_1(x_1^2 - x_2) + 2(x_1 - 1), \tag{1.6}$$

$$\frac{\partial f(\mathbf{x})}{\partial x_2} = 200(x_2 - x_1^2). \tag{1.7}$$

1.1.3.3 Differentiability

A continuous objective function may be non-differentiable or differentiable of order n (≥ 1). A continuous non-differentiable function is shown in Figure 1.3.

Figure 1.3 A continuous non-differentiable one-dimensional objective function

1.1.3.4 Modality

An objective function $f(\mathbf{x})$ is unimodal if there is some path from every point $\mathbf{x}$ to the optimal solution point $\mathbf{x}^*$ along which it is monotonous ([2], p. 106). Otherwise, it is multimodal. Figure 1.1 shows a one-dimensional multimodal objective function and Figure 1.4 shows a two-dimensional multimodal objective function where the dot is the optimal point $\mathbf{x}^*$.

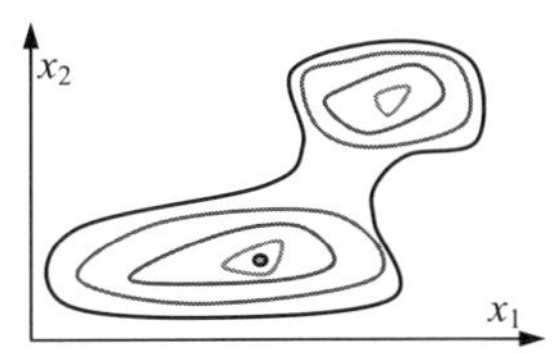

Figure 1.4 A two-dimensional multimodal objective function

A one-dimensional unimodal objective function is drawn in Figure 1.5 while a two-dimensional unimodal objective function is drawn in Figure 1.6.

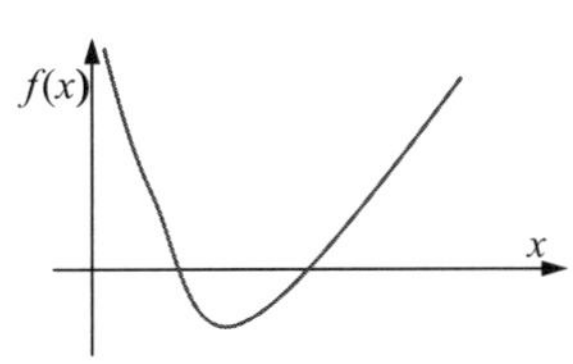

Figure 1.5 A one-dimensional unimodal objective function

1.1.3.5 Noise

Some optimization problems are dynamic or even noisy. In another word, their objective and/or constraint functions are random to a certain extent. Noise is most often used to represent randomness.

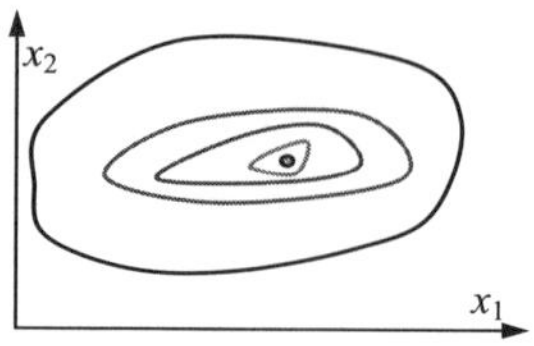

Figure 1.6 A two-dimensional unimodal function

1.1.3.6 Scalability

An objective function $f(\mathbf{x})$ is not scalable if its number of parameters, or dimension, is fixed. On the other hand, scalable objective functions can be scaled to any dimension. As dimension increases, the search space size also increases exponentially. The difficulty of finding an optimal solution increases accordingly.

1.1.3.7 Symmetry

A symmetric objective function $f(\mathbf{x})$ does not change its value if $\mathbf{x}$ is replaced by any of its $N!$ permutations. The aforementioned sphere function and Chung–Reynolds function are symmetric while the Rosenbrock saddle function is non-symmetric or asymmetric.

1.1.3.8 Uniqueness

An optimal solution point $\mathbf{x}^*$ is unique if there is only one optimal solution. Some objective functions, for example, $f(x)=\sin(x)$, $x\in[-2\pi, 2\pi]$, may have multiple optimal solution points. Objective function values at all optimal solution points are equal. In this regard, no optimal solution is distinguishable from others.

Some people confuse uniqueness with modality.

1.1.4 Constraint Functions

Constraints allow the optimization parameters to take on certain values but exclude others. For the investment fund management problem, the amount of money invested must not exceed the available money.

Constraints are not absolutely necessary for an optimization problem. In fact, the field of unconstrained optimization is a large and important one for which a lot of algorithms and software are available. However, it has been argued that almost all problems really do have constraints. For example, any optimization parameter denoting the number of objects in a system can only be meaningful if it is less than the number of elementary particles in the known universe! Nevertheless, in practice, answers that make good sense in terms of the underlying physics can often be obtained without putting constraints on the optimization problems.

Sometimes, constraint functions and objective functions for an optimization problem are exchangeable, depending on the priority of the desirable properties. In fact, later in this book, we avoid distinguishing objective functions and constraint functions whenever possible. Moreover, without loss of generality, in the rest of this book, we are exclusively concerned with minimization problems unless explicitly stated otherwise. As a matter of fact, it is very easy to convert a maximization problem into minimization problem.

Many people regard the search space for an optimization problem as a constraint. However, in this book, we do not take this approach.

1.1.5 Applications

Optimization has a consistent track record across a wide range of science, engineering, industry and commerce. In fact, many optimization problems come directly from real-world applications.

A simple search on Google with the keywords "optimization" and "application" will get numerous hits. Publications on optimization that do not mention applications are very rare. There is no need for us to prove the usefulness of optimization by presenting a long list of fields of application in which optimization has been involved. Space considerations also do not permit us to give an exhaustive and in-depth review on applications of optimization. Therefore, no further discussion on applications of optimization will be given here.

1.1.6 Optimization Algorithms

Optimization has long been the subject of intensive study. Numerous optimization algorithms have been proposed. In general, these algorithms can be divided into two major categories, deterministic and stochastic. Hybrid algorithms which combine deterministic and stochastic features are stochastic in essence and regarded as such. However, it is acceptable to treat them as a third category from the point of view of purity.

1.2 Deterministic Optimization Algorithms

A deterministic optimization algorithm will always get the same solution with the same number of objective function evaluations regardless of the time it is started, if the search space, starting-point, and termination conditions are unchanged. If the algorithm is run multiple times on the same computer, the search time for each run will be exactly the same. In other words, deterministic optimization is clonable.

Dimension is a good criterion for classifying deterministic optimization algorithms. Deterministic optimization algorithms are accordingly divided into one-dimensional and multi-dimensional deterministic optimization algorithms. Some multi-dimensional deterministic algorithms need the help of one-dimensional deterministic optimization algorithms.

1.2.1 One-Dimensional Deterministic Optimization Algorithms

Use of the derivative is a good choice for distinguishing one-dimensional optimization algorithms.

1.2.1.1 Zeroth-Order One-Dimensional Deterministic Optimization Algorithms

Zeroth-order algorithms involve objective function evaluation and comparison only. Prominent algorithms include the exhaustive search algorithm, dichotomous algorithms, the parabolic interpolation algorithm, and the Brent algorithm. If the minimum of the objective function $f(x)$ is known, a nonlinear equation can be formulated. In this case, the Secant algorithm for nonlinear equations is applicable.

1.2.1.1.1 Exhaustive Search Algorithm The exhaustive search algorithm ([2], pp. 137–138) samples the search space $[a, b]$ at m points. Usually, the sample points are equally spaced within $[a, b]$. The minimum value of the objective function at each and every sample point is regarded as the optimum and the corresponding sample point is regarded as the optimal solution.

The exhaustive search algorithm is also known as enumeration algorithm or brute force algorithm ([3], p. 13).

1.2.1.1.2 Dichotomous Algorithms For minimizing the objective function $f(x)$, the Fortran-style pseudo-code for dichotomous algorithms ([2], p. 141) is shown in Figure 1.7.

```
n = 0
b_L = a
b_U = b
do while (termination conditions not satisfied)
  n = n + 1
  get y1 within [b_L, b_U]
  get y2 within [b_L, b_U]
  f1 = f(y1)
  f2 = f(y2)
  if (f2 < f1) then
    b_L = y1
  else
    b_U = y2
  end if
end do
```

Figure 1.7 Fortran-style pseudo-code for dichotomous algorithms

There are many different schemes for obtaining y_1 and y_2 within $[b_L, b_U]$, such as the equal interval, Fibonacci, and golden section schemes.

1.2.1.1.3 Parabolic Interpolation Algorithm A local quadratic approximation to the objective function $f(x)$ is useful because the minimum of a quadratic is easy to compute. The parabolic interpolation algorithm ([4], p. 185, [5], pp. 51, 72) interpolates the objective function by a quadratic polynomial which fits the objective function values at three points. The minimum point of the parabola, a new estimate of the minimum point of the objective function, will replace one of the three previous points. This process is repeated until termination conditions are fulfilled.

1.2.1.1.4 Brent Algorithm The Brent algorithm [6] is a hybrid of the parabolic interpolation algorithm and the golden section algorithm. The objective function in each iteration is approximated by an interpolating parabola through three existing points. The minimum point of the parabola is taken as a guess for the minimum point. It is accepted and used to generate a smaller interval if it lies within the bounds of the current interval. Otherwise, the algorithm falls back to an ordinary golden section step.

1.2.1.1.5 Secant Algorithm for Nonlinear Equation For an objective function $f(x)$ with known optimum, it is straightforward to formulate a nonlinear equation $f(x)=0$. Its truncated Taylor series $f(x+h)\approx f(x)+f'(x)\,h$ is a linear function of h that approximates $f(x)$ near a given x. Assuming $f'(x)\neq 0$ and $f(x+h)=0$, we get $h=-f(x)/f'(x)$. Because the truncated Taylor series is only an approximation to the nonlinear function $f(x)$, its root $x+h$ does not equal to the root of $f(x)$. Therefore, this process has to be repeated until an acceptable root is located. This motivates the update scheme of the Newton algorithm for nonlinear equation ([4], pp. 156–158) as $x^{n+1}=x^n-f(x^n)/f'(x^n)\,f'(x^n)$. However, the derivative $f'(x^n)$ may not be available. In this case, it is approximated by finite difference. The Newton update scheme accordingly becomes the Secant update scheme ([4], p. 158), expressed as

$$x^{n+1}=x^n-f(x^n)\frac{x^n-x^{n-1}}{f(x^n)-f(x^{n-1})}. \tag{1.8}$$

1.2.1.2 Higher-Order One-Dimensional Deterministic Optimization Algorithms

Besides objective function evaluation and comparison, higher-order algorithms directly make use of derivatives. Three algorithms in this category are commonly used.

1.2.1.2.1 Newton Algorithm Another way to obtain a local quadratic approximation to the objective function $f(x)$ is to use a truncated Taylor series expansion ([4], pp. 185–186)

$$f(x+h)\approx f(x)+f'(x)h+\frac{f''(x)}{2}h^2. \tag{1.9}$$

The minimum point of this quadratic function of h is given by

$$h=-\frac{f'(x)}{f''(x)}. \tag{1.10}$$

This result motivates the iteration scheme for the Newton algorithm

$$x^{n+1}=x^n-\frac{f'(x^n)}{f''(x^n)}. \tag{1.11}$$

Obviously, the Newton algorithm here is equivalent to the Newton algorithm for the nonlinear equation $f'(x)=0$.

1.2.1.2.2 Secant Algorithm The Secant algorithm ([7], p. 55) here is equivalent to the Secant algorithm for nonlinear equation $f'(x)=0$. Finite difference is applied to approximate the second-order derivative, $f''(x_k)$, in Equation 1.11. Its update scheme is

$$x^{n+1}=x^n-f'(x^n)\frac{x^n-x^{n-1}}{f'(x^n)-f'(x^{n-1})}. \tag{1.12}$$

1.2.1.2.3 Cubic Interpolation Algorithm This is another polynomial approximation algorithm in which the objective function $f(x)$ is approximated by a local third-order polynomial $p_3(x)$ ([7], pp. 56–58). The basic logic is similar to that of the parabolic interpolation algorithm. However, in this instance, evaluation of both objective function and its derivative at each point is required. Consequently, the approximation polynomial can be constructed using fewer points.

A third-order polynomial fitting the objective function $f(x)$ and its derivative $f'(x)$ at points a and b is given by $p_3(x)=c_0+c_1(x-a)+c_2(x-a)(x-b)+c_3(x-a)^2(x-b)$, where $f(a)=c_0, f(b)=c_0+c_1(b-a), f'(a)=c_1+c_2(a-b), f'(b)=c_1+c_2(b-a)+c_3(b-a)^2$. The minimum point of $p_3(x)$ is

$$x^*=\begin{cases} b & \lambda<0,\\ b-\lambda(b-a) & 0\le\lambda\le1,\\ a & \lambda>1.\end{cases} \tag{1.13}$$

where $\lambda=[f'(b)+\alpha-\beta]/[f'(b)-f'(a)+2\alpha]$, $\alpha=\sqrt{\beta^2-f'(a)f'(b)}$, $\beta=3[f(a)-f(b)]/(b-a)+f'(a)+f'(b)$. Either a or b will be replaced by x^*, depending on both function values and the values of derivatives. The Fortran-style pseudo-code for one implementation is given in Figure 1.8 ([5], p. 82).

1.2.2 Multi-Dimensional Deterministic Optimization Algorithms

Similarly, the use of gradient, Hessian matrix or even higher-order partial derivatives is a good way to distinguish multi-dimensional optimization algorithms.

1.2.2.1 Zeroth-Order Multi-Dimensional Deterministic Optimization Algorithms

Likewise, zeroth-order algorithms here also involve objective function evaluation and comparison only. Prominent algorithms include the grid search algorithm, univariate search algorithms, the pattern search algorithm, Powell's conjugate direction algorithm, and the downhill simplex algorithm. If the minimum of the objective function $f(\mathbf{x})$ is known, a nonlinear equation can be formulated. In this case, Broyden's algorithm for nonlinear equation is applicable.

1.2.2.1.1 Grid Search Algorithm The grid search algorithm ([2], pp. 156–158) is the simplest algorithm for finding the minimum and the corresponding minimum solution point of objective function $f(\mathbf{x})$. D_i, the search space of x_i, is sampled at m_i points to form a grid of $\prod_{i=1}^{N} m_i$ points. The minimum value of the objective function at each and every grid point is regarded as the minimum and the corresponding grid point is regarded as the optimal

```
compute f(a)
compute f(b)
compute f'(a)
compute f'(b)
do while (termination conditions not satisfied)
  β = 3[f(a)–f(b)] / (b – a) + f'(a) + f'(b)
  α = sqrt(β² – f'(a)f'(b))
  λ = [f'(b) + α – β] / [f'(b) – f'(a) + 2 α]
```

$$x^* = \begin{cases} a & \lambda < 0 \\ b - \lambda(b-a) & 0 \le \lambda \le 1 \\ b & \lambda > 1 \end{cases}$$

```
  compute f(x*)
  compute f'(x*)
  if( f'(a) < 0 and (f'(x*) > 0 or f(x*) ≥ f(a)) ) then
    b = x*
    f(b) = f(x*)
    f'(b) = f'(x*)
  else
    a = x*
    f(a) = f(x*)
    f'(a) = f'(x*)
  end if
end do
```

Figure 1.8 Fortran-style pseudo-code for one implementation of the cubic interpolation algorithm

solution point. Obviously this approach soon becomes prohibitive as the dimension N and the number of sample points m_i for each dimension increase. A more efficient approach starts from a grid point, evaluates the objective function values at the surrounding $3^N - 1$ grid points, selects the grid point with the smallest objective function value as the new starting-point, and repeats this process until termination conditions are fulfilled.

1.2.2.1.2 Univariate Search Algorithms The guiding idea behind univariate search algorithms ([2], pp. 158–159) is to change one optimization parameter at a time so that the function is optimized in each of the coordinate directions $\mathbf{e}^i$, $1 \le i \le N$. The Fortran-style pseudo-code is given in Figure 1.9.

Univariate search algorithms are regarded as zeroth-order optimization algorithms for multi-dimensional optimization problems because the gradient of the objective function is not explicitly involved, although the derivative of $f(\mathbf{x}^{n-1} + \lambda \mathbf{e}^i)$ with respect to λ might be involved in the inner one-dimensional optimization algorithm.

```
choose starting point x^0
n = 1
do while (termination conditions not fulfilled)
   i = mod(n, N)
   if (i = 0) i = N
   find λ^n to minimize f(x^(n-1) + λ^n e^i)
   x^n = x^(n-1) + λ^n e^i
   n = n + 1
end do
```

Figure 1.9 Fortran-style pseudo-code for univariate search algorithms

1.2.2.1.3 Pattern Search Algorithm The pattern search algorithm is also known as the Hooke–Jeeves algorithm [8]. The essential idea behind it is to move from one solution point to the next.

There are two kinds of moves in the pattern search algorithm: the exploratory move and the pattern move. An exploratory move consists of a series of univariate searches. Each univariate search moves a little along its coordinate direction. The pattern move is larger in size.

Fortran-style pseudo-code is given in Figure 1.10.

```
choose starting point x
choose exploratory move steps δ_i, 1 ≤ i ≤ N
choose initial pattern move steps Δ_i, 1 ≤ i ≤ N
do while (termination conditions not fulfilled)
   do i = 1, N
     if( f(x + δ_i e^i) < f(x))
        x = x + δ_i e^i
        Δ_i = Δ_i + δ_i
     else if( f(x − δ_i e^i) < f(x))
        x = x − δ_i e^i
        Δ_i = Δ_i − δ_i
     end if
   end do
   if( f(x + Δ) < f(x))
     x = x + Δ
   end if
end do
```

Figure 1.10 Fortran-style pseudo-code for pattern search algorithms

1.2.2.1.4 Powell's Conjugate Direction Algorithms The univariate search algorithm is very appealing due to its simplicity. However, it may not converge to the optimal solution point in the case of non-decomposable objective functions. Powell proposed a simple but vastly superior variation. It gets the name Powell's conjugate direction algorithm ([2], pp. 162–163) because it chooses conjugate directions to move when applied to an objective function of quadratic form, $f(\mathbf{x}) = \mathbf{x} \cdot \mathbf{A} \cdot \mathbf{x} + \mathbf{b} \cdot \mathbf{x} + c$, where $\cdot$ denotes the dot product of vectors. Directions $\mathbf{p}$ and $\mathbf{q}$ are conjugate if $\mathbf{p} \cdot \mathbf{A} \cdot \mathbf{q} = 0$.

The Fortran-style pseudo-code for Powell's conjugate direction algorithm is given in Figure 1.11.

```
choose starting point x^0
do i = 1, N
   p^i = e^i
end do
do while (termination conditions not fulfilled)
   do i = 1, N
      find λ^i to minimize f(x^(i-1) + λ_i p^i)
      x^i = x^(i-1) + λ_i p^i
   end do
   do i = 1, N - 1
      p^i = p^(i+1)
   end do
   p^N = x^N - x^0
   find λ to minimize f(x^N + λ (x^N - x^0))
   x^0 = x^N + λ (x^N - x^0)
end do
```

Figure 1.11 Fortran-style pseudo-code for powell's conjugate direction algorithm

1.2.2.1.5 Downhill Simplex Algorithm The downhill simplex algorithm for minimizing $f(\mathbf{x})$ is due to Nelder and Mead [9]. A non-degenerate N-dimensional simplex is a geometrical figure consisting of $N + 1$ distinct vertices. For example, a two-dimensional simplex is a triangle and three-dimensional simplex is a tetrahedron.

Suppose we have obtained an N-dimensional simplex with vertices $\mathbf{x}^i$, $i = 1, \ldots, N + 1$. Identify the vertices with the highest, second highest, and lowest objective function value, $\mathbf{x}^{\mathrm{H}}$, $\mathbf{x}^{\mathrm{S}}$, and $\mathbf{x}^{\mathrm{L}}$. Let $\mathbf{x}^{\mathrm{G}}$ be the geometric centroid of all the vertices except $\mathbf{x}^{\mathrm{H}}$, that is,

$$\mathbf{x}^{\mathrm{G}} = \frac{1}{n}\left(\sum_{i=1}^{N+1} \mathbf{x}^i - \mathbf{x}^{\mathrm{H}}\right). \tag{1.14}$$

Several key operations, including reflection, contraction, and expansion, are involved in determining a new vertex to replace $\mathbf{x}^{\mathrm{H}}$. If a vertex better than $\mathbf{x}^{\mathrm{H}}$ is not generated by all these operations, the simplex will contract toward $\mathbf{x}^{\mathrm{L}}$ through a simultaneous contraction $\mathbf{x}^{i} = \mathbf{x}^{\mathrm{L}} + \eta\,(\mathbf{x}^{i} - \mathbf{x}^{\mathrm{L}})$ where $0 < \eta < 1$, $1 \le i \le N+1$, $i \ne \mathrm{L}$, as shown in Figure 1.12.

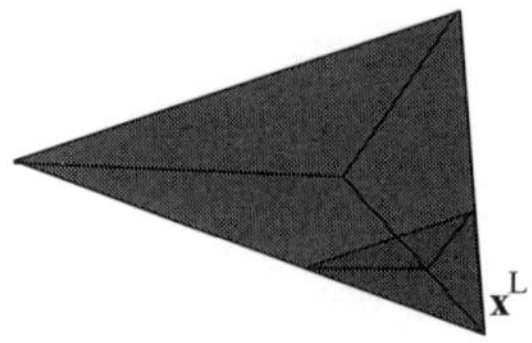

Figure 1.12 Simultaneous contraction

The mirror vertex $\mathbf{x}^{\mathrm{R}}$ of $\mathbf{x}^{\mathrm{H}}$ with respect to $\mathbf{x}^{\mathrm{G}}$ is obtained through reflection $\mathbf{x}^{\mathrm{R}} = \mathbf{x}^{\mathrm{G}} + \alpha\,(\mathbf{x}^{\mathrm{G}} - \mathbf{x}^{\mathrm{H}})$ where $\alpha > 0$. A schematic three-dimensional reflection is drawn in Figure 1.13.

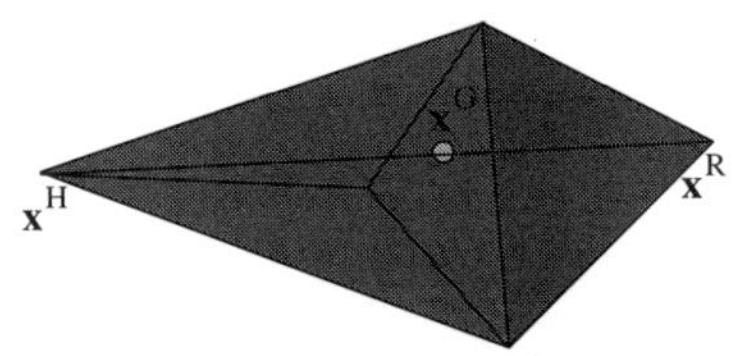

Figure 1.13 Reflection

An expansion as shown in Figure 1.14, $\mathbf{x}^{\mathrm{E}} = \mathbf{x}^{\mathrm{R}} + \beta\,(\mathbf{x}^{\mathrm{R}} - \mathbf{x}^{\mathrm{G}})$, where $\beta > 0$, will be made to search for better point if $f(\mathbf{x}^{\mathrm{R}}) < f(\mathbf{x}^{\mathrm{L}})$. $\mathbf{x}^{\mathrm{E}}$ will be accepted to replace $\mathbf{x}^{\mathrm{H}}$ if $f(\mathbf{x}^{\mathrm{E}}) < f(\mathbf{x}^{\mathrm{R}})$. Otherwise, $\mathbf{x}^{\mathrm{R}}$ will replace $\mathbf{x}^{\mathrm{H}}$.

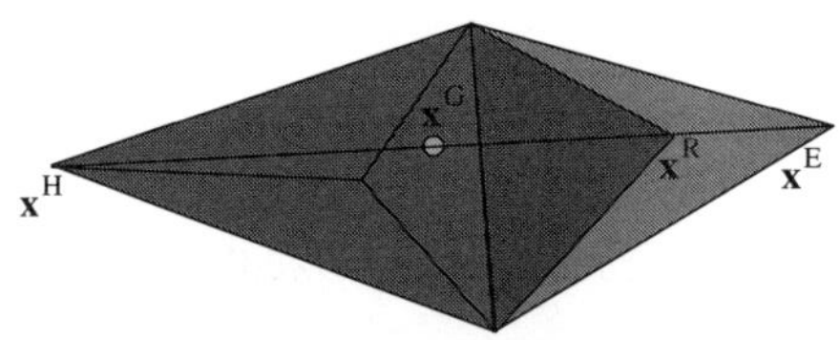

Figure 1.14 Expansion

$\mathbf{x}^{\mathrm{R}}$ will also replace $\mathbf{x}^{\mathrm{H}}$ if $f(\mathbf{x}^{\mathrm{R}}) < f(\mathbf{x}^{\mathrm{S}})$. Otherwise, a contraction as shown in Figure 1.15 is made to obtain a new point $\mathbf{x}^{\mathrm{C}} = \mathbf{x}^{\mathrm{H}} + \lambda\,(\mathbf{x}^{\mathrm{R}} - \mathbf{x}^{\mathrm{H}})$, where $0 < \lambda < 1$. $\mathbf{x}^{\mathrm{C}}$ will be accepted to replace $\mathbf{x}^{\mathrm{H}}$ if $f(\mathbf{x}^{\mathrm{C}}) < f(\mathbf{x}^{\mathrm{H}})$.

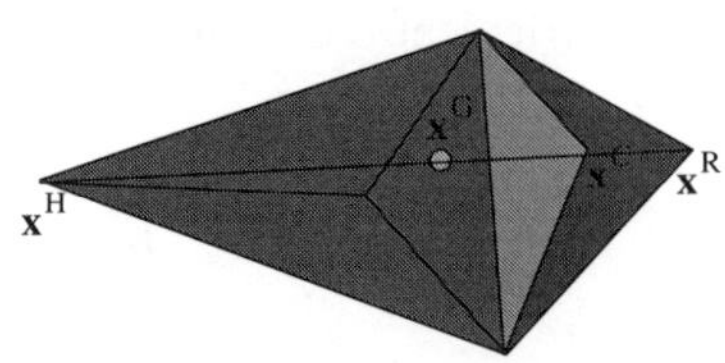

Figure 1.15 Contraction

The Fortran-style pseudo-code for the above downhill simplex algorithm is given in Figure 1.16.

```
set α, β, λ, and η
choose starting simplex x^i, 1 ≤ i ≤ N + 1
do while (termination conditions not fulfilled)
  get x^H, x^S, x^L, and x^G
  x^R = x^G + α (x^G – x^H)
  if (f(x^R) < f(x^L))
    x^E = x^R + β (x^R – x^G)
    if (f(x^E) < f(x^R))
      x^H = x^E
    else
      x^H = x^R
    end if
  else
    if (f(x^R) < f(x^S))
      x^H = x^R
    else
      x^C = x^H + λ (x^R – x^H)
      if (f(x^C) < f(x^H))
        x^H = x^C
      else
        do i = 1, N + 1
          if( i ≠ L) x^i = x^L + η (x^i – x^L)
        end do
      end if
    end if
  end if
end do
```

Figure 1.16 Fortran-style pseudo-code for downhill simplex algorithm

1.2.2.1.6 Broyden's Algorithm for Nonlinear Equation It is straightforward to generalize the one-dimensional Newton algorithm for nonlinear equations to multiple dimensions as $\mathbf{x}^{n+1} = \mathbf{x}^n + \mathbf{s}^n$, where $\mathbf{J}(\mathbf{x}^n) \cdot \mathbf{s}^n = -f(\mathbf{x}^n)$ and $J_i(\mathbf{x}^n) = \partial f(\mathbf{x}^n)/\partial x_i$ is the Jacobian (partial derivative) with respect to x_i. It is also straightforward to generalize the corresponding one-dimensional Secant algorithm for nonlinear equation to multiple dimensions ([4], pp. 167–168). One of the simplest and most effective Secant algorithms for multi-dimensional nonlinear equations is Broyden's algorithm, whose Fortran-style pseudo-code is given in Figure 1.17.

```
choose starting point x^0
choose starting Jacobian vector J^0
n = 0
do while (termination conditions not fulfilled)
   get s^n from J^n (x^n) · s^n = -f(x^n)
   x^(n+1) = x^n + s^n
   J^(n+1) = J^n + f(x^(n+1)) / (s^n · s^n) s^n
   n = n + 1
end do
```

Figure 1.17 Fortran-style pseudo-code for broyden's algorithm for multi-dimensional nonlinear equation

1.2.2.2 Higher-Order Multi-Dimensional Deterministic Optimization Algorithms

Besides objective function evaluation and comparison, higher-order algorithms make direct use of partial derivatives. Four algorithms in this category are commonly used.

1.2.2.2.1 Steepest Descent Algorithm It is well known that the negative gradient $-\nabla f(\mathbf{x})$ at a given point $\mathbf{x}$ is locally the steepest descent direction in the sense that the function value decreases more rapidly along it than along any other direction. This fact leads to one of the oldest algorithms for multi-dimensional optimization, the steepest descent algorithm ([2], pp. 166–168, [4], pp. 187–188, [5], pp. 100–101) whose update scheme is $\mathbf{x}^{n+1} = \mathbf{x}^n - \alpha^n \nabla f(\mathbf{x}^n)$, where the step α^n is usually obtained by applying any of the aforementioned one-dimensional minimization algorithms for the function $f(\mathbf{x}^n - \alpha \nabla f(\mathbf{x}^n))$ of α.

The steepest descent algorithm is also referred to as the Cauchy algorithm.

1.2.2.2.2 Newton Algorithm The Newton algorithm ([7], pp. 103–105, [4], pp. 189–190) for minimizing the multi-dimensional function $f(\mathbf{x})$ is a straightforward generalization of its one-dimensional counterpart. The straightforward generalization of the update scheme of the Newton algorithm for one-dimensional minimization problems would be $\mathbf{x}^{n+1} = \mathbf{x}^n - \mathbf{H}^{-1}(\mathbf{x}^n) \nabla f(\mathbf{x}^n)$, where $\mathbf{H}(\mathbf{x})$ is the Hessian matrix and $H_{ij}(\mathbf{x}) = \partial^2 f(\mathbf{x})/\partial x_i \partial x_j$. However, to avoid matrix inversion, the update scheme is slightly modified as $\mathbf{x}^{n+1} = \mathbf{x}^n + \mathbf{s}^n$, where $\mathbf{H}(\mathbf{x}^n) \cdot \mathbf{s}^n = -\nabla f(\mathbf{x}^n)$.

Sometimes, the Newton algorithm is modified ([7], p. 105) by introducing a line search as $\mathbf{x}^{n+1} = \mathbf{x}^n + \alpha^n \mathbf{s}^n$ where the step α^n is obtained by applying any of the aforementioned one-dimensional minimization algorithms for the function $f(\mathbf{x}^n + \alpha\, \mathbf{s}^n)$ of α.

Marquardt ([7], p. 105) even combines the steepest descent algorithm and the Newton algorithm as $\mathbf{x}^{n+1} = \mathbf{x}^n + \mathbf{s}^n$, where $[\mathbf{H}(\mathbf{x}^n) + \alpha^n \mathbf{I}] \cdot \mathbf{s}^n = -\nabla f(\mathbf{x}^n)$ and $\mathbf{I}$ is the identity matrix.

1.2.2.2.3 Quasi-Newton Algorithm Quasi-Newton algorithms ([7], pp. 112–114, [4], pp. 191–193, [5], pp. 151–157) are used to improve the reliability and reduce the overhead of the Newton algorithm. In general, an update scheme $\mathbf{x}^{n+1} = \mathbf{x}^n + \alpha^n \mathbf{s}^n$ is used, where

$\mathbf{B}(\mathbf{x}^n)\cdot\mathbf{s}^n = -\nabla f(\mathbf{x}^n)$. Here $\mathbf{B}(\mathbf{x}^n)$ is an $N \times N$ matrix approximating $\mathbf{H}(\mathbf{x}^n)$. $\mathbf{B}(\mathbf{x}^n)$ is called a metric. Thus, they are often called variable metric algorithms. The Davidon–Fletcher–Powell (DFP) algorithm and the Broyden–Fletcher–Goldfarb–Shanno (BFGS) algorithms are two of the most famous quasi-Newton algorithms.

The Fortran-style pseudo-code for BFGS algorithm is given in Figure 1.18, where the symbol $\otimes$ stands for the outer or direct product of two vectors $\mathbf{u}$ and $\mathbf{v}$. $\mathbf{u}\otimes\mathbf{v}$ is a matrix. Its element $(\mathbf{u}\otimes\mathbf{v})_{ij}$ is $= u_i v_j$.

```
choose starting point x^0
choose starting Jacobian vector B^0
n = 0
do while (termination conditions not fulfilled)
   get s^n from B_n(x^n) · s^n = −∇f(x^n)
   x^(n+1) = x^n + α^n s^n
   y^n = ∇f(x^(n+1)) − ∇f(x^n)
   B^(n+1) = B^n + (y^n ⊗ y^n) / (y^n · s^n) − (B^n · s^n) ⊗ (B^n · s^n) / (s^n · B^n · s^n)
   n = n + 1
end do
```

Figure 1.18 Fortran-style pseudo-code for BFGS algorithm

1.2.2.2.4 Conjugate Gradient Method The conjugate gradient algorithm ([7], pp. 107–110, [4], pp. 193–195, [5], pp. 109–115) was initially proposed by Hestenes and Stiefel [10]. Later, Fletcher and Reeves [11] proved its quadratic convergence and extended it to nonquadratic functions. Therefore, it is also referred to as the Fletcher–Reeves algorithm ([2], pp. 169–173).

The conjugate gradient algorithm is another alternative to Newton's algorithm that does not require explicit second derivatives. In fact, the conjugate gradient does not even store an approximation to the Hessian matrix. As its name suggests, the conjugate gradient algorithm also uses the gradient, but it searches for the solution along a sequence of conjugate (i.e., orthogonal in some inner product) directions. The Fortran-style pseudo-code for the conjugate gradient algorithms is given in Figure 1.19.

1.2.3 Randomizing Deterministic Optimization Algorithms

Some deterministic algorithms can be randomized. Random multi-start is the simplest and most common approach. It is applicable to all deterministic algorithms requiring starting-points. The other approach is to randomize quantities such as the line search step. It applies to most of multi-dimensional deterministic optimization algorithms.

Some people regard randomized deterministic optimization algorithms as stochastic algorithms. This approach causes unnecessary confusion and should be abandoned. The original algorithms are strictly followed here unless explicitly stated otherwise.

```
choose starting point x^0
g^0 = ∇f(x^0)
s^0 = – g^0
n = 0
do while (termination conditions not fulfilled)
   get step α^n by minimizing function f(x^n + α^n s^n) of α
   x^(n+1) = x^n + α^n s^n
   g^(n+1) = ∇f(x^(n+1))
   B^(n+1) = B^n + (g^(n+1) ⊗ g^(n+1)) / (g^n · g^n)
   s^(n+1) = –g^(n+1) + B^(n+1) · s^n
   n = n + 1
end do
```

Figure 1.19 Fortran-style pseudo-code for conjugate gradient algorithm

1.3 Stochastic Optimization Algorithms

1.3.1 Motivation

Deterministic optimization algorithms have undergone many decades of intensive development. Nowadays, many deterministic optimization algorithms have appeared in textbooks of different levels. Some of them have even been implemented as intrinsic subroutines in commercial software packages. They have enjoyed considerable success.

Nevertheless, deterministic optimization algorithms have been facing more and more serious challenges arising from diverse real-world applications. Failure cases have been rapidly accumulating. Their inherent weaknesses have been frequently exposed.

Most deterministic optimization algorithms are mathematically elegant. However, they are never user-friendly. A user may be required to provide not only the objective and constraint functions but also their derivatives. This requirement is invariably tedious and may prove prohibitive. Computation of derivatives may impose serious overhead. The overhead becomes even worse when computation of derivatives has to be done through approximation such as finite difference.

Most deterministic optimization algorithms imply excessive restrictions on optimization parameters, objective functions, and constraint functions. Most deterministic optimization algorithms apply to real continuous optimization parameters only. Convexity, continuity, and differentiability of objective and constraint functions are the most common implications assumed by deterministic optimization algorithms. Unfortunately, many real-world application problems do not satisfy even one of these assumptions.

It is obvious that most deterministic optimization algorithms demand one or more starting-points. Good starting-points are critical for the success of deterministic optimization algorithms. Poor starting-points may have a significant adverse effect on deterministic optimization algorithms' efficiency, or even cause them to fail.

Most often, good starting-points cannot be chosen by chance. A good choice relies heavily on a priori knowledge which may require years of experience. When such a priori knowledge is not available, people usually have to use trial and error to locate starting-points before starting the optimization process. The aforementioned random multi-start approach is the other choice for this problem.

Computational efficiency is the most promising feature of most deterministic optimization algorithms. However, as computer technology advances speedily with each passing day, people's expectations with regard to efficiency have risen significantly.

As such, in recent decades, people have increasingly turned their attention to stochastic optimization algorithms, especially the evolutionary algorithms. An exponential growth in the use of stochastic optimization has been seen.

1.3.2 Outstanding Features

Stochastic optimization algorithms have many interesting features. Some of these features are controversial. Nevertheless, stochastic optimization has been gaining more and more popularity and acceptance.

1.3.2.1 Randomness

As mentioned earlier, deterministic optimization is clonable. In contrast, as the name indicates, results obtained from a stochastic optimization algorithm are in general unpredictable due to randomness. In practice, one may never be able to get identical optimal solutions, although the solutions obtained may differ only very slightly.

However, from the point of view of practical application, two mathematically different results are regarded as identical if both of them meet the tolerance requirement imposed by the practical application.

A controversy accompanying stochastic optimization algorithms is their proof of absolute success, either theoretically or numerically. No stochastic optimization algorithm guarantees absolute success, although the failure percentage might be very small. A search by a stochastic optimization algorithm may miss the optimal solution. This constitutes a major challenge for the entire stochastic optimization community. Mathematicians working on it are a long way from a successful conclusion.

1.3.2.2 Simplicity

Stochastic optimization algorithms are in general mathematically simpler than deterministic algorithms. Usually, neither an exact nor an approximate derivative is involved. Most stochastic optimization algorithms generate their initial solution through an inherent initialization process, and thus avoid being troubled by choosing a starting-point. Relief from heavy reliance on trial and error or a priori knowledge in guessing starting-points is a tremendous advantage in the eyes of many optimization practitioners.

Another controversy as regards stochastic optimization algorithms is their rigorous mathematical foundation. Most of the stochastic optimization algorithms are inspired by natural phenomena which mankind has been struggling to understand.

Each stochastic optimization algorithm has at least one intrinsic control parameter. The performance of stochastic optimization algorithms more or less depends on these intrinsic control parameters. It is well known that tuning these intrinsic control parameters for better performance is usually very hard. In this sense, stochastic optimization algorithms are not as simple as people have believed.

1.3.2.3 Efficiency

Stochastic optimization algorithms usually require more objective function evaluations to find the optimal solution than deterministic optimization algorithms, given that both of them succeed. In other words, they are computationally more expensive or less efficient.

1.3.2.4 Robustness

This is the third controversy as far as stochastic optimization algorithms are concerned. Stochastic optimization algorithms may occasionally miss the optimal solution even if everything is favorable. On the other hand, stochastic optimization algorithms are usually capable of bracketing a quasi-optimal solution within a wide search space, while deterministic optimization algorithms usually fail to do so in the same situation. In most practical applications, a quasi-optimal solution is welcome. Very often, it is immediately accepted. In the event that it is not acceptable, certain measures can be taken to refine it.

1.3.2.5 Versatility

Most stochastic optimization algorithms do not impose restrictions on optimization problems. In addition, many stochastic optimization algorithms apply to discrete or even symbolic optimization parameters as well as real ones. In this regard, stochastic optimization optimizations are versatile.

1.3.3 Classification

Stochastic optimization algorithms can be divided into two major categories according to their origins: physical algorithms and evolutionary algorithms. Some people regard artificial neural networks and artificial immune systems as stochastic optimization algorithms. Indeed, they can be used to solve certain optimization problems. However, before solving the optimization problem at hand, training has to be carried out. These algorithms cannot accomplish the optimization by themselves without the help of training sets. For this reason, this author personally would not regard them as stochastic optimization algorithms.

Once again, we do not regard hybrids of two or more stochastic optimization algorithms as a separate category.

1.3.4 Physical Algorithms

Stochastic optimization algorithms in this category are inspired by physical phenomena. The Monte Carlo algorithm [12] and the simulated annealing algorithm [13,14] are two of the most prominent algorithms in this category.

1.3.4.1 Monte Carlo Algorithm

The Monte Carlo algorithm, named for a famous casino city in Monaco, relies on repeated random sampling to find the optimal solution. The use of randomness and the repetitive nature of the process are analogous to the activities conducted at a casino. It was originally practiced under more generic names such as statistical sampling. In the 1940s, physicists working on nuclear weapons projects at the Los Alamos National Laboratory coined the present name.

The Fortran-style pseudo-code for the Monte Carlo algorithm for minimizing $f(\mathbf{x})$ is given in Figure 1.20.

```
n = 1
do while (termination conditions not fulfilled)
   randomly generate x^n
   compute f(x^n)
   n = n + 1
end do
```

Figure 1.20 Fortran-style pesudo-code for the Monte Carlo algorithm

1.3.4.2 Simulated Annealing Algorithm

The simulated annealing algorithm imitates the annealing process in metallurgy, a technique involving heating and controlled cooling of a material to increase the size of its crystals and reduce defect. By analogy with this physical process, each step of the simulated annealing algorithm replaces the current solution by a new solution with a probability that depends on the difference of objective function values at the two solution points and the temperature. The most often implemented probability distribution function is the Boltzmann probability distribution which is $p(\mathbf{x}, \mathbf{y}, \mathrm{T}) = \exp\{-[f(\mathbf{y}) - f(\mathbf{x})]/(k\,T)\}$ where k is the Boltzmann constant.

The Fortran-style pseudo-code for a generic simulated annealing algorithm is given in Figure 1.21. Here $p(\mathbf{x}^{n-1}, \mathbf{y}^n, T^{n-1})$ is positive value when $f(\mathbf{y}^n) > f(\mathbf{x}^{n-1})$. This means that the new solution $\mathbf{y}^n$ may be accepted even if it is worse than $\mathbf{x}^{n-1}$. It is this feature that prevents the simulated annealing algorithm from being trapped in a locally minimum solution point.

1.4 Evolutionary Algorithms

Evolutionary algorithms were inspired by Darwin's theory of evolution. Natural selection is the foundation of Darwin's theory of evolution.

The study of evolutionary algorithms began in the 1960s. A number of creative researchers independently came up with the idea of mimicking the biological evolution mechanism and developed three mainstream evolutionary algorithms, namely, genetic algorithms [15–17], evolutionary programming [18,19], and evolution strategies [20,21]. Other evolutionary algorithms include memetic algorithms [22,23], scatter search ([22,24], [25], pp. 183–193, [26], pp. 519–537, [27]), self-organizing migrating [28], and tabu search [29–32].

```
generate x^0
set initial temperature T^0
n = 1
do while (termination conditions not fulfilled)
  get new solution y^n
  get f(y^n)
  if(p(x^(n-1), y^n, T^(n-1)) > rand(0, 1))
    x^n = y^n
    f(x^n) = f(y^n)
  end if
  get new temperature T^n
  n = n + 1
end do
```

Figure 1.21 Fortran-style pseudo-code for simulated annealing algorithm

In recent decades, swarm algorithms [33] including ant colony optimization [22,34], bees algorithm [35], cultural algorithm [22,36], particle swarm optimization [22,37], have emerged within the evolutionary computation community. The swarm algorithms, as their name implies, imitate human or animal behaviors.

1.4.1 Evolutionary Terminologies

For the convenience of the following description, several essential terminologies frequently encountered in evolutionary computation are defined.

1.4.1.1 Gene

The gene is the basic building block of all evolutionary algorithms. There are usually two classes of genes: real, where a gene is a real number; and alphabetic, where a gene takes a value from an alphabet set. Common alphabet sets are the binary, octal, decimal, and hexadecimal sets.

1.4.1.2 Chromosome

The chromosome is another essential building block of all evolutionary algorithms. It is a symbolic representation of optimization parameters **x**.

Genes are concatenated into chromosomes in the following form

$$\mathbf{g} = \overbrace{g_{11}\cdots g_{1L_1}}^{x_1}\ \overbrace{g_{21}\cdots g_{1L_2}}^{x_2}\cdots\ \overbrace{g_{N1}\cdots g_{NL_N}}^{x_N}$$

where g_{ij} is the *j*th gene representing x_i, and L_i is the number of genes in the substring for x_i.

A two-way mapping exists between an actual optimization parameter and the corresponding substring of genes in a chromosome. Such mapping is usually referred to as an encoding/decoding operation in evolutionary computation.

Prominent mappings include the following:

- Natural mapping. The optimization parameter itself is the gene, that is, $\mathbf{x}=\mathbf{g}$. In this case, optimization parameters and chromosome will be used interchangeably.
- Digitizer. Digitize an optimization parameter as follows

$$x_i \approx b_i^L + (b_i^U - b_i^L) \sum_{j=1}^{L_i} g_{ij} B_i^{-j}, \tag{1.15}$$

where B_i is the base of the alphabet set for x_i, $D_i = [b_i^L, b_i^U]$ is the search space of x_i.

1.4.1.3 Fitness

Fitness is the measure of goodness of a chromosome. It is directly related to objective and constraint function values through a scaling operation. It is not absolutely necessary for some evolutionary algorithms such as differential evolution.

1.4.1.4 Individual

In a real society, an individual is a living creature. However, in the community of evolutionary computation, an individual **p** is an aggregate of a chromosome **g**, optimization parameter values **x**, and objective function (including constraint) values **f**. Additional attributes of an individual may include fitness value, generation, velocity, age, gender, and even memory.

The similarity of two individuals **p** and **q** is quantitatively defined as

$$s(\mathbf{p}, \mathbf{q}) = e(\mathbf{x}^p, \mathbf{x}^q) + \sum_{i=1}^{N_f} e[f_i(\mathbf{x}^p), f_i(\mathbf{x}^q)], \tag{1.16}$$

where

$$e(\mathbf{x}, \mathbf{y}) = \begin{cases} ||\mathbf{x}-\mathbf{y}||/\min(||\mathbf{x}||, ||\mathbf{y}||), & ||\mathbf{x}|| \neq 0 \cap ||\mathbf{y}|| \neq 0, \\ ||\mathbf{x}-\mathbf{y}||, & \text{otherwise}, \end{cases}$$

$$||\mathbf{x}|| = \sqrt{\sum_{j=1}^{N} x_j^2},$$

$$e(f, g) = \begin{cases} |f-g|/\min[|f|, |g|], & |f| \neq 0 \cap |g| \neq 0, \\ |f-g|, & \text{otherwise}. \end{cases}$$

To quantitatively compare two individuals, a logic dominance function is defined as

$$d(\mathbf{p},\mathbf{q}) = \begin{cases} \text{true}, & \begin{array}{l} \forall 1 \le i \le N_{o^{\min}} : o_i^{\min}(\mathbf{x}^p) \le o_i^{\min}(\mathbf{x}^q) \cap \\ \forall 1 \le i \le N_{o^{\max}} : o_i^{\max}(\mathbf{x}^p) \ge o_i^{\max}(\mathbf{x}^q) \cap \\ \forall 1 \le i \le N_{c^=} : |c_i^=(\mathbf{x}^p)| \le |c_i^=(\mathbf{x}^q)| \cap \\ \forall 1 \le i \le N_{c^-} : c_i^-(\mathbf{x}^p) \le c_i^-(\mathbf{x}^q) \cap \\ \forall 1 \le i \le N_{c^+} : c_i^+(\mathbf{x}^p) \ge c_i^+(\mathbf{x}^q), \end{array} \\ \text{false}, & \text{otherwise.} \end{cases} \tag{1.17}$$

1.4.1.5 Population

A population is a congregation of individuals. An important characteristic feature of a population is its age, expressed in terms of generation. It is the object all evolutionary algorithms work on. On average, statistically speaking, a later population is better than an earlier population.

The most important feature of a population $\mathbf{P}^n$ of generation n containing N_p individuals is its diversity which can be defined as

$$d(\mathbf{P}^n) = \min_{1 \le i \ne j \le N_p} s(\mathbf{p}^{n,i}, \mathbf{p}^{n,j}). \tag{1.18}$$

If the best individual of population $\mathbf{P}^n$, $\mathbf{p}^{n,\text{best}}$, and the worst individual, $\mathbf{p}^{n,\text{worst}}$, are identifiable, it is computationally more efficient to redefine the population diversity as

$$d(\mathbf{P}^n) = s(\mathbf{p}^{n,\text{best}}, \mathbf{p}^{n,\text{worst}}). \tag{1.19}$$

In practice, both diversity functions must be applied with care. Mathematically, the former definition is more reasonable. It is also more beneficial for diversity preservation. However, it is computationally much more expensive. On the other hand, the latter definition may overemphasize an individual's performance and discriminate an individual's genetic features.

1.4.2 Prominent Evolutionary Algorithms

1.4.2.1 Genetic Algorithms

Genetic algorithms, a class of search techniques inspired by evolutionary biology, were originated by J.H. Holland. The general flow chart of genetic algorithms is presented in Figure 1.22.

As seen in Figure 1.22, genetic algorithms involve two stages: initialization and evolution. Initialization generates an initial population $\mathbf{P}^0$. Then $\mathbf{P}^0$ evolves into $\mathbf{P}^1$, $\mathbf{P}^1$ evolves into $\mathbf{P}^2$, . . ., until the termination conditions are fulfilled. While evolving from $\mathbf{P}^n$ to $\mathbf{P}^{n+1}$, the three evolutionary operations – selection, crossover, and mutation – are executed in sequence.

Originally, optimization parameters were represented by binary chromosomes regardless of their actual types. Accordingly, crossover and mutation are Boolean operations. This applies to the following description unless otherwise explicitly stated. In addition, here, maximizing

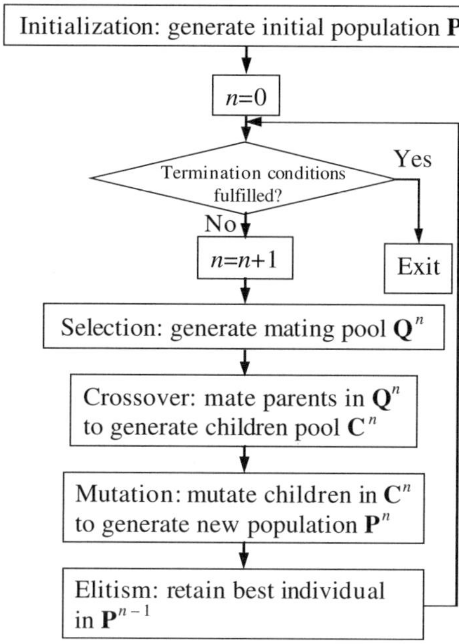

Figure 1.22 General flow chart of genetic algorithms

$f(\mathbf{x})$ is assumed for convenience and consistency with the original standard binary genetic algorithm. In this case, we use the objective function $f(\mathbf{x})$ directly as a fitness function.

1.4.2.1.1 Initialization Initialization generates an initial population $\mathbf{P}^0$ which contains N_p individuals $\mathbf{p}^{0,i}$, $1 \le i \le N_p$. Individual $\mathbf{p}^{0,i}$ is generated in three steps. A binary chromosome $\mathbf{g}^{0,i}$ is first generated. Then the binary chromosome is decoded to obtain the actual optimization parameter values $\mathbf{x}^{0,i}$. The optimization parameters are finally used to evaluate the objective function value $f(\mathbf{x}^{0,i})$. The Fortran-style pseudo pseudo-code for this process is given in Figure 1.23.

do $i = 1, N_p$

  do $j = 1, N$

    do $k = 1, L_j$

      $g_{j,k}^{0,i} = \text{int}(2 * \text{rand}(0, 1))$

    end do

    $$x_j^{0,i} = b_j^L + \left(b_j^U - b_j^L\right)\sum_{k=1}^{L_j} g_{j,k}^{0,i} 2^{-k}$$

  end do

  evaluate $f(\mathbf{x}^{0,i})$

end do

Figure 1.23 Fortran-style pseudo-code for initialization of genetic algorithms

1.4.2.1.2 Selection The selection operator is believed to be responsible for the convergence of genetic algorithms. It selects good individuals on the basis of their fitness values and produces a temporary population, namely, the mating pool $\mathbf{Q}^n$ with member individuals $\mathbf{q}^{n,i}$. The mating pool is usually, but not necessarily, the same size as $\mathbf{P}^{n-1}$. This can be achieved by many different schemes, but the most common methods are roulette-wheel, ranking and tournament selection.

1.4.2.1.2.1 Roulette-Wheel Selection The essential idea behind roulette-wheel selection is that the probability of selection of an individual $\mathbf{p}^{n-1,i}$ in the parent population $\mathbf{P}^{n-1}$ is proportional to its fitness value, that is, $p^{n-1,i} \propto f(\mathbf{x}^{n-1,i})/\sum_{i=1}^{N_p} f(\mathbf{x}^{n-1,i})$. It gets its name due to its analogy to the roulette-wheel in a casino.

1.4.2.1.2.2 Ranking Selection The idea of ranking selection is very simple. All individuals $\mathbf{p}^{n-1,i}$ in the parent population $\mathbf{P}^{n-1}$ are ranked according to their fitness. Individuals with higher rank (higher fitness value) are admitted into $\mathbf{Q}^n$. Usually, the better half of the individuals in $\mathbf{P}^{n-1}$ are copied and duplicated to fill $\mathbf{Q}^n$.

1.4.2.1.2.3 Tournament Selection Tournament selection implements a tournament to decide the membership of an individual $\mathbf{p}^{n-1,i}$ in parent population $\mathbf{P}^{n-1}$ in $\mathbf{Q}^n$. The winner of an m-participant tournament is selected. The Fortran-style pseudo-code for the binary, that is, two-participant, tournament selection procedure is given in Figure 1.24.

```
do i = 1, Np
  j = int(Np * rand(0, 1)) + 1
  k = int(Np * rand(0, 1)) + 1
  do while (k = j)
    k = int(Np * rand(0, 1)) + 1
  end do
  if(f(x^(n-1,j)) > f(x^(n-1,k)))
    q^(n,i) = p^(n-1,j)
  else
    q^(n,i) = p^(n-1,k)
  end if
end do
```

Figure 1.24 Fortran-style pseudo-code for binary tournament selection

1.4.2.1.3 Crossover The crossover operator has been believed to be the main search tool of genetic algorithms. It mates individuals $\mathbf{q}^{n,i}$ and $\mathbf{q}^{n,j}$ in the mating pool by pairs and generates children by crossing over the mated pairs with probability p_c, one of the intrinsic control parameters of genetic algorithms. Analogously to human society, the two mating partners are defined as parents, among which individual $\mathbf{q}^{n,i}$ is the mother and individual $\mathbf{q}^{n,j}$ is the father. Usually, but not necessarily, every pair of parents delivers two children.

Many crossover schemes have been developed. Four of them are introduced here.

1.4.2.1.3.1 One-Point Crossover One-point crossover randomly selects a single crossover point r ($1 < r \leq N$). The two children, $\mathbf{c}^{n,a}$ and $\mathbf{c}^{n,b}$, are delivered by swapping genes of the two mating partners, $\mathbf{q}^{n,i}$ and $\mathbf{q}^{n,j}$, as shown in Figure 1.25. More exactly, child $\mathbf{c}^{n,a}$ inherits genes of $\mathbf{q}^{n,i}$ to the left of the crossover point and genes of $\mathbf{q}^{n,j}$ from the crossover point (including) to the right end of the chromosomes. $\mathbf{c}^{n,b}$ inherits the remaining genes of $\mathbf{q}^{n,i}$ and $\mathbf{q}^{n,j}$.

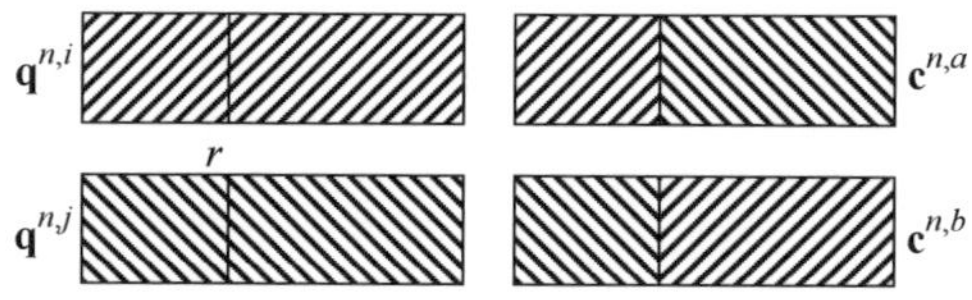

Figure 1.25 One-point crossover

1.4.2.1.3.2 Multi-Point Crossover In the multi-point crossover scheme, the chromosomes of both $\mathbf{q}^{n,i}$ and $\mathbf{q}^{n,j}$ are cut into $m + 1$ partitions. Child $\mathbf{c}^{n,a}$ inherits odd-partition genes of $\mathbf{q}^{n,i}$ and even-partition genes of $\mathbf{q}^{n,j}$. Similarly, $\mathbf{c}^{n,b}$ inherits the remaining genes of $\mathbf{q}^{n,i}$ and $\mathbf{q}^{n,j}$. An example of a two-point crossover ($m = 2$) is shown in Figure 1.26.

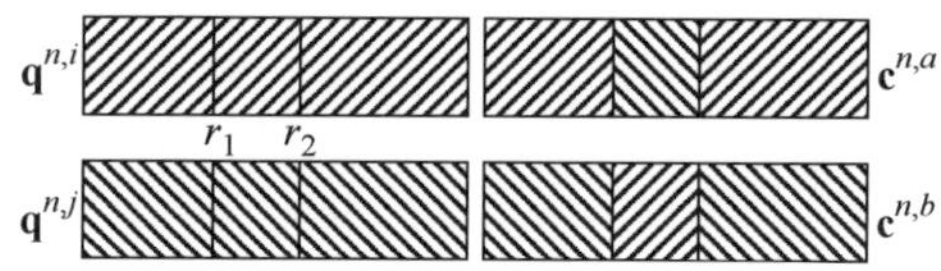

Figure 1.26 Two-point crossover

1.4.2.1.3.3 Binomial Crossover The Fortran-style pseudo-code for the binomial crossover scheme is given in Figure 1.27, where L is the chromosome length and p_c is the probability of crossover. One extreme case is that $\mathbf{q}^{n,i}$ and $\mathbf{q}^{n,j}$ do not exchange any genes. In this case, they will be forced to exchange one gene at a randomly selected site.

```
do k = 1, L
  if(rand(0, 1) < p_c)
    g_k^{n,c,a} = g_k^{n,q,j}
    g_k^{n,c,b} = g_k^{n,q,i}
  else
    g_k^{n,c,a} = g_k^{n,q,i}
    g_k^{n,c,b} = g_k^{n,q,j}
  end if
end do
```

Figure 1.27 Fortran-style pseudo-code for binomial crossover

1.4.2.1.3.4 Exponential Crossover Exponential crossover is a cyclic two-point crossover. The Fortran-style pseudo-code is given in Figure 1.28, where p_e is the probability of exponential crossover which controls the crossover length. A possible crossover result is given in Figure 1.29, where $r + M > L$.

```
do k = 1, L

    g_k^{n,c,a} = g_k^{n,q,i}
    g_k^{n,c,b} = g_k^{n,q,j}
end do
M = 1
β = rand(0, 1)
do while (β ≤ p_c and M < L – 2)
   M = M + 1
   β = rand(0, 1)
end do
r = L * rand(0, 1) + 1
do m = 0, M – 1
   k = mod(r + m, L)
   if (k = 0) k = L
   g_k^{n,c,a} = g_k^{n,q,j}
   g_k^{n,c,b} = g_k^{n,q,i}
end do
```

Figure 1.28 Fortran-style pseudo-code for exponential crossover

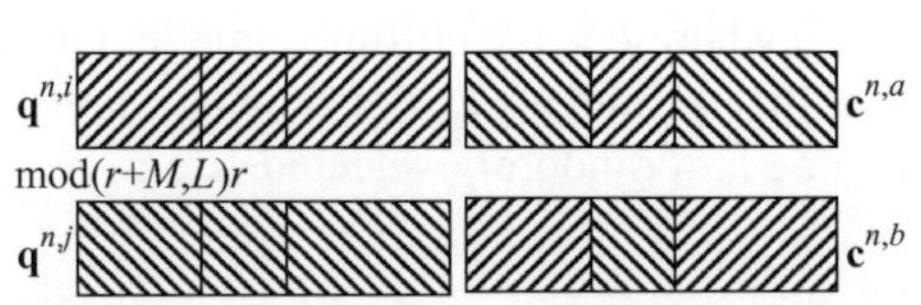

Figure 1.29 Exponential crossover

1.4.2.1.4 Mutation After crossover, some of the genes in the child chromosome are inverted with probability p_m, another intrinsic control parameter of genetic algorithms. The mutation operator is included to prevent premature convergence by ensuring the population diversity.

1.4.2.1.5 Elitism Obviously, there is no direct comparison between $\mathbf{P}^{n-1}$ and $\mathbf{P}^n$. All children are unconditionally admitted to $\mathbf{P}^n$ and all parents in $\mathbf{P}^{n-1}$ are unconditionally abandoned. Consequently, the best individual (in terms of fitness) in $\mathbf{P}^n$ may be worse than that in $\mathbf{P}^{n-1}$. In this case, elitism will be implemented by replacing the worst individual in $\mathbf{P}^n$ with the best individual in $\mathbf{P}^{n-1}$.

1.4.2.1.6 Real-Coded Genetic Algorithm The standard binary genetic algorithm encodes optimization parameters into binary chromosomes regardless of the nature of the optimization parameters. This makes the standard binary genetic algorithm flexible in that it can handle almost all kinds of optimization parameters. However, staircase error will inevitably be introduced when encoding a real number. The encoding and decoding operations also make the algorithm more computationally expensive for problems with real optimization parameters. A less often mentioned serious pitfall is the increased difficulty in controlling the algorithm due to the introduction of at least one more intrinsic control parameter, the chromosome length. Working directly on real optimization parameters can remove all these overheads. This leads to the real-coded genetic algorithm [38], in which natural real code is adopted, that is, $\mathbf{x} = \mathbf{g}$. Besides the aforementioned Boolean crossover operators, other schemes are also applicable. Arithmetic crossover is most often applied.

1.4.2.1.6.1 Arithmetic One-Point Crossover Arithmetic one-point crossover randomly selects a single crossover point r $(1 < i < N)$. Child $\mathbf{c}^{n,a}$ inherits genes of $\mathbf{q}^{n,i}$ to the left of the crossover point while $\mathbf{c}^{n,b}$ inherits genes of $\mathbf{q}^{n,j}$ to the left of the crossover point. The remaining genes of $\mathbf{c}^{n,a}$ and $\mathbf{c}^{n,b}$ are generated by arithmetic swapping

$$\left.\begin{aligned} x_k^{n,c,a} &= hx_k^{n,q,i} + (1-h)x_k^{n,q,j} \\ x_k^{n,c,b} &= (1-h)x_k^{n,q,i} + hx_k^{n,q,j} \end{aligned}\right\} \quad k \geq r. \tag{1.20}$$

where h is the arithmetic swapping intensity. An example of schematic crossover is shown in Figure 1.30.

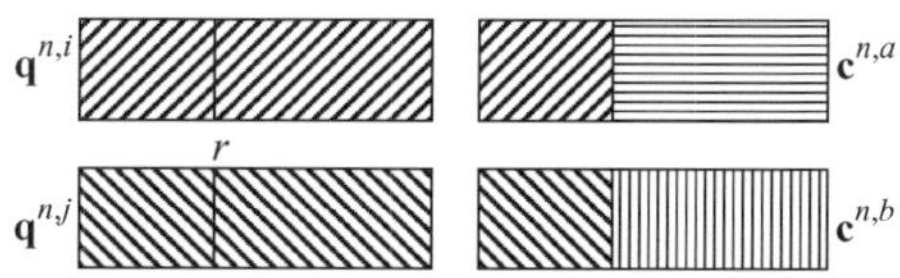

Figure 1.30 Arithmetic one-point crossover

1.4.2.1.6.2 Arithmetic Multi-Point Crossover In arithmetic multi-point crossover the optimization parameters of both $\mathbf{q}^{n,i}$ and $\mathbf{q}^{n,j}$ are likewise cut into m + 1 partitions. Child $\mathbf{c}^{n,a}$ inherits odd-partition genes from $\mathbf{q}^{n,i}\mathbf{c}^{n,b}$ inherits odd-partition genes from $\mathbf{q}^{n,j}$. The remaining genes of $\mathbf{c}^{n,a}$ and $\mathbf{c}^{n,b}$ are generated by arithmetic swapping. As a example, an arithmetic two-point crossover $(m = 2)$ is shown in Figure 1.31.

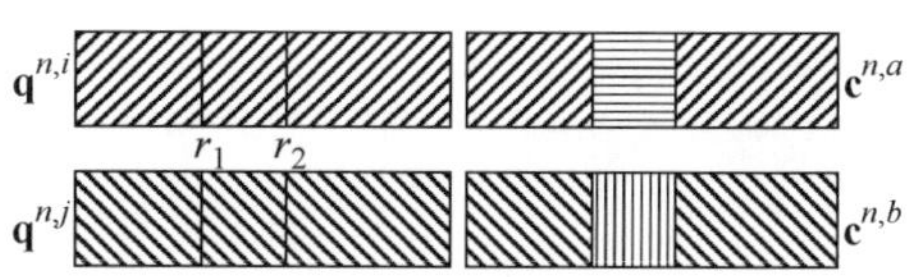

Figure 1.31 Arithmetic two-point crossover

1.4.2.1.6.3 Arithmetic Binomial Crossover The Fortran-style pseudo-code for arithmetic binomial crossover is given in Figure 1.32. One extreme case is where $\mathbf{q}^{n,i}$ and $\mathbf{q}^{n,j}$ do not exchange any parameters. In this case, they will be forced to exchange one parameter at a randomly selected site.

```
do k = 1, N
  if(rand(0, 1) < p_c)
    x_k^{n,c,a} = h x_k^{n,q,i} + (1 - h) x_k^{n,q,j}
    x_k^{n,c,b} = (1 - h) x_k^{n,q,i} + h x_k^{n,q,j}
  else
    x_k^{n,c,a} = x_k^{n,q,i}
    x_k^{n,c,b} = x_k^{n,q,j}
  end if
end do
```

Figure 1.32 Fortran-style pseudo-code for arithmetic binomial crossover

1.4.2.1.6.4 Arithmetic Exponential Crossover The Fortran-style pseudo-code for arithmetic exponential crossover is given in Figure 1.33. The only difference with exponential crossover happens at lines 15 and 16 as highlighted. A possible crossover result is given in Figure 1.34, where $r + M > N$.

```
do k = 1, N
  x_k^{n,c,a} = x_k^{n,q,i}
  x_k^{n,c,b} = x_k^{n,q,j}
end do
M = 1
β = rand(0, 1)
do while (β ≤ p_c and M < N - 2)
  M = M + 1
  β = rand(0, 1)
end do
r = N * rand(0, 1) + 1
do m = 0, M - 1
  k = mod(r + m, N)
  if (k = 0) k = N
    x_k^{n,c,a} = h x_k^{n,q,i} + (1 - h) x_k^{n,q,j}
    x_k^{n,c,b} = (1 - h) x_k^{n,q,i} + h x_k^{n,q,j}
end do
```

Figure 1.33 Fortran-style pseudo-code for arithmetic exponential crossover

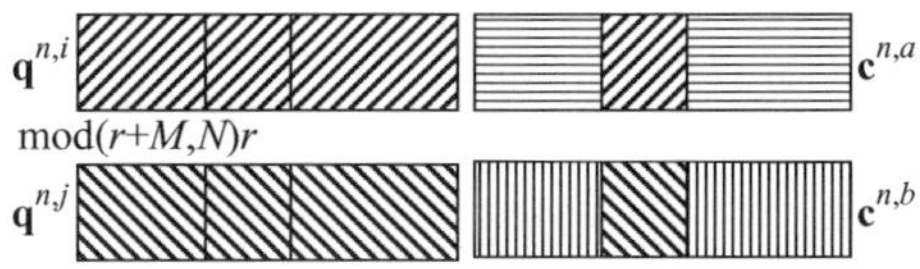

Figure 1.34 Arithmetic exponential crossover

1.4.2.1.6.5 Non-Uniform Arithmetic One-Point Crossover Non-uniform arithmetic one-point crossover randomly selects a single crossover point r ($1 < i < N$). Child $\mathbf{c}^{n,a}$ inherits genes of $\mathbf{q}^{n,i}$ to the left of the crossover point while $\mathbf{c}^{n,b}$ inherits genes of $\mathbf{q}^{n,j}$ to the left of the crossover point. Then, a crossover intensity, $h_k^{n,i}, k \geq r$, is randomly generated for each gene from the crossover point (including) to the right end of chromosomes. $h_k^{n,i}$ usually but not necessarily lies in [0,1]. Finally, the remaining genes of $\mathbf{c}^{n,a}$ and $\mathbf{c}^{n,b}$ are generated by non-uniform arithmetic swapping

$$\left.\begin{aligned} x_k^{n,i} &= h_k^{n,i} x_k^{n,q,i} + (1-h_k^{n,i}) x_k^{n,q,j} \\ x_k^{n,i+1} &= (1-h_k^{n,i}) x_j^{n,q,i} + h_k^{n,i} x_k^{n,q,j} \end{aligned}\right\} \quad k \geq r. \tag{1.21}$$

A schematic crossover result is shown in Figure 1.35.

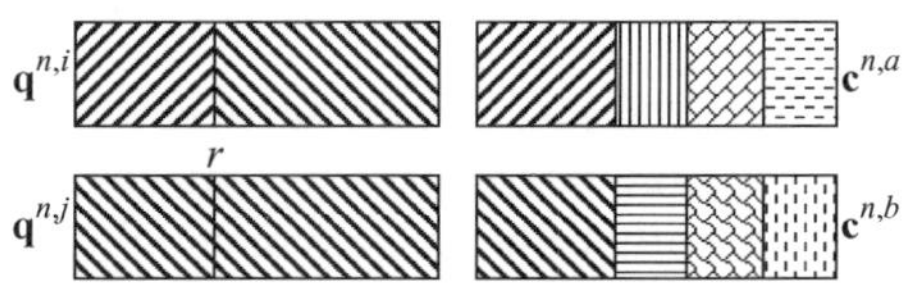

Figure 1.35 Non-uniform arithmetic one-point crossover

It is apparent that non-uniform arithmetic one-point crossover looks very similar with arithmetic one-point crossover. However, the crossover intensity $h_k^{n,i}$ for each child parameter $h_k^{n,i}$ to the right of the crossover point ($k \geq r$) is in general different. In this regard, it is non-uniform.

1.4.2.1.6.6 Non-Uniform Arithmetic Multi-Point Crossover In non-uniform arithmetic multi-point crossover the optimization parameters of both $\mathbf{q}^{n,i}$ and $\mathbf{q}^{n,j}$ are likewise cut into $m + 1$ partitions. Child $\mathbf{c}^{n,a}$ inherits odd-partition genes from $\mathbf{q}^{n,i}$, while $\mathbf{c}^{n,b}$ inherits odd-partition genes from $\mathbf{q}^{n,j}$. The remaining genes of $\mathbf{c}^{n,a}$ and $\mathbf{c}^{n,b}$ are generated by non-uniform arithmetic swapping.

A schematic crossover result is shown in Figure 1.36.

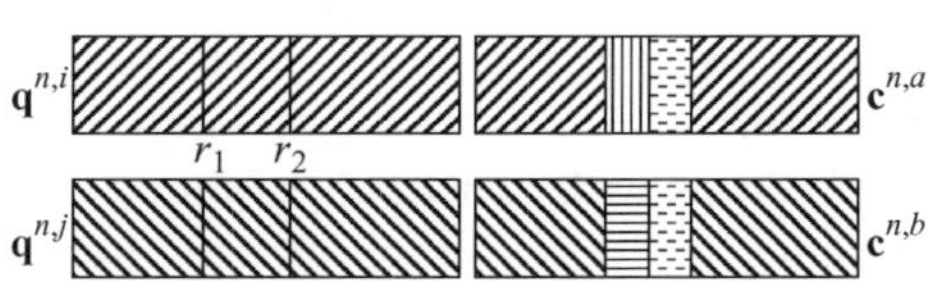

Figure 1.36 Non-uniform arithmetic multi-point crossover

1.4.2.1.6.7 Non-Uniform Arithmetic Binomial Crossover The Fortran-style pseudo-code for non-uniform arithmetic binomial crossover is given in Figure 1.37. The difference with arithmetic binomial crossover is highlighted.

```
do k = 1, N
  if(rand(0, 1) < p_c)
    h_k^{n,i} = rand(0, 1)
    x_k^{n,c,a} = h_k^{n,i} x_k^{n,q,i} + (1 - h_k^{n,i}) x_k^{n,q,j}
    x_k^{n,c,b} = (1 - h_k^{n,i}) x_k^{n,q,i} + h_k^{n,i} x_k^{n,q,j}
  else
    x_k^{n,c,a} = x_k^{n,q,i}
    x_k^{n,c,b} = x_k^{n,q,j}
  end if
end do
```

Figure 1.37 Fortran-style pseudo-code for non-uniform arithmetic binomial crossover

One extreme case is where $\mathbf{q}^{n,i}$ and $\mathbf{q}^{n,j}$ do not exchange any parameters. In this case, they will be forced to exchange one parameter at a randomly selected site.

1.4.2.1.6.8 Non-Uniform Arithmetic Exponential Crossover Non-uniform arithmetic exponential crossover is a cyclic two-point crossover. The Fortran-style pseudo-code is given in Figure 1.38, where the difference with arithmetic exponential crossover is highlighted.

A possible crossover result is given in Figure 1.39, where $r + M > N$. However, the Boolean inversion mutation operator is not applicable any more. Schemes working on real optimization parameters directly have to be implemented. Random perturbation mutation is most often applied.

Random perturbation mutation alters an optimization parameter of child $\mathbf{c}^{n,a}$ by adding a random perturbation term as $x_k^{n,c,a} = x_k^{n,c,a} + \alpha_k^{n,a} B_k$ where $-1 \le \alpha_k^{n,a} \le 1$ is a random number and B_k is the perturbation amplitude of the kth optimization parameter. B_k is usually no more than 10% of the corresponding search space.

1.4.2.2 Evolution Strategies

Evolution strategies were proposed by I. Rechenberg and H.P. Schwefel. There are two types: the (μ, λ)-strategy and the $(\mu + \lambda)$-strategy. Both generate a child population $\mathbf{C}^n$ of size λ through mutation from parent population $\mathbf{P}^{n-1}$ of size μ. The (μ, λ)-strategy generates $\mathbf{P}^n$ of size μ by selecting μ individuals from $\mathbf{C}^n$, while the $(\mu + \lambda)$-strategy generates $\mathbf{P}^n$ of size μ by selecting μ individuals from the union of $\mathbf{C}^n$ and $\mathbf{P}^{n-1}$. The general flow chart for evolution strategies is shown in Figure 1.40.

There are two notable differences with genetic algorithms. One is the reversed order of selection and mutation. The other one is the disappearance of crossover.

An individual in evolution strategies has an additional attribute: strategy parameters, that is, variances and covariances of a generalized N-dimensional normal distribution for mutation,

```
do k = 1, N
   x_k^{n, c, a} = x_k^{n, q, i}
   x_k^{n, c, b} = x_k^{n, q, j}
end do
M = 1
β = rand(0, 1)
do while (β ≤ p_c and M < N – 2)
   M = M + 1
   β = rand(0, 1)
end do

r = N * rand(0, 1) + 1
do m = 0, M – 1
   k = mod(r + m, N)
   if (k = 0) k = N
      h_k^{n, i} = rand(0, 1)
      x_k^{n, c, a} = h_k^{n, i} x_k^{n, q, i} + (1 – h_k^{n, i}) x_k^{n, q, j}
      x_k^{n, c, b} = (1 – h_k^{n, i}) x_k^{n, q, i} + h_k^{n, i} x_k^{n, q, j}
end do
```

Figure 1.38 Fortran-style pseudo-code for non-uniform arithmetic exponential crossover

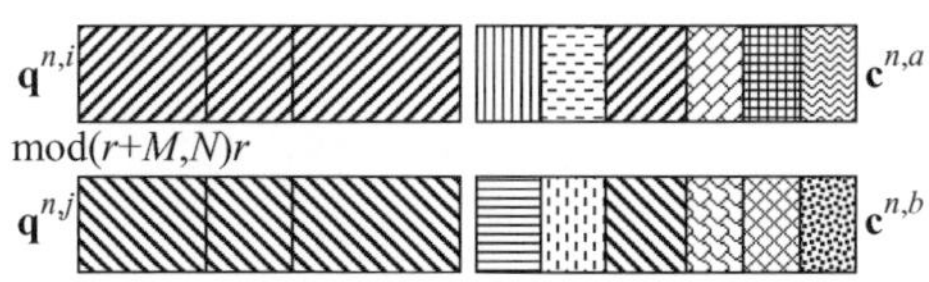

Figure 1.39 Non-uniform arithmetic exponential crossover

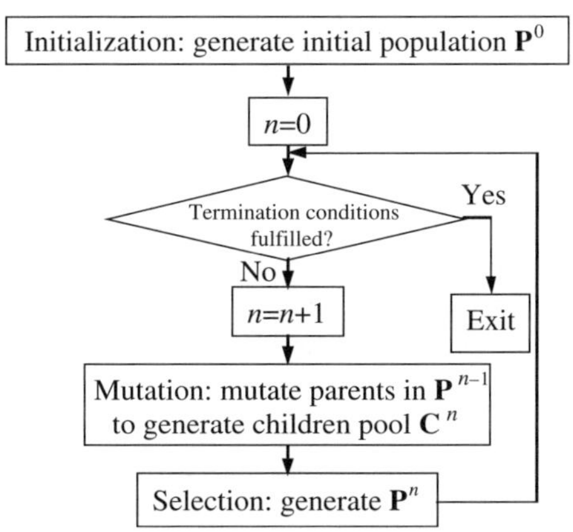

Figure 1.40 General flow chart of evolution strategies

$\mathbf{p} = (\mathbf{x}, \boldsymbol{\sigma}, \boldsymbol{\eta})$. $\boldsymbol{\sigma}$ is an N_σ-dimensional variance vector and $\boldsymbol{\eta}$ is an N_η-dimensional covariance vector, $\eta_j \in [-\pi, \pi]$, $1 \le j \le N_\eta$, $N_\sigma \ge 0$, $N_\eta \ge 0$.

N_σ and N_η play an essential role in mutation. Therefore, mutation schemes of evolution strategies are expressed as (N_σ, N_η)-mutation. In what follows we consider three representative mutation schemes.

1.4.2.2.1 (1, 0)-Mutation In this scheme, a child individual **c** is generated from its parent individual **p** as follows:

$$
\begin{aligned}
\sigma^c &= \sigma^p e^{\tau N(0,1)}, \\
x_j^c &= x_j^p + \sigma^c N_j(0,1),
\end{aligned}
\tag{1.22}
$$

where $\tau \propto \sqrt{1/N}$, $N(0, 1)$ denotes a normally distributed one-dimensional random number with mean zero and standard deviation one. $N_j(0, 1)$ is also a normally distributed one-dimensional random number with mean zero and standard deviation one but applies to σ_j^c only.

1.4.2.2.2 (N, 0)-Mutation In this scheme, a child individual **c** is generated from its parent individual **p** as follows:

$$
\begin{aligned}
\sigma_j^c &= \sigma_j^p e^{\tau N(0,1) + \tau' N_j(0,1)}, \\
x_j^c &= x_j^p + \sigma_j^c N_j(0,1),
\end{aligned}
\tag{1.23}
$$

where $\tau \propto \sqrt{1/(2N)}$, $\tau' \propto \sqrt{1/(2\sqrt{N})}$, and $N(0, 1)$ applies to all σ_j^c, $1 \le j \le N$.

1.4.2.2.3 (N, N (N − 1)/2)-Mutation In this scheme, a child individual **c** is generated from its parent individual **p** as follows:

$$
\begin{aligned}
\sigma_j^c &= \sigma_j^p e^{\tau N(0,1) + \tau' N_j(0,1)}, \\
\eta_j^c &= \eta_j^p + \beta N_j(0,1), \\
x_j^c &= x_j^p + N_j(\mathbf{0}, \boldsymbol{\sigma}^c, \boldsymbol{\eta}^c),
\end{aligned}
\tag{1.24}
$$

where $\tau \propto \sqrt{1/(2N)}$, $\tau' \propto \sqrt{1/(2\sqrt{N})}$, $\beta \approx 0.0873$, $N_j(\mathbf{0}, \boldsymbol{\sigma}, \boldsymbol{\eta})$ is the jth element of the N-dimensional correlated mutation vector $\mathbf{N}(\mathbf{0}, \boldsymbol{\sigma}, \boldsymbol{\eta})$ with mean zero, variances $\boldsymbol{\sigma}$, and covariances $\boldsymbol{\eta}$.

1.4.2.3 Evolutionary Programming

Evolutionary programming was proposed by L.J. Fogel. The original evolutionary programming was intended to operate on finite-state machines and the corresponding discrete representations. However, the present variants developed by L.J. Fogel are utilized for optimization problems with real optimization parameters. These variants have much in common with evolution strategies. A minor difference between them lies in selection. Each individual in the union of $\mathbf{C}^n$ and $\mathbf{P}^{n-1}$ is compared with q (>1) randomly chosen individuals, or opponents, which are also from the union. The individual is assigned a score which is the number of wins, or the number of individuals among its opponents dominated by it. The μ individuals with the highest scores survive.

1.4.2.4 Particle Swarm Optimization

Particle swarm optimization was introduced by Kennedy and Eberhart in 1995 [37]. As its name implies, it was inspired by the movement and intelligence of swarms. A swarm is a structured collection of interacting organisms such as bees, ants, or birds. Each organism in a swarm is a particle, or agent. Particles and swarms in particle swarm optimization are equivalent to individuals and populations in other evolutionary algorithms. The position, or site, of a particle in a swarm is the vector of optimization parameters **x** in particle swarm optimization.

Besides the attributes of an individual in other evolutionary algorithms, a particle in a swarm has two additional attributes: its velocity $\mathbf{v}^i$ and its memory $\mathbf{p}^{i,\text{ best}}$ for the best site it has ever visited. Usually the velocity of $\mathbf{p}^{i,\text{ best}}$ is not part of the memory. Accordingly, a swarm also has in its memory $\mathbf{p}^{\text{best}}$, the best site all the particles in the swarm have ever visited. Similarly, the velocity of $\mathbf{p}^{\text{best}}$ is not part of the memory.

Particles in a swarm cooperate by sharing knowledge. This has been shown to be the critical idea behind the success of the particle swarm optimization algoirthm. The block diagram for particle swarm optimization is shown in Figure 1.41.

The movement of a particle is accomplished in four steps: velocity update; position update; memory update; and swarm memory update.

1.4.2.4.1 Velocity Update The velocity $\mathbf{v}^{n,i}$ of particle $\mathbf{p}^{n,i}$ is updated by

$$v_j^{n+1,i} = wv_j^{n,i} + c_p\alpha_j(p_j^{i,\text{best}} - x_j^{n,i}) + c_s\beta_j(g_j^{\text{best}} - x_j^{n,i}), \tag{1.25}$$

where w, c_p, and c_s are the inertial weight, cognitive rate, and social rate which are intrinsic control parameters of particle swarm optimization.

1.4.2.4.2 Position Update The position $\mathbf{x}^{n,i}$ is updated by

$$x_j^{n+1,i} = x_j^{n,i} + \Delta t v_j^{n,i}, \tag{1.26}$$

where Δt is the time step. It usually is assumed 1 since it can always be absorbed by w, c_p, and c_s.

1.4.2.4.3 Memory Update The personal memory $\mathbf{p}^{i,\text{ best}}$ of particle $\mathbf{p}^{n,i}$ will be updated if the updated position $\mathbf{x}^{n,i}$ outperforms the current best position $\mathbf{x}^{i,\text{ best}}$ in its memory.

1.4.2.4.4 Swarm Memory Update The swarm memory $\mathbf{p}^{\text{best}}$ will be updated if the updated position of particle $\mathbf{p}^{n,i}$ outperforms the best position $\mathbf{x}^{\text{best}}$ in the swarm's memory.

1.4.3 Evolutionary Crimes

After years of involvement in the evolutionary computation community, this author has noticed various acts of misconduct within it. Some of these acts have led to serious misconceptions. By analogy with inverse crimes in the inverse problems community, this author refers these acts of misconduct as evolutionary crimes.

In fact, although it is embarrassing to admit, this author is both a criminal and a victim of evolutionary crimes.

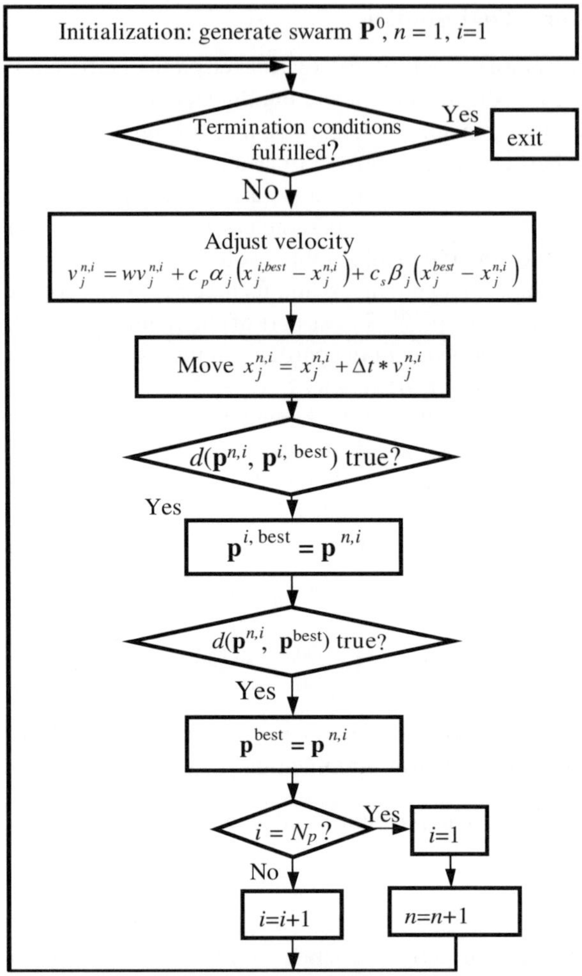

Figure 1.41 Block diagram for particle swarm optimization

1.4.3.1 Definition

An evolutionary crime is the exploitation of inadequate numerical evidence to claim advantage of an optimization algorithm over an evolutionary algorithm and/or the goodness of certain intrinsic control parameters of an evolutionary algorithm. Four kinds of evolutionary crime have been observed: ignorance of randomness (type A); baseless generalization (type B); biased comparison (type C); and transfer of problem (type D).

1.4.3.1.1 Ignorance of Randomness The stochastic nature of evolutionary algorithms means that the result of a single simulation run is much less trustworthy because of randomness. Instead, comparison should be based on averaged results over multiple runs. The higher the number of runs, the more trustworthy the average simulation result is.

Babu and co-workers reported two studies on intrinsic control parameters of differential evolution in 2000 ([39,40]). Evolutionary crime mentioned here is committed.

1.4.3.1.2 Baseless Generalization Many people have realized the sensitivity of differential evolution to its intrinsic control parameters. Therefore, they have carried out case studies on intrinsic control parameters. These case studies are conducted on very limited cases of intrinsic control parameters. The span of the intrinsic control parameters studied only partially covers the intrinsic control parameter domain. However, the researchers claim that the observed best intrinsic control parameters are the optimal intrinsic control parameters of differential evolution. Such a claim is unconvincing and constitutes an evolutionary crime.

Another baseless generalization is concerned with the test bed. Sometimes, the test bed implemented is not well designed. Problem features are not carefully taken care of. Only limited problem features are addressed. However, the final conclusion is unintentionally or purposely generalized.

1.4.3.1.3 Biased Comparison When a new optimization algorithm is developed, its proposer needs to prove its capability. Comparison with other existent optimization algorithms is the most common approach. This crime is usually evident when at least one evolutionary algorithm is chosen as competitor. It can be found throughout the evolutionary computation literature.

Every evolutionary algorithm has its specific intrinsic control parameters. For example, population size, mutation intensity, and crossover probability are the intrinsic control parameters for some strategies of differential evolution, while population size, crossover probability, and mutation probability are the three intrinsic control parameters for the real-coded genetic algorithm.

It is well known that the performance of an evolutionary algorithm is more or less sensitive to its intrinsic control parameters. Inappropriate choice of intrinsic control parameter values may make its performance deteriorate significantly or even cause it to fail.

Bias is due to fixing intrinsic control parameters of the competing evolutionary algorithms so as claim that the new optimization algorithm has advantages over competing evolutionary algorithms. It is appropriate to claim advantage only if the best intrinsic control parameters of the competing evolutionary algorithm are known. In this case, the new optimization algorithm would be remarkable because it absolutely outperforms its competitors. Unfortunately, very often, the best intrinsic control parameter values and the corresponding best performance of the competing evolutionary algorithms are not known.

In fact, bias, and thus evolutionary crime, may still be evident even if more sets of intrinsic control parameter values are investigated, if the sets are not carefully chosen.

1.4.3.1.4 Transfer of Problem This crime has been committed by most practitioners of evolutionary algorithms working on adaptation of intrinsic control parameters. They have realized the sensitivity of evolutionary algorithms to their intrinsic control parameters and the difficulty of choosing optimal intrinsic control parameters which implies that they do not know the best intrinsic control parameter values of the evolutionary algorithms concerned. They have made various proposals to adapt the intrinsic control parameters and usually claimed performance improvement over the original evolutionary algorithm. However, these proposals have their own specific intrinsic control parameters. Therefore, the problem changes merely from choosing optimal intrinsic control parameters for the evolutionary algorithms concerned to choosing intrinsic control parameters for a specific proposal to adapt those intrinsic control parameters. In the worst scenario, a proposal to adapt intrinsic control parameters for an evolutionary algorithm introduces more intrinsic control parameters and builds a hierarchy.

1.4.3.2 Damages

Evolutionary crimes have caused serious damages, most notoriously the overestimation of crossover and the underestimation of mutation in genetic algorithms. One aspect of the damage to differential evolution is the three misconceptions widely circulated within the differential evolution community.

1.4.3.2.1 There is No Dramatic Difference in Performance Between the One- and Two-Array Methods More exactly, the one-array method is the classic differential evolution while the two-array method is its dynamic counterpart. We will show later that dynamic differential evolution significantly outperforms classic differential evolution.

1.4.3.2.2 DE/best/1/* is More Prone to Being Trapped in a Local Optimum than DE/rand/1/* Indeed, fixing the best individual in a population as the base for mutant generation is greedier than using a random base. However, differential evolution has its own balance mechanism.

In fact, this misconception is against the secular administration. It is a kind of anarchism in differential evolution. At all levels of secular administration, a leader guides the whole community. A good national leader will bring stability and prosperity to the whole country while anarchism brings disastrous suffering in the form of poverty, starvation, or even war.

1.4.3.2.3 Crossover is Not so Important The role of crossover in differential evolution has been underestimated since its inception. The success of the innovative idea of differential mutation may be to blame for this conception. However, evolutionary crime is definitely responsible for it too.

1.4.3.3 Remedies

Evolutionary crimes have caused huge damage to the evolutionary computation community. It is expected to cause more damage, at least in the near future, since many evolutionary computation practitioners have not even realized that they are committing evolutionary crimes. It is therefore imperative to promote the concept within the whole community. Awareness of evolutionary crimes is only the first step. Measures have to be taken to avoid future commitment.

1.4.3.3.1 Application Instead of Comparison It is understandable that the computational cost of multiple runs may not be affordable, especially for computationally expensive application problems. In this case, most often, the first priority is to prove the potential of the evolutionary algorithm concerned by successfully capturing the solution of the problem. Hence, it may not be absolutely necessary to compare the performance of the evolutionary algorithm concerned with other optimization algorithms, which requires averaged performance over multiple runs.

1.4.3.3.2 Comprehensive Parametric Study A proposer of a new optimization algorithm might want to claim advantages over its competing evolutionary algorithms. In this case, besides multiple runs, a comprehensive parametric study of intrinsic control parameters of at least the competitors has to be conducted to locate the best intrinsic control parameter values and the corresponding best performance.

As already observed, combinations of intrinsic control parameters are infinite. It is computationally neither practical nor necessary to try all combinations of intrinsic control parameters. A more reasonable approach is to replace the infinite set of intrinsic control parameters with a finite set containing representative values of intrinsic control parameters.

References

[1] Salomon, R. (1996) Re-evaluating genetic algorithm performance under coordinate rotation of benchmark functions: a survey of some theoretical and practical aspects of genetic algorithms. *Bio Systems*, **39**(3), 263–278.

[2] Cooper, L. and Steinberg, D. (1970) *Introduction to Methods of Optimization*, W.B. Saunders, Philadelphia.

[3] Price, K.V., Storn, R.M. and Lampinen, J.A. (2005) *Differential Evolution: A Practical Approach to Global Optimization*, Springer, Berlin.

[4] Heath, M.T. (1997) *Scientific Computing: An Introductory Survey*, McGraw-Hill, Boston.

[5] Joshi, M.C. and Moudgalya, K.M. (2004) *Optimization: Theory and Practice*, Alpha Science International, Harrow.

[6] Brent, R.P. (1973) *Algorithms for Minimization without Derivatives*, Prentice-Hall, Englewood Cliffs, NJ.

[7] Reklaitis, G.V., Ravindran, A. and Ragsdell, K.M. (1983) *Engineering Optimization: Methods and Applications*, John Wiley & Sons, Inc., New York.

[8] Hooke, R. and Jeeves, T.A. (1966) Direct search of numerical and statistical problems. *Journal of the ACM*, **8**, 212–229.

[9] Nelder, J.N. and Mead, R. (1964) A simplex method for function minimization. *Computer Journal*, **7**(4), 308–313.

[10] Hestenes, M.R. and Stiefel, E. (1952) Methods of conjugate gradients for solving linear systems. *NBS Research Journal*, **49**, 409–436.

[11] Fletcher, R. and Reeves, C.M. (1964) Function minimization by conjugate grdients. *Computer Journal*, **7**(4), 149–154.

[12] Metropolis, N., Rosenbluth, A.W., Rosenbluth, M.N., Teller, A.H. and Teller, E. (1953) Equations of state calculations by fast computing machines. *Journal of Chemical Physics*, **21**(6), 1087–1092.

[13] Kirkpatrick, S., Gelatt, C.D. and Vecchi, M.P. (1983) Optimization by simulated annealing. *Science*, **220**(4598), 671–680.

[14] Davis, L. (ed.) (1987) *Genetic Algorithms and Simulated Annealing*, Pitman, London.

[15] Holland, J.H. (1975) *Adaptation in Natural and Artificial Systems*, University of Michigan Press, Ann Arbor.

[16] Goldberg, D.E. (1989) *Genetic Algorithms in Search, Optimization and Machine Learning*, Addison-Wesley, Reading, MA.

[17] Davis, L. (ed.) (1991) *The Handbook of Genetic Algorithms*, Van Nostrand Reinhold, New York.

[18] Fogel, L.J. (1962) Autonomous automata. *Industrial Research*, **4**, 14–19.

[19] Fogel, L.J., Owens, A.J. and Walsh, M.J. (1966) *Artificial Intelligence through Simulated Evolution*, John Wiley & Sons, Inc., New York.

[20] Schwefel, H.P. (1995) *Evolution and Optimum Seeking*, John Wiley & Sons, Inc., New York.

[21] Beyer, H.G. and Schwefel, H.P. (2002) Evolution strategy – a comprehensive introduction. *Natural Computing*, **1**(1), 3–52.

[22] Corne, D., Dorigo, M. and Glover, F. (eds) (1999) *New Ideas in Optimization*, McGraw-Hill, London.

[23] Caorsi, S., Massa, A., Pastorino, M. and Randazzo, A. (2004) Detection of PEC elliptic cylinders by a memetic algorithm using real data. *Microwave and Optical Technology Letters*, **43**(4), 271–273.

[24] Glover, F., Laguna, M. and Marti, R. (2000) Fundamentals of scatter search and path relinking. *Control Cybernetics*, **39**(3), 653–684.

[25] Pardalos, P.M. and Resende, M.G.C. (eds) (2002) *Handbook of Applied Optimization*, Oxford University Press, Oxford.

[26] Ghosh, A. and Tsutsui, S. (eds) (2003) *Advances in Evolutionary Computation: Theory and Applications*, Springer, New York.

[27] Laguna, M. and Marti, R. (2003) *Scatter Search: Methodology and Implementations in C*, Kluwer Academic, Boston.

[28] Onwubolu, G.C. and Babu, B.V. (eds) (2004) *New Optimization Techniques in Engineering*, Springer, Berlin.

[29] Glover, F. (1989) Tabu search, part I. *ORSA Journal on Computing*, **1**(3), 190–206.

[30] Glover, F. (1990) Tabu search, part II. *ORSA Journal on Computing*, **2**(1), 4–32.

[31] Cvijovic, D. and Klinowski, J. (1995) Taboo search – an approach to the multiple minima problem. *Science*, **267**, 664–666.

[32] Glover, F. and Laguna, M. (1997) *Tabu Search*, Kluwer, Norwell, MA.

[33] Kennedy, J., Eberhart, R.C. and Shi, Y. (2001) *Swarm Intelligence*, Morgan Kaufmann, San Francisco.

[34] Dorigo, M. (1992) Optimization, Learning and Natural Algorithms, PhD thesis, Politecnico di Milano, Italy.

[35] Pham, D.T., Ghanbarzadeh, A., Koc, E. *et al.* (2006) The bees algorithm – a novel tool for complex optimisation problems, in *Intelligent Production Machines and Systems: 2nd I*PROMS Virtual Conference* (eds. D.T. Pham, E.E. Eldukhri and A.J. Soroka), Elsevier, Oxford, pp. 454–461.

[36] Reynolds, R.G. (1994) An introduction to cultural algorithms, in *Proceedings of the Third Annual Conference on Evolutionary Programming*, World Scientific, Singapore, pp. 131–139.

[37] Kennedy, J. and Eberhart, R.C. (1995) Particle swarm optimization. IEEE International Conference on Neural Networks, Perth, WA, 27 Nov.–1 Dec. 1995, vol. 4, pp. 1942–1948.

[38] Qing, A., Lee, C.K. and Jen, L. (2001) Electromagnetic inverse scattering of two-dimensional perfectly conducting objects by real-coded genetic algorithm. *IEEE Transactions on Geoscience and Remote. Sensing*, **39**(3), 665–676.

[39] Babu, B.V. and Chaturvedi, G. (2000) Evolutionary computation strategy for optimization of an alkylation reaction. International Symposium 53rd Annual Session IIChE, Science City, Calcutta, December 18–21.

[40] Babu, B.V. and Munawar, S.A. (2000) Differential evolution for the optimal design of heat exchangers. All India Seminar Chemical Engineering Progress Resource Development: A Vision 2010 Beyond, Orissa State Centre Bhuvaneshwar, March 13.

2

Fundamentals of Differential Evolution

2.1 Differential Evolution at a Glimpse

2.1.1 History

Differential evolution ([1], [2], pp. 79–108, [3]) was proposed by Kenneth V. Price and R. Storn in 1995 while trying to solve the Chebyshev polynomial fitting problem. It stems from the *genetic annealing algorithm* which was also developed by Kenneth V. Price. Troubled by the slow convergence of the genetic annealing algorithm and the difficulties it faced in determining effective control parameters, Price modified the genetic annealing algorithm by using real code with arithmetic operations instead of binary code with Boolean operations. During this process, he discussed the *differential mutation operator* which was later shown to be the key to the success of differential evolution. Thus, differential evolution came to being.

Differential evolution has been the subject of intensive performance evaluation since its appearance. Apart from the performance evaluation conducted by its originators, to this author's knowledge, the first performance evaluation of differential evolution by other researchers was carried out in 1995 on a test bed of 15 test functions [4].

Differential evolution soon earned its reputation by ranking third among all entries in the first international contest on evolutionary optimization held in Nagoya, Japan, in May 1996 and best among all qualified entries in the second international contest on evolutionary optimization held in Indianapolis, Indiana, USA, in April 1997 (the actual contest was cancelled due to a lack of valid entries).

In late 1997, Storn established a website (http://www.icsi.berkeley.edu/∼storn/code.html) to publicize differential evolution. Since then we have witnessed explosive growth in differential evolution research.

2.1.2 Applications

Differential evolution is a very simple but very powerful stochastic global optimizer. Since its inception, it has proved to be a very efficient and robust technique for function optimization and has been used to solve problems in many scientific and engineering fields.

Differential Evolution Anyong Qing

Soon after differential evolution was proposed, Storn and Price used it to assign optimal redundancies for the priority encoding transmission technology for the special case of Moving Picture Experts Group-1 (MPEG) encoded video sequences [5], to design an infinite impulse response (IIR) filter [6,7], to design a howling removal unit [8], and to design switched capacitor filter [9].

Pioneering work in differential evolution was also carried out by Joshi and Sanderson. They applied differential evolution to solve a multi-sensor fusion problem commonly encountered in robotics [10]. As time passes, more and more researchers are becoming aware of differential evolution's potential.

A literature search on differential evolution has been carried out using Web of Science, Engineering Village 2, IEEE Xplore, and Google. Publications on differential evolution available from Web of Science, Engineering Village 2, and IEEE Xplore have been fully (as at January 2008) collected. In addition, publications on differential evolution available from Google have been selectively collected. These publications have been summarized as a bibliography on differential evolution and can be found on the companion website for this book. Based on this work, the following fields where differential evolution has been applied have been identified.

Applications of differential evolution in electrical and electronic engineering will be discussed separately in detail in Chapters 9–16 of this book. Therefore, they are not dealt with here.

Due to the vast number of publications involved in this survey, it is impossible to provide a full set of references for this chapter. Interested readers should refer to the companion website for details.

2.1.2.1 Acoustics

The applications of differential evolution in acoustics are summarized in Table 2.1.

At the Delft Institute of Earth Observation and Space Systems, Faculty of Aerospace Engineering, Delft University of Technology, The Netherlands, research on acoustic remote

Table 2.1 Applications of differential evolution in acoustics

Year	Researchers	Application
2000	Ainslie *et al.*	Estimation of geoacoustic seabed parameters in shallow water
2004	Ganchev *et al.*	Text-independent speaker verification
2005	Karasalo and P. Skogqvist	Identification of submerged or buried objects
	Zou *et al.*	Matched field inversion with synthetic data
2006	Cheng and Yang	Distinguishing multi-sources of vector transducer
	Perichon *et al.*	Synthesis of woodwind sound in real time
	Svensson	Inversion of acoustic communication signals for the sound speed profile
	Witos, Gacek, and Paduch	Decomposition of frequency characteristics of acoustic emission signals for different types of partial discharge sources

sensing and seafloor mapping has been going on under the Mathematical Geodesy and Positioning program. Researchers are trying to extract information about the acoustic medium (either the seafloor or the human body) using differential evolution from received acoustic signals that have traveled through the medium.

2.1.2.2 Aerodynamics, Aeronautics and Aerospace

Most of the applications of differential evolution in aerodynamics are concerned with the design of blades, turbine stages, fins, airfoils, engines, and so on. It is also used to calibrate multihole aerodynamic pressure probes. In aeronautics, it is most often used for flight control. In aerospace, differential evolution is applied in scheduling and evaluating spacecraft launches.

2.1.2.3 Agriculture and Pasturage

A farmer has to maximize his profit while at the same time assuring the quality of his products and conserving the environment for sustainable development. Hence, some researchers are using differential evolution to help make strategic decisions.

Another important application of differential evolution in agriculture and pasturage is in the daily management of farms.

2.1.2.4 Automobile and Automotive Industry

The applications of differential evolution in the automotive industry are summarized in Table 2.2.

Table 2.2 Applications of differential evolution in the automobile and automotive industry

Year	Researchers	Application
2000	Kyprianou *et al.*	Identify the parameters of an automotive hydraulic engine mount model
2003	Pedchote and Purdy	Estimate parameters of a single wheel station of a medium sized family car
2005	Yang *et al.*	Design of disk brake

2.1.2.5 Biological Science and Engineering

The first report of the use of differential evolution in biology appeared in 1998. It was applied to optimize fed-batch bioreactors and to solve a time-optimal drug displacement problem [11].

Other problems in biology solved using differential evolution are characterizing benzodiazepine receptors in the cerebellum, investigating a flexible ligand docking protocol, estimating the parameters model of the fermentation bioprocess, online detection of the physiological state of strains in a bioreactor, biological crystallography, optimization of indentation creep and stress relaxation of biological brain tissue, determining kinetic parameters in fixed-film bio-reactors for waste water purification, inference of gene regulatory networks, estimating parameters of a mathematical model for biological systems, metabolic flux analysis, estimating parameters of a mathematical model for biofilters, automatic structural determination of active sites of metalloproteins, and molecular image registration.

2.1.2.6 Chemical Science and Engineering

Wang has pioneered the application of differential evolution in chemistry. He has used differential evolution to study optimal temperature control of batch styrene polymerization initiated with a mixed initiator system, to study optimal feed control and optimal parameter selection for a fed-batch fermentation process, to estimate the parameters of the Monod model of a recombinant fermentation process, to synthesize the nonsharp distillation sequence and heat exchanger network, to optimize low pressure chemical vapor deposition reactors and a two-stage fermentation process for lactic acid production, and to estimate the kinetic parameters of batch polymerization described by differential-algebraic equations.

Another pioneer, Babu, has been actively applying differential evolution to the estimation of heat transfer parameters, the management of unexpected events in chemical process industries, optimum fuel allocation in power plants and optimization of the drying process for a through-circulation dryer, optimization of water pumping systems, optimization of extraction processes, design of a gas transmission network, estimation of optimal time of pyrolysis and heating rate, design of an auto-thermal ammonia synthesis reactor, a two-reactor problem, optimal operation of an alkylation unit in the petroleum industry, optimization of an adiabatic styrene reactor, optimal feed control and optimal parameter selection for a fed-batch fermentation process, design of a heat exchanger network, design of a reactor network, and optimization of an isothermal continuous stirred tank reactor.

Chemical problems solved by other practitioners using differential evolution include optimization of fed-batch bioreactors and time optimal drug displacement, gas carburizing, induced aeration in agitated vessels, minimization of transition time for a continuous methyl methacrylate-vinyl acetate copolymerization reactor, and optimal control of pH, characterization of benzodiazepine receptors in the cerebellum, binding of Si–H clusters, optimal feed control and optimal parameter selection for a fed-batch fermentation process, a chemical process converting methylcyclopentane to benzene in a tubular reactor, heat exchanger network design, reactor network design, optimization of an isothermal continuous stirred tank reactor, Pb–S–O vapor system, nonlinear parameter estimation of the model of low temperature SO_2 oxidation with Cs–Rb–V sulfuric acid catalyst, optimization of pure terephthalic acid crystallization process, selective product enhancement in *Aspergillus niger* fermentation, and aromatic hydrocarbon isomerization process optimization.

2.1.2.7 Climatology

Differential evolution is used for climate prediction and downscaling [12].

2.1.2.8 Defense

Applications of differential evolution in defense are summarized in Table 2.3.

Table 2.3 Applications of differential evolution in defense

Year	Researchers	Application
2005	Nikolos and Brintaki	Unmanned aerial vehicle path planning
2005	Robins and Thomas	Estimating releases of chemical, biological, radiological or nuclear atmospheric material
2005	Starkey, Rankins, and Pines	Optimize the hypersonic cruise trajectories of air-breathing, waverider-based vehicles

2.1.2.9 Economics

In many differential evolution applications, cost reduction is one of the objectives. From this point of view, differential evolution has been extensively applied to solve economic problems. However, such applications can only be regarded as indirect.

Differential evolution has also been directly applied to solve economic problems as summarized in Table 2.4.

Table 2.4 Applications of differential evolution in economics

Year	Researchers	Application
2003	Abbass	Australian credit card assessment problem
2005	Beynon	Credit rating problem
2006	Ali	Data network investment planning
2007	Pavlidis, Vrahatis, and Mossay	Economic geography

2.1.2.10 Environmental Science and Engineering

Environment conservation has become a more and more urgent problem. It is one of the most influential factors in many decision-making processes, such as the location and management of chemical and power plants. Differential evolution has been implemented to help address the environmental concerns involved in many application problems as summarized in Table 2.5.

Table 2.5 Applications of differential evolution in environmental science and engineering

Year	Researchers	Application
2001	Booty *et al.*	RAISON Environmental Decision Support System
2003	Banga *et al.*	Food industry
2004	DeVoil, Rossing, and Hammer	Cropping system
2004	Moles, Banga, and Keller	Climate change
2005	Carpio *et al.*	Cement industry
2006	Olamaie and Niknam	Power industry
2007	Groot *et al.*	Design and assessment of multifunctional agricultural landscapes
2007	Karterakis	Coastal subsurface water management

2.1.2.11 Food Engineering and Food Industry

Banga has been actively implementing differential evolution in food-related research. He summarized computer-aided optimization tools to improve food processing in a review article [13]. Application examples including thermal processing, contact cooking, food drying, microwave heating, as well as future trends and research needs are outlined.

2.1.2.12 Forestry

In 2000, Ochi and Cao [14] applied differential evolution to model forest tree growth.

2.1.2.13 Gas, Oil, and Petroleum Industry

Applications of differential evolution in the gas, oil, and petroleum industry are summarized in Table 2.6.

Table 2.6 Applications of differential evolution in the gas, oil, and petroleum industry

Year	Researchers	Application
2000	Babu and Chaturvedi	Alkylation reaction
2002	Chen, Chen, and Cao	Boiling point and effect of pressure on entropy of crude oil
2003, 2005	Babu *et al.*	Gas transmission network
2006	Chu and Wu	Pipeline for transmitting petroleum products
2006, 2008	Nobakhti and Wang	ALSTOM gasifier

2.1.2.14 Geoscience

Differential evolution has been used to identify short-run equilibria in economic geography [15].

2.1.2.15 Hydroscience

Water is one of the most precious natural resources for human survival on the earth. Efficient and proper use of water is essential for sustainable development. Differential evolution has been intensively applied to solve problems in this aspect as summarized in Table 2.7.

Other water-related applications of differential evolution including identification of soil physical parameters of the sustainable grazing systems pasture model [16], optimization water pumping system [17,18], and 2-chloropheol oxidation in supercritical water [19].

2.1.2.16 Industry

Chakraborti is a pioneer in applying differential evolution to solve optimization problems encountered in the iron and steel industry. Problems solved by him include the design of an integrated steel plant bloom reheating furnace, optimization of continuous casting

Table 2.7 Applications of differential evolution in water management

Year	Researchers	Application
2001	Booty *et al.*	Decision support system for 1994 Canada–Ontario Agreement to reduce pollutants in the Great Lakes of North America
2004	Kisi	Estimation of sediment concentration from streamflow
2005, 2006	Kiranmai, Jyothirmai, and Murty, Bhat *et al.*	Waste water purification
2005, 2006	Kremer and Dolinar; Lakshminarasimman and Subramanian	Scheduling of cascaded hydro- and thermo-power plants
2006	El-Telbany, Konsowa, and El-Adawy	Forecasting the Nile river flood data to improve the water management polices in Egypt
2006	Liu, Liu, and Yang	Monitoring dam deformation
2007	Cannon	Hydro-climatology
2007	Karterakis	Coastal subsurface water management
2007	Reddy and Kumar	Operation of Hirakud reservoir in Orissa State, India
2007	Shoemaker, Regis, and Fleming	Watershed calibration

mould parameters, scheduling of a reversing strip mill (minimizing the hot rolling time of an ingot, from a given initial thickness to a prescribed final one, subject to a number of system constraints), and optimizing surface profiles during hot rolling of an integrated steel plant.

Due to the importance of the iron and steel industry, many other researchers have also shown interest in solving optimization problems encountered in this area using differential evolution as summarized in Table 2.8.

Table 2.8 Applications of differential evolution in the iron and steel industry

Year	Researchers	Application
2001	Kilkki, Lampinen, and Martikka	Optimization of cross-section of a welded or edged steel column
2003	Nath and Mitra	Optimization of coke rate and sinter quality for two-layer sintering of iron ore
2004	Lyden, Kalviainen, and Nykanen	Surface roughness estimation of shot-blasted steel bars
2004	Sahay and Mitra	Carburizing operation
2005	Zhao, Liu, and Luo	Blast furnace hot metal silicon content prediction
2006, 2007	Filipič, Tušar, and Laitinen	Optimizing secondary coolant flows on a steel casting machine
2006	Pal, Datta, and Sahay	Batch annealing

2.1.2.17 Logistics and Enterprise Management

Differential evolution is inherently a good decision-making aid. It provides great relief for decision makers in logistics and enterprise managers as summarized in Table 2.9.

Table 2.9 Applications of differential evolution in enterprise and management

Year	Researchers	Application
2000	Lin, Hwang, and Wang	Plant scheduling and planning
2006	Onwubolu and Davendra	
2006	Qian, Wang, Huang, and Wang	
2001	Babu and Gautam	Management of unexpected events in chemical process industry
2003, 2005	Xue, Sanderson, and Graves 2003	Enterprise design, supplier and manufacturing planning
2005, 2007	Routroy Kodali, and Sanisetty	
2006	Coelho and Lopes	Optimization of supply chain
2007	Lieckens and Vandaele	Design of reverse logistics network
2007	Liu and Li	Logistics delivery routing scheduling

2.1.2.18 Maritime

Nikolos applied differential evolution to unmanned aerial vehicle path planning [21,22]. Differential evolution is also used in optimize port scheduling [23].

2.1.2.19 Materials Science and Engineering

Besides the iron and steel industry, Chakraborti is also very active in applying differential evolution to solve problems in materials science and engineering. In 2004, he presented a review on the application of differential evolution in materials science [24].

Many differential evolution applications in materials science and engineering involve characterization, classification and modeling of materials. Differential evolution also plays an important role in materials manufacturing.

2.1.2.20 Mathematics

Differential evolution was applied to compute the Nash equilibria of finite strategic games in game theory [25].

2.1.2.21 Mechanics

Researchers from the mechanics community are very interested in differential evolution. Lampinen has pioneered the use of differential evolution to solve mechanical problems. The intensive use of differential evolution in mechanics can be seen in Table 2.10.

Table 2.10 Applications of differential evolution in mechanics

Year	Researchers	Application
1999	Lin, Wang, and Hwang	Gear train and pressure vessel
2001	DiSilvestro and Suh	Articular cartilage
2001	Kyprianou, Worden, and Panet	Hysteretic or memory-dependent vibrations
2002	Tin-Loi and Que	Cohesive crack fracture
2003	Hrstka *et al.*	Reinforced concrete beam layout
2004	Han *et al.*	Thermo-mechanical reliability of land grid array solder joints
2005	Alvarez-Gallegos, Villar, and Flores	Optimization of a continuously variable transmission
2003–2006	Babu *et al.*	Cantilever design
2005	Hickey *et al.*	Force appropriation for nonlinear systems
2005–2006	Ng, Ajavakom, and Ma	Prediction of degrading structures
2005–2006	Yousefi and Handroos	Position control of a flexible servohydraulic system
2006	Giambanco and Scimemi	Failure analysis of bonded joints
2006	Park and Sohn	Structural health monitoring

In 2005, a survey of structural optimization in mechanical product development was published to further enhance acceptance of structural optimization within the industry [26].

2.1.2.22 Medicine and Pharmacology

Most medical applications of differential evolution focus on the diagnosis, classification and treatment of cancer. Other applications are summarized in Table 2.11.

2.1.2.23 Optics

Differential evolution has been intensively implemented in the optics community, as shown in Table 2.12.

Table 2.11 Applications of differential evolution in medicine and pharmacology

Year	Researchers	Application
2001	Salomon *et al.*	Medical image registration
2002	He and Narayana	
2004	Dony and Xu	
2003–2006	Li *et al.*	Medical imaging of brain section based on real head model
2004	Magoulas, Plagianakos, and Vrahatis	Colonoscopic diagnosis
2006	Koutsojannis and Hatzilygeroudis	Intelligent diagnosis and treatment of acid-base disturbances based on blood gas analysis data
2004–2005	Saastamoinen, Ketola, and Turunen	Sport medicine

Table 2.12 Applications of differential evolution in optics

Year	Researchers	Application
1999, 2001	Doyle, Corcoran, and Connell	Design luminaire reflectors
2005	Patow and Pueyo	
1999	Kasemir and Betzler	Characterize materials
2004	Al-Kuzee, Matsuura, and Goodyear	Optimize plasma etch processes
2004	Zhang and Zhong	Calibrate camera
2005	Chan, Toader, and John	PBG design
2006	Akdagli and Yuksel	Laser diode nonlinearity
2006	Bluszcz and Adamiec	Optical stimulated luminescence decay
2006	Ling, Wu, Yang, and Wan	Design holographic grating
2007	Pan and Xie	Deformation measurement

2.1.2.24 Physics

Most of the applications of differential evolution in physics focus on stellarator design and plasma. Other reported applications include chaos control [27], and optimization of the structure of atomic and molecular clusters [28].

2.1.2.25 Seismology

Seismologists are working intensively towards the early accurate prediction of earthquakes. Differential evolution has joined this effort as summarized in Table 2.13.

Table 2.13 Applications of differential evolution in seismology

Year	Researchers	Application
1998, 2000	Bartal *et al.*	Optimize the seismic networks in Israel
2001	Ruzek and Kvasnicka	Earthquake hypocenter location
2007	Ruzek *et al.*	Find seismic velocity models yielding travel times consistent with observed experimental data

2.1.2.26 Thermal Engineering

Thermal problems arise from many engineering processes such as chemical and materials processing. Thermal problems solved by using differential evolution are summarized in Table 2.14.

2.1.2.27 Transportation

Chang has pioneered the implemention of differential evolution in mass transit systems. He has used differential evolution for optimal train movement, layout design of signaling blocks, train schedule optimization, and the effect of train movement and scheduling on harmonic distortions in a.c. supply systems.

Table 2.14 Applications of differential evolution in thermal engineering

Year	Researchers	Application
1998, 2000	Chakraborti, Deb, and Jha	Modeling of an integrated steel plant bloom reheating furnace
1999	Babu and Sastry	Heating system identification
2006	Coelho and Mariani 2006	
2000, 2001, 2007	Babu and Munawar	Design of heat exchangers
2001	Liao, Tzeng, and Wang	Synthesis problems of the nonsharp distillation sequence and heat exchanger network
2006	Coelho	Modeling of a thermal system

Other applications include design optimization for high-speed train suspension systems [29].

2.1.3 Differential Evolution Strategies

2.1.3.1 Classification

There are many differential evolution strategies. Each strategy has its unique features. These features can be used to classify differential evolution strategies. Consequently, there are many alternative ways to classify differential evolution strategies.

As mentioned earlier, differential mutation is the key to the success of differential evolution. Thus, it is reasonable to classify differential evolution strategies according to the evolution mechanism implemented.

In general, from the point of view of evolution mechanism, there are four classes of differential evolution strategies:

(a) classic differential evolution (CDE);
(b) dynamic differential evolution (DDE);
(c) modified differential evolution (MDE);
(d) hybrid differential evolution (HDE).

Classic differential evolution refers to the first strategy proposed by the originators of differential evolution. The other three categories were developed from classic differential evolution and will be discussed in detail in Chapter 3 with other advances in differential evolution.

2.1.3.2 Notation

Differential evolution is initially proposed to minimize a single objective function $f(\mathbf{x})$ with real optimization parameters $\mathbf{x}$ without any constraint. Natural real code is adopted. It operates on a population with N_p individuals. Each individual is a symbolic representation of the N optimization parameters.

Strategies of differential evolution are denoted by DE/x/y/z where x indicates how the base is chosen, $y \geq 1$ is the number of vector differences added to the base vector, and z is the law which the number of parameters donated by the mutant follows.

2.2 Classic Differential Evolution

2.2.1 Evolution Mechanism

Classic differential evolution involves two stages: *initialization* and *evolution*. Initialization generates an initial population $\mathbf{P}^0$. Then $\mathbf{P}^0$ evolves to $\mathbf{P}^1$, $\mathbf{P}^1$ evolves to $\mathbf{P}^2$, ..., until the termination conditions are fulfilled. While evolving from $\mathbf{P}^n$ to $\mathbf{P}^{n+1}$, the three evolutionary operations, namely, *differential mutation*, *crossover* and *selection* are executed in sequence.

The flow chart of classic differential evolution is shown in Figure 2.1.

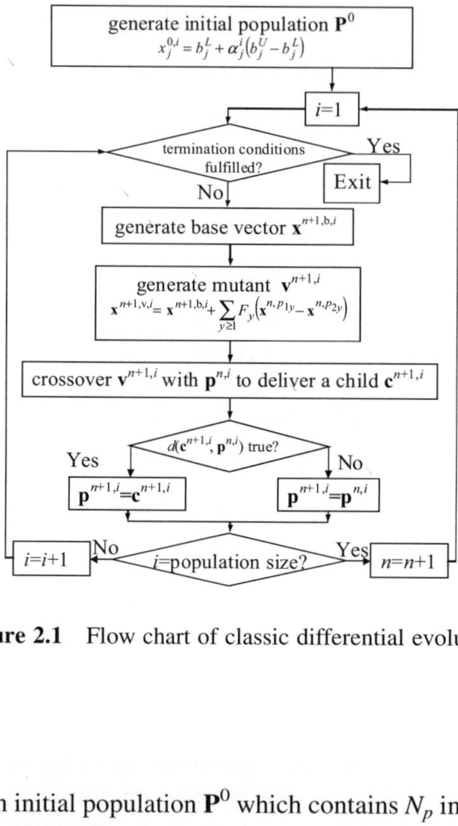

Figure 2.1 Flow chart of classic differential evolution

2.2.2 Initialization

Initialization generates an initial population $\mathbf{P}^0$ which contains N_p individuals $\mathbf{p}^{0,i}$, $1 \leq i \leq N_p$.

$$x_j^{0,i} = b_j^L + \alpha_j^i(b_j^U - b_j^L), \quad 1 \leq j \leq N, \tag{2.1}$$

where α_j^i is a real random number usually, but not necessarily, uniformly distributed in [0, 1].

2.2.3 Differential Mutation

2.2.3.1 General Formulation

Mutation, or more precisely, *differential mutation*, generates a mutant $\mathbf{v}^{n+1,i}$ with mutant vector $\mathbf{x}^{n+1,v,i}$, for each $\mathbf{p}^{n,i}$ (also known as the mother) in $\mathbf{P}^n$:

$$x^{n+1,v,i} = x^{n,b,i} + \sum_{y \geq 1} F_y(x^{n,p_{1y}} - x^{n,p_{2y}}), \quad 1 \leq i \neq p_{1y} \neq p_{2y} \leq N_p, \tag{2.2}$$

where $\mathbf{x}^{n,\mathrm{b},i}$ is the vector of optimization parameters of *differential mutation base* $\mathbf{b}^{n,i}$, p_{1y} and p_{2y} are integers (usually random), F_y, a constant usually in [0, 1], is the mutation intensity for the yth vector difference $\mathbf{x}^{n,p_{1y}} - \mathbf{x}^{n,p_{2y}}$, and $\mathbf{x}^{n,p_{1y}}$ and $\mathbf{x}^{n,p_{2y}}$ are difference vectors of individuals $\mathbf{p}^{n,p_{1y}}$ and $\mathbf{p}^{n,p_{2y}}$ from $\mathbf{P}^n$.

2.2.3.2 Differential Mutation Base

There are various base generators.

(a) Current: $\mathbf{b}^{n,i} = \mathbf{p}^{n,i}$.
(b) Best: $\mathbf{b}^{n,i} = \mathbf{p}^{n,\mathrm{best}}$, where $\mathbf{p}^{n,\mathrm{best}}$ is the best or most dominant individual in $\mathbf{P}^n$, that is, $d(\mathbf{p}^{n,\mathrm{best}}, \mathbf{p}^{n,i}) = \mathrm{true}$ for $1 \leq i \neq \mathrm{best} \leq N_p$.
(c) Better: $\mathbf{b}^{n,i}$ is an individual in $\mathbf{P}^n$ better or more dominant than $\mathbf{p}^{n,i}$, that is, $\mathbf{b}^{n,i} \in \{\mathbf{p}^{n,i} | d(\mathbf{b}^{n,i}, \mathbf{p}^{n,i}) = \mathrm{true}\}$. Usually, $\mathbf{b}^{n,i}$ is randomly chosen from qualified individuals in $\mathbf{P}^n$. *Mutation with better base* is also called *mutation with selection pressure* [30].
(d) Random: $\mathbf{b}^{n,i}$ is randomly chosen from $\mathbf{P}^n$.

Sometimes, current-to-random and current-to-best are also regarded as basic base generators where the base vector is generated as follows:

$$\mathbf{x}^{n,b,i} = \mathbf{x}^{n,i} + K(\mathbf{x}^{n,p_3} - \mathbf{x}^{n,i}), \quad 1 \leq p_3 \neq p_{1y} \neq p_{2y} \leq N_p, \tag{2.3}$$

where K, a constant in [0, 1], is the coefficient of combination, and $\mathbf{x}^{n,p_3}$ is the vector of optimization parameters of $\mathbf{p}^{n,p_3}$. $\mathbf{p}^{n,p_3}$ is randomly chosen from $\mathbf{P}^n$ for current-to-random, and $\mathbf{p}^{n,\mathrm{best}}$ for current-to-best.

However, it is confusing to regard current-to-random and current-to-best as basic differential mutation base generators. In fact, both DE/current-to-random/y/* and DE/current-to-best/y/* are special cases of DE/current/y + 1/*. Therefore, it might be more appropriate to exclude both current-to-random and current-to-best as basic base generators.

2.2.3.3 Regularization of Infeasible Mutant

Values of optimization parameters of $\mathbf{v}^{n+1,i}$ generated through Equation 2.2 may exceed the search space $[b_i^L, b_i^U]$. This is helpful when the search space is improperly preset due to lack of a priori knowledge. It gives differential evolution some freedom to find its solution outside the search space. However, it is harmful and dangerous in other cases. For example, a negative value of the control point for the benchmark electromagnetic inverse scattering problem [31–34], which is absolutely inadmissible, may be generated through 2.2 although $b_j^L = 0$.

Some researchers treat the search space as a boundary constraint and penalize infeasible mutant vectors by imposing a penalty on their objective function value. The *brick wall penalty* and the *adaptive penalty* discussed in ([3], pp. 203) are two typical penalty functions. However, this approach is not applicable for problems such as the aforementioned benchmark electromagnetic inverse scattering problems. No objective function value is available for a vector containing infeasible optimization parameter values.

There are two approaches to regularization of infeasible mutant vectors. The first approach is *random reinitialization*. Any infeasible optimization parameter value of the mutant vector, $x_j^{n+1,v,i} \notin [b_j^L, b_j^U]$, is replaced by a value randomly generated within the search space. The Fortran-style pseudo-code for random reinitialization is given in Figure 2.2.

```
do j = 1, N
  do while ( x_j^{n+1,v,i} < b_j^L  or  x_j^{n+1,v,i} > b_j^U )
    x_j^{n+1,v,i} = b_j^L + α_j^i (b_j^U − b_j^L)
  end do
end do
```

Figure 2.2 Fortran-style pseudo-code for random reinitialization

Bounce-back is the other approach to regularization of infeasible mutant vectora. One implementation of the bounce-back approach is given in ([3], pp. 204–205). However, the regularized mutant may remain infeasible because the potentially infeasible base is involved.

In our practice, a similar but slightly different bounce-back approach is implemented as shown in Figure 2.3.

```
doj = 1, N
  do while ( x_j^{n+1,v,i} < b_j^L )
    x_j^{n+1,v,i} = x_j^{n+1,v,i} + b_j^U − b_j^L
  end do
  do while ( x_j^{n+1,v,i} > b_j^U )
    x_j^{n+1,v,i} = x_j^{n+1,v,i} − b_j^U + b_j^L
  end do
end do
```

Figure 2.3 Fortran-style pseudo-code for bounce-back

2.2.4 Crossover

$\mathbf{v}^{n+1,i}$ is then mated with $\mathbf{p}^{n,i}$ to deliver a child $\mathbf{c}^{n+1,i}$. In principle, all crossover schemes mentioned in Chapter 1 are applicable. $\mathbf{p}^{n,i}$ serves as mother while $\mathbf{v}^{n+1,i}$ serves as father. However, in classic differential evolution, each couple is allowed one child only. Either of the two potential children is acceptable. In practice, the first child, child $\mathbf{c}^{n,a}$, is adopted.

2.2.5 Selection

Selection is also called *mother–child competition*. As its name implies, mother $\mathbf{p}^{n,i}$ and child $\mathbf{c}^{n+1,i}$ compete with each other to survive in the next generation. Mother–child competition is expressed mathematically as follows:

$$\mathbf{p}^{n+1,i} = \begin{cases} \mathbf{c}^{n+1,i}, & d(\mathbf{c}^{n+1,i}, \mathbf{p}^{n,i}) \text{ is true}, \\ \mathbf{p}^{n,i}, & \text{otherwise}. \end{cases} \tag{2.4}$$

2.2.6 Termination Conditions

Six termination conditions are given in ([3], pp. 128–130). In our practice, the following three termination conditions are applied simultaneously.

2.2.6.1 Objective Met

The *objective met* termination condition is only applicable to optimization problems whose optima are known. Such optimization problems include multi-dimensional toy functions and benchmark application problems for unconstrained single-objective optimization. Usually, they are involved in algorithm evaluation.

Mathematically, this condition can be expressed as

$$\begin{aligned} &\exists 1 \le p \le N_p \\ &\qquad \forall 1 \le i \le N_{f^{\min}} : f_i^{\min}(\mathbf{x}^{n,p}) - f_i^{\min}(\mathbf{x}^*) \le \varepsilon_{fi}^{\min} \cap \\ &\qquad \forall 1 \le i \le N_{f^{\max}} : f_i^{\max}(\mathbf{x}^*) - f_i^{\max}(\mathbf{x}^{n,p}) \le \varepsilon_{fi}^{\max} \cap \\ &\qquad \forall 1 \le i \le N_{c^=} : |c_i^=(\mathbf{x}^{n,p})| \le \varepsilon_{ci}^= \cap \\ &\qquad \forall 1 \le i \le N_{c^-} : c_i^-(\mathbf{x}^{n,p}) < 0 \cap \\ &\qquad \forall 1 \le i \le N_{c^+} : c_i^+(\mathbf{x}^{n,p}) > 0 \end{aligned}$$

where $\varepsilon_{oi}^{\min}$, $\varepsilon_{oi}^{\max}$, and $\varepsilon_{ci}^=$ are the preset values to reach (VTR).

2.2.6.2 Limit on Number of Objective Function Evaluations

It is not possible in practice to allow unlimited time to search for an optimal solution. The search process has to be terminated when the time taken exceeds a certain limit. Although such a limit on search time is straightforward, it does not capture the essence of optimization, that is, objective function evaluation. Instead, a limit on the number of objective function evaluations is a good marker for time taken.

Most people use an alternative to this, a limit on the number of generations. However, this is not applicable to optimization algorithms without a clear concept of generation, for example, deterministic optimization algorithms. It is also inappropriate for evolutionary algorithms with dynamic evolution mechanism, for example, the dynamic differential evolution discussed in Chapter 3.

2.2.6.3 Limit of Population Diversity

The *limit of population diversity* termination condition is implemented to address the problem of premature populations in evolutionary computation. Individuals in a premature population differ very little from each other. In other words, the population diversity is very small. Usually, the search for the optimum is highly unlikely to succeed in a premature population. Therefore, it is practically pointless to continue when the population is premature. The search process will be terminated when the population diversity $d(\mathbf{P}^n)$ is below a preset threshold, ε_d.

2.3 Intrinsic Control Parameters of Differential Evolution

2.3.1 Introduction

According to the above description, there are at least three kinds of intrinsic control parameters for classic differential evolution:

(a) population size N_p;
(b) mutation intensities F_y $(y \geq 1)$;
(c) crossover probability p_c.

There is more than one mutation intensity if two or more vector differences are used to generate the mutation. However, I personally do not enjoy more than one mutation intensity. Accordingly, F, instead of F_y, will be used for mutation intensity from now on unless specified otherwise.

If any of the uniform arithmetic crossover schemes is chosen, the crossover intensity h will be the fourth kind of intrinsic control parameter. However, most people in the differential evolution community do not implement arithmetic crossover, and so it is not mentioned further in this book.

2.3.2 Originators' Rules of Thumb

The originators of differential evolution have recommended some rules of thumb ([2,8], [35], pp. 79–108) for the choice of intrinsic control parameters.

It is observed by the originators that differential evolution is much more sensitive to the choice of mutation intensity than it is to the choice of crossover probability. Crossover probability is more like a fine-tuning element.

Price claims that binomial is never worse than exponential. However, Storn claims that crossover is not so important.

2.3.2.1 First Choice

The originators recommend $N_p/N = 10$, $F = 0.8$, and $p_c = 0.9$. However, $F = 0.5$ and $p_c = 0.1$ are also claimed to be a good first choice.

2.3.2.2 Range of Choice

The originators claim that population size, mutation intensity and crossover probability are not difficult to choose in order to obtain good results. The value of N_p/N varies from as low as 1.2

to as high as 30, although it is claimed that "a reasonable choice" is between 5 and 10. As to the mutation intensity F, highly diverse or even contradictory ranges, such as [0, 2], [0, 1], [−1, 1], [0.4, 1], [0.5, 1], have been reported. In most of the originators' publications, [0, 1] is referred to as the range of crossover probability. However, narrower ranges such as [0.5, 1] or even [0.8, 1] have also been mentioned.

2.3.2.3 Adjusting Intrinsic Control Parameters

Different problems usually require different intrinsic control parameters. In addition, different differential evolution strategies may have their own preference for intrinsic control parameters. For example, Storn mentions that the crossover probability for binomial crossover is usually higher than that for exponential crossover. Therefore, the above first choice of intrinsic control parameters may not work for certain problems. Consequently, they have to be adjusted. A tradeoff between efficiency and robustness is necessary while adjusting the intrinsic control parameters.

Increasing the population size is suggested if differential evolution does not converge. However, this only helps to a certain extent.

Mutation intensity has to be adjusted a little lower or higher than 0.8 if population size is increased. It is claimed that convergence is more likely to occur but generally takes longer with larger population and weaker mutation (smaller F).

Most often the crossover probability must be considerably lower than one (e.g., 0.3). High values of crossover probability such as $p_c = 1$ give faster convergence if convergence occurs. Sometimes, however, crossover probability needs to be as low as $p_c = 0$ to make differential evolution robust enough for a particular problem.

2.3.3 *Other Recommendations*

Differential evolution is sensitive to intrinsic control parameters. More importantly, optimal intrinsic control parameters are problem- and strategy-specific. Therefore, other differential evolution practitioners have also made recommendations based on their own experience.

Users of differential evolution may use values of intrinsic control parameters recommended by predecessors, or choose their own values according to certain criteria. Consequently, a huge variety of combinations of values of intrinsic control parameters have been reported. It is interesting to observe that some users have employed a population as small as 5, or as big as 1000, 2000 or even 3000 [36], although N_p/N tends to vary between 2 and 40.

In [37], the value of N_p/N varies from 2 to 40, and the range [0.4, 0.95] is recommended for F. Surprisingly, the range [0, 0.2] is recommended for p_c in order to optimize separable objective functions.

Zaharie [38] obtained the following empirical relationship between population size, mutation intensity, and crossover probability:

$$2F^2 - 2/N_p + p_c/N_p = 0. \tag{2.5}$$

The test bed includes three 30-dimensional test functions: sphere, Griewank, and Rastrigin. In addition, $N_p = 50$, $p_c = 0.2$.

2.4 Differential Evolution as an Evolutionary Algorithm

Differential evolution is one of the evolutionary algorithms. It soon established its reputation as an efficient and robust optimizer. Its applications have been increasingly expanding in both breadth and depth. Consequently, more and more people who have not used differential evolution before are thinking about implementing it to solve their optimization problems.

2.4.1 Common Evolutionary Ingredients

As an evolutionary algorithm, differential evolution has all the essential evolutionary ingredients: gene, chromosome, individual, population, and evolutionary operations. In this regard, differential evolution is a firmly established member of the evolutionary algorithm family.

2.4.2 Distinctive Evolutionary Ingredients

Differential evolution cannot achieve its success without cause. Differential mutation is the key to its success. Besides, its constructive evolutionary mechanism and creative mechanism of cooperation and competition also contribute to its success.

2.4.2.1 Constructive Evolution Mechanism

In each evolution loop, the three evolutionary operations, mutation, crossover, and selection, are executed in turn. Unlike genetic algorithms, in differential evolution, a child is accepted only if it dominates its direct mother. In this regard, its evolution mechanism is *constructive* instead of destructive.

2.4.2.2 Creative Mechanism of Cooperation and Competition

Cooperation and *competition* coexist in nature everywhere. Natural selection enables complex creatures evolve from more simplistic ancestors naturally over time. At the same time, cooperation or sharing (e.g., among creatures such as ants, bees, and birds) makes unbelievable things happen. Differential evolution implements this concept creatively. Parents compete to serve as base and donors of difference vectors in differential mutation. However, once chosen, difference vectors cooperate with each other by donating vector differences. The mutant vector and mother cooperate to deliver a child. Each mother has equal right to deliver. The child competes with its mother for survival.

2.4.3 Strength

Differential evolution has been widely recognized as an efficient and robust optimizer. Many comparisons have been made with other optimization algorithms, especially evolutionary algorithms. Most often, differential evolution outperforms its counterparts in efficiency and robustness. Later, in Chapter 4, the results of a comprehensive comparative study with genetic algorithms and a biased comparative study with particle swarm optimization will be given.

Differential evolution's popularity also attributes to its simplicity. The C code for DE/rand/1/bin given in ([2], pp. 107–108) has only 24 lines. Amazing!

User-friendliness is another important advantage of differential evolution. Differential evolution has only three intrinsic control parameters. More importantly, its performance is more weakly dependent on its intrinsic control parameters. Differential evolution may work fairly well using reported intrinsic control parameter values in relevant literatures.

2.4.4 Weakness

To my knowledge, two critics have pointed to the poor performance of differential evolution, due to noise or dynamic problems [39] and epistatic problems ([3], p. 102). At present, I do not have any experience of solving dynamic problems. However, it may be a dubious claim that differential evolution is not good for epistatic problems since all evolutionary algorithms perform equally badly on epistatic problems.

References

[1] Storn, R. and Price, K.V. (1995) Differential Evolution – A Simple and Efficient Adaptive Scheme for Global Optimization over Continuous Spaces. *TR-95-012*, International Computer Science Insitute, Berkeley, CA.

[2] Corne, D., Dorigo, M. and Glover, F. (eds) (1999) *New Ideas in Optimization*, McGraw-Hill, London.

[3] Price, K.V., Storn, R.M. and Lampinen, J.A. (2005) *Differential Evolution: A Practical Approach to Global Optimization*, Springer, Berlin.

[4] Brutovský, B., Ulicný, J. and Miškovský, P. (1995) Application of genetic algorithms based techniques in the theoretical analysis of molecular vibrations. 1st Int. Conf. Genetic Algorithms Occasion 130th Anniversary Mendel's Laws in Brno, Brno, Czech Republic, September 26–28, 1995, pp. 29–33.

[5] Storn, R. (1995a) Modeling and Optimization of PET-Redundancy Assignment for MPEG-Sequences. *Technical Report. TR-95-018*, International Computer Science Insitute, May.

[6] Storn, R. (1995b) Differential Evolution Design of an IIR-Filter with Requirements for Magnitude and Group Delay. *Technical Report TR-95-026*, International Computer Science Insitute, June.

[7] Storn, R. (1996a) Differential evolution design of an IIR-filter. 1996 IEEE Int. Conf. Evolutionary Computation, Nagoya, May 20–22, pp. 268–273.

[8] Storn, R. (1996b) On the usage of differential evolution for function optimization. 1996 Biennial Conf. North American Fuzzy Information Processing Society, Berkeley, CA, 19–22 June, pp. 519–523.

[9] Storn, R. (1996c) System Design by Constraint Adaptation and Differential Evolution. *Technical Report TR-96-039*, International Computer Science Institute, November.

[10] Joshi, R. and Sanderson, A.C. (1996) Multisensor fusion and model selection using a minimal representation size framework. 1996 IEEE/SICE/RSJ Int. Conf. Multisensor Fusion Integration Intelligent Systems, Washington, DC, December 8–11, 1996, pp. 25–32.

[11] Balsa-Canto, E., Alonso, A.A. and Banga, J.R. (1998) Dynamic optimization of bioprocesses: deterministic and stochastic strategies. Automatic Control of Food and Biological Processes, Gothenburg, Sweden, September 21–23.

[12] Cannon, A.J. (2007) Nonlinear analog predictor analysis: a coupled neural network/analog model for climate downscaling. *Neural Networks*, **20**(4), 444–453.

[13] Banga, J.R., Balsa-Canto, E., Moles, C.G. and Alonso, A.A. (2003) Improving food processing using modern optimization methods. *Trends in Food Science and Technology*, **14**(4), 131–144.

[14] Ochi, N. and Cao, Q.V. (2000) Application of differential evolution strategy in modeling forest tree growth. *Proceedings of the Louisiana Academy Science*, Jan.

[15] Pavlidis, N.G., Vrahatis, M.N. and Mossay, P. (2007) Existence and computation of short-run equilibria in economic geography. *Applied Mathematics Computation*, **184**(1), 93–103.

[16] Johnson, I.R., Kinghorn, B.P., Murphy, S.R., Lodge, G.M. and Meszaros, S.A. (2002) Estimating soil physical parameters using simulation differential evolution. IASTED Int. Conf. Applied Simulation Modelling, Crete, Greece, June 25–28, pp. 274–279.

[17] Babu, B.V. and Angira, R. (2003) Optimization of water pumping system using differential evolution strategies. 2nd Int. Conf. Computational Intelligence Robotics Autonomous Systems, Singapore, December 15–18.

[18] Babu, B.V. and Angira, R. (2004) Optimization using hybrid differential evolution algorithms. Int. Symp. 57th Annual Session IIChE, Mumbai, December 27–30.

[19] Yan, X.F., Yu, J., Qian, F. and Ding, J.W. (2006) Kinetic parameter estimation of oxidation in supercritical water based on modified differential evolution. Huadong Ligong Daxue Xuebao/J. East China University Science Technology, **32**(1), 94–97 and 124 (in Chinese).

[20] Derksen, R.W. and Hancox, E. (2005) The optimization of an industrial pneumatic supply system. 9th Int. Conf. Computer Aided Optimum Design Engineering, Skiathos, Greece, May 23–25, pp. 413–422.

[21] Nikolos, I.K. (2005) Coordinated UAV path planning using an ANN assisted differential evolution algorithm. EUROGEN 2005, September 12–14.

[22] Nikolos, I.K. and Brintaki, A.N. (2005) Coordinated UAV path planning using differential evolution. 20th IEEE Int. Symp. Intelligent Control, Limassol, Cyprus, June 27–29, pp. 549–556.

[23] Lorenzoni, L.L., Ahonen, H. and de Alvarenga, A.G. (2006) A multi-mode resource-constrained scheduling problem in the context of port operations. *Computers and Industrial Engineering*, **50**(1–2), 55–65.

[24] Chakraborti, N. (2004) Genetic algorithms in materials design and processing. *International Materials Reviews*, **49**(3–4), 246–260.

[25] Pavlidis, N.G., Parsopoulos, K. E. and Vrahatis, M.N. (2005) Computing Nash equilibria through computational intelligence methods. *Journal of Computational and Applied Mathematics*, **175**(1), 113–136.

[26] Saitou, K., Izui, K., Nishiwaki, S. and Papalambros, P. (2005) A survey of structural optimization in mechanical product development. *Journal of Computing and Information Science in Engineering*, **5**, 214–226.

[27] Zelinka, I. (2005) Investigation on evolutionary deterministic chaos control – extended study. 19th European Conf. Modelling Simulation, Riga, Latvia, June 1–4, pp. 51–58.

[28] Ali, M.M., Smith, R. and Hobday, S. (2006) The structure of atomic and molecular clusters, optimised using classical potentials. *Computer Physics Communications*, **175**(7), 451–464.

[29] Kim, Y. G., Park, C.K., Hwang, H.S. and Park, T.W. (2003) Design optimization for suspension system of high speed train using neural network. *JSME International Journal Series C, Mechanical Systems, Machine Elements and Manufacturing*, **46**(2), 727–735.

[30] Bergey, P.K. and Ragsdale, C. (2005) Modified differential evolution: a greedy random strategy for genetic recombination. *OMEGA – The International Journal of Management Science*, **33**(3), 255–265.

[31] Qing, A., Lee, C.K. and Jen, L. (2001) Electromagnetic inverse scattering of two-dimensional perfectly conducting objects by real-coded genetic algorithm. *IEEE Transactions on Geoscience and Remote. Sensing*, **39**(3), 665–676.

[32] Qing, A. (2003) Electromagnetic inverse scattering of multiple two-dimensional perfectly conducting objects by the differential evolution strategy. *IEEE Transactions on Antennas & Propagation*, **51**(6), 1251–1262.

[33] Qing, A. and Gan, Y.B. (2005) Electromagnetic inverse problems (invited), in *Encyclopedia of RF and Microwave Engineering*, vol. **2** (ed. K. Chang), John Wiley & Sons, Inc., Hoboken, NJ, pp. 1200–1216.

[34] Qing, A. (2006) Dynamic differential evolution strategy and applications in electromagnetic inverse scattering problems. *IEEE Transactions on Geoscience and Remote Sensing*, **44**(1), 116–125.

[35] Storn, R. and Price, K.V. (1997) Differential evolution – a simple and efficient heuristic for global optimization over continuous spaces. *Journal of Global Optimization*, **11**(4), 341–359.

[36] Massa, A., Pastorino, M. and Randazzo, A. (2004) Reconstruction of two-dimensional buried objects by a differential evolution method. *Inverse Problems*, **20**(6), S135–S150.

[37] Rönkkönen, J., Kukkonen, S. and Price, K.V. (2005) Real-parameter optimization with differential evolution. 2005 IEEE Congress Evolutionary Computation, Edinburgh, UK, Sept. 2–5, 2005, vol. 1, pp. 506–513.

[38] Zaharie, D. (2002) Critical values for the control parameters of differential evolution algorithms. 8th Int. Mendel Conference Soft Computing, Brno, Czech Republic, June 5–7, pp. 62–67.

[39] Krink, T., Filipič, B., Fogel, G.B. and Thomsen, R. (2004) Noisy optimization problems – a particular challenge for differential evolution? 2004 IEEE Congress Evolutionary Computation, Portland, OR, June 19–23, 2004, vol. 1, pp. 332–339.

3

Advances in Differential Evolution

3.1 Handling Mixed Optimization Parameters

The initial differential evolution aims to minimize a single objective function $f(\mathbf{x})$ of real optimization parameters $\mathbf{x}$. Accordingly, it adopts natural real code.

However, optimization problems of mixed optimization parameters arise in many practical applications. For example, in the design of microwave absorbers, material properties have to be chosen from a limited real materials database. Each material is represented by an integer material index in the real materials database. In this case, both real and integer optimization parameters may be involved. Without loss of generality, integer optimization parameters are considered here.

It is natural to implement natural integer code to represent integer optimization parameters. However, the differential mutation will definitely result in real mutant vectors. Arithmetic crossover will also lead to real child vectors. One may convert the real mutant vector and real child vector by truncating at increased computational cost.

Although implementing natural integer code for integer optimization parameter looks natural, it is disadvantageous in practice. It significantly reduces population diversity, increases programming complexity, and requires additional real-to-integer conversion.

A more advantageous approach uses real code regardless of type of optimization parameters. The objective function evaluator receives real parameters, truncates them to get the required integer optimization parameters, and evaluates the objective function value.

This approach requires no modification to the developed differential evolution program for real optimization parameters since it is usually the end user's responsibility to code the objective function according to the differential evolution developer's template. A more diverse population and consequently a less premature convergence probability will result from applying this approach.

Another interesting modification to parameter representation in differential evolution uses wavelet code [1]. Evolutionary operators for wavelet code differential evolution have been defined accordingly.

3.2 Advanced Differential Evolution Strategies

3.2.1 Dynamic Differential Evolution

3.2.1.1 Motivations

The evolution mechanism of classic differential evolution implies two inherent defects, namely, slow response to update of population status and extra memory requirement.

In nature, from the point of view of population updating, classic differential evolution is static, that is, the whole population $\mathbf{P}^n$ of generation n remains unchanged until it is replaced by population $\mathbf{P}^{n+1}$ of generation $n+1$. In evolving from $\mathbf{P}^n$ to $\mathbf{P}^{n+1}$, classic differential evolution does not make use of any improvement taking place in the current evolution loop. The mutation operator keeps using $\mathbf{P}^n$ to produce base and mutant vectors until the current evolution loop completes. For example, DE/best/*/* keeps using $\mathbf{p}^{n,\text{best}}$ even if $\mathbf{p}^{n,\text{best}}$ has already been replaced by her own more dominant child in $\mathbf{P}^{n+1}$, or another individual $\mathbf{p}^{n,i}$ has given birth to a more dominant (than $\mathbf{p}^{n,\text{best}}$) child $\mathbf{c}^{n+1,i}$ or equivalently $\mathbf{p}^{n,i+1}$. At the same time, the mutation operator chooses difference vectors $\mathbf{x}^{n,p_{1y}}$ and $x^{n,p_{2y}}$ always from population $\mathbf{P}^n$, regardless of the fact that $\mathbf{p}^{n,i}$ may have been replaced by her more dominant child $\mathbf{c}^{n+1,i}$ or equivalently $\mathbf{p}^{n,i+1}$. Therefore, classic differential evolution does not follow the progress of the population status immediately. Instead, it responds to the population progress after a time lag.

The evolution mechanism of classic differential evolution also implies extra memory requirement since memory should be allocated for both $\mathbf{P}^n$ and $\mathbf{P}^{n+1}$.

3.2.1.2 Inspiration

For a set of linear equations $\mathbf{A}\cdot\mathbf{x}=\mathbf{b}$, the Jacobi method updates its solution as

$$x_i^{n+1}=\frac{b_i-\sum_{j\neq i}A_{ij}x_j^n}{A_{ii}}. \tag{3.1}$$

No component of the solution $\mathbf{x}^n$ can be overwritten before the new solution $\mathbf{x}^{n+1}$ is ready. It is well known that the convergence rate of the Jacobi method is usually unacceptably slow. One reason for the slow convergence is that it does not make use of the latest information available. New component values of $\mathbf{x}^{n+1}$ are used only after the entire sweep has been completed.

The Gauss–Seidel method remedies this drawback by introducing a dynamic updating scheme in which each new component of the solution $\mathbf{x}^{n+1}$ is used as soon as it has been updated:

$$x_i^{n+1}=\frac{b_i-\sum_{j<i}A_{ij}x_j^{n+1}-\sum_{j>i}A_{ij}x_j^n}{A_{ii}}. \tag{3.2}$$

Although iteration index n is still present in the above formulation, it is used to highlight the dynamic updating scheme only. It is unnecessary in actual programming. Consequently, there is no need to allocate memory for both $\mathbf{x}^n$ and $\mathbf{x}^{n+1}$.

This brings two remarkable advantages: faster guaranteed convergence under weaker conditions and less memory requirement. The dynamic updating scheme plays an essential role.

The situation for classic differential evolution is quite similar. It is observed that all evolutionary operations of classic differential evolution are workable if one or more individuals in the population is updated. The dynamic counterpart of classic differential evolution is accordingly developed.

3.2.1.3 Key Innovations

The flow chart of dynamic differential evolution [2] is shown in Figure 3.1. It is readily apparent that dynamic differential evolution looks very similar to classic differential evolution. A slight but very significant modification is the disappearance of the generation index n which implies that there is only one population **P** whose individuals are continuously updated by their more dominant children. Such modification results in a significantly different evolution mechanism, the dynamic evolution mechanism, and halving of memory requirement.

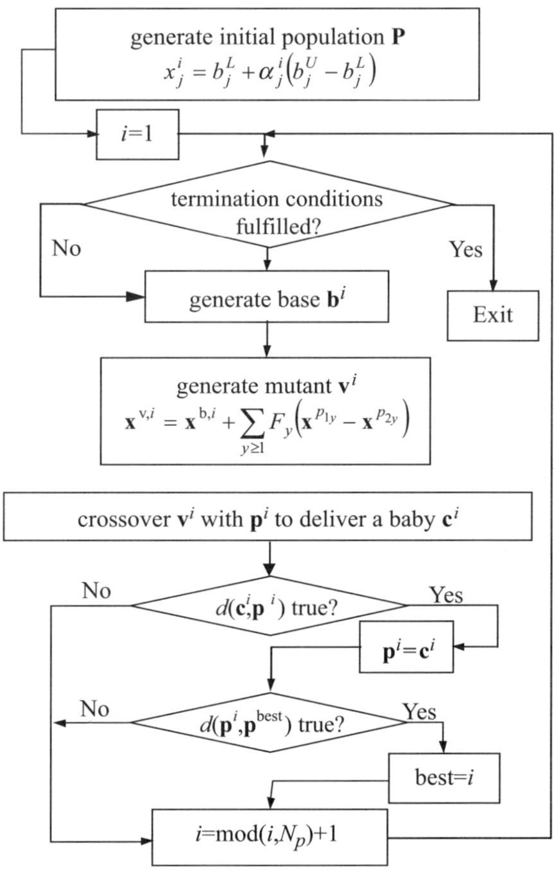

Figure 3.1 Flow chart of dynamic differential evolution

One of the significant differences between dynamic differential evolution and classic differential evolution is the dynamic update of the most dominant individual $\mathbf{p}^{\text{best}}$. As shown in Figure 3.1, there is an additional competition between a new-born more dominant (than $\mathbf{p}^i$) child $\mathbf{c}^i$ and $\mathbf{p}^{\text{best}}$. $\mathbf{p}^{\text{best}}$ will be replaced by $\mathbf{c}^i$ if $\mathbf{c}^i$ is more dominant than $\mathbf{p}^{\text{best}}$. The updated $\mathbf{p}^{\text{best}}$ will be used in the following evolutions immediately.

In addition, non-optimal individuals $\mathbf{p}^i$ are also updated dynamically. This is the other significant difference between dynamic differential evolution and classic differential evolution. A non-optimal individual $\mathbf{p}^i$ will be replaced by her new-born child $\mathbf{c}^i$ if $\mathbf{c}^i$ is more dominant than she. The new-born child $\mathbf{c}^i$, or equivalently the updated $\mathbf{p}^i$, will also be used immediately in the following evolutions.

3.2.1.4 Advantages

Dynamic differential evolution has been subject to extensive performance evaluation. As expected, dynamic differential evolution significantly outperforms classic differential evolution, in the same way as the Gauss–Seidel method does the Jacobi method – it is faster and more robust. Detailed results will be given in later chapters.

As mentioned earlier, the dynamic updating scheme of the Gauss–Seidel method requires less memory since $\mathbf{x}^n$ and $\mathbf{x}^{n+1}$ are now indistinguishable. This observation also applies to dynamic differential evolution. In fact, memory is almost halved.

The dynamic evolution mechanism also allows a slightly different mother–child competition scheme as

$$\mathbf{p}^{\text{worst}} = \mathbf{c}^i, \quad d(\mathbf{c}^i, \mathbf{p}^i) \text{ is true}\,. \tag{3.3}$$

It is also possible to replace a randomly chosen worse mother with the dominant child.

3.2.1.5 Disadvantage

The dynamic differential evolution is not without disadvantage. It cannot be parallelized any more.

3.2.1.6 Similar Ideas

K.V. Price observed that one population can be used if execution of differential evolution is sequential ([3], p. 85). Only parallel execution requires two populations. Sequential execution of differential evolution with one population only is one of the key ideas leading to dynamic differential evolution. However, Price claims "no dramatic difference in performance between the one- and two-array methods".

A similar idea was independently proposed by Angira and Babu [4–8] and Omran *et al.* [9].

3.2.2 Modified Differential Evolution

3.2.2.1 Introduction

The performance of classic differential evolution is not satisfactory for some problems. Researchers have proposed various modifications to classic differential evolution to enhance its performance.

As stated earlier, classic differential evolution works on a population with N_p individuals and involves two stages: initialization and evolution. Differential mutation, crossover and selection are involved in each evolution loop. Objective and constraint functions for each new individual have to be evaluated.

Accordingly, modifications to differential evolution are classified with respect to population, initialization, differential mutation, crossover, objective and constraint function evaluation, and selection.

3.2.2.2 Multi-population Differential Evolution

Population is the object on which differential evolution works. In actual programming of classic differential evolution, two populations, $\mathbf{P}^n$ and $\mathbf{P}^{n+1}$, are implemented. However, both $\mathbf{P}^n$ and $\mathbf{P}^{n+1}$ are the same evolving population at different stages. Therefore, in nature, classic differential evolution is a one-population strategy.

Various multi-population differential evolution strategies have been proposed.

3.2.2.2.1 Auxiliary Population In each evolution loop of classic differential evolution, a child $\mathbf{c}^{n+1,i}$ is rejected if it is dominated by its mother $\mathbf{p}^{n,i}$. In the worst scenario, the mother is the most dominant individual in $\mathbf{P}^n$ and the child is dominant over all other individuals in $\mathbf{P}^n$. To avoid this problem, an auxiliary population [10] containing rejected children is introduced. Good individuals from the auxiliary population are used to replace bad individuals in the principal population periodically.

3.2.2.2.2 Differential Evolution with Individual in Groups For some engineering problems such as the benchmark electromagnetic inverse scattering problem [11–13], due to lack of a priori information, the actual problem dimension is unknown. Instead, a finite set of problem dimensions including the actual problem dimension is available. Intuitively, one may try each problem dimension in the set one by one. However, it is extremely time-consuming. Differential evolution with individuals in groups [14,15] is therefore proposed to deal with such problems. The key idea of this is to organize the entire evolving population into different groups. Each group takes care of one of the dimensions and searches for its optimal solution through differential evolution. Groups compete to adjust group sizes dynamically.

3.2.2.2.3 Subpopulation Classic differential evolution uses one evolving population and identifies one solution once. However, non-unique functions have multiple solutions. Searching through all possible solutions one by one is extremely inefficient. The concept of subpopulations [16–28] is therefore implemented to find multiple solutions simultaneously. An evolving population is subdivided into multiple subpopulations according to distances between individuals. Each subpopulation evolves by itself. However, from time to time, subpopulations are forced to talk with each other by implementing either reorganization or migration to break the isolation between subpopulation.

The subpopulation approach is also used for parallelization and diversity preservation.

3.2.2.2.4 Opposition-Based Differential Evolution The opposite number $\breve{x}$ of a number $x \in [a, b]$ is defined as $\breve{x} = a + b - x$. A vector of opposite optimization parameters $\breve{\mathbf{x}}^{n,i}$ of vector of optimization parameters $\mathbf{x}^{n,i}$, an opposite individual $\breve{\mathbf{p}}^{n,i}$ of individual $\mathbf{p}^{n,i}$, and an opposite population $\breve{\mathbf{p}}^{n}$ of population $\mathbf{P}^n$, are accordingly defined for differential evolution.

Apparently, $\breve{\mathbf{p}}^{n,i}$ and $\mathbf{p}^{n,i}$ have an equal chance of being more dominant. Inspired by this observation, S. Rahnamayan, H.R. Tizhoosh, and M.M.A. Salama introduced opposite populations into differential evolution and developed opposition-based differential evolution [28–33] in which opposite numbers play a crucial role.

Opposition-based differential evolution implements an ordinary evolving population $\mathbf{P}^n$ as well as an opposite population $\breve{\mathbf{p}}^n$. However, the opposite population $\breve{\mathbf{p}}^n$ does not evolve by itself. The Fortran-style pseudo-code of opposition-based differential evolution is given in Figure 3.2, where p_j is the jumping rate, an extra intrinsic control parameter of opposition-based differential evolution.

```
get initial population P'^0
get initial opposite population P̆^0
get initial population P^0 from the union of P'^0 and P̆^0
n = 1
do while (termination conditions not fulfilled)
  get population P'^n from P^(n-1) through differential evolution
  if(rand(0, 1) < p_j)
    get opposite population P̆^n of P^(n-1)
    get population P^n from the union of P'^n and P̆^n
  else
    P^n = P'^n
  end if
end do
```

Figure 3.2 Fortran-style pseudo-code of opposition-based differential evolution

3.2.2.3 Biased Initialization

Usually, the initial population is generated randomly and uniformly in the search space. However, when a priori information is available, a biased initial population might be more favorable [34,35]. Usually, the biased initialization is realized by implementing a non-uniform probability density function to generate the initial optimization parameters as shown in Equation 2.1 of Chapter 2.

As shown in Figure 3.2, the initial population $\mathbf{P}^0$ of opposition-based differential evolution is chosen from the union of $\mathbf{P}'^0$ and $\breve{\mathbf{p}}^0$. It is the other approach to biasing the initial population.

3.2.2.4 Variants of Differential Mutation

It is well known that differential mutation is the key to the success of differential evolution. Some form of differential mutation is most likely to lead to improvements in differential mutation. Thus, a large amount of effort has been devoted to it.

3.2.2.4.1 Perturbation The perturbation mutation [34,35] is introduced to prevent premature convergence by introducing an independent noise term to the mutant vector $\mathbf{x}^{n+1,v,i}$:

$$\mathbf{x}^{n+1,v,i} = \mathbf{x}^{n,b,i} + \sum_{y \geq 1} F_y(\mathbf{x}^{n,p_{1y}} - \mathbf{x}^{n,p_{2y}}) + \text{noise}, \quad 1 \leq i \neq p_{1y} \neq p_{2y} \leq N_p. \tag{3.4}$$

3.2.2.4.2 Trigonometric Mutation The trigonometric mutation was proposed by H.Y. Fan and J. Lampinen in 2003 [36,37]. It aims to bias the differential mutation towards the most dominant individual among the three individuals involved in the differential mutation by making use of the individuals' objective function values:

$$\mathbf{x}^{n+1,v,i} = \frac{\mathbf{x}^{n,p_1} + \mathbf{x}^{n,p_2} + \mathbf{x}^{n,p_3}}{3} + \frac{\sum_{j=1}^{3}(|f(\mathbf{x}^{n,p_k})| - |f(\mathbf{x}^{n,p_j})|)(\mathbf{x}^{n,p_j} - \mathbf{x}^{n,p_k})}{|f(\mathbf{x}^{n,p_1})| + |f(\mathbf{x}^{n,p_2})| + |f(\mathbf{x}^{n,p_3})|}, \quad k = \text{mod}(j,3)+1. \tag{3.5}$$

3.2.2.4.3 Multiple Mutations In this approach [38–40], multiple mutations are implemented to deliver multiple children. The mutations after the first mutation are either unconditional or conditional. An unconditional follow-up mutation delivers a child. The child is then compared with the children born by previous mutations. The most dominant child survives. On the other hand, a conditional follow-up mutation is carried out only if the children born by preceding mutations are not satisfactory.

As a matter of fact, the mutation involved in the opposition-based differential evolution is of this type.

3.2.2.4.4 Non-uniform Mutation The non-uniformity here is related to the mutation probability of individuals in a population. The mutation probability depends on the individual's performance.

The non-uniform mutation proposed by H. Sarimveis and A. Nikolakopoulos in 2005 [39] gives better individuals a higher probability of moving forward or backward, leading to a better solution. Its Fortran-style pseudo-code is given in Figure 3.3, where d^n is the dynamically adjusted step. Individuals in $\mathbf{P}^n$ have been arranged in descending order.

3.2.2.4.5 Binomial Mutation Usually, all genes of the mutant are generated in the same way. In 2003, B.V. Babu and A. Angira [41] proposed a new mutation scheme which borrows the concept of binomial crossover. Hence, we call it binomial mutation here.

Binomial mutation involves two different mutation schemes. Each gene of the mutant is generated by one of these schemes. The choice of a specific mutation scheme for a gene is usually the result of a Bernoulli experiment.

Binomial mutation is hybrid of two different mutation schemes. In this sense, it may also be called a hybrid mutation.

3.2.2.4.6 Mutation with Selection Pressure In mutation with selection pressure [42–44], the mutant is generated in the same way as in classic differential evolution. The difference lies in the way base and difference vectors are chosen. As its name implies, individuals $\mathbf{p}^{n,i}$ in population

```
do i = 1, N_p
  p_i = (N_p - i + 1) / N_p
  if(rand(0, 1) < p_i) then
    do j = 1, N
      if(int(2 * rand(0, 1)) + 1 = 0) then
        c_j^{n+1,i} = x_j^{n,i} + (b_j^U - x_j^{n,i}) * rand(0, 1) * d^n
      else
        c_j^{n+1,i} = x_j^{n,i} - (x_j^{n,i} - b_j^L) * rand(0, 1) * d^n
      end if
    end do
    if(d(c^{n+1,i}, p^{n,i}) true) then
      p^{n+1,i} = c^{n,i}
    end if
  end if
end do
```

Figure 3.3 Fortran-style pseudo-code for non-uniform mutation

$\mathbf{P}^n$ might be restricted to participate in the mutation. Usually, such restriction is inversely proportional to individuals' performance. Better individuals have a better chance of participating.

Best base and better base are two of the special cases of this mutation scheme.

3.2.2.4.7 Differential-Free Mutation Ali and colleagues observed that calculating vector differences for each mutant is time-consuming and may limit the scope for exploration of differential evolution. Consequently, the differential-free point generation scheme [10,45] is proposed. Instead of calculating vector differences for each mutant $\mathbf{v}^{n,i}$, differential-free mutation chooses vector differences from an array of difference vectors. These difference vectors are updated from time to time.

3.2.2.5 Peculiar Crossovers

As mentioned before, crossover is the main search tool of genetic algorithms. It has been extensively studied in genetic algorithms. Various schemes have been proposed. Unfortunately, its role in genetic algorithms has been overemphasized. In differential evolution, however, its role has suffered significant neglect and few people have been working on it.

In classic differential evolution, the mutant $\mathbf{v}^{n+1,i}$ mates with mother $\mathbf{p}^{n,i}$ to deliver a child $\mathbf{c}^{n+1,i}$. This scheme can be modified in two ways: by changing the mating partner and the number of children. The mutant $\mathbf{v}^{n+1,i}$ must be one of the mating partners. However, using $\mathbf{p}^{n,i}$ as the other mating partner is not mandatory. Many other choices are also feasible. For example, in [41], $\mathbf{p}^{n,\text{best}}$ is mated with $\mathbf{v}^{n+1,i}$. The two mating parents to have more than one child, and indeed the original crossover schemes presented in Chapter 1 all deliver two children.

3.2.2.6 Approximating Objective and Constraint Functions

For most practical problems, evaluating objective and constraint functions is very time-consuming. Thus, some researchers have also proposed to approximate objective and constraint functions at certain stages of evolution so that the computation cost can be significantly reduced without sacrificing solution accuracy [46,47].

3.2.2.7 Alternative Selections

In the mother–child competition in classic differential evolution, each child $\mathbf{c}^{n+1,i}$ competes with its direct mother $\mathbf{p}^{n,i}$. Neither inter-parent nor inter-child competition is enforced. Consequently, a child $\mathbf{c}^{n+1,i}$ is rejected if it is not dominant over its mother $\mathbf{p}^{n,I}$, even if the child is dominant over all other individuals in $\mathbf{P}^n$ except its direct mother and the mother is the best individual in $\mathbf{P}^n$. Similarly, a dominant child replaces its direct mother even if the mother is the best individual in $\mathbf{P}^n$. In addition, very similar children may be generated. Population diversity is thus at risk. On account of this observation, many alternative selection schemes have been proposed.

3.2.2.7.1 Cross-selection In the cross-selection scheme, a child $\mathbf{c}^{n+1,i}$ competes with but does not replace its direct mother $\mathbf{p}^{n,i}$. Instead, it replaces other parent individuals. The eliminated parent individual can be chosen randomly from the parent population $\mathbf{P}^n$, randomly chosen from parent individuals dominated by $\mathbf{c}^{n+1,i}$, or the worst individual in $\mathbf{P}^n$.

3.2.2.7.2 Group Selection Group selection [48] is proposed to keep the dominating individuals in the union of $\mathbf{C}^{n+1}$ and $\mathbf{P}^n$. Such a scheme has been widely implemented in other evolutionary algorithms. For example, the selection scheme in the $(\mu + \lambda)$-evolution strategy is just a generalized group selection.

Group selection requires ranking all individuals in the union of $\mathbf{C}^{n+1}$ and $\mathbf{P}^n$ according to specified criteria. It is well known that ranking – and therefore group selection – is computationally very expensive, especially when the population size is large.

3.2.2.7.3 Similarity Selection The aim of similarity selection [49] is to enhance population diversity. A child $\mathbf{c}^{n+1,i}$ competes with the most similar parent individual instead of its direct mother in $\mathbf{P}^n$. If the child dominates that parent individual, the latter is replaced.

Similarity selection is also computationally expensive since the similarities (or distances) between the child and all parent individuals in $\mathbf{P}^n$ have to be computed.

3.2.2.7.4 Threshold Margin Selection Noisy problems have been a challenge to differential evolution [50]. Threshold margin selection [51] is designed to address the noise in objective and/or constraint functions. A child $\mathbf{c}^{n+1,i}$ replaces its mother in $\mathbf{P}^n$ if it dominates its mother by a threshold margin which is proportional to the noise strength.

The aforementioned selection schemes can also be generalized by using the notion of threshold margin selection.

3.2.3 Hybrid Differential Evolution

3.2.3.1 Introduction

Although differential evolution has proved to be a very efficient global optimizer, its local search ability has long been questioned. In addition, there is still considerable scope to improve its global search ability. Hence, many researchers have been working to enhance differential evolution by combining it with deterministic and/or stochastic optimizers.

In general, there are two mechanisms to hybridize differential evolution: embedding and sequential. In the embedding mechanism, one optimizer is embedded in another optimizer. Differential evolution serves as the host optimizer if the embedded optimizer is deterministic. Most often, the best individual of each generation of differential evolution is used as the starting point for the deterministic optimizer. However, in the sequential mechanism, the optimizers execute in series, that is, one optimizer starts when the other optimizer stops. Solutions obtained by the first optimizer are used as seed for the second optimizer.

Each optimization algorithm has certain operators. For example, differential evolution involves differential mutation, crossover and mother–child competition. An interesting hybridization approach implements some operators borrowed from other optimization algorithms in an optimization algorithm. The borrowed operators are supposed to complement those in the host in order to enhance the host. However, due to the large number of varieties of this approach, it will not be discussed here.

3.2.3.2 Enhancing Differential Evolution with Deterministic Optimization Algorithms

In general, most deterministic optimizers are good local optimizers. Consequently, it is the first choice to hybridize differential evolution with deterministic optimizers to enhance the local search capability of differential evolution.

Wang and co-workers have been pioneering in this area. They proposed a gradient-based hybrid differential evolution [52,53] in 1998. After the ordinary evolutionary operations, acceleration, a local search process based on the steepest descent algorithm, is introduced to push the best individual forward to obtain a better solution. In addition, to avoid premature convergence, the migrating operation, a reinitialization around the current best individual, is introduced.

Many other deterministic algorithms have also been hybridized with differential evolution to improve the efficiency of differential evolution. Some representative schemes are summarized in Table 3.1. The references cited there and in Tables 3.2 and 3.3 below are not listed at the end of this chapter as there are too many. Interested readers are advised to obtain them from the companion website for this book.

3.2.3.3 Enhancing Differential Evolution with Stochastic Optimization Algorithms

The evolution mechanism of differential evolution differs from that of other stochastic optimizers. For example, in genetic algorithms, the three evolutionary operations are executed

Table 3.1 Deterministic optimization algorithms hybridized with differential evolution

Deterministic Optimization Algorithm	References
Adjustable control weight gradient method	(Lopez Cruz, van Willigenburg, and van Straten, 2003)
Combined feasible direction method	(Chang and Wu, 2004)
Davidon-Fletcher-Powell method	(Colaco, Dulikravich, and Martin, 2003a; Colaco, Dulikravich, and Martin, 2003b; Colaco, Dulikravich, Orlande, and Rodrigues, 2003; Colaco, Dulikravich, and Martin, 2004)
Dividing rectangle algorithm	(He and Narayana, 2002)
Gaussian-Newton algorithm	(Xing and Xue, 2007)
Implicit filtering algorithm	(Coelho and Mariani, 2006c)
Levenberg–Marquardt descent strategy	(Bluszcz and Adamiec, 2006; Chen, Chen and Cao, 2002)
Markov chain particle filtering	(Du and Guan, 2006)
Multi-mode left shift	(Lorenzoni, Ahonen, and de Alvarenga, 2006)
Modified Newton-Raphson method	(Li, Rao, He, Xu, Wu, Yan, Dong, and Yang, 2005)
Newton-Raphson method	(Crutchley and Zwolinski, 2002)
Powell's direction set method	(Dony and Xu, 2004; Xu and Dony, 2004)
Quasi-Newton algorithm	(Babu and Angira, 2004)
Rough sets theory	(Hernandez-Diaz, Santana-Quintero, Coello, Caballero, and Monina, 2006)
Sequential quadratic programming	(Fung and Yau, 2005; Coelho and Mariani, 2006a; Menon, Kim, Bates, and Postlethwaite, 2006; Menon, Bates, and Postlethwaite, 2007)
Simplex method	(Rogalsky and Derksen, 2000; Niu, Wang, and Gu, 2005; Zou, Ma, Yang, and Li, 2005; Bhat, Venkataramani, Ravi, and Murty, 2006; Nasimul and Hitoshi, 2008)
Variable neighborhood search method	(Qian, Wang, Huang, and Wang, 2006)

in the order selection–crossover–mutation. Crossover plays a more important role in genetic algorithms. In differential evolution, by constrast, the execution order of genetic operations is differential mutation–crossover–selection. Differential mutation is believed more critical for differential evolution.

Generally speaking, although differential evolution works on a population, it encourages individualism and puts social behavior in a subordinate position. In addition, individuals in differential evolution have neither gender nor long-term memory, which have been proved to be essential contributors to the efficiency and robustness of some other stochastic optimizers.

On account of these fundamental differences, the potential to improve the efficiency and robustness of differential evolution by hybridizing differential evolution with other stochastic optimizers has been realized by many evolutionary researchers. Many hybrid differential evolution strategies have been proposed. Some representative hybridization schemes are summarized in Table 3.2.

Table 3.2 Other stochastic algorithms hybridized with differential evolution

Algorithm	References
Ant colony optimisation	(Chiou, Chang, and Su, 2004; Korošec and Šilc, 2006; Wang, Liu, and Chiou, 2006)
Artificial immune system	(Han and He, 2006)
Condensation algorithm	(Yal ın and Gökmen, 2006)
Cultural algorithm	(Becerra and Coello, 2004a; Becerra and Coello, 2004b; Becerra and Coello, 2006a; Becerra and Coello, 2006b; Chong and Tremayne, 2006)
Estimation of distribution Algorithm	(Sun, Zhang, and Tsang, 2005)
Evolution strategy	(Yang, Horng, and Kao, 2001)
Extreme learning machine	(Zhu, Qin, Suganthan, and Huang, 2005)
Genetic algorithms	(Abbass, 2002a; Abbass, 2002b; Abbass, 2002c; Lii, Chiang, Su, and Hwung, 2003; Das, Konar, and Chakraborty, 2005a; De, Rai, Konar, and Chatterjee, 2005; Kaelo and Ali, 2007)
Monte Carlo method	(Weber and Burgi, 2002; Weber, 2005; Ter Braak, 2006)
Multi-start method	(Laskari, Parsopoulos, and Vrahatis, 2003)
Particle swarm optimization	(Hendtlass, 2001; Parsopoulos and Vrahatis, 2002; Zhang and Xie, 2003; Kannan, Slochanal, Subbaraj, and Padhy, 2004; Phan, Lech, and Nguyen, 2004; Talbi and Batouche, 2004; Das, Konar, and Chakraborty, 2005b; Das, Konar, and Chakraborty, 2005c; De, Ray, Konar, and Chatterjee, 2005b; Liu, Wang, Jin, and Huang, 2005; Moore and Venayagamoorthy, 2006; Zheng and Qian, 2006)
Scatter search	(Davendra and G. Onwubolu, 2007)
Simulated annealing	(Zou, Ma, Yang, and Zhang, 2005; Kong, Xu, and Liu, 2006; Subramanian, Slochanal, Subramanian, and Padhy, 2006; Yan, Ling, and Sun, 2006)
Stochastic gradient descent	(Magoulas, Plagianakos, and Vrahatis, 2001)
Threshold accepting algorithm	(Schmidt and Thierauf, 2002; Schmidt and Thierauf, 2005)

3.3 Multi-objective Differential Evolution

3.3.1 Introduction

Hitherto we have been talking about minimizing a single objective function $f(\mathbf{x})$ without constraint functions. However, in practice, one is more likely to have to deal with multi-objective problems as stated in Chapter 1.

Differential evolution has been extended to solve multi-objective problems. In general, there are three approaches to solve constrained multi-objective problems.

(1) weighted sum method;
(2) generalized differential evolution;
(3) Pareto differential evolution.

3.3.2 Weighted Sum Approach

3.3.2.1 General Formulation

Mathematically, the weighted sum method converts a multi-objective optimization problem into an unconstrained single objective optimization problem by defining a single objective function through a weighted sum of all objective and constraint functions:

$$f(\mathbf{x}) = \sum_{i=1}^{N_{o\min}} w_i^{\min} f_i^{\min}(\mathbf{x}) - \sum_{i=1}^{N_{o\max}} w_i^{\max} f_i^{\max}(\mathbf{x}) - \sum_{i=1}^{N_{c+}} w_i^{+} C_i^{+}(\mathbf{x}) + \sum_{i=1}^{N_{c-}} w_i^{-} C_i^{-}(\mathbf{x}) + \sum_{i=1}^{N_{c=}} w_i^{=} |C_i^{=}(\mathbf{x})|. \tag{3.6}$$

The single objective optimization problem can then be solved as normal.

Although the weighted sum method is pervasive in multi-objective optimization due to its simplicity, there is a high risk that the solution obtained is local. Furthermore, the selection of weights strongly depends on the user's experience and may significantly slow down differential evolution's convergence.

3.3.2.2 Weighted Chebysheff

Alternatively, the single objective function can be constructed through the weighted Chebysheff method [54],

$$\begin{aligned} f(\mathbf{x}) = \max\big[& w_i^{\min} |f_i^{\min}(\mathbf{x}) - f_i^{\min}(\mathbf{x}^*)| w_i^{\max} |f_i^{\max}(\mathbf{x}) - f_i^{\max}(\mathbf{x}^*)| \\ & - w_i^{+} C_i^{+}(\mathbf{x}) w_i^{-} C_i^{-}(\mathbf{x}) w_i^{=} |C_i^{=}(\mathbf{x})| \big], \end{aligned} \tag{3.7}$$

which is a special case of the weighted sum approach.

3.3.2.3 Fuzzy Logic Weighted Sum

An interesting modification of the weighted sum approach to multi-objective optimization applies fuzzy logic [53]. The approach is as follows:

(1) Define a threshold interval for each objective and constraint function.
(2) Define a member function for each objective and constraint function:

$$\begin{aligned} & 0 \le \mu_i^{\min}[f_i^{\min}(\mathbf{x}_1)] \le \mu_i^{\min}[f_i^{\min}(\mathbf{x}_2)] \le 1 \quad f_i^{\min}(\mathbf{x}_1) \le f_i^{\min}(\mathbf{x}_2), \\ & 0 \le \mu_i^{\max}[f_i^{\max}(\mathbf{x}_1)] \le \mu_i^{\max}[f_i^{\max}(\mathbf{x}_2)] \le 1 \quad f_i^{\max}(\mathbf{x}_1) \ge f_i^{\max}(\mathbf{x}_2), \\ & 0 \le \mu_i^{=}[|C_i^{=}(\mathbf{x}_1)|] \le \mu_i^{=}[|C_i^{=}(\mathbf{x}_2)|] \le 1 \quad |C_i^{=}(\mathbf{x}_1)| \le |C_i^{=}(\mathbf{x}_2)|, \\ & 0 \le \mu_i^{-}[C_i^{-}(\mathbf{x}_1)] \le \mu_i^{-}[C_i^{-}(\mathbf{x}_2)] \le 1 \quad C_i^{-}(\mathbf{x}_1) \le C_i^{-}(\mathbf{x}_2), \\ & 0 \le \mu_i^{+}[C_i^{+}(\mathbf{x}_1)] \le \mu_i^{+}[C_i^{+}(\mathbf{x}_2)] \le 1 \quad C_i^{+}(\mathbf{x}_1) \ge C_i^{+}(\mathbf{x}_2). \end{aligned}$$

(3) Define a single objective function:

$$f(\mathbf{x}) = \max\{ \mu_i^{\min}[f_i^{\min}(\mathbf{x})] \quad \mu_i^{\max}[f_i^{\max}(\mathbf{x})] \quad \mu_i^{=}[C_i^{=}(\mathbf{x})] \quad \mu_i^{-}[C_i^{-}(\mathbf{x})] \quad \mu_i^{+}[C_i^{+}(\mathbf{x})] \}. \tag{3.8}$$

In essence, the membership functions take the role of weights for objective and constraint functions.

3.3.3 Generalized Differential Evolution

To ease the pressure of choosing weights for objective and constraint functions in the weighted sum approach, Lampinen [55–58] suggests the use of generalized differential evolution for optimizing a single objective function with multiple constraint functions. Generalized differential evolution was later extended to optimize multiple objective functions with multiple constraint functions [59,60].

The generalized differential evolution generalizes the selection (or competition) operator to replace the mother $\mathbf{p}^{n,i}$ with her dominant child $\mathbf{c}^{n+1,i}$. Child $\mathbf{c}^{n+1,i}$ is dominant over mother $\mathbf{p}^{n,i}$ if one or more of the following conditions are satisfied:

(1) Both $\mathbf{c}^{n+1,i}$ and $\mathbf{p}^{n,i}$ are infeasible. However, $\mathbf{c}^{n+1,i}$ violates fewer constraints than $\mathbf{p}^{n,i}$.
(2) $\mathbf{c}^{n+1,i}$ is feasible while $\mathbf{p}^{n,i}$ is infeasible.
(3) Both $\mathbf{c}^{n+1,i}$ and $\mathbf{p}^{n,i}$ are feasible. However, $\mathbf{c}^{n+1,i}$ is better than $\mathbf{p}^{n,i}$ in terms of objective function values, or $\mathbf{c}^{n+1,i}$ is worse than $\mathbf{p}^{n,i}$ in terms of objective function values, but the neighborhood of $\mathbf{c}^{n+1,i}$ is larger than that of $\mathbf{p}^{n,i}$.

The logic dominance function is accordingly defined as

$$d(\mathbf{p},\mathbf{q}) = \begin{cases} & \{\forall 1 \le i \le N_{c^=} : |c_i^=(\mathbf{x}^p)| \le |c_i^=(\mathbf{x}^q)| \cap \\ & \quad \forall 1 \le i \le N_{c^-} : c_i^-(\mathbf{x}^p) \le c_i^-(\mathbf{x}^q) \cap \\ & \quad \forall 1 \le i \le N_{c^+} : c_i^+(\mathbf{x}^p) \ge c_i^+(\mathbf{x}^q)\} \cup \\ \text{true}, & \{\forall 1 \le i \le N_{c^=} : [|c_i^=(\mathbf{x}^p)| = 0 \cap |c_i^=(\mathbf{x}^q)| = 0] \cap \\ & \quad \forall 1 \le i \le N_{c^-} : [c_i^-(\mathbf{x}^p) \le 0 \cap c_i^-(\mathbf{x}^q) \le 0] \cap \\ & \quad \forall 1 \le i \le N_{c^+} : [c_i^+(\mathbf{x}^p) \ge 0 \cap c_i^+(\mathbf{x}^q) \ge 0] \cap \\ & \quad \forall 1 \le i \le N_{o^{\min}} : [f_i^{\min}(\mathbf{x}^p) \le f_i^{\min}(\mathbf{x}^q)] \cap \\ \text{false}, & \quad \forall 1 \le i \le N_{o^{\max}} : [f_i^{\max}(\mathbf{x}^p) \ge f_i^{\max}(\mathbf{x}^q)]\}, \\ & \quad \text{otherwise}. \end{cases} \tag{3.9}$$

Generalized differential evolution gives priority to constraint satisfaction. In practice, sometimes, different objective and constraint functions may have to be considered with different priorities. In this case, child $\mathbf{c}^{n+1,i}$ can be regarded as dominant over mother $\mathbf{p}^{n,i}$ if any improvement of higher-order objective/constraint functions takes place.

3.3.4 Pareto Differential Evolution

Almost all the differential evolution strategies for unconstrained single-objective optimization can be directly generalized to solve multi-objective optimization problems by applying the general logic dominance function defined in Equation 1.17 in Chapter 1. However, such a straightforward generalization may not be efficient because multi-objective optimization problems have many distinctive features. Algorithms addressing these features properly may yield very good results.

3.3.4.1 Pareto Front

Objective functions and constraint functions may be commensurate or incommensurate. For problems with incommensurate objective and constraint functions, a dominant optimal solution may not exist. Instead, it is more practical to provide a set of competing solutions for the decision maker to choose according to specific preference criteria which may be dependent on the real-time situation. The competing solution set forms a Pareto front, which notion was introduced by the engineer/economist Vilfredo Pareto ([61], p. 246). No Pareto solution dominates any other Pareto solutions. When differential evolution tries to obtain a Pareto front for a multi-objective optimization problem it is generally referred to as Pareto differential evolution.

3.3.4.2 Auxiliary Pareto Population

Most often, simple generalization of differential evolution strategies can only capture one solution instead of a Pareto front. One solution for this situation is to introduce an auxiliary Pareto population in which all Pareto solutions found are restored. The auxiliary Pareto population does not participate in differential evolution. It accommodates new Pareto solutions generated by differential evolution. Of course, members of the present auxiliary Pareto population will be deleted if they are dominated by any of the new Pareto solutions since they are not Pareto solutions any more.

In practice, due to memory limitations and programming difficulty, the size of the auxiliary Pareto population is fixed. If the auxiliary Pareto population is full and a new Pareto solution is found, a Pareto solution in the auxiliary Pareto population will be replaced by the new Pareto solution according to certain criteria. Similarity or distance between the two Pareto solutions is usually the criterion most often implemented.

3.3.4.3 Pareto Set Differential Evolution

Pareto set differential evolution was proposed by H.A. Abbass, R. Sarker, and C. Newton in 2001 [62]. The crucial idea behind it is the Pareto set $\mathbf{b}^n$ of a population $\mathbf{P}^n$. The Pareto set of a population contains all Pareto individuals in the population. The general evolution mechanism of Pareto set differential evolution, as shown in Figure 3.4, is identical with that of classic differential evolution. What makes it unique is that only Pareto set members are allowed to donate to the mutant.

In this approach, the Pareto set members do not evolve. They are simply duplicated to $\mathbf{P}^{n+1}$. It is unreasonable to deprive the birth right of the Pareto solutions. All individuals should have equal birth right. Therefore, it might be more reasonable to remove the step shaded in Figure 3.4.

3.3.4.4 Co-evolution Pareto Differential Evolution

In auxiliary Pareto population differential evolution, the auxiliary Pareto population does not interfere with or confer any benefit on the evolution process of differential evolution.

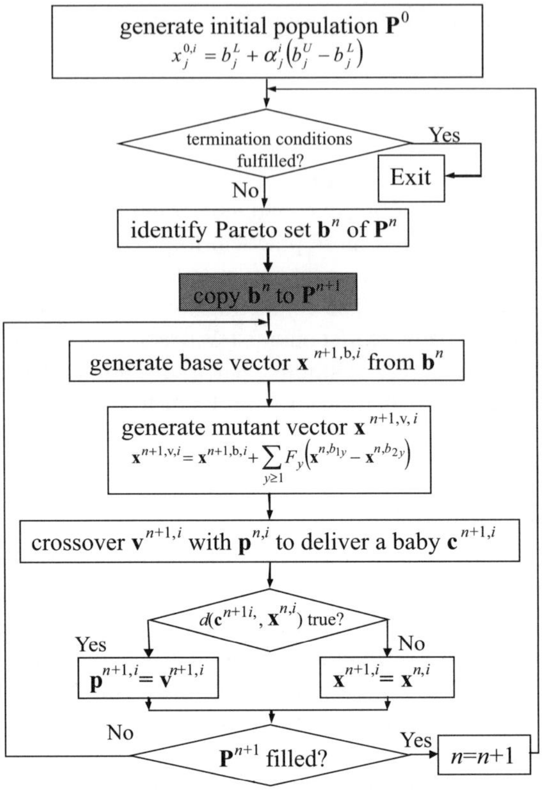

Figure 3.4 Flow chart of Pareto set Pareto differential evolution. Note that $\mathbf{p}^{n,b_{1y}}$ and $\mathbf{p}^{n,b_{2y}}$ belong to $\mathbf{b}^n$

Furthermore, the Pareto population does not evolve by itself. It is updated only when either of the two aforementioned updating conditions is fulfilled. Consequently, a member of the auxiliary Pareto population may never evolve if it is absent in the evolving population.

On the other hand, Pareto set differential evolution abandons Pareto children. It is also dangerous because it may not be able to capture many Pareto solutions.

In this regard, a more meaningful approach is to make the auxiliary Pareto population evolve too. The main population and the Pareto population may evolve separately, cooperatively, or in a unified manner.

3.3.4.4.1 Separate Co-evolution In separate co-evolution, the main population and the Pareto population evolve by themselves. Each co-member of the main population and the Pareto population will deliver two children, one by the main population evolution and the other by the Pareto population. New Pareto solutions generated will be added to the Pareto population.

3.3.4.4.2 Cooperative Co-evolution In cooperative co-evolution, the main population and the Pareto population also evolve separately. However, each co-member of the main population and the Pareto population delivers one child only. They may also work closely by sharing the Pareto population as base and difference vectors provider.

3.3.4.4.3 Unified Co-evolution In unified co-evolution, the main population and the Pareto population form a union. The union evolves once. New Pareto solutions are joined to the Pareto population. The main difference with separate co-evolution lies with the evolution of co-member of the main population and the Pareto population: each co-member delivers one child only. In addition, most often, the Pareto population provides base and difference vectors.

3.3.4.5 Non-dominated Sorting Differential Evolution

The concept of non-dominated sorting was proposed by N. Srinivas and K. Deb in 1994 [63]. It was first implemented in genetic algorithms for solving multi-objective optimization problems and adopted by the differential evolution community from 2002 onwards [64,65].

Non-dominated sorting differential evolution, as shown in Figure 3.5, makes use of a buffer population $\mathbf{B}^{n+1}$ of size $2 \times N_p$. The buffer population contains all individuals of $\mathbf{P}^n$, $\mathbf{p}^{n,i}$,

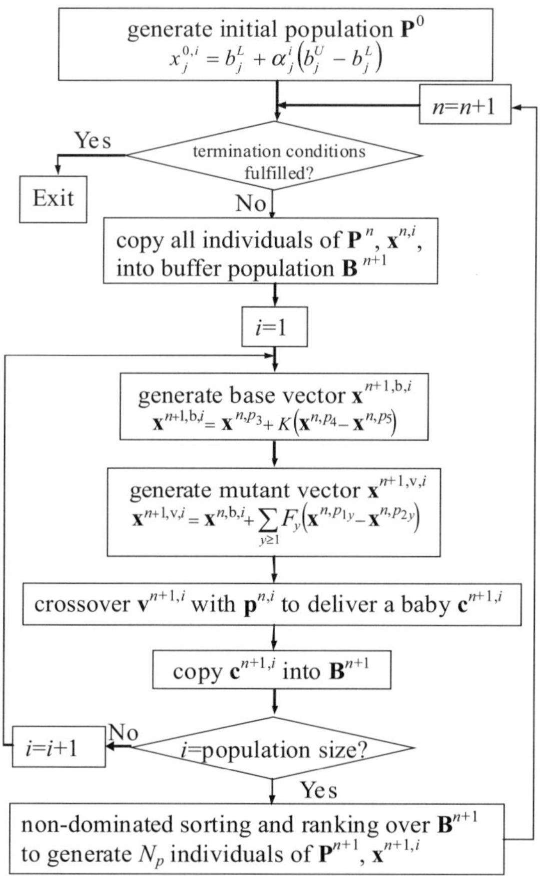

Figure 3.5 Flow chart of non-dominated sorting differential evolution

and all children of $\mathbf{P}^n$, $\mathbf{c}^{n+1,i}$. Non-dominated sorting and ranking are then applied to $\mathbf{B}^{n+1}$ to generate the N_p individuals of $\mathbf{P}^{n+1}$, $\mathbf{x}^{n+1,i}$. In other words, non-dominated sorting differential evolution replaces one-to-one mother–child competition with group competition.

3.4 Parametric Study on Differential Evolution

3.4.1 Motivations

The sensitivity of differential evolution to its intrinsic control parameters has been a challenge to practitioners since its inception. Quantifying the sensitivity of differential evolution to its intrinsic control parameters and subsequently extracting empirical rules for choosing optimal intrinsic control parameters for future applications is of great importance. Consequently, it has been a hot topics in the community of differential evolution.

Generally speaking, the intrinsic control parameters investigated in each study form a finite set $\mathbf{S}_{\text{icp}} = \mathbf{S}_{\text{ps}} \cup \mathbf{S}_{\text{mi}} \cup \mathbf{S}_{\text{cp}}$, where $\mathbf{S}_{\text{ps}}$ is the population size set, $\mathbf{S}_{\text{mi}}$ is the mutation intensity set, $\mathbf{S}_{\text{cp}}$ is the crossover probability set.

Different practitioners have applied different intrinsic control parameter sets $\mathbf{S}_{\text{icp}}$, or more specifically, $\mathbf{S}_{\text{ps}}$, $\mathbf{S}_{\text{mi}}$, and $\mathbf{S}_{\text{cp}}$. Although the single element set $\mathbf{S}_{\text{icp}}$ can be regarded as a special case, it is of little interest to the general public. Hence, it is not discussed here.

3.4.2 Comprehensive Case Studies

Babu and co-workers reported two comprehensive case studies in 2000. In the first study [66], differential evolution, or more precisely, CDE/rand/1/bin with weighted sum of objective and constraint functions, was applied to determine the optimal operating conditions for the alkylation process. The intrinsic control parameter set $\mathbf{S}_{\text{icp}}$ involved is $\mathbf{S}_{\text{ps}} = \{N_p | N_p = 5j, 1 \leq j \leq 20\}$, $\mathbf{S}_{\text{mi}} = \{0.4, 0.5, 0.6, 0.7\}$, $\mathbf{S}_{\text{cp}} = \{0.3, 0.4, 0.5, 0.6, 0.7, 0.8, 0.9\}$. The observed optimal intrinsic control parameters are $N_p = 20$, $F = 0.4$, $p_c = 0.6$. Obviously, type B evolutionary crime (see Section 1.4.3.1) is committed here because mutation intensity and crossover probability are sampled with bias.

The design of a shell-and-tube heat exchanger is the target problem involved in the second comprehensive case study [67]. The intrinsic control parameter set $\mathbf{S}_{\text{icp}}$ is $\mathbf{S}_{\text{ps}} = \{N_p | N_p = 10 + 5j, 0 \leq j \leq 28\}$, $\mathbf{S}_{\text{mi}} = \{F | F = 0.1j, 0 \leq j \leq 10\}$, $\mathbf{S}_{\text{cp}} = \{p_c | p_c = 0.1j, 0 \leq j \leq 10\}$. Comparison between differential evolution and the standard binary genetic algorithm (population size varies from 72 to 224, crossover probability 0.8, and mutation probability 0.1) is carried out. It is very clear that type C evolutionary crime is committed here by fixing the intrinsic control parameters of the competing standard binary genetic algorithm.

It should be pointed out that each combination of intrinsic control parameters is executed only once in both studies. Type A evolutionary crime is committed. Consequently, the reported results and recommended optimal intrinsic control parameters should be accepted with extreme care due to the stochastic nature of both differential evolution and genetic algorithms.

3.4.3 Biased Case Studies

Besides the above two comprehensive case studies on intrinsic control parameters, there are many more biased case studies on intrinsic control parameters. In these studies, at least one intrinsic control parameter is fixed. Type B evolutionary crime is therefore committed in all biased cased studies.

Usually, the fixed value applies to all combinations of variable intrinsic control parameters. Interestingly, some researchers [68,69] apply Zaharie's empirical rule [70,71] to determine the value of the fixed intrinsic control parameter.

There is another interesting variant of the fixed intrinsic control parameter. It is applied when the fixed intrinsic control parameter is mutation intensity or crossover probability. It is either assigned a fixed random number before evolution [72], or a pure random number during evolution [73–78].

3.4.3.1 Biased Case Studies with One Intrinsic Control Parameter Fixed

Most of the case studies here fix population size [41,69,79–85]. It is interesting to note that a very wide range of mutation intensity, [0, 3], was studied in [84]. The effect of population size has also been studied [68,86].

3.4.3.2 Biased Case Studies with Two Intrinsic Control Parameters Fixed

All three intrinsic control parameters have been studied by fixing the other two as summarized in Table 3.3. This is the most common approach implemented to investigate the effect of intrinsic control parameters on optimal values.

Table 3.3 Biased case studies with one intrinsic control parameters flexible

Flexible Intrinsic Control Parameter	References
Population size	(Fischer, Hlavackova-Schindler, and Reismann, 1999; Fischer, Reismann, and Hlavackova-Schindler, 1999; Lampinen and Zelinka, 2000; Babu and Angira, 2001; Salomon, Perrin, and Heitz, 2001; Tin-Loi and Que, 2002; Ilonen, Kamarainen, and Lampinen, 2003; Karaboğa and Ökdem, 2004; Li, 2005; Noman and Iba, 2006)
Mutation intensity	(Fischer, Hlavackova-Schindler, and Reismann, 1999; Ruzek and Kvasnicka, 2001; Gämperle, Müller, and Koumoutsakos, 2002; Tremayne, Seaton, and Glidewell, 2002; Liu and Lampinen, 2005; Yan, 2006)
Crossover probability	(Chiou and Wang, 1999; Fischer, Hlavackova-Schindler, and Reismann, 1999; Schmitz and Aldrich, 1999; Ruzek and Kvasnicka, 2001; Abbass and Sarker, 2002; Gämperle, Müller, and Koumoutsakos, 2002; Karaboğa and Ökdem, 2004; Sarker and Abbass, 2004; Xie, Zhang, Guo, and Yang, 2004; Liu and Lampinen, 2005; Mezura-Montes, Velázquez-Reyes, and Coello, 2006)

3.4.4 *Applicability*

It is understood that it takes a lot of time for researchers to carry out case studies of intrinsic control parameters. Unfortunately, the reported results are more or less unreliable as a result of the evolutionary crimes committed.

3.5 Adaptation of Intrinsic Control Parameters of Differential Evolution

3.5.1 Motivation

We have observed that differential evolution is sensitive to intrinsic control parameters. Choosing appropriate intrinsic control parameters has been a challenge to researchers from the outset.

Many researchers have also pointed out that fixed intrinsic control parameters throughout the evolution process may not be advantageous. At early stages of evolution, population diversity and convergence are more important. Therefore, a larger population, stronger mutation, and lower crossover probability are more favorable. In contrast, at later stages of evolution, more attention should be paid to efficiency, or convergence speed. Accordingly, a smaller population, weaker mutation, and higher crossover probability are more beneficial.

Various approaches to ease the pressure of choosing appropriate intrinsic control parameters for differential evolution have been proposed. Generally speaking, these approaches can be classified into four categories

(1) random;
(2) deterministic;
(3) adaptive;
(4) self-adaptive.

3.5.2 Random Adaptation

Random adaptation is the easiest way to ease the pressure of choosing appropriate intrinsic control parameters. It has only been applied to adapt mutation intensity and crossover probability due to the difficulty of programming differential evolution using a population of random size. Values of mutation intensity and/or crossover probability in each evolution loop are chosen randomly from a pre-specified range. Random numbers may follow a uniform or Gaussian distribution, or some other distribution with specified probability density function.

For the uniform distribution, although the benchmark range recommended by Storn is most often adopted, the reported range is quite diverse. For example, some researchers use a wide range, such as $[-2, 2]$, for mutation intensity while others use a narrow range, such as $[0.5, 1]$.

Similarly, the reported mean value and the standard deviation of the Gaussian distribution are also very diverse. Some researchers even suggest a discrete range for mutation intensity and/or crossover probability.

An interesting variation has been proposed by Orman, Engelbrecht and Salman [87–89]. Differential mutation is implemented to adapt mutation intensity.

Unfortunately, all these approaches commit more or less evolutionary crime of at least type D.

3.5.3 Deterministic Adaptation

In deterministic adaptation, the intrinsic control parameters concerned are adjusted according to some deterministic rules which are independent of the status of the evolving population.

3.5.3.1 Population Size

To reduce computational cost, Michalski [90] proposed using an initial population of size 50 for the first nine generations and then restarting using a shrunken population of size 25. The shrunken population is initialized by perturbing the best individual so far with normally distributed random deviations.

3.5.3.2 Mutation Intensity

Various deterministic adaptation rules have been proposed.

3.5.3.2.1 Linear Adaptation The general formulation of the linear adaptation scheme [38,91–94] is

$$F^{n+1} = F^{n} - a, \tag{3.10}$$

where the constant a is usually dependent on the maximum number of generations and the variation range of the mutation intensity.

3.5.3.2.2 Power Adaptation In the power adaptation scheme [95,96], the adaptation obeys the power law

$$F^{n+1} = aF^{n} \tag{3.11}$$

where the power constant a is usually dependent on the maximum number of generations and on the variation range of the mutation intensity.

3.5.3.2.3 Combined Adaptation In the combined adaptation scheme, the linear and power adaptation rules can be unified to give

$$F^{n+1} = aF^{n} - b. \tag{3.12}$$

3.5.3.2.4 Chaotic Adaptation In chaotic adaptation, the mutation intensity is given by a chaotic sequence generated by iterator logistic map [97,98].

3.5.3.2.5 Directed Adaptation Usually, only one mutation intensity and one crossover probability are applied while evolving from parent population $\mathbf{P}^{n}$ to child population $\mathbf{P}^{n+1}$. This approach does not bestow any privilege on the dominating individuals in population $\mathbf{P}^{n}$ and may therefore not be beneficial in directing the evolution toward the optimal solution. One implementation to provide such guidance is to apply mutation intensity proportional to the performance difference of individuals donating difference vectors.

H.Y. Fan and co-workers proposed three approaches aiming to bias the differential mutation by making use of the individuals' objective function values information. Trigonometric mutation is one of them. The other two are directed mutation [99] and direct mutation [100].

3.5.3.2.5.1 Directed Mutation Directed mutation was proposed to bias the differential mutation towards the most dominant individual among the three individuals involved in the differential mutation:

$$\mathbf{x}^{n+1,v,i} = \mathbf{x}^{n,p_1} + \sum_{j=2}^{3} \frac{f(\mathbf{x}^{n,p_j}) - f(\mathbf{x}^{n,p_1})}{f(\mathbf{x}^{n,p_j})} (\mathbf{x}^{n,p_1} - \mathbf{x}^{n,p_j}) \tag{3.13}$$

where $\mathbf{p}^{n,p_1}$ dominates both $\mathbf{p}^{n,p_2}$ and $\mathbf{p}^{n,p_3}$.

3.5.3.2.5.2 Direct Mutation The direct mutation works by following the formula

$$\mathbf{x}^{n+1,v,i} = \mathbf{x}^{n,b,i} + \frac{|f(\mathbf{x}^{n,p_1}) - f(\mathbf{x}^{n,p_2})|(\mathbf{x}^{n,p_1} - \mathbf{x}^{n,p_2})}{|\mathbf{x}^{n,b,i}|}. \tag{3.14}$$

3.5.3.3 Crossover Probability

In principle, all the aforementioned deterministic adaptation rules for mutation intensity except directed adaptation can be used to adjust the crossover probability.

3.5.4 Adaptive Adaptation

In adaptive adaptation, the intrinsic control parameters are adjusted dynamically depending on the status of the evolving population. The aforementioned Zaharie empirical rule [70,71] is one of the proposed adaptive rules.

3.5.4.1 Population Size

In 2004, Feoktistov and Janaqi [101] proposed to decrease population size according to an energetic barrier function which depends on the generation number. The value of this function acts as an energetic filter, through which only individuals with lower fitness can pass.

In 2006, dynamic population size differential evolution [102] was proposed. In this method, the variance of samples of population is defined to monitor population status and adjust population size accordingly. The worst individual is removed from the population when the population status goes to mode 1, while a randomly initialized new individual is added to population if the population status shifts to mode 2.

3.5.4.2 Mutation Intensity

In 2002, Liu and Lampinen proposed fuzzy adaptive differential evolution [103–105] to adjust mutation intensity and crossover probability using a fuzzy logic controller.

Ali and Törn [10,69] proposed to adjust mutation intensity according to the ratio of minimum and maximum objective function values of individuals in the current population. However, the formulation presented might be incorrect.

The ratio of successful mutations is implemented as an indicator of population status by some researchers [95,96,106,107]. Mutation intensity is adapted accordingly.

In 2006, Guo and co-workers [108] proposed to adjust mutation intensity according to the ratio between fitness of the best individual in the current population and the average fitness of the current population.

3.5.4.3 Crossover Probability

Guo and co-workers [108] proposed to adjust crossover probability according to the ratio between fitness of the best individual in the current population and that in the preceding population. Nearchou and Omirou [69] proposed to adjust crossover probability according to the ratio between minimal and average cost of the individual in the current population.

3.5.5 Self-Adaptive Adaptation

In the self-adaptive adaptation approach, the intrinsic control parameters are optimized simultaneously with the target problem. In general, there are two approaches to optimizing the intrinsic control parameters: joint and nested.

3.5.5.1 Joint Self-Adaptive

In this approach [79,80,109–118], some or all of the intrinsic control parameters are treated as extra optimization parameters. They are joined with the optimization parameters of the target problem and are simultaneously optimized by differential evolution.

When the population size is adapted by this approach [110,111,115], additional operations, such as deletion and cloning, have to be included to adjust the population size.

3.5.5.2 Nested Self-Adaptive

Babu, Angira, and Nilekar [119,120] proposed nested differential evolution to adapt intrinsic control parameters of differential evolution. The outer loop takes care of intrinsic control parameters of differential evolution while the inner loop takes care of the target problem.

References

[1] Watson, G.H., Gilholm, K. and Jones, J.G. (1999) A wavelet-based method for finding inputs of given energy which maximize the outputs of nonlinear systems. *International Journal of Systems Science*, **30** (12), 1297–1307.

[2] Qing, A. (2006) Dynamic differential evolution strategy and applications in electromagnetic inverse scattering problems. *IEEE Transactions on Geoscience and Remote Sensing*, **44** (1), 116–125.

[3] Corne, D., Dorigo, M. and Glover, F. (eds) (1999) *New Ideas in Optimization*, McGraw-Hill, London.

[4] Angira, R. and Babu, B.V. (2005) Process synthesis and design using modified differential evolution (MDE). Int. Symp. 58th Annual Session IIChE, New Delhi, December 14–17.

[5] Angira, R. and Babu, B.V. (2005) Optimization of non-linear chemical processes using modified differential evolution (MDE). 2nd Indian Int. Conf. Artificial Intelligence, Pune, India, December 20–22, pp. 911–923.

[6] Angira, R. and Babu, B.V. (2006) Optimization of process synthesis and design problems: a modified differential evolution approach. *Chemical Engineering Science*, **61** (14), 4707–4721.

[7] Angira, R. and Babu, B.V. (2006) Performance of modified differential evolution for optimal design of complex and non-linear chemical processes. *Journal of Experimental Theoretical Artificial Intelligence*, **18** (4), 501–512.

[8] Babu, B.V. and Angira, R. (2006) Modified differential evolution (MDE) for optimization of non-linear chemical processes. *Computers and Chemical Engineering*, **30** (6–7), 989–1002.

[9] Omran, M.G.H., Engelbrecht, A.P. and Salman, A. (2005) Differential evolution methods for unsupervised image classification. 2005 IEEE Congress Evolutionary Computation, Edinburgh, UK, Sept. 2–5, vol. 2, pp. 966–973.

[10] Ali, M.M. and Törn, A. (2004) Population set-based global optimization algorithms: some modifications and numerical studies. *Computers and Operations Research*, **31** (10), 1703–1725.

[11] Qing, A., Lee, C.K. and Jen, L. (2001) Electromagnetic inverse scattering of two-dimensional perfectly conducting objects by real-coded genetic algorithm. *IEEE Transactions on Geoscience and Remote Sensing*, **39** (3), 665–676.

[12] Qing, A. (2003) Electromagnetic inverse scattering of multiple two-dimensional perfectly conducting objects by the differential evolution strategy. *IEEE Transactions on Antennas and Propagation*, **51** (6), 1251–1262.

[13] Qing, A. and Gan, Y.B. (2005) Electromagnetic inverse problems (invited), in *Encyclopedia of RF and Microwave Engineering*, vol. **2** (ed. K. Chang), John Wiley & Sons, Inc., Hoboken, NJ, pp. 1200–1216.

[14] Qing, A. (2003) Electromagnetic inverse scattering of multiple perfectly conducting cylinders by differential evolution strategy with individuals in groups (GDES). 2003 IEEE AP-S Int. Symp., Columbus, Ohio, June 22–27, vol. 1, pp. 519–522.

[15] Qing, A. (2004) Electromagnetic inverse scattering of multiple perfectly conducting cylinders by differential evolution strategy with individuals in groups (GDES). *IEEE Transactions on Antennas and Propagation*, **52** (5), 1223–1229.

[16] Rigling, B.D. and Moore, F.W. (1999) Exploitation of sub-populations in evolution strategies for improved numerical optimisation. 10th Midwest Artificial Intelligence Cognitive Science Conf., Bloomington, Indiana, April 23–25, pp. 80–88.

[17] Rumpler, J.A. and Moore, F.W. (2001) Automatic selection of sub-populations and minimal spanning distances for improved numerical optimisation. 2001 Congress Evolutionary Computation, Seoul, South Korea, May 27–30, vol. 1, pp. 38–43.

[18] Zaharie, D. (2004) A multipopulation differential evolution algorithm for multimodal optimisation. 10th Int. Mendel Conf. Soft Computing, Brno, Czech Republic, June, pp. 17–22.

[19] Zaharie, D. (2004) Extensions of differential evolution algorithms for multimodal optimisation. 6th Int. Symp. Symbolic Numeric Algorithms Scientific Computing, Timişoara, September 26–30, pp. 523–534.

[20] Li, X. (2005) Efficient differential evolution using speciation for multimodal function optimisation. Genetic Evolutionary Computation Conf., Washington DC, June 25–29, pp. 873–880.

[21] Mendes, R. and Mohais, A.S. (2005) DynDE: a differential evolution for dynamic optimization problems. 2005 IEEE Congress Evolutionary Computation, Pasadena, CA, June 8–10, vol. 3, pp. 2808–2815.

[22] Shi, Y.J., Teng, H.F. and Li, Z.Q. (2005) Cooperative co-evolutionary differential evolution for function optimization, in *Advances in Natural Computation, Part II* (eds. L. Wang, K. Chen, and Y.S. Ong), Lecture Notes in Computer Science, **3611**, Springer, Berlin, pp. 1080–1088.

[23] Ali, M.M., Smith, R. and Hobday, S. (2006) The structure of atomic and molecular clusters, optimised using classical potentials. *Computer Physics Communications*, **175** (7), 451–464.

[24] Tasgetiren, M.F. and Suganthan, P.N. (2006) A Multi-Populated Differential Evolution Algorithm for Solving Constrained Optimization Problem. 2006 IEEE Congress Evolutionary Computation, Vancouver, Canada, July 16–21, pp. 33–40.

[25] Zaharie, D. (2006) Distributed clustering based on representatives evolved by crowding differential evolution. 12th Int. Mendel Conf. Soft Computing, Brno, Czech Republic, May 31–June 2, pp. 51–56.

[26] Liang, C.H., Chung, C.Y., Wong, K.P. and Duan, X.Z. (2007) Parallel optimal reactive power flow based on cooperative co-evolutionary differential evolution and power system decomposition. *IEEE Transactions on Power Systems*, **22** (1), 249–257.

[27] Singh, L. and Kumar, S. (2007) Parallel evolutionary asymmetric subsethood product fuzzy-neural inference system: an island model approach. Int. Conf. Computing Theory Applications, March, pp. 282–286.

[28] Rahnamayan, S., Tizhoosh, H.R. and Salama, M.M.A. (2008) Opposition-based differential evolution. *IEEE Transactions on Evolutionary Computation*, **12** (1), 64–79.

[29] Rahnamayan, S., Tizhoosh, H.R. and Salama, M.M.A. (2006) Opposition-based differential evolution for optimization of noisy problems. 2006 IEEE Congress Evolutionary Computation, Vancouver, Canada, July 16–21, pp. 1865–1872.

[30] Rahnamayan, S., Tizhoosh, H.R. and Salama, M.M.A. (2006) Opposition-based differential evolution algorithms. 2006 IEEE Congress Evolutionary Computation, Vancouver, Canada, July 16–21, pp. 2010–2017.

[31] Rahnamayan, S., Tizhoosh, H.R. and Salama, N.M.M. (2007) Opposition-based differential evolution (ODE) with variable jumping rate. 2007 IEEE Symp. Foundations Computation Intelligence, Honolulu, HI, April 1–5, pp. 81–88.

[32] Rahnamayan, S., Tizhoosh, H.R. and Salama, M.M.A. (2007) A novel population initialization method for accelerating evolutionary algorithms. *Computers and Mathematics with Applications*, **53** (10), 1605–1614.
[33] Rahnamayan, S., Tizhoosh, H.R. and Salama, N.M.M. (2007) Quasi-oppositional differential evolution. 2007 IEEE Congress Evolutionary Computation, Singapore, Sept. 25–28, pp. 2229–2236.
[34] Chang, T.T. and Chang, H.C. (1998) Application of differential evolution to passive shunt harmonic filter planning. 8th Int. Conf. Harmonics Quality Power, Athens, Greece, Oct. 14–16, vol. 1, pp. 149–153.
[35] Chen, C.W., Chen, D.Z. and Cao, G.Z. (2002) An improved differential evolution algorithm in training and encoding prior knowledge into feedforward networks with application in chemistry. *Chemometrics and Intelligent Laboratory Systems*, **64** (1), 27–43.
[36] Fan, H.Y. and Lampinen, J. (2003) A trigonometric mutation operation to differential evolution. *Journal of Global Optimization*, **27**, 105–129.
[37] Beynon, M.J. (2005) Optimizing object classification under ambiguity/ignorance: application to the credit rating problem. *International Journal of Intelligent Systems in Accounting, Finance and Management*, **13** (2), 113–130.
[38] Lii, G.R., Chiang, C.L., Su, C.T. and Hwung, H.R. (2003) An induction motor position controller optimally designed with fuzzy phase-plane control and genetic algorithms. *Electric Power Systems Research*, **68** (2), 103–112.
[39] Sarimveis, H. and Nikolakopoulos, A. (2005) A line up evolutionary algorithm for solving nonlinear constrained optimization problems. *Computers and Operations Research*, **32** (6), 1499–1514.
[40] Wu, L.H., Wang, Y.N., Yuan, X.F. and Shou, S.W. (2006) Differential evolution algorithm with adaptive second mutation. *Control and Decision*, **21** (8), 898–902 (in Chinese).
[41] Babu, B.V. and Angira, R. (2003) New strategies of differential evolution for optimization of extraction process. Int Symp 56th Annual Session IIChE, Bhubaneswar, December 19–22.
[42] Feoktistov, V. and Janaqi, S. (2004) Generalization of the strategies in differential evolution. 18th Int. Parallel Distributed Processing Symp., April 26–30, pp. 2341–2346.
[43] Bergey, P.K. and Ragsdale, C. (2005) Modified differential evolution: a greedy random strategy for genetic recombination. *Omega – International Journal of Management Science*, **33** (3), 255–265.
[44] Du, J.X., Huang, D.S., Wang, X.F. and Gu, X. (2005) Neural network-based shape recognition using generalized differential evolution training algorithm. 2005 IEEE Int. Joint Conf. Neural Networks, July 31-August 4, vol. 4, pp. 2012–2017.
[45] Ali, M.M. and Fatti, L.P. (2006) A differential free point generation scheme in the differential evolution algorithm. *Journal of Global Optimization*, **35** (4), 551–572.
[46] Pahner, U. and Hameyer, K. (2000) Adaptive coupling of differential evolution and multiquadrics approximation for the tuning of the optimization process. *IEEE Transactions on Magnetics*, **36** (4), 1047–1051.
[47] Vancorenland, P., De Ranter, C., Steyaert, M. and Gielen, G. (2000) Optimal RF design using smart evolutionary algorithms. 37th Design Automation Conf., Los Angeles, CA, June 5–9, pp. 7–10.
[48] Noman, N. and Iba, H. (2006) A new generation alternation model for differential evolution. Genetic Evolutionary Computation, Seattle, WA, July 8–12, vol. 2, pp. 1265–1272.
[49] Thomsen, R. (2004) Multimodal optimization using crowding-based differential evolution. 2004 IEEE Congress Evolutionary Computation, Portland, OR, June 19–23, vol. 2, pp. 1382–1389.
[50] Krink, T., Filipič, B., Fogel, G.B. and Thomsen, R. (2004) Noisy optimization problems – a particular challenge for differential evolution? 2004 IEEE Congress Evolutionary Computation, Portland, OR, June 19–23, vol. 1, pp. 332–339.
[51] Das, S., Konar, A. and Chakraborty, U.K. (2005) Improved differential evolution algorithms for handling noisy optimization problems. 2005 IEEE Congress Evolutionary Computation, Edinburgh, September 2–5, vol. 2, pp. 1691–1698.
[52] Chiou, J.P. and Wang, F.S. (1998) A hybrid method of differential evolution with application to optimal control problems of a bioprocess system. IEEE Congress Computational Intelligence, Anchorage, Alaska, May 4–9, pp. 627–632.
[53] Wang, F.S., Jing, C.H. and Tsao, G.T. (1998) Fuzzy-decision-making problems of fuel ethanol production using a genetically engineered yeast. *Industrial and Engineering Chemistry Research*, **37** (8), 3434–3443.
[54] Li, H. and Zhang, Q.F. (2006) A multiobjective differential evolution based on decomposition for multiobjective optimization with variable linkages, in *Parallel Problem Solving from Nature – PPSN IX* (eds T.P. Runarsson*et al.*), Lecture Notes in Computer Science, **4193**, Springer, Berlin, pp. 583–592.

[55] Lampinen, J. (2001) Solving problems subject to multiple nonlinear constraints by the differential evolution. 7th Int. Conf. Soft Computing, Brno, Czech Republic, June 6–8, pp. 50–57.

[56] Lampinen, J. (2001) Multi-constrained optimization by the differential evolution. IASTED Int. Conf. Artificial Intelligence Applications, Marbella, Spain, September 4–7, pp. 177–184.

[57] Lampinen, J. (2001) Multi-constrained nonlinear optimization by the differential evolution algorithm. 6th On-line World Conf. Soft Computing Industrial Applications, September 10–24.

[58] Lampinen, J. (2002) A constraint handling approach for the differential evolution algorithm. 2002 Congress Evolutionary Computation, Honolulu, Hawaii, May 12–17, vol. 2, pp. 1468–1473.

[59] Kukkonen, S. and Lampinen, J. (2004) An extension of generalized differential evolution for multi-objective optimization with constraints, in *Parallel Problem Solving from Nature – PPSN VIII* (eds X. Yao, E. Burke, J.A. Lozano*et al.*) Lecture Notes in Computer Science, **3242**, Springer, Berlin, pp. 752–761.

[60] Kukkonen, S. and Lampinen, J. (2004) A differential evolution algorithm for constrained multi-objective optimization: initial assessment. IASTED Int. Conf. Applied Informatics, Innsbruck, Austria, Feb. 16–18, pp. 96–102.

[61] Price, K.V., Storn, R.M. and Lampinen, J.A. (2005) *Differential Evolution: A Practical Approach to Global Optimization*, Springer, Berlin.

[62] Abbass, H.A., Sarker, R. and Newton, C. (2001) PDE: a Pareto-frontier differential evolution approach for multi-objective optimization problems. IEEE Congress Evolutionary Computation, May 27–30, Seoul, South Korea, vol. 2, pp. 971–978.

[63] Srinivas, N. and Deb, K. (1995) Multi-objective optimization using nondominated sorting in genetic algorithms. *Evolutionary Computation*, **2**, 221–248.

[64] Madavan, N.K. (2002) Turbomachinery airfoil design optimization using differential evolution. 2nd Int. Conf. Computational Fluid Dynamics, Sydney, Australia, July 15–19.

[65] Angira, R. and Babu, B.V. (2005) Non-dominated sorting differential evolution (NSDE): an extension of differential evolution for multi-objective optimisation. 2nd Indian Int. Conf Artificial Intelligence, Pune, India, December 20–22, pp. 1428–1443.

[66] Babu, B.V. and Chaturvedi, G. (2000) Evolutionary computation strategy for optimization of an alkylation reaction. Int. Symp. 53rd Annual Session IIChE, Science City, Calcutta, December 18–21.

[67] Babu, B.V. and Munawar, S.A. (2000) Differential evolution for the optimal design of heat exchangers. All India Seminar Chemical Engineering Progress Resource Development: A Vision 2010 Beyond, Orissa State Centre Bhuvaneshwar, March 13.

[68] Nearchou, A.C. (2006) Meta-heuristics from nature for the loop layout design problem. *International Journal of Production Economics*, **101** (2), 312–328.

[69] Nearchou, A.C. and Omirou, S.L. (2006) Differential evolution for sequencing and scheduling optimization. *Journal of Heuristics*, **12** (6), 395–411.

[70] Zaharie, D. (2002) Critical values for the control parameters of differential evolution algorithms. 8th Int. Mendel Conference Soft Computing, Brno, Czech Republic, June 5–7, pp. 62–67.

[71] Zaharie, D. (2002) Parameter adaption in differential evolution by controlling the population diversity. 4th Int. Workshop Symbolic Numeric Algorithms Scientific Computing, Timişoara, Romania, October 9–12, 2002, Analele Universitaţii din Timiş oara, vol. XL, Fasc. special, 2002, Seria Matematică–Informatică, pp. 281–295.

[72] Mezura-Montes, E., Coello, C.A.C. and Tun-Morales, E.I. (2004) Simple feasibility rules and differential evolution for constrained optimization, in *MICAI 2004: Advances in Artificial Intelligence* (eds. R. Monroy, G. Arroyo-Figueroa, L.E. Sucar, and H. Sossa), Lecture Notes in Computer Science, **2972**, Springer, Berlin, pp. 707–716.

[73] Abbass, H.A. and Sarker, R. (2002) The Pareto differential evolution algorithm. *International Journal on Artificial Intelligence Tools (Architectures, Languages, Algorithms)*, **11** (4), 531–552.

[74] Karaboğa, D. and Ökdem, S. (2004) A simple and global optimization algorithm for engineering problems: differential evolution algorithm. *Turkish Journal of Electrical Engineering*, **12** (1), 53–60.

[75] Sarker, R. and Abbass, H.A. (2004) Differential evolution for solving multiobjective optimization problems. *Asia-Pacific Journal of Operational Research*, **21** (2), 225–240.

[76] Xie, X.F., Zhang, W.J., Guo, R. and Yang, Z.L. (2004) Empirical study of differential evolution. *Control Decision*, **19** (1), 49–52 & 56 (in Chinese).

[77] Mezura-Montes, E., Velázquez-Reyes, J. and Coello, C.A.C. (2006) A comparative study of differential evolution variants for global optimisation. GECCO'06, July 8–12, 2006, Seattle, Washington, USA, pp. 485–492.

[78] Yan, X. (2006) Aromatic hydrocarbon isomerization process optimization based on IDE and AOS. 6th World Congress Intelligent Control Automation, Dalian, China, June 21–23, pp. 7692–7696, (in Chinese).

[79] Abbass, H.A. (2002) The self-adaptive Pareto differential evolution algorithm. 2002 IEEE Congress Evolutionary Computation, Honolulu, Hawaii, May 12–17, vol. 1, pp. 831–836.

[80] Abbass, H.A. (2002) A memetic Pareto evolutionary approach to artificial neural networks, in *AI 2001: Advances in Artificial Intelligence* (eds. M. Brooks, D. Corbett and M. Stumptner), Lecture Notes in Artificial Intelligence, **2256**. Springer, Berlin, pp. 1–12.

[81] Chang, C.S. and Du, D.Y. (1999) Further improvement of optimisation method for mass transit signalling block-layout design using differential evolution. *IEE Proceedings B, Electric Power Applications*, **146** (5), 559–569.

[82] Babu, B.V. and Gautam, K. (2001) Evolutionary computation for scenario-integrated optimization of dynamic systems. Int. Symp. 54th Annual Session IIChE, CLRI, Chennai, December 19–22.

[83] Babu, B.V. and Munawar, S.A. (2001) Optimal design of shell-and-tube heat exchangers using different strategies of differential evolution. PreJournal.org, The Faculty Lounge, Article No. 003873, posted on website Journal, http://www.prejournal.org, March 2003.

[84] Kukkonen, S. and Lampinen, J. (2005) An empirical study of control parameters for generalized differential evolution. EUROGEN 2005, Munich, Germany, September 12–14.

[85] Zielinski, K., Weitkemper, P., Laur, R. and Kammeyer, K.D. (2006) Parameter study for differential evolution using a power allocation problem including interference cancellation. 2006 IEEE Congress Evolutionary Computation, Vancouver, Canada, July 16–21, pp. 1857–1864 & pp. 6748–6755.

[86] Price, K.V. and Ronkkonen, J.I. (2006) Comparing the uni-modal scaling performance of global and local selection in a mutation-only differential evolution algorithm. 2006 IEEE Congress Evolutionary Computation, Vancouver, Canada, July 16–21, pp. 2034–2041.

[87] Omran, M.G.H. and Engelbrecht, A.P. (2006) Self-adaptive differential evolution methods for unsupervised image classification. 2006 IEEE Conf. Cybernetics Intelligent Systems, Bangkok, Thailand, June 7–9, pp. 1–6.

[88] Omran, M.G.H., Engelbrecht, A.P. and Salman, A. (2006) Using the ring neighborhood topology with self-adaptive differential evolution, in *Advances in Natural Comnputation* (eds. L. Jiao *et al.*), Lecture Notes in Computer Science, **4221**, Springer, Berlin, 976–979.

[89] Salman, A., Engelbrecht, A.P. and Omran, M.G. (2007) Computing, artificial intelligence and information management – empirical analysis of self-adaptive differential evolution. *European Journal of Operational Research*, **183** (2), 785–804.

[90] Michalski, K.A. (2001) Electromagnetic imaging of elliptical-cylindrical conductors and tunnels using a differential evolution algorithm. *Microwave and Optical Technology Letters*, **28** (3), 164–169.

[91] Chang, C.S. and Du, D.Y. (2000) Differential evolution based tuning of fuzzy automatic train operation for mass rapid transit system. *IEE Proceedings B-Electric Power Applications*, **147** (3), 206–212.

[92] Chang, C.S. and Kwan, C.M. (2004) Evaluation of evolutionary algorithms for multi-objective train schedule optimization, in *AI 2004: Advances in Artificial Intelligence* (eds. G. Webb and X. Yu), Lecture Notes in Computer Science, Springer, Berlin, **3339**, pp. 803–815.

[93] Chang, C.S., Lu, L.R. and Wang, F. (2004) Application of differential evolution in harmonic worst-case identification of mass rapid transit power supply system. *International Journal of Systems Science*, **35** (13–14), 731–739.

[94] Mayer, D.G., Kinghorn, B.P. and Archer, A.A. (2005) Differential evolution – an easy and efficient evolutionary algorithm for model optimisation. *Agricultural Systems*, **83** (3), 315–328.

[95] Chiou, J.P., Chang, C.F. and Su, C.T. (2005) Variable scaling hybrid differential evolution for solving network reconfiguration of distribution systems. *IEEE Transactions on Power Systems*, **20** (2), 668–674.

[96] Chiou, J.P., Chang, C.F. and Su, C.T. (2006) Capacitor placement in large-scale distribution systems using variable scaling hybrid differential evolution. *International Journal of Electrical Power Energy Systems*, **28** (10), 739–745.

[97] Coelho, L.S. and Alotto, P. (2006) Electromagnetic device optimization using improved differential evolution methods. 12th Biennial IEEE Conf. Electromagnetic Field Computation, Miami, April 30–May 3, p. 295.

[98] Coelho, L.S. and Mariani, V.C. (2006) Chaotic differential evolution applied to electromagnetics optimisation. 12th Biennial IEEE Conf. Electromagnetic Field Computation, Miami, April 30–May 3, p. 230.

[99] Fan, H.Y. and Lampinen, J. (2003) A directed mutation operation for the differential evolution algorithm. *International Journal of Industrial Engineering*, **10** (1), 6–15.

[100] Fan, H.Y., Lu, W.Z. and Xi, G. (2003) An improved neural-network-based calibration method for aerodynamic pressure probes. *Journal of Fluids Engineering*, **125** (1), 113–120.

[101] Feoktistov, V. and Janaqi, S. (2004) New energetic selection principle in differential evolution. 6th International Conf. Enterprise Information Systems, Porto, Portugal, April 14–17, pp. 151–157.

[102] Huang, F., Wang, L. and Liu, B. (2006) Improved differential evolution with dynamic population size, in *Intelligent Computing: International Conference on Intelligent Computing, ICIC 2006, Kunming, China, August 16–19, 2006: Proceedings, Part I* (eds. D.-S. Huang, K. Li and G.W. Irwin), Lecture Notes in Computer Science, **4113**, Springer, Berlin, pp. 725–730.

[103] Liu, J. and Lampinen, J. (2005) A fuzzy differential evolution algorithm. *Soft Computing*, **9** (6), 448–462.

[104] Liu, J. and Lampinen, J. (2002) Adaptive parameter control of differential evolution. 8th Int. MENDEL Conf. Soft Computing, Brno, Czech Republic, June 5–7, pp. 19–26.

[105] Liu, J. and Lampinen, J. (2002) A fuzzy adaptive differential evolution algorithm. 2002 IEEE Region 10 Conf. Computers Communications Control Power Engineering, Oct. 28–31, vol. 1, pp. 606–611.

[106] Nobakhti, A. and Wang, H. (2006) A self-adaptive differential evolution with application on the ALSTOM gasifier. American Control Conf., June 14–16, pp. 4489–4494.

[107] Nobakhti, A. and Wang, H. (2006) Co-evolutionary self-adaptive differential evolution with a uniform-distribution update rule. IEEE Int. Symp. Intelligent Control, Munich, Germany, October 4–6, pp. 1264–1269.

[108] Guo, Z.Y., Kang, L.Y., Cheng, B. *et al.* (2006) Chaos differential evolution algorithm with dynamically changing weighting factor and crossover factor. *Journal of Harbin Engineering University*, **27**, 523–526 (in Chinese).

[109] Abbass, H.A. (2002) An evolutionary artificial neural networks approach for breast cancer diagnosis. *Artificial Intelligence in Medicine*, **25** (3), 265–281.

[110] Teo, J. (2005) Differential evolution with self-adaptive populations, in *Knowledge-Based Intelligent Information and Engineering System* (eds. R. Khosla, R.J. Howlett and L.C. Jain), Lecture Notes in Computer Science, **3681**, Springer, Berlin, pp. 1284–1290.

[111] Teo, J. and Hamid, M.Y. (2005) Investigating the search quality, population dynamics and evolutionary dynamics of a parameterless differential evolution optimizer. *WSEAS Transactions Systems*, **4** (11), 1993–2000.

[112] Brest, J., Greiner, S., Bošković, B. *et al.* (2006) Self-adapting control parameters in differential evolution: a comparative study on numerical benchmark problems. *IEEE Transactions on Evolutionary Computation*, **10** (6), 646–657.

[113] Brest, J., Žumer, V. and Maučec, M.S. (2006) Self-adaptive differential evolution algorithm in constrained real-parameter optimisation. 2006 IEEE Congress Evolutionary Computation, Vancouver, Canada, July 16–21, pp. 215–222.

[114] Brest, J., Žumer, V. and Maučec, M.S. (2006) Control parameters in self-adaptive differential evolution. 2nd Int. Conf. Bioinspired Optimization Methods Applications, Ljubljana, Slovenia, October 9–10, pp. 33–44.

[115] Teo, J. (2006) Exploring dynamic self-adaptive populations in differential evolution. *Soft Computing*, **10** (8), 673–686.

[116] Thitithamrongchai, C. and Eua-Arporn, B. (2006) Economic load dispatch for piecewise quadratic cost function using hybrid self-adaptive differential evolution with augmented Lagrange multiplier method. 2006 Int. Conf. Power System Technology, Chongqing, China, October 22–26, pp. 1–8.

[117] Thitithamrongchai, C. and Eua-Arporn, B. (2006) Hybrid self-adaptive differential evolution method with augmented Lagrange multiplier for power economic dispatch of units with valve-point effects and multiple fuels. 2006 IEEE PES Power Systems Conf., Atlanta, Georgia, October 29–November 1.

[118] Brest, J., Bošković, B., Greiner, S. *et al.* (2007) Performance comparison of self-adaptive and adaptive differential evolution algorithms. *Soft Computing*, **11** (7), 617–629.

[119] Babu, B.V. and Jehan, M.M.L. (2003) Differential evolution for multi-objective optimisation. 2003 Congress Evolutionary Computation, Canberra, Australia, December 8–12, pp. 2696–2703.

[120] Babu, B.V., Angira, R. and Nilekar, A. (2004) Optimal design of an auto-thermal ammonia synthesis reactor using differential evolution. 8th World Multi-Conference Systemics Cybernetics Informatics, Orlando, FL, July 18–21, vol. 16, pp. 266–271.

4

Configuring a Parametric Study on Differential Evolution

4.1 Motivations

As mentioned in Chapter 1, evolutionary crimes have caused serious damage to the evolutionary computation community. Misconceptions on the evolution mechanism, differential mutation base, and crossover are widespread in the community. It is therefore a matter of urgency to clear these misconceptions and find the best strategy of differential evolution and the corresponding optimal intrinsic control parameter values.

As already mentioned, differential evolution is sensitive to its intrinsic control parameters. It is common practice to choose intrinsic control parameters according to existing recommendations and/or personally accumulated experience. If the chosen intrinsic control parameters do not work for a specific problem, a trial-and-error approach is usually employed to adjust intrinsic control parameters until the problem is successfully solved. Such practice is computationally very inefficient and may even be impractical for large problems.

In this regard, guidelines for the choice of appropriate intrinsic control parameters for differential evolution to solve future optimization problems efficiently and robustly are desperately needed. Evolutionary computation practitioners have realized that obtaining the relationship between an evolutionary algorithm's performance and its intrinsic control parameters analytically is scarcely possible. A more practical approach is to conduct a comprehensive parametric study on the evolutionary algorithm of interest and extract empirical rules for choosing intrinsic control parameters.

4.2 Objectives

In accordance with the need mentioned above, the objective of the parametric study is to look for best differential evolution strategy and its optimal intrinsic control parameters, or rules of thumb for choosing intrinsic control parameters among selected differential evolution strategies to solve future optimization problems robustly and efficiently. More specifically, the

Differential Evolution Anyong Qing

parametric study should answer the following questions:

(1) Which strategy performs best?
(2) When does the strategy perform best?
(3) Why does the strategy perform best?

4.3 Scope

4.3.1 *Differential Evolution Strategies*

There are many variants of differential evolution as discussed in Chapters 2 and 3. This parametric study aims to provide general guidelines for a differential evolution user to choose the most appropriate strategy for his or her specific application problem. This section addresses the evolution mechanism, differential mutation base, and crossover schemes.

4.3.1.1 Evolution Mechanism

As discussed in Chapter 2, differential evolution strategies are classified into four categories according to the evolution mechanism implemented. Classic differential evolution is fundamental and has been widely applied by many practitioners. Dynamic differential evolution differs from classic differential evolution slightly but significantly from the point of view of the evolution mechanism. Moreover, it has been observed that dynamic differential evolution outperforms classic differential evolution over a very limited test bed, including the benchmark electromagnetic inverse scattering problem and two toy functions defined by this author. It is very beneficial if the advantage of dynamic differential evolution persists over a bigger test bed with all important problem features fully addressed. On the other hand, it is equally helpful if dynamic differential evolution is observed performing worse than classic differential evolution over test problems with certain problem features. Thus, classic differential evolution and dynamic differential evolution are chosen as target strategies here.

Modified differential evolution and hybrid differential evolution are of less importance, so they are not considered in the parametric study.

At present, optimization of a single objective function without constraint is our only concern. Therefore, multi-objective differential evolution is also not considered in the parametric study. It may be involved in future parametric study if there is sufficient demand from practical applications.

4.3.1.2 Differential Mutation Base

Best and random bases are the two extreme cases of differential mutation base. In the inventive version of differential evolution, the differential mutation base is randomly chosen among the current population. Historically, this is the most widely employed approach. However, strategies of differential evolution implementing random base do suffer loss of efficiency.

In real-life societies, leaders play a crucial role. They provide guidance to the society, although such guidance may be potentially harmful in a non-democratic or dictatorial society. The best (or the most dominant) individual in a differential evolution population is a bionic leader. A good bionic leader is expected to guide the whole population for better efficiency and robustness. However, a misconception in this community claims that best base vector leads to trapping in local minima, or equivalently, poorer robustness.

With this in mind, we examine best and random bases here. Other bases mentioned in Chapter 2 are not considered.

4.3.1.3 Crossover Scheme

Crossover is a very important evolutionary operator for many evolutionary algorithms. However, in differential evolution, crossover has been thought to play a less critical role. A view prevailing in the differential evolution community is that differential evolution is insensitive to crossover, and therefore very little effort has been made to study it seriously. Taking account of the inexact or even false claims with respect to evolution mechanism and differential mutation base due to inappropriate parametric study, it is reasonable to doubt this prevailing view.

Many crossover schemes have been discussed in Chapter 1, among which binomial and exponential crossover are most commonly implemented. It has been claimed that binomial is never worse than exponential, even though solid theoretical or numerical evidence is lacking. Accordingly, these two crossover schemes are investigated here.

The following eight strategies are considered:

(1) DDE/best/1/bin;
(2) CDE/best/1/bin;
(3) DDE/rand/1/bin;
(4) CDE/rand/1/bin;
(5) DDE/best/1/exp;
(6) CDE/best/1/exp;
(7) DDE/rand/1/exp;
(8) CDE/rand/1/exp.

4.3.2 Intrinsic Control Parameters

This is the main concern of the parametric study. The relationship between performance and intrinsic control parameters of differential evolution will be extensively studied. Optimal intrinsic control parameters, or rules of thumb to choose intrinsic control parameters, are to be extracted. Interaction between intrinsic control parameters is also of great interest.

The intrinsic control parameters form an infinite set. It is practically impossible to study infinite combinations of intrinsic control parameters. A finite set $\mathbf{S}_{icp} = \mathbf{S}_{ps} \cup \mathbf{S}_{mi} \cup \mathbf{S}_{cp}$ containing representative values of intrinsic control parameters is a practical alternative.

4.3.2.1 Population Size

Four different sets of population sizes are used:

$$\mathbf{S}_{ps} = \{6, 12, 18, 24, 30, 36, 42, 48, 54, 60, 90, 120, 150, 180, 360\},$$
$$\mathbf{S}_{ps} = \{8, 16, 24, 32, 40, 48, 56, 64, 72, 80, 120, 160, 200, 400\},$$
$$\mathbf{S}_{ps} = \{9, 18, 27, 36, 45, 54, 63, 72, 81, 90, 120, 150, 180, 360\},$$
$$\mathbf{S}_{ps} = \{10, 20, 30, 40, 50, 60, 70, 80, 90, 100, 150, 200, 250, 500\}.$$

Choosing a set of population sizes for a specific test problem depends only on its dimension.

4.3.2.2 Mutation Intensity

The mutation intensity set tested takes the form $\mathbf{S}_{mi} = \{F|F = j \times \Delta_{mi}, 0 \leq j \leq m/\Delta_{mi}\}$, $m = 1$, unless specified otherwise. Δ_{mi} is determined according to the computational cost of a test problem. For computationally cheap test problems, $\Delta_{mi} = 0.025$. However, if the test problem is computationally very expensive, $\Delta_{mi} = 0.1$.

4.3.2.3 Crossover Probability

The crossover probability set tested is $\mathbf{S}_{cp} = \{p_c|p_c = j \times \Delta_{cp}, 0 \leq j \leq 1/\Delta_{cp}\}$. Δ_{cp} is similarly determined according to the computational cost of a test problem and is usually equal to Δ_{mi}.

4.3.3 Non-intrinsic Control Parameters

Unlimited time to search the solution of an application problem is not feasible in practice. Thus, non-intrinsic control parameters have to be used to terminate differential evolution. The effect of non-intrinsic control parameters on differential evolution and the interaction between intrinsic and non-intrinsic control parameters have to be studied.

In this study, three non-intrinsic control parameters are applied: value to reach (or accuracy, tolerance), limit of number of objective function evaluations, and limit of population diversity. The default value of the value to reach is $\varepsilon_o = 1 \times 10^{-2}$. The default limit of number of objective function evaluations is 2000 times the number of optimization parameters, while the default limit of population diversity is $\varepsilon_d = 1 \times 10^{-5}$. Values of non-intrinsic control parameters different from the default settings will be pointed out explicitly.

It is worthwhile to point out that many researchers do not use value to reach to terminate their simulation. Instead, they keep their program running until the limit of number of objective function evaluations is reached. The accuracy of the final result is then used to evaluate their algorithm. From the point view of engineering applications, this is completely unnecessary: 1% or even 5% accuracy is usually acceptable.

4.4 Implementation Terminologies

4.4.1 Search

A search is an execution of a strategy of differential evolution with fixed intrinsic control parameters. It is successful if it finds the solution. The number of objective function evaluations of a successful search is recorded.

4.4.2 Trial

To reduce randomness, each search with fixed intrinsic control parameters is repeated 100 times, each repetition constituting a trial. The number of successful searches (NSS) in a trial is recorded. A trial with at least one successful search is a partially successful trial. The average number of objective function evaluations (ANOFE) of all successful searches in a partially successful trial is accordingly computed. If all of the 100 searches in a trial are successful, it is a successful trial. Total number of successful searches (TNSS) and number of successful trials (NST) of all trials at fixed population sizes are computed accordingly.

4.5 Performance Indicators

The performance of any evolutionary algorithm includes efficiency and robustness. Efficiency is a measure of how fast the optimizer finds the desired optima, while robustness (or stability, sensitivity) indicates how the optimizer's performance changes with respect to its intrinsic control parameters.

The ANOFE of a trial indicates the efficiency of a differential evolution strategy at fixed intrinsic control parameters. The NSS in the trial and the distribution of number of objective function evaluations of all successful searches in the trial indicate the robustness of the differential evolution strategy at fixed intrinsic control parameters. Efficiency and robustness defined in this way are functions of the three intrinsic control parameters. It is therefore not possible to view these functions graphically. More importantly, application engineers are not interested in these functions. Instead, overall optimal intrinsic control parameters at which best performance is achieved are their primary concern. Therefore, these functions will not be examined unless necessary.

Alternatively, two performance indicators are defined to quantitatively represent the performance of differential evolution.

4.5.1 Robustness Indicator

Intuitively, the success ratio of all searches, that is, the ratio of successful searches to the total number of searches, can be used to represent the robustness of differential evolution. Although such an indicator provides a general idea of the robustness of differential evolution, it does not meet the most essential demand to locate the optimal intrinsic control parameters. The result may also be potentially misleading since it is highly dependent on the finite set of intrinsic control parameters which may be intentionally or unintentionally biased. Therefore, it should not be used.

Instead, the highest number of successful trials (HNST) is defined as the robustness indicator. The population size at which the HNST is achieved is the overall optimal population size in terms of robustness. Overall optimal mutation intensity and crossover probability in terms of robustness correspond to the successful trial with minimal average number of objective function evaluations (MANOFE) at the overall optimal population size in terms of robustness.

The highest total number of successful searches (HTNSS) can be defined as the alternative robustness indicator of differential evolution for those who are not concerned by occasional failure.

4.5.2 Efficiency Indicator

The lowest average number of objective function evaluations (LANOFE) of all successful trials is defined as the efficiency indicator. The overall optimal intrinsic control parameters in terms of efficiency correspond to the successful trial with the LANOFE.

It must be emphasized that the trial must be fully successful, that is, all of the 100 searches in the trial must find the optimal solution. It has occasionally been observed that a partially successful trial may involve a lower ANOFE.

4.5.3 Priority of Performance Indicators

Sometimes, the overall optimal intrinsic control parameters in terms of efficiency and those in terms of robustness are not consistent with each other. In this case, the overall optimal intrinsic control parameters in terms of robustness are more important since it is pointless to talk about number of objective function evaluations when a trial or a search is not successful.

4.5.4 Solution for Insufficient Sampling of Intrinsic Control Parameters

Insufficient sampling of intrinsic control parameters may result in the absence of actual overall optimal intrinsic control parameters. On the other hand, too many sampling points of intrinsic control parameters are computationally unaffordable. A remedy for this issue is to look at the variation of NST (or TNSS) and MANOFE at fixed population size with respect to population size whenever necessary.

4.6 Test Bed

The parametric study on differential evolution has to be carried out on a test bed. Although many test beds are available, none of them is satisfactory. The benchmarking of a test bed for the evaluation of evolutionary algorithms especially for parametric study on differential evolution has been ongoing since 2004. It will be discussed separately in Chapter 5 of this book.

4.7 Similar Works

4.7.1 Introduction

Due to the desperate need for optimal intrinsic control parameters for differential evolution, and of course, other evolutionary algorithms, many researchers have been working in this area. All the case studies mentioned in Chapter 3 have the same objective as the present parametric study. However, all of them have committed evolutionary crimes, whether fully or partially. In addition, several critical issues have been neglected. Consequently, the case study results are somewhat questionable.

4.7.2 Neglected Critical Issues

4.7.2.1 Performance Indicators

Various performance indicators have been employed by different researchers. Efficiency indicators include the number of objective function evaluations, number of generations, CPU time, and even solution accuracy. Some people show several convergence curves simultaneously as a proof of robustness. Others count the number of successes in all simulations and use the success ratio as a proof of robustness. The inconsistency and arbitrariness of these performance indicators seriously compromise the practicability of the case study results.

It should be pointed out that it is absurd to apply solution accuracy as a performance indicator because it makes no sense for an application engineer to impose unachievable requirements on solution accuracy.

4.7.2.2 Test Bed

So far, no effort to benchmark a test bed for unconstrained single-objective optimization has been found with problem features fully and properly addressed. While evaluating an evolutionary algorithm, people usually choose test problems from the available literature. Very often, insufficient attention is given to problem features.

4.7.2.3 Non-intrinsic Control Parameters

Most of the time, people focus their attention on intrinsic control parameters. Little consideration is given to non-intrinsic control parameters. Values of non-intrinsic control parameters are taken from the available literature. The effect of non-intrinsic control parameters has not been adequately addressed.

4.8 A Comparative Study

4.8.1 Motivations

Differential evolution has established its reputation as an efficient evolutionary algorithm. It has been compared with numerous optimization algorithms. Usually, these comparisons commit evolutionary crimes, leading to questionable conclusions.

Right at the beginning of the parametric study, a very limited comparative study is carried out between the standard binary genetic algorithm, real-coded genetic algorithm, differential evolution, and particle swarm optimization. Test bed members participating in the comparative study are the Griewank function, the Qing function, the Rastrigin function, the Rosenbrock function, the translated sphere function, and the benchmark electromagnetic inverse scattering problem. The comparative study results, although limited, clearly demonstrate differential evolution's characteristics.

However, it is not the main concern of this book to compare differential evolution with other optimization algorithms.

4.8.2 Settings of Competing Optimization Algorithms

4.8.2.1 Standard Binary Genetic Algorithm

Tournament selection and one-point crossover are implemented, as is elitism. Two valid decimal digits are guaranteed for each optimization parameter. The crossover probability is sampled in [0, 1] in steps of 0.025 and the mutation probability is sampled in [0, 0.5] in steps of 0.005.

4.8.2.2 Real-Coded Genetic Algorithm

Tournament selection, arithmetic one-point crossover, and perturbation mutation are implemented, as is elitism. The cdrossover probability is sampled in [0, 1] in steps of 0.025. The perturbation intensity is empirically chosen as 0.5% of the dynamic range of optimization parameters in line with personal experience. The mutation probability is sampled in [0, 0.5] in steps of 0.005.

4.8.2.3 Particle Swarm Optimization

Two sets of intrinsic control parameters reported in the literature are used:

(a) $w = 0.729$, $c_p = c_s = 1.494$, $\Delta t = 1$;
(b) $w = 0.729$, $c_p = 2.0412$, $c_s = 0.9477$, $\Delta t = 1$.

Therefore, the results reported here may not be the best particle swarm optimization results.

4.8.3 Numerical Results

4.8.3.1 Translated Sphere Function

4.8.3.1.1 8-Dimensional The performance of the standard binary genetic algorithm is shown in Figure 4.1. Only a few searches are successful, and there is no successful trial.

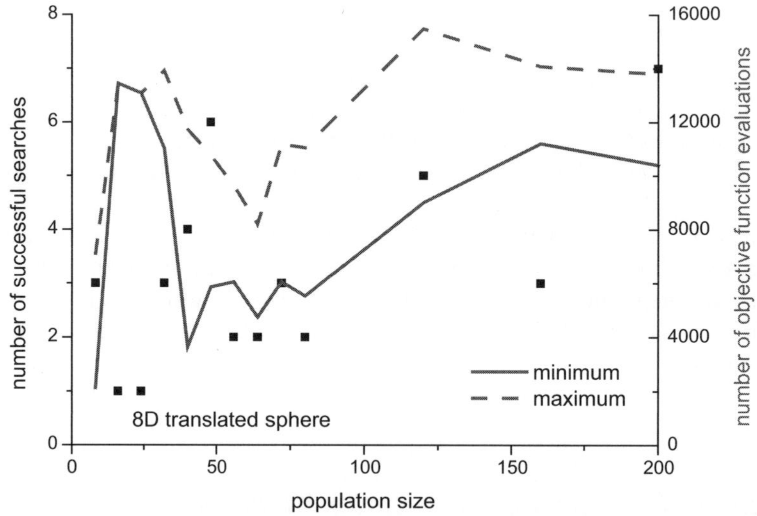

Figure 4.1 Performance of standard binary genetic algorithm for 8-dimensional translated sphere function

The efficiency of other algorithms is shown in Figure 4.2. It is very clear that DDE/best/1/bin is the most efficient algorithm. Classic differential evolution outperforms other optimization algorithms. Their robustness is shown in Figure 4.3. Dynamic differential evolution outperforms all other optimization algorithms.

A very attractive characteristics of DDE/best/1/bin observed in Figure 4.2 is that the MANOFE is almost flat within a wide range of population size. Robustness is therefore in harmony with efficiency. This harmony allows greater freedom to choose population size for robustness.

In contrast, the MANOFE of other optimization algorithms increases linearly with population size. Efficiency is in conflict with robustness. One must always seek a tradeoff between efficiency and robustness.

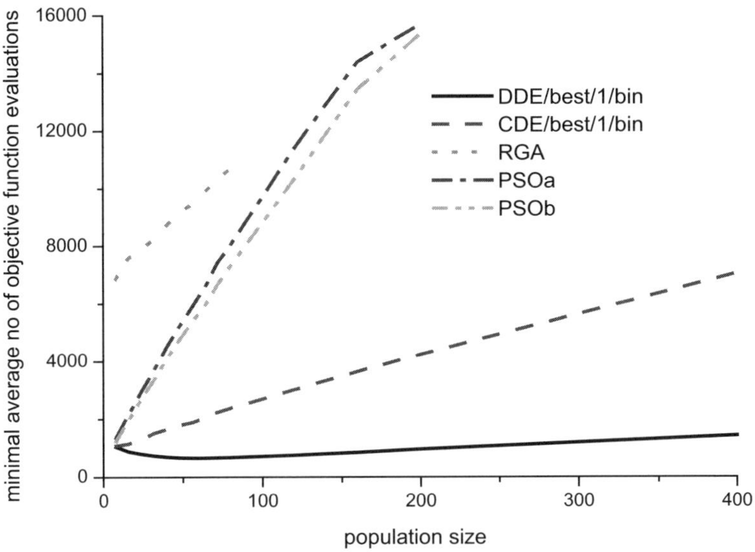

Figure 4.2 Efficiency of differential evolution, real-coded genetic algorithm and particle swarm optimization for 8-dimensional translated sphere function

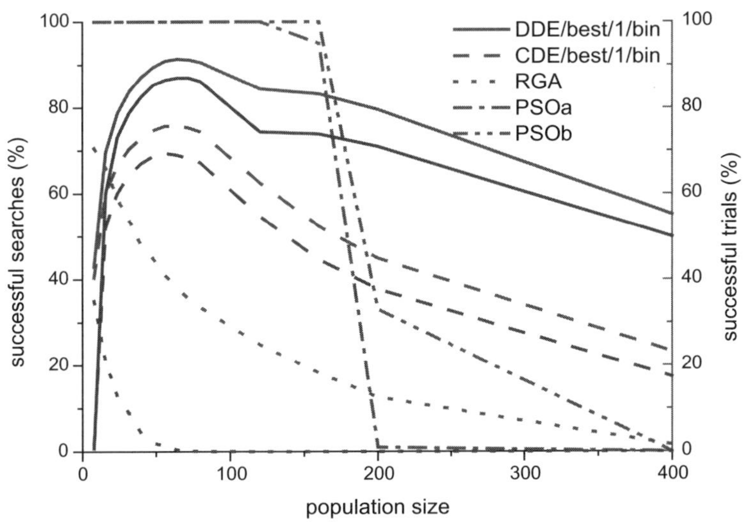

Figure 4.3 Robustness of differential evolution, real-coded genetic algorithm and particle swarm optimization for 8-dimensional translated sphere function

4.8.3.1.2 16-Dimensional No successful search is observed for the standard binary genetic algorithm; this observation applies to higher dimensions.

The efficiency and robustness of other participating optimization algorithms are shown in Figures 4.4 and 4.5, respectively. All observations mentioned above still hold.

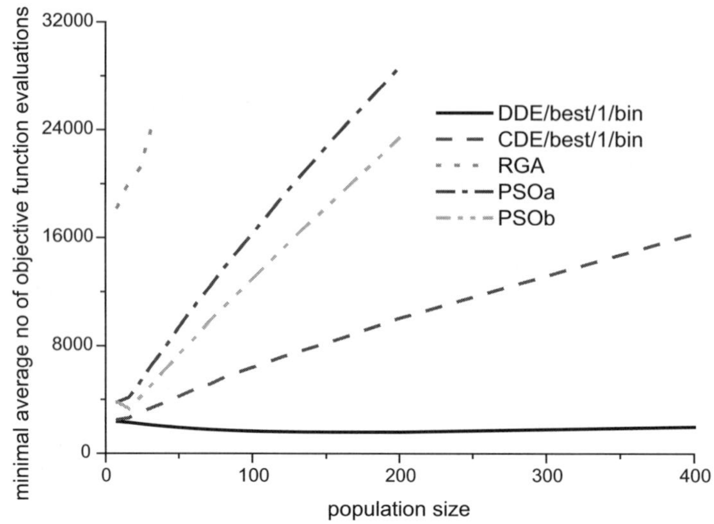

Figure 4.4 Efficiency of differential evolution, real-coded genetic algorithm and particle swarm optimization for 16-dimensional translated sphere function

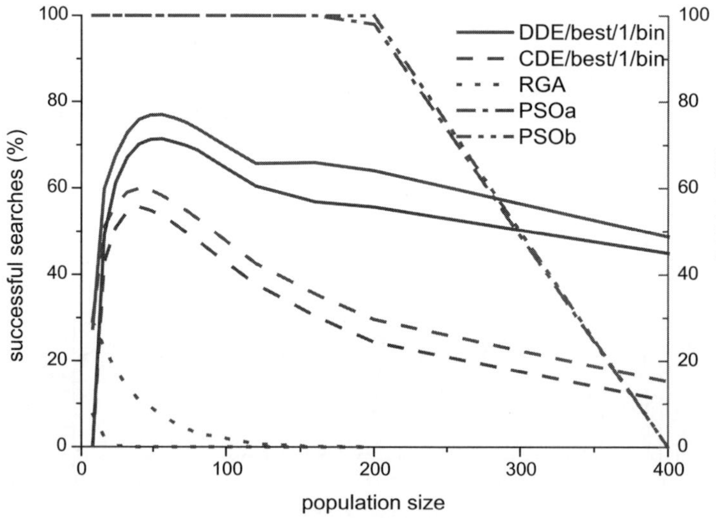

Figure 4.5 Robustness of differential evolution, real-coded genetic algorithm and particle swarm optimization for 16-dimensional translated sphere function

It is further observed that the success rate of all algorithms has fallen. This is to be expected because the search space increases exponentially with problem dimension, while the limit of number of objective function evaluations is linearly proportional to the problem dimension.

4.8.3.1.3 24-Dimensional The efficiency and robustness of participating optimization algorithms are shown in Figures 4.6 and 4.7, respectively. As expected, the success rate of all algorithms has fallen further.

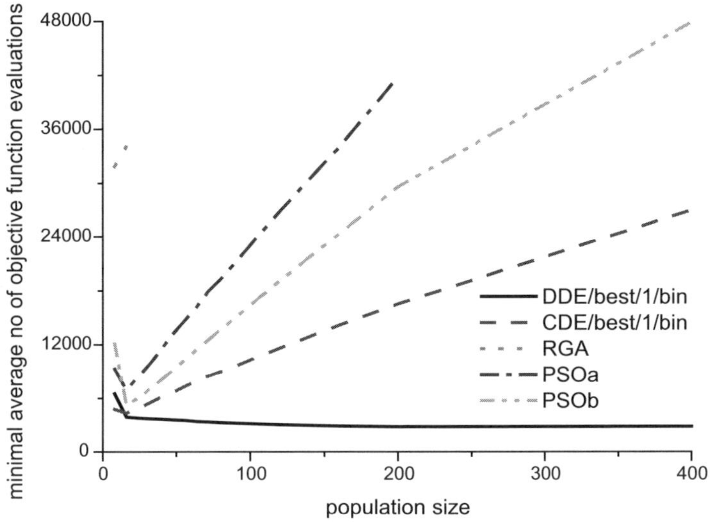

Figure 4.6 Efficiency of differential evolution, real-coded genetic algorithm and particle swarm optimization for 24-dimensional translated sphere function

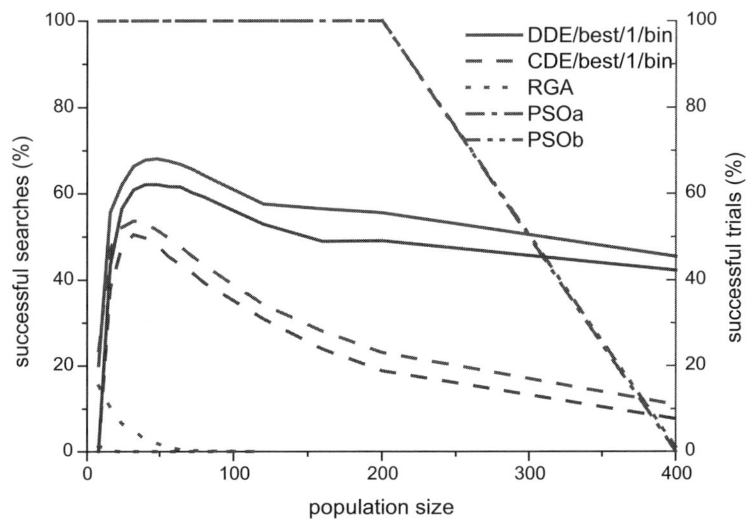

Figure 4.7 Robustness of differential evolution, real-coded genetic algorithm and particle swarm optimization for 24-dimensional translated sphere function

4.8.3.1.4 50-Dimensional The efficiency and robustness of participating optimization algorithms are shown in Figures 4.8 and 4.9, respectively. It is observed that the real-coded

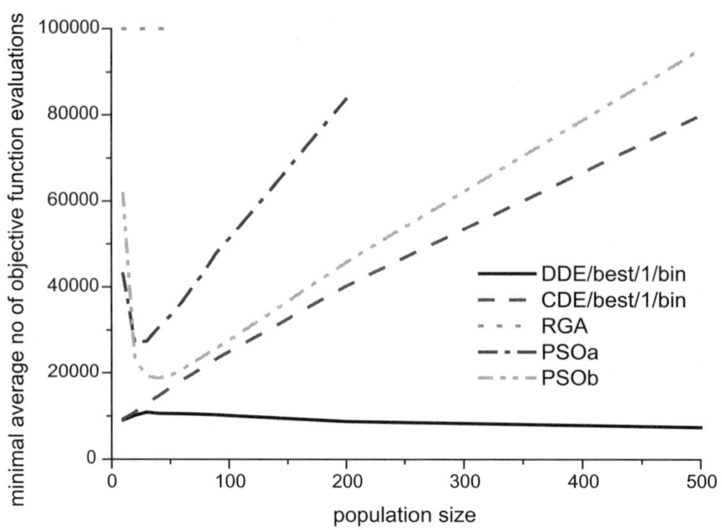

Figure 4.8 Efficiency of differential evolution, real-coded genetic algorithm and particle swarm optimization for 50-dimensional translated sphere function

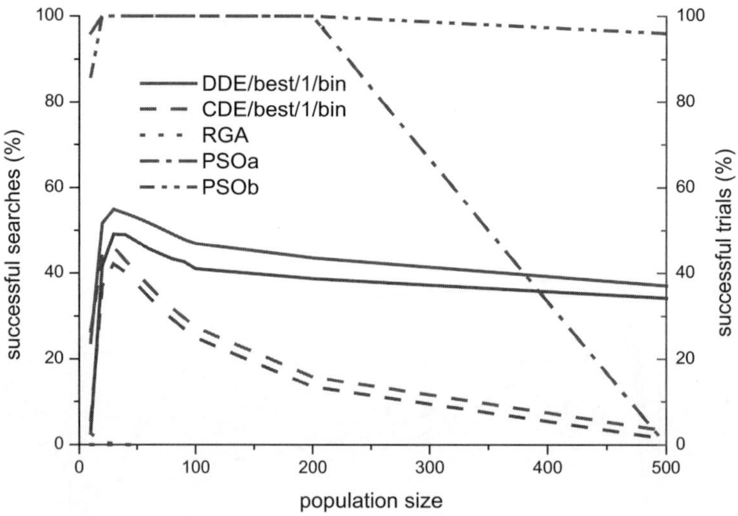

Figure 4.9 Robustness of differential evolution, real-coded genetic algorithm and particle swarm optimization for 50-dimensional translated sphere function

genetic algorithm is unable to solve this problem robustly, although limited searches are still successful.

4.8.3.1.5 100-Dimensional The efficiency and robustness of participating optimization algorithms are shown in Figures 4.10 and 4.11, respectively. It is interesting to observe

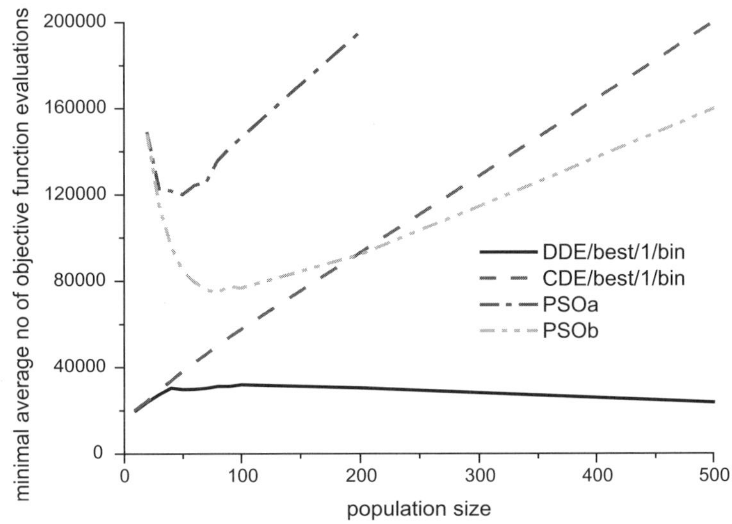

Figure 4.10 Efficiency of differential evolution, real-coded genetic algorithm and particle swarm optimization for 100-dimensional translated sphere function

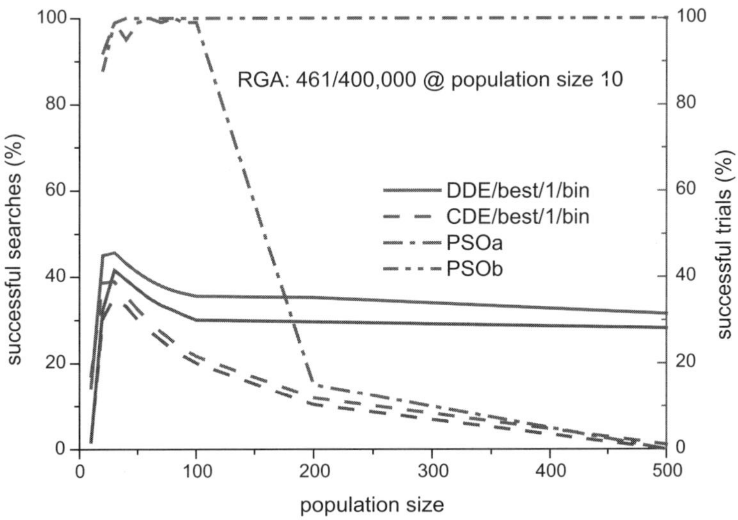

Figure 4.11 Robustness of differential evolution, real-coded genetic algorithm and particle swarm optimization for 100-dimensional translated sphere function

that classic differential evolution does not consistently outperform particle swarm optimization any more.

4.8.3.1.6 Effect of Dimension The effect of dimension on differential evolution is shown in Figure 4.12. The values shown correspond to the most robust cases of differential evolution. It is

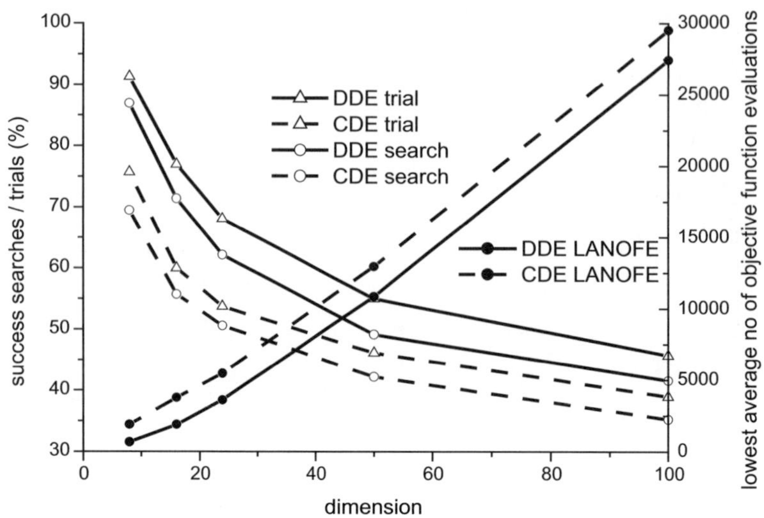

Figure 4.12 Effect of dimension on differential evolution

observed that although the search space increases exponentially with dimension, the performance of differential evolution falls almost linearly.

4.8.3.2 Qing Function

4.8.3.2.1 8-Dimensional The efficiency and robustness of participating optimization algorithms are shown in Figures 4.13 and 4.14, respectively. It is observed that more

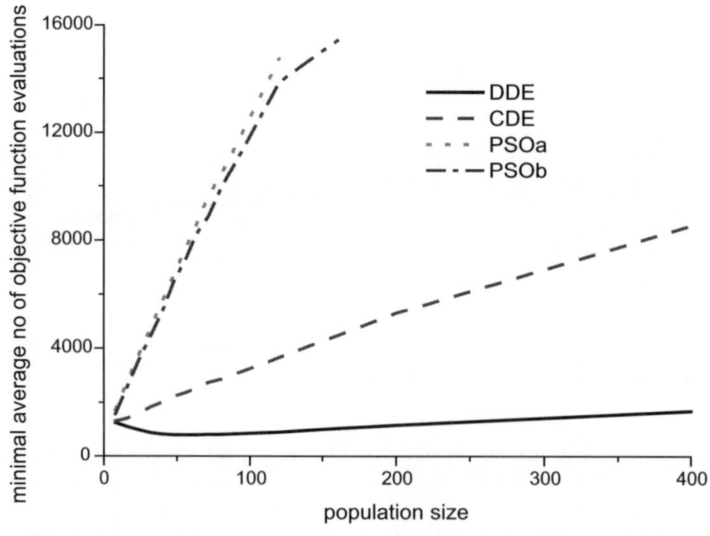

Figure 4.13 Efficiency of differential evolution, real-coded genetic algorithm and particle swarm optimization for 8-dimensional Qing function

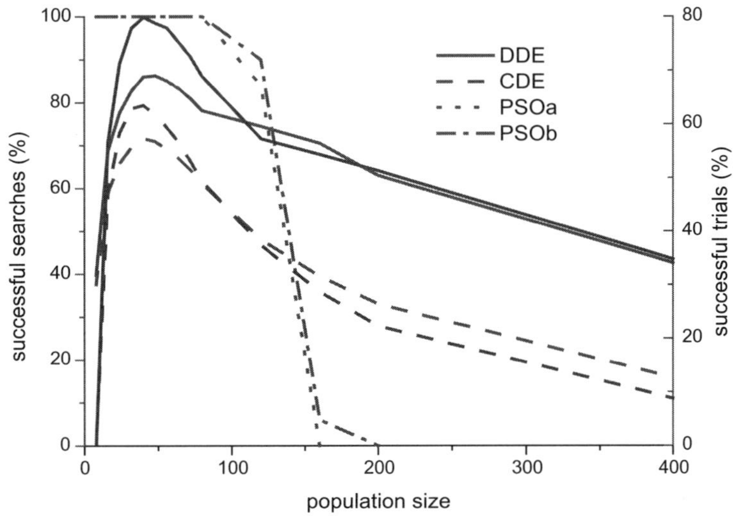

Figure 4.14 Robustness of differential evolution, real-coded genetic algorithm and particle swarm optimization for 8-dimensional Qing function

objective function evaluations are required in order to reach the optimal solution. This is due to the non-uniqueness of solutions.

4.8.3.2.2 16-Dimensional The efficiency and robustness of other participating optimization algorithms are shown in Figures 4.15 and 4.16, respectively.

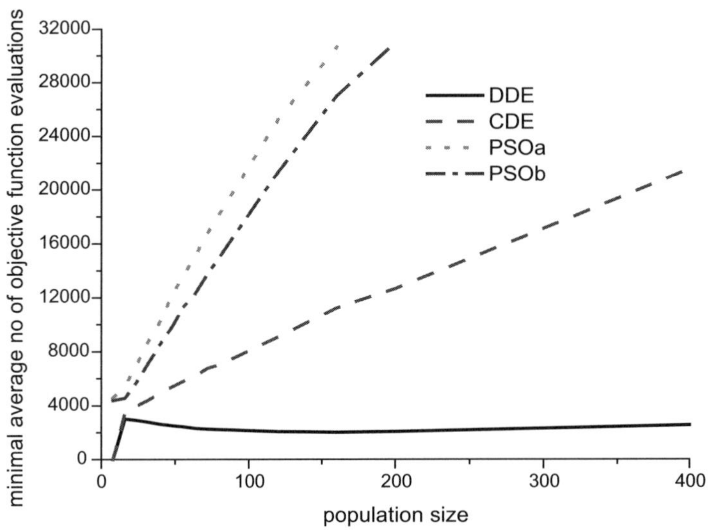

Figure 4.15 Efficiency of differential evolution, real-coded genetic algorithm and particle swarm optimization for 16-dimensional Qing function

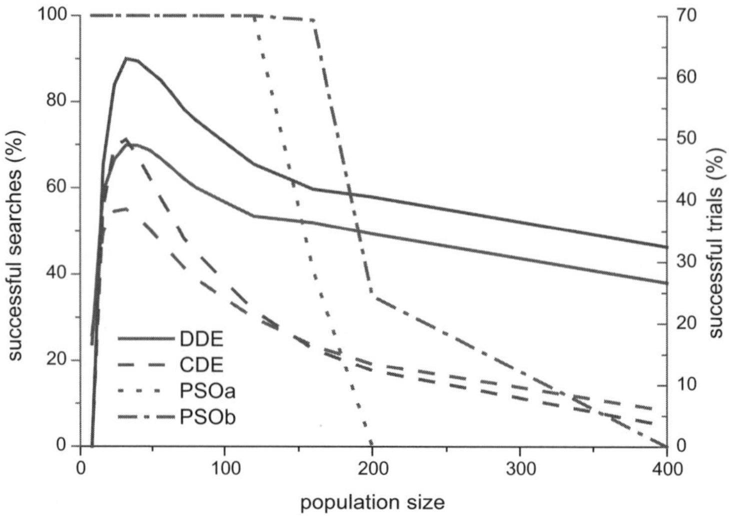

Figure 4.16 Robustness of differential evolution, real-coded genetic algorithm and particle swarm optimization for 16-dimensional Qing function

4.8.4 Conclusions

The results of the comparative study show the advantage of using differential evolution, especially dynamic differential evolution, over the standard genetic algorithm and the real-coded genetic algorithm. Although it is observed from the results presented that differential evolution outperforms particle swarm optimization, it is inappropriate to conclude that differential evolution outperforms particle swarm optimization, because three of the four intrinsic control parameters of the particle swarm optimization are fixed during our numerical simulation.

5

Benchmarking a Single-Objective Optimization Test Bed for Parametric Study on Differential Evolution

5.1 Motivation

The parametric study on differential evolution must be carried out on a test bed. When the parametric study was initiated in 2004, the initial test bed only included the benchmark electromagnetic inverse scattering problem and two toy functions defined by this author. As such it was seriously inadequate from the point of view of problem features. The features of an application problem may affect the choice of strategy, intrinsic and non-intrinsic control parameters. The relationship between problem features, strategy, intrinsic and non-intrinsic control parameters is of great significance. It is therefore necessary to build a test bed with all representative problem features fully and appropriately addressed.

Although many test beds are available, none of them is satisfactory. No effort to benchmark a test bed for unconstrained single-objective optimization has been found with problem features fully and properly addressed. When evaluating an evolutionary algorithm, people usually choose test problems from the available literature. Very often, insufficient attention is given to problem features.

5.2 A Survey on Test Problems

5.2.1 Sources

There are two sources for test problems. Optimization publications including books, journals, conference proceedings, technical reports, degree thesis, and so on, contain numerous test problems. In general, test problems from this source are more trustworthy because of the careful reviewing and proofreading that precede publication. However, using this source is very tedious and time-consuming. Moreover, access to optimization publications is usually limited.

The internet provides an easy and instant complementary source. A great deal of information on test problems for constrained and unconstrained optimization is available from this source. Usually, uploading and downloading such information is free. However, there is no guarantee that the available information is accurate. Information from the internet has to be treated with extreme care.

The internet is also increasingly the medium chosen by researchers to showcase their contributions. Many have set up their own homepages to advertise their achievements and share their collections. Rich relevant information has been observed on these homepages. The homepages of some outstanding researchers providing information on test problems are listed in Table 5.1.

Table 5.1 Homepages of some outstanding researchers

Researcher	Homepage
Dr Ernesto P. Adorio	http://www.geocities.com/eadorio/
Dr Abdel-Rahman Hedar	http://www-optima.amp.i.kyoto-u.ac.jp/member/student/hedar/Hedar.html
Prof. Kaj Madsen	http://www2.imm.dtu.dk/~km/GlobOpt/testex/
Dr Jorge Moré	http://www-unix.mcs.anl.gov/~more/
Prof. Arnold Neumaier	http://www.mat.univie.ac.at/~neum/
Prof. Klaus Schittkowski	http://www.math.uni-bayreuth.de/~kschittkowski/home.htm
Prof. Darrel Whitley	http://www.cs.colostate.edu/~whitley/
Prof. P.N. Suganthan	http://www.ntu.edu.sg/home/EPNSugan/

5.2.2 Prominent Test Beds

Various test beds differing in size and problem features are scattered throughout the literature. These test beds have been more or less accepted by researchers. Some of the test beds are discussed here. The test beds mentioned are more representative from the point of view of size and problem features and have gained more publicity and acceptance accordingly.

5.2.2.1 Conventional Test Beds

5.2.2.1.1 Branin Test Bed This test bed [1] is used to investigate the extraneous singularities suffered by Branin's method for simultaneous nonlinear equations and optimizations. Almost all test problems included have been accepted as benchmark.

5.2.2.1.2 CEC 2005 Test Bed This test bed [2] was prepared for the special session on real-parameter optimization of the 2005 IEEE Congress on Evolutionary Computation. It consists of 25 test problems. Test problems are formulated in a more general form, taking account of biasing, translation, rotation, expansion, and composition. Documentation and source code are freely downloadable (http://www.ntu.edu.sg/home/EPNSugan/).

5.2.2.1.3 Chung–Reynolds Test Bed Full consideration has been given to representative problem features while designing this test bed [3], which consists of 34 test problems are collected.

However, there may be some typographic errors with the Michalewicz function. In addition, no information about vectors **a** and **b** for the Kowalik function and coefficients a and e for the Branin RCOS function is given. Although such information is available from other references, it may cause unnecessary inconvenience and confusion.

5.2.2.1.4 De Jong Test Bed De Jong is one of the pioneers of evolutionary computation. He is one of the originators of genetic algorithms. The De Jong test bed [4] was established to evaluate genetic algorithms and has been well accepted by the evolutionary computation community. Five test problems are included.

The minimum of the step function (De Jong function 3) is dependent on the search space.

5.2.2.1.5 Dekkers–Aarts Test Bed Dekkers and Aarts [5] compiled eight test problems from two sources [6,7] to test their simulated annealing algorithm for continuous optimization problems. Attention is focused on function modality to demonstrate the capability of the proposed simulated annealing algorithm to locate global optima.

5.2.2.1.6 Dixon–Szegö Test Bed Dixon and Szegö edited papers presented in three workshops held in 1974–1977 [6,8]. Many test problems are reported in these two books. However, the Branin RCOS function and Hartman functions are incorrectly formulated, and these errors remain in place.

5.2.2.1.7 Levy Montalvo Test Bed This test bed ([9], pp. 18–33, [10]) focuses on modality. It consists of 16 test problems, widely implemented for evaluation and comparison of optimization algorithms.

It must be pointed out that numbering of the test problems in ([9], pp. 18–33) and [10] is inconsistent. Such inconsistency causes unnecessary confusion for later researchers.

5.2.2.1.8 MINPACK-2 Test Bed The distinctive feature of this test bed [11] is that every test problem included comes from a real application. Eighteen test problems from such diverse fields as fluid dynamics, medicine, combustion, nondestructive testing, chemical kinetics, lubrication, mathematics, and superconductivity, are presented. A brief description of test problems, detailed documentation, and Fortran source codes are all available from http://www-unix.mcs.anl.gov/~more/tprobs/.

It is usually very hard for a researcher to understand and solve a test problem without a solid background knowledge of the relevant field, unless the problem is mathematically formulated independent of that knowledge. In this regard, some test problems in this test bed are not suitable for the benchmark test bed here.

5.2.2.1.9 Moré–Garbow–Hillstrom Test Bed This test bed [12] is the first comprehensive collection of test problems for unconstrained single-objective optimization to avoid criticism from cynical observers. In considering test problems for inclusion, the efficiency and robustness of algorithms were given equal consideration.

Thirty-five test problems are included in this test bed. Three problem areas – systems of nonlinear equations, nonlinear least squares, and unconstrained minimization – are considered.

The formulation of the Gulf Research and Development function may be incorrect. Unfortunately, the original reference by R.A. Cox is not available.

5.2.2.1.10 Powell Test Bed M.J.D. Powell is one of the giants in optimization and contributed various well-known optimization algorithms such as the conjugate direction algorithm and quasi-Newton algorithm. The test problems reported in his publications [13–15] form this test bed. These test problems have been accepted as benchmark for optimization algorithm testing and comparison.

5.2.2.1.11 Price–Storn–Lampinen Test Bed This test bed [16] consists of 20 test problems. It evolved from the ICEO test problems and appears as an appendix to their monograph. Codes in various programming languages are also available.

The odd square function and Katsuura function are incorrectly formulated.

5.2.2.1.12 Salomon Test Bed To investigate the effect of parameter dependence on genetic algorithms, Salomon [17] introduced parameter dependence through coordinate rotation. The coordinate rotation was implemented on 12 widely used test problems. It was found that parameter dependence causes significant performance loss to genetic algorithms.

5.2.2.1.13 Schwefel Test Bed H.P. Schwefel is one of the originators of evolution strategies. He built his test bed of 62 problems [18,19] for strategy comparison. Due to his established position in the evolutionary computation community, this test bed is widely accepted.

5.2.2.1.14 Whitley–Rana–Dzubera–Mathias Test Bed This test bed came about as a byproduct of the study by these authors [20]. Limitations of previous test suites regarding problem features such as computational cost, decomposability, linearity, modality, scalability, and symmetry are discussed. Guidelines for designing test functions as well as strategies to build new test functions are given.

5.2.2.1.15 Yao–Liu–Lin Test Bed This test bed [21] is a collection of 23 test problems. It has been accepted as a benchmark test bed for evaluating evolutionary algorithms. Many researchers have followed the practice of these authors in evaluating evolutionary algorithms.

Unfortunately, this test bed carries the mistakes by Dixon and Szegö [6]. It also makes new mistakes regarding Hartman's functions.

5.2.2.2 Online Test Beds

5.2.2.2.1 CET The Cross-Entropy Toolbox website (http://www.maths.uq.edu.au/CEToolBox/) was set up on December 17, 2004. Its initial purpose was to make a comprehensive toolbox to compare cross-entropy method with other methods. Twenty-eight test problems are gathered together. There may be some typographical errors.

5.2.2.2.2 CUTEr CUTEr (http://hsl.rl.ac.uk/cuter-www/) evolved from the constrained and unconstrained testing environment, or CUTE [22]. It is a versatile testing environment for optimization and linear algebra solvers. The package contains a collection of test problems to help developers design, compare and improve new and existing solvers. Established on December 15, 1994, it gathered together 738 unconstrained and constrained test problems. These problems span a wide spectrum of difficulty, ranging from small-scale differentiable unconstrained minimization to large-scale equality and inequality-constrained dense and sparse problems, systems of nonlinear equations, network problems and so on.

5.2.2.2.3 GEATbx The genetic and evolutionary algorithm toolbox (http://www.geatbx.com/) implements a wide range of genetic and evolutionary algorithms to solve large and complex real-world problems. Sixteen scalable test problems are included.

5.2.2.2.4 ICEO Differential evolution earned its reputation by ranking the third among all entries in the first international contest on evolutionary optimization and the best among all qualified entries in the second international contest on evolutionary optimization. The ICEO test problems are still available from the website http://iridia.ulb.ac.be/~aroli/ICEO/Functions/Functions.html set up on November 27, 1996. However, many typographical errors have been found.

5.2.2.2.5 MVF The MVF test bed (http://www.geocities.com/eadorio/mvf.pdf) consists of at least 58 test problems. Mathematical formulation and C++ source code are given. However, it can only be used with extreme care because there are so many mistakes in it.

5.2.2.2.6 Netlib Netlib (http://netlib.sandia.gov/) is a repository of mathematical software, data, documents, address lists, and other useful items. Test problems for unconstrained optimization and nonlinear least squares can be found in the uncon subdirectory (http://netlib.sandia.gov/uncon/data/index.html).

5.2.2.2.7 NLS NLS (http://people.scs.fsu.edu/~burkardt/f_src/test_nls/test_nls.html) is a library of Fortran 90 routines defining test problems for least squares minimization.

5.2.2.2.8 SOMA SOMA, the self-organizing migrating algorithm, was proposed in 2000. Zelinka, one of its originators, constructed a website (http://www.ft.utb.cz/people/zelinka/soma/) to disseminate it. Test functions to evaluate SOMA and problems solved by SOMA, as well as other interesting information, are presented.

5.2.2.2.9 Tracer TRACER is a coordinated research project aimed at performing research in computer science to solve complex problems with modern optimization, search and learning tools. Test problems can be found on the website (http://tracer.lcc.uma.es/).

5.2.3 Latest Collections

5.2.3.1 Classification

In general, the test problems collected can be classified into two categories: toy functions and practical application problems. Some test problems are pure mathematical functions formulated by mathematicians for a specific purpose. They are usually nicknamed toy functions. Each toy function possesses specific features. Understanding these toy functions usually does not require knowledge in any field other than mathematics. The computation cost of these functions is usually low.

The application problems extracted from practical applications, on the other hand, are more attractive to application engineers. It would be useful to application engineers if consistency of parametric study results were observed between a toy function and an application problem with identical features. However, solving these application problems usually requires solid knowledge of the field concerned. In addition, most of the application problems are very time-consuming. This makes it difficult even to know the problem's features.

5.2.3.2 Toy Functions

By means of a survey and analysis of the literature, the following toy functions have been identified.

(T1) Ackley function [23]
(T2) Ackley function 1.4 ([24], pp. 10–12)
(T3) Ackley function 1.6 ([243], pp. 13–14)
(T4) Ackley function 1.7 ([24], pp. 16–17)
(T5) Adjiman function [25]
(T6) Alpine function [26]
(T7) Alpine function 2 [27]
(T8) Bard function [28]
(T9) Barrodale–Roberts function ([29], pp. 449, 463)
(T10) Bartels–Conn function ([9], p. 58)
(T11) Beale function [30]
(T12) Biggs EXP2 function [31]
(T13) Biggs EXP3 function [31]
(T14) Biggs EXP4 function [31]
(T15) Biggs EXP5 function [31]
(T16) Biggs EXP6 function [31]
(T17) Bird function [32]
(T18) Bohachevsky function [33]
(T19) Bohachevsky function 2 [33]
(T20) Bohachevsky function 3 [34]
(T21) Booth function [34,35]
(T22) Box two-dimensional function [36]
(T23) Box three-dimensional function [36]
(T24) Box three-dimensional function 2 [36]
(T25) Branin function [1]
(T26) Branin function 2 [1]
(T27) Branin RCOS function [1]
(T28) Branin RCOS function 2 [37]
(T29) Brent function [1]
(T30) Brent function 2 ([8], pp. 148)
(T31) Brown function [38]
(T32) Brown function 2 [38]
(T33) Brown almost-linear function [38]
(T34) Brown badly scaled function [12]
(T35) Brown–Conte function [39]
(T36) Brown–Dennis function [12]
(T37) Brown–Gearheardt function [40]
(T38) Brown–Gearheardt function 2 [40]
(T39) Brown–Gearheardt function 3 [40]
(T40) Brown–Gearheardt function 4 [40]
(T41) Broyden banded function [41]

(T42) Broyden tridiagonal function [41,42]
(T43) Bukin function 2 [43]
(T44) Bukin function 4 [43]
(T45) Bukin function 6 [43]
(T46) Camel-back function (Three-hump) [1]
(T47) Camel-back function (six-hump) [1]
(T48) Chebyquad function [44]
(T49) Chen–Bird function [45]
(T50) Chen *V* function [45]
(T51) Chichinadze function [46]
(T52) Chung–Reynolds function [3]
(T53) Cola function [47]
(T54) Corana function [48]
(T55) Cosine mixture function [49]
(T56) Cross function [32]
(T57) Cube function ([50], p. 33)
(T58) Damped cosine wave function [51]
(T59) Damped sinus function [51]
(T60) de Villers–Glasser function [52]
(T61) de Villers–Glasser function 2 [52]
(T62) Deceptive function [53]
(T63) Deflected corrugated spring function [32]
(T64) Dennis function ([54], p. 29)
(T65) Discrete boundary value function [55]
(T66) Discrete integral equation function [55]
(T67) Dixon–Price function [56]
(T68) Dolan function (http://www.aridolan.com/default.aspx)
(T69) Easom function [3]
(T70) El-Attar–Vidyasagar–Dutta function [57]
(T71) El-Attar–Vidyasagar–Dutta function 2 [57]
(T72) El-Attar–Vidyasagar–Dutta function 3 [57]
(T73) El-Attar–Vidyasagar–Dutta function 4a [57]
(T74) El-Attar–Vidyasagar–Dutta function 4b [57]
(T75) El-Attar–Vidyasagar–Dutta function 4c [57]
(T76) Engvall function ([54], p. 78)
(T77) Engvall function 2 ([54], p. 82)
(T78) Evtushenko function [58]
(T79) EX1 function [59]
(T80) EX3 function [59]
(T81) Exponential function [60]
(T82) Exponential data fitting function [61]
(T83) F101 function [20]
(T84) F102 function [20]
(T85) F102 function expanded [20]
(T86) F102 function expanded and weighted [20]
(T87) F103 function [20]

(T88) F103 function expanded and weighted [20]
(T89) F8F2 function a [20]
(T90) F8F2 function b ([16], pp. 522–523)
(T91) F8F2 function c [20]
(T92) F8F2 function d [20]
(T93) F8F2 function e [20]
(T94) F8F2 function f [20]
(T95) Ferraris–Tronconi function [39]
(T96) Ferraris–Tronconi function 2 [39]
(T97) Ferraris–Tronconi function 3 [39]
(T98) Ferraris–Tronconi function 4 [39]
(T99) Fletcher–Powell function [14]
(T100) Fletcher–Powell function 2 [14]
(T101) Floudas function ([62], p. 354)
(T102) Fractal function ([63], pp. 144–148)
(T103) Freudenstein–Roth function [64]
(T104) Gaussian function [12]
(T105) Gheri–Mancino function [65]
(T106) Giunta function [32]
(T107) Goldstein–Price function [66]
(T108) Goldstein–Price function 2 [66]
(T109) Griewank function [67]
(T110) Griewank function 2 [67]
(T111) Gulf Research Development function [68]
(T112) Hansen function [69]
(T113) Hanson function [70]
(T114) Hanson function 2 [70]
(T115) Hartman function [71]
(T116) Helical valley function [14]
(T117) Helmholtz energy function [61]
(T118) Henon map function [72]
(T119) Hesse function ([73], p. 186)
(T120) Hiebert function [74]
(T121) Hiebert function 2 [74]
(T122) Hilbert function ([16], p. 526)
(T123) Hilbert quadratic function ([6], p. 212)
(T124) Himmelblau function ([75], p. 6)
(T125) Himmelblau function 2 ([75], p. 7)
(T126) Himmelblau function 3 ([75], p. 67)
(T127) Himmelblau function 4 ([75], p. 117)
(T128) Himmelblau function 5 ([75], p. 145)
(T129) Himmelblau function 6 ([75], p. 168)
(T130) Himmelblau function 7 ([75], p. 427)
(T131) Himmelblau function 8 ([75], p. 428)
(T132) Himmelblau function 9 ([75], p. 428)
(T133) Himmelblau function 10 ([75], pp. 428–429)

(T134) Himmelblau function 11 ([75], p. 429)
(T135) Himmelblau function 12 ([75], p. 430)
(T136) Himmelblau function 13 ([75], p. 430–431)
(T137) Himmelblau function 14 ([54], p. 76)
(T138) Himmelblau function 15 ([54], p. 88)
(T139) Himmelblau function 16 ([76], pp. 326–327)
(T140) Hosaki function [77]
(T141) Hyper-ellipsoid function ([2], [16], p. 515, [17])
(T142) Hyper-ellipsoid function 2 [78]
(T143) Hyper-ellipsoid function 3 ([54], pp. 101, 120)
(T144) Hyper-ellipsoid function 4 [47]
(T145) Jennrich–Sampson function [79]
(T146) Jennrich–Sampson function 2 [79]
(T147) Jensen function [51]
(T148) Katsuura function [78,80]
(T149) Kjellström function [81]
(T150) Koon–Sebald function 3 ([82], p. 491)
(T151) Koon–Sebald function 4 ([82], p. 492)
(T152) Koon–Sebald function 5 ([82], p. 493)
(T153) Langerman function [83]
(T154) Leon function ([50], p. 46)
(T155) Levy function 3 ([9], p. 33)
(T156) Levy function 3 penalized [7]
(T157) Levy function 7 ([9], p. 33)
(T158) Levy function 7 penalized [7]
(T159) Levy function 8 [10]
(T160) Levy function 8 penalized [7]
(T161) Levy function 11 [10]
(T162) Levy function 13 ([9], p. 33)
(T163) Levy function 13 penalized [7]
(T164) Linear function – full rank [12]
(T165) Linear function – rank 1 [12]
(T166) Linear function – rank 1 with zero columns and rows [12]
(T167) Luenberger function ([84], p. 111)
(T168) Luenberger function 2 ([84], p. 113)
(T169) Luenberger function 3 ([30], [84], pp. 149–154, 199)
(T170) Luenberger function 4 ([30], [84], pp. 149–154, 199)
(T171) Makino–Berz function [85]
(T172) Masters cosine wave function [26]
(T173) Matrix square root function – rank 2 [69]
(T174) Matrix square root function – rank 3 [69]
(T175) Matyas function [86]
(T176) Mazzoleni function ([8], p. 355)
(T177) Mazzoleni function 2 ([8], pp. 356–357)
(T178) McCormick function ([54], pp. 209–221, [87])
(T179) McKeown function ([8], pp. 243–247, [88])

(T180) McKeown function 2 ([8], pp. 243–247, [88])
(T181) McKeown function 3 ([8], pp. 243–247, [88])
(T182) Michalewicz function [83]
(T183) Michalewicz function 2 ([16], pp. 531–532)
(T184) Michalewicz function 3 ([89], p. 34)
(T185) Michalewicz function 4 ([89], p. 154)
(T186) Michalewicz function 5 (http://www.maths.uq.edu.au/CEToolBox/)
(T187) Miele function [90]
(T188) Mishra function [91]
(T189) Mishra function 2 [91]
(T190) Mishra function 3 [32]
(T191) Mishra function 4 [32]
(T192) Mishra function 5 [32]
(T193) Mishra function 6 [32]
(T194) Mishra function 7 [32]
(T195) Mishra function 8 [32]
(T196) Mishra function 9 [32]
(T197) Mishra function 10 [32]
(T198) Mishra function 11 [32]
(T199) Mishra function 12 [32]
(T200) Mladineo function ([62], pp. 379)
(T201) Moon function [81]
(T202) Needle eye function [32]
(T203) Neumaier function [47]
(T204) Neumaier function 2 [47]
(T205) Neumaier function 3 ([47], [16], p. 517)
(T206) Neumaier function 4 [47]
(T207) Neumaier function 5 ([92], p. 281)
(T208) Odd square function ([93], [16], pp. 529–530)
(T209) Odd square function 2
(T210) Osborne–Watson function [94]
(T211) Pahner–Hameyer function [95]
(T212) Parkinson–Hutchinson function ([54], p. 100)
(T213) Parkinson–Hutchinson function 2 ([54], p. 100)
(T214) Parkinson–Hutchinson function 3 ([54], p. 100)
(T215) Parkinson–Hutchinson function 4 ([54], p. 100)
(T216) Parkinson–Hutchinson function 5 ([54], p. 100)
(T217) Pathological function [26,96]
(T218) Paviani function ([75], p. 416)
(T219) Peak function ([16], p. 16)
(T220) Pen function ([54], p. 163)
(T221) Pen holder function [32]
(T222) Penalty function [12]
(T223) Penalty function 2 [12]
(T224) Pintér function ([97], p. 183)
(T225) Pintér function 2 ([97], p. 187)

(T226) Pintér function 3 ([97], p. 248)
(T227) Powell function [15]
(T228) Powell function 2 ([98], p. 90)
(T229) Powell function 3 ([98], p. 118)
(T230) Powell badly scaled function ([98], p. 146)
(T231) Powell singular function [13]
(T232) Powell singular function 2 [99]
(T233) Power sum function [26]
(T234) Price function [100]
(T235) Price function 2 [100]
(T236) Price function 3 [100]
(T237) Price function 4 ([6], p. 80)
(T238) PseudoDirac function (http://www.ft.utb.cz/people/zelinka/soma/)
(T239) Qing function [101]
(T240) Quadratic function [102]
(T241) Quadratic function 2 ([54], p. 67)
(T242) Quadratic power function ([6], p. 212)
(T243) Quartic function [4]
(T244) Quintic function [32]
(T245) Rana function ([16], pp. 532–533, [47])
(T246) Rastrigin function [103]
(T247) Rastrigin function 2 [104]
(T248) Rastrigin function 3 ([82], p. 494)
(T249) RECIP function ([54], p. 163)
(T250) Rosenbrock function [35]
(T251) Rosenbrock function 2 ([6], p. 213)
(T252) Rosenbrock function 3 ([54], p. 163)
(T253) Rosenbrock function 4 (http://www.ft.utb.cz/people/zelinka/soma/)
(T254) Rotated ellipse function (http://www.krellinst.org/UCES/archive/resources/conics/node66.html)
(T255) Rotated ellipse function 2 ([16], p. 84)
(T256) Rump function ([92], p. 110)
(T257) Salomon function [17]
(T258) Sargan function ([6], p. 259)
(T259) Schaffer function [105]
(T260) Schaffer function 2 [105]
(T261) Schaffer function 3 [105]
(T262) Schaffer function 4 [105]
(T263) Schmidt–Vetters functions ([54], p. 81)
(T264) Schumer–Steiglitz function [102]
(T265) Schwefel function ([18], p. 96)
(T266) Schwefel function 1.2 ([18], p. 292)
(T267) Schwefel function 2.4 ([18], p. 295)
(T268) Schwefel function 2.6 ([18], p. 297)
(T269) Schwefel function 2.20 ([18], p. 305)
(T270) Schwefel function 2.21 ([18], p. 306)

(T271) Schwefel function 2.22 ([18], p. 306)
(T272) Schwefel function 2.23 ([18], p. 306)
(T273) Schwefel function 2.25 ([18], p. 309)
(T274) Schwefel function 2.26 ([18], p. 309)
(T275) Schwefel function 2.36 ([18], p. 314)
(T276) Sekaj function [106]
(T277) Shekel function [107]
(T278) Shekel foxholes function ([16], pp. 528–529, [83])
(T279) Shekel foxholes function 2 [17,108]
(T280) Shubert function [109]
(T281) Shubert function 2 ([9], p. 33)
(T282) Shubert function 2 penalized [7]
(T283) Shubert function 3 [47]
(T284) Shubert function 4 [47]
(T285) Sine envelope sine wave function [110]
(T286) Sphere function [102]
(T287) Step function [4]
(T288) Step function 2 [23]
(T289) Step function 3
(T290) Storn–Chebyshev function ([16], pp. 523–525)
(T291) Stretched *V* sine wave function [110]
(T292) Styblinski–Tang function [43]
(T293) Table function [32]
(T294) Treccani function ([8], p. 141)
(T295) Trefethen function [47]
(T296) Trigonometric function ([6], p. 212)
(T297) Trigonometric function 2 [99]
(T298) Tripod function [26,96]
(T299) Tube Holder function [32]
(T300) Variably-dimensioned function [12]
(T301) Watson function ([111], pp. 104–105)
(T302) Watson function 2 [112]
(T303) Wavy function [113]
(T304) Wayburn–Seader function [114]
(T305) Wayburn–Seader function 2 [114]
(T306) Wayburn–Seader function 3 [114]
(T307) Weierstrass function [2]
(T308) Wolfe function ([18], pp. 309, 324)
(T309) Wood function [90]
(T310) Zakharov function [26]
(T311) Zangwill function ([54], p. 74)
(T312) Zangwill function 2 ([54], p. 80)
(T313) Zero Sum function [32]
(T314) Zettl function ([19], p. 344)
(T315) Zimmerman function [47]
(T316) Zirilli function [7]

(T317) Zirilli function 2 [7]
(T318) Zirilli function 3 [7]
(T319) Zirilli function 4 [7].

All these toy functions have been coded in Fortran 90/95 in a separate module. The module will be part of the source code package downloadable from the companion website for this book.

Documentation for the above toy functions will be given separately and is also downloadable from the companion website. The mathematical formulation from the earliest available literature is strictly followed unless it is explicitly pointed out that the formulation is incorrect. To avoid potential confusion, modifications and observed incorrect citations will be pointed out explicitly.

5.2.3.3 Application Problems

Besides the application problems mentioned in Chapters 2 and 9 of this book, through our literature survey and analysis, we have collected, documented, and coded the following application problems. Documentation for these application problems can be downloaded from the companion website. The corresponding Fortran 90/95 source code is also part of the electronic package.

(A1) Andersen data fitting problem [115]
(A2) Benchmark electromagnetic inverse scattering problem [101,116]
(A3) Coating thickness standardization problem [11]
(A4) EX2 problem ([73], p. 213, [117,118], [119], p. 192, [59])
(A5) Gear design problem ([120], p. 134, [47])
(A6) Hiebert propane combustion problem [11,74,121,122]
(A7) Hougen problem ([123], [124], pp. 271–272, [125])
(A8) Human heart dipole problem [11,69,126,127]
(A9) Kowalik enzyme reaction problem ([111], p. 104, [12,128,129])
(A10) Lennard-Jones potential problem ([76], pp. 188–193, [47], [16], p. 525)
(A11) Meyer thermistor resistance problem ([29], pp. 60, 483–484, [12,74])
(A12) Osborne problem ([54], pp. 185–186, [11,12,69,74])
(A13) Osborne problem 2 ([54], pp. 186–188, [11,12,69,74])
(A14) Photoelectron spectroscopy problem ([62], pp. 337–340)
(A15) XOR problem ([130–132], [133], chapters 26–28, [134])

We have also collected the following application problems. However, documentation and Fortran 90/95 source code are not yet ready.

(A16) Alkylation reaction problem http://www.icsi.berkeley.edu/~storn/code.html, [135]
(A17) Alphabetic font learning problem ([133], chapter 27)
(A18) Antenna problem
(A19) Ausubel auction problem ([136–138], http://tracer.lcc.uma.es/)
(A20) Avionics problem [139]
(A21) Bin packing problem (http://gplus2004.tripod.com/)
(A22) Biomass pyrolysis problem [140]
(A23) Bolding–Andersen potential problem ([76], pp. 202–204)

(A24) Brenner potential problem ([76], pp. 199–202)
(A25) Cantilever design problem [141–143]
(A26) Carbon materials physiochemistry problem (http://www.icsi.berkeley.edu/~storn/code.html, http://www.chem.uni.torun.pl/~gaudip/wegiel/index.html)
(A27) Chemical equilibrium problem ([144], pp. 46–49, [145], chapter 5, [75]), pp. 395–396, [122])
(A28) Chemical equilibrium problem 2 [122]
(A29) Chrystallorgraphic [*sic*] characterization problem (http://www.icsi.berkeley.edu/~storn/code.html)
(A30) Circles packing problem [146,147]
(A31) Circuit design problem ([6], pp. 81–83, [100,148,149])
(A32) Coil spring design problem ([120], pp. 141–143)
(A33) Colors design problem ([73], pp. 198–201, 204–211)
(A34) Computer game problem (http://tracer.lcc.uma.es/)
(A35) Computer network design problem ([73], p. 216)
(A36) Cropping patterns planning problem http://www.icsi.berkeley.edu/~storn/code.html)
(A37) Dataset clustering problem (http://www.maths.uq.edu.au/CEToolBox/)
(A38) Diffraction grating design problem (http://www.icsi.berkeley.edu/~storn/code.html, http://www.gsolver.com/gsprod.html)
(A39) Disjoint paths problem (http://tracer.lcc.uma.es/)
(A40) Dynamic systems scenario-integrated optimization problem ([150], http://www.icsi.berkeley.edu/~storn/code.html)
(A41) Edge-2 connectivity augmentation problem (http://tracer.lcc.uma.es/)
(A42) Effort estimation problem (http://tracer.lcc.uma.es/)
(A43) Electricity market simulation problem (http://www.icsi.berkeley.edu/~storn/code.html, http://www.draytonanalytics.com/)
(A44) Electroencephalogram problem (http://tracer.lcc.uma.es/)
(A45) Electronics problem ([73], p. 215)
(A46) Error correcting code design problem (http://tracer.lcc.uma.es/)
(A47) Filter design problem ([81], [120], chapter 7)
(A48) Frequency modulation sounds problem (http://tracer.lcc.uma.es/)
(A49) Gas transmission network design problem ([151], http://www.icsi.berkeley.edu/~storn/code.html)
(A50) Gear train problem ([152], p. 123)
(A51) Hash function design problem (http://tracer.lcc.uma.es/)
(A52) Helicopter rotor blade design problem [153]
(A53) High tide prediction problem (http://tracer.lcc.uma.es/)
(A54) Howling removal unit design problem [154]
(A55) Initial public offer problem (http://tracer.lcc.uma.es/)
(A56) Judge economics problem ([155], chapter 24, [156], pp. 956–957, [32])
(A57) Knapsack problem (http://gplus2004.tripod.com/, http://tracer.lcc.uma.es/)
(A58) Kubíček problem ([76], pp. 332–333, [157])
(A59) Loney solenoids problem [158–160]
(A60) Magnetic deflection system problem ([73], pp. 195–198, 202–204)

(A61) Magnetizer design problem [161]
(A62) Maximum Boolean K-satisfiability problem (http://tracer.lcc.uma.es/)
(A63) Maximum graph cut problem (http://tracer.lcc.uma.es/)
(A64) Mechanical engineering design problem ([120], chapter 8)
(A65) Median filter problem (http://tracer.lcc.uma.es/)
(A66) Meyer–Roth problem [93]
(A67) Minimum graph bisection problem (http://tracer.lcc.uma.es/)
(A68) Minimum job scheduling problem (http://tracer.lcc.uma.es/)
(A69) Minimum linear arrangement problem (http://tracer.lcc.uma.es/)
(A70) Minimum weighted K-tree subgraph problem (http://tracer.lcc.uma.es/)
(A71) Modulator synthesis problem (http://www.icsi.berkeley.edu/~storn/code.html, [162])
(A72) Morse potential problem ([76], pp. 193–197)
(A73) Multi-Gaussian problem [93,96]
(A74) Multiprocessor synthesis problem (http://www.icsi.berkeley.edu/~storn/code.html, [163])
(A75) Multi-sensor fusion problem [164–168]
(A76) Network capacity planning problem (http://www.maths.uq.edu.au/CEToolBox/)
(A77) Neural network learning problem [130]
(A78) Neurotransmitter problem [33,169]
(A79) Non-linear chemical processes optimization problem ([170], http://www.icsi.berkeley.edu/~storn/code.html)
(A80) Number partitioning problem (http://tracer.lcc.uma.es/)
(A81) Numeric font generalization problem [131]
(A82) Numeric font learning problem ([131,132], [133], chapter 27)
(A83) OneMax problem (http://tracer.lcc.uma.es/)
(A84) Optical filter design problem ([6], pp. 179–188)
(A85) Permutation flow shop problem (http://www.maths.uq.edu.au/CEToolBox/)
(A86) Polling problem (http://tracer.lcc.uma.es/)
(A87) Polynomial fitting problem (http://tracer.lcc.uma.es/)
(A88) pressure vessel design problem ([120], pp. 136–138)
(A89) Project scheduling problem (http://tracer.lcc.uma.es/)
(A90) Quadratic assignment problem (http://tracer.lcc.uma.es/)
(A91) Radio network design problem (http://www.icsi.berkeley.edu/~storn/code.html, [171–174])
(A92) Radiotherapy problem [175–177]
(A93) Robosoccer problem (http://tracer.lcc.uma.es/)
(A94) Robot kinematics problem ([76], pp. 329–331)
(A95) Royal trees problem (http://tracer.lcc.uma.es/)
(A96) Semiconductor boundary condition problem [139,178]
(A97) Shacham reaction rate equation problem [122,179]
(A98) Shell-and-tube heat exchangers design problem ([180], http://www.icsi.berkeley.edu/~storn/code.html,)
(A99) Simpleton problem (http://www.aridolan.com/default.aspx)

(A100) Single neuron training problem [131]
(A101) Software testing problem (http://tracer.lcc.uma.es/)
(A102) Statistical natural language tagging problem (http://tracer.lcc.uma.es/)
(A103) Steiner problem (http://gplus2004.tripod.com/, http://www-optima.amp.i.kyoto-u.ac.jp/member/student/hedar/Hedar.html)
(A104) Strongin problem ([73], p. 186, [181])
(A105) Swaney–Wilhelm problem [122]
(A106) Swaney–Wilhelm problem 2 [122]
(A107) Table design problem (http://tracer.lcc.uma.es/)
(A108) Team 22 problem [182–184]
(A109) Tersoff potential problem ([76], pp. 197–199)
(A110) Texture classification problem [131,132]
(A111) Thermal cracker operation optimization problem ([185], http://www.icsi.berkeley.edu/~storn/code.html)
(A112) Thomson problem (http://tracer.lcc.uma.es/)
(A113) Three pit parity problem ([133], chapter 26)
(A114) Time series identification problem (http://tracer.lcc.uma.es/)
(A115) Traveling salesman problem (http://tracer.lcc.uma.es/)
(A116) Water pumping system optimization problem (http://www.icsi.berkeley.edu/~storn/code.html)
(A117) Wolff problem [7].

5.3 Generating New Test Problems

Existing test problems may not have all desirable features. To remove this limitation, various approaches to generating new test problems with desirable features using existent test problems have been proposed.

5.3.1 *Biasing*

Biasing is the simplest way to generate a new test problem. The new test problem is generated from the existing test problem by the displacement

$$f(\mathbf{x}) = g(\mathbf{x}) + \text{bias}. \tag{5.1}$$

5.3.2 *Cascading*

Cascading, or wrapping, aims to generate a new test problem exhibiting simultaneously all interesting features of the test problems involved. The cascading operation can be expressed as

$$f(\mathbf{x}) = g[h_1(\mathbf{x}), \ldots, h_M(\mathbf{x})], \tag{5.2}$$

where $g(\mathbf{x})$ and $h_i(\mathbf{x})$ may have different features.

5.3.3 Perturbation

Noise in a test problem is also regarded as a problem feature. Optimizing noisy problems has been a challenge to differential evolution [186–189]. A noisy test problem can be generated by perturbation of its noiseless counterpart:

$$f(\mathbf{x}) = g(\mathbf{x}) + \text{noise}. \tag{5.3}$$

5.3.4 Translation

Coordinate translation is one way to break the symmetry of a test problem. The new test problem is generated from the existing test problem by coordinate translation

$$f(\mathbf{x}) = g(\mathbf{x}-\mathbf{o}), \tag{5.4}$$

where $\mathbf{o}$ is the translation vector.

5.3.5 Rotation

Coordinate rotation is usually implemented to transform a separable test problem into a non-separable problem. A new test problem is generated by

$$f(\mathbf{x}) = g(\mathbf{R} \cdot \mathbf{x}), \tag{5.5}$$

where $\mathbf{R}$ is the rotation matrix.

5.3.6 Weighted Sum

Weighted sum is another approach which generates a new test problem with all the interesting features of the test problems of interest. A new function is obtained by a weighted sum of multiple functions:

$$f(\mathbf{x}) = \sum_i w_i g_i(\mathbf{x}). \tag{5.6}$$

5.3.7 Scaling

Lower-dimensional test problems can be expanded to obtain higher-dimensional test problems through scaling:

$$f(\mathbf{x}) = \sum_i g(\mathbf{x}_i) \; \mathbf{x}_i \subseteq \mathbf{x}. \tag{5.7}$$

Usually, this approach is implemented to introduce scalability into a non-scalable lower-dimensional test problem.

5.3.8 Composition

A new test problem can be generated by composition, that is, imposing some or all of the above operations.

5.4 Tentative Benchmark Test Bed

5.4.1 Admission Criteria

In benchmarking a test bed for single-objective optimization, each member of the benchmark test bed satisfies basic admission criteria and has its own unique merits.

5.4.1.1 Mathematical Features

A test problem in the benchmark test bed has to have representative features. Non-scalable toy functions and application problems are excluded in order to investigate the effect of dimensionality. In general, composite test problems are excluded since it is difficult to identify the effect of a specific problem feature using composite problems.

5.4.1.2 Computational Expense

Time-consuming test problems are excluded from the benchmark test bed unless specified otherwise, since a comprehensive parametric study involves huge number of objective function evaluations.

5.4.1.3 Optimum

The primary purpose of developing the benchmark test bed is the parametric study on differential evolution. From the point of view of the parametric study, a test problem is either examinatorial or exploratory, depending on whether its optimum is known. The optimum of an examinatorial member is known while that of an exploratory member is yet to be determined.

Known optima play a critical role in parametric study. Without a known optimum, it will not be possible to implement the first termination condition, that is, that the objective has been met. Accordingly, it is impossible to identify a search as successful or unsuccessful. Values of performance indicators would not be available. An examinatorial test problem is suitable for parametric study because at least its optimum is known.

5.4.1.4 Search Space

The search space is regarded as an extra non-intrinsic control parameter. The effect of non-intrinsic control parameters and the relationship between intrinsic and non-intrinsic control parameters are also of concern. It is therefore necessary to standardize the search space. A basic requirement on search space is that it is independent of problem dimension.

One of the advantages of evolutionary algorithms is their strong global search ability. In another word, evolutionary algorithms are more capable of locating global optimal solutions within a reasonably wide search space. In our practice, $[-500, 500]$ is implemented if the highest number of successful trials of the best strategy exceeds 20%. Otherwise, it will be adjusted.

5.4.2 Members of Tentative Benchmark Test Bed

5.4.2.1 Ackley function

The Ackley function is continuous, differentiable, non-separable, scalable, multimodal, and symmetric. It is therefore suitable for investigating the effect of modality and separability. It is given by

$$f(\mathbf{x}) = 20 - 20\exp\left(-0.2\sqrt{\frac{1}{N}\sum_{i=1}^{N}x_i^2}\right) + \exp(1) - \exp\left(\frac{1}{N}\sum_{i=1}^{N}\cos(2\pi x_i)\right). \tag{5.8}$$

Its minimum is $f(\mathbf{x}^*) = 0$ at $\mathbf{x}^* = \mathbf{0}$, and the recommended search space is $\mathbf{x} \in [-30,\ 30]^N$.

5.4.2.2 F8F2 function b

The F8F2 function b is a composite function. It is continuous, differentiable, non-separable, scalable, multimodal, and symmetric. It is very difficult to optimize. It is also used to study the effect of modality and separability. The F8F2 function b is given by

$$f(x) = \sum_{i=1}^{N}\sum_{j=1}^{N} F_8^1[F_2^2(x_i, x_j)] = \sum_{i=1}^{N}\sum_{j=1}^{N}\left[\frac{y_{ji}^2}{4000} - \cos y_{ji} + 1\right], \tag{5.9}$$

where $y_{ji} = 100(x_i - x_j^2)^2 + (x_j - 1)^2$. Its minimum is $f(\mathbf{x}^*) = 0$ at $\mathbf{x}^* = \mathbf{1}$. The recommended search space is $\mathbf{x} \in [-100,\ 100]^N$.

5.4.2.3 Griewank function

The Griewank function is continuous, differentiable, non-separable, scalable, multimodal, and asymmetric. It is also difficult to optimize. Accordingly, it is a good candidate for studying the effect of modality, separability, and symmetry. The Griewank function is given by

$$f(\mathbf{x}) = \frac{1}{4000}\sum_{i=1}^{N}x_i^2 - \prod_{i=1}^{N}\cos\left(\frac{x_i}{\sqrt{i}}\right) + 1. \tag{5.10}$$

Its minimum is $f(\mathbf{x}^*) = 0$ at $\mathbf{x}^* = \mathbf{0}$. The recommended search space is $\mathbf{x} \in [-500,\ 500]^N$.

5.4.2.4 Hyper-ellipsoid function

The hyper-ellipsoid function is continuous, differentiable, separable, scalable, unimodal, and asymmetric. It is usually implemented to study the effect of symmetry. It is given by

$$f(\mathbf{x}) = \sum_{i=1}^{N} b^{i-1}x_i^2, \tag{5.11}$$

where b is the base. Commonly used bases include 2 ([16], p. 515), 10 [17], and $10^{6/(N-1)}$ [2]. Its minimum is $f(\mathbf{x}^*) = 0$ at $\mathbf{x}^* = \mathbf{0}$. The recommended search space is $\mathbf{x} \in [-500, 500]^N$.

5.4.2.5 Qing function

The Qing function (Figure 5.1) was originally designed to study the effect of non-uniqueness. It is continuous, differentiable, separable, scalable, unimodal, and asymmetric. It is given by

$$f(\mathbf{x}) = \sum_{i=1}^{N} (x_i^2 - i)^2. \tag{5.12}$$

It has minima $f(\mathbf{x}^*) = 0$ at $x_i^* = \pm\sqrt{i}$. The recommended search space is $\mathbf{x} \in [-500, 500]^N$.

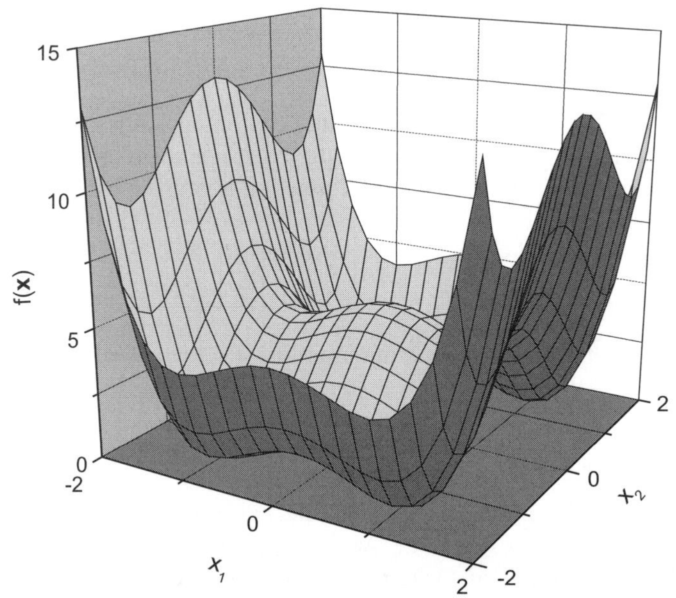

Figure 5.1 Landscape of Qing function

5.4.2.6 Rastrigin function

The Rastrigin function is continuous, differentiable, separable, scalable, multimodal, and symmetric. It is very suitable for investigating the effect of modality. It is given by

$$f(\mathbf{x}) = \sum_{i=1}^{N} [x_i^2 - 10\cos(2\pi x_i) + 10]. \tag{5.13}$$

Its minimum is $f(\mathbf{x}^*) = 0$ at $\mathbf{x}^* = \mathbf{0}$. The recommended search space is $\mathbf{x} \in [-100, 100]^N$.

5.4.2.7 Rosenbrock function

The Rosenbrock function is continuous, differentiable, non-separable, scalable, unimodal, and asymmetric. It is extremely difficult to optimize. It is used to study the effect of separability, symmetry, and non-intrinsic control parameters. It is also a good candidate to study the effect of convexity because its landscape is non-convex. The Rosenbrock function is given by

$$f(\mathbf{x}) = \sum_{i=1}^{N-1}[a(x_{i+1}-x_i^2)^2 + b(x_i-1)^2], \tag{5.14}$$

where $(a, b) = (100, 1), (2, 1), (1, 1), (0.5, 1)$, or $(1, 100)$. Its minimum is $f(\mathbf{x}^*) = 0$ at $\mathbf{x}^* = \mathbf{1}$. The recommended search space is $\mathbf{x} \in [-100, 100]^N$.

5.4.2.8 Salomon function

The Salomon function is continuous, differentiable, non-separable, scalable, multimodal, and symmetric. It is also extremely difficult to optimize. It is used to study the effect of separability and modality. It is given by

$$f(\mathbf{x}) = 1-\cos(2\pi||\mathbf{x}||) + 0.1||\mathbf{x}||, \tag{5.15}$$

where

$$||\mathbf{x}|| = \sqrt{\sum_{i=1}^{N} x_i^2}.$$

Its minimum is $f(\mathbf{x}^*) = 0$ at $\mathbf{x}^* = \mathbf{0}$. The recommended search space is $\mathbf{x} \in [-100, 100]^N$.

5.4.2.9 Schwefel function 1.2

The Schwefel function 1.2 is a rotated sphere function. It is continuous, differentiable, partially separable, scalable, unimodal, and asymmetric. It can be used to investigate the effect of symmetry and rotation, and is given by

$$f(\mathbf{x}) = \sum_{i=1}^{N}\left(\sum_{j=1}^{i} x_j\right)^2. \tag{5.16}$$

Its minimum is $f(\mathbf{x}^* = \mathbf{0}) = 0$. The recommended search space is $\mathbf{x} \in [-500, 500]^N$.

5.4.2.10 Schwefel function 2.22

The Schwefel function 2.22 is continuous, non-differentiable, separable, scalable, unimodal, and symmetric. It can be implemented to examine the effect of convexity because its landscape is non-convex. It is given by

$$f(\mathbf{x}) = \sum_{i=1}^{N} |x_i| + \prod_{i=1}^{N} |x_i|. \tag{5.17}$$

Its minimum is $f(\mathbf{x}^*) = 0$ at $\mathbf{x}^* = \mathbf{0}$. The recommended search space is $\mathbf{x} \in [-500, 500]^N$.

5.4.2.11 Schwefel function 2.26

The Schwefel function 2.26 is continuous, non-differentiable, separable, scalable, multimodal, and symmetric. It can be used to examine the effect of differentiability and modality, and is given by

$$f(\mathbf{x}) = -\frac{1}{N}\sum_{i=1}^{N} x_i \sin\sqrt{|x_i|}. \tag{5.18}$$

Its minimum is $f(\mathbf{x}^*) = -418.983$ at $x_i^* = \pm[\pi(0.5+k)]^2$. The recommended search space is $\mathbf{x} \in [-500, 500]^N$.

5.4.2.12 Sphere function

The sphere function is continuous, differentiable, separable, scalable, unimodal, and symmetric. It is the benchmark for studying problem features, and is given by

$$f(\mathbf{x}) = \sum_{i=1}^{N} x_i^2 \tag{5.19}$$

Its minimum is $f(\mathbf{x}^*) = 0$ at $\mathbf{x}^* = \mathbf{0}$. The recommended search space is $\mathbf{x} \in [-500, 500]^N$.

5.4.2.13 Step function 2

Step function 2 is discontinuous, non-differentiable, separable, scalable, unimodal, and symmetric. It is designed for studying the effect of continuity, and is given by

$$f(\mathbf{x}) = \sum_{i=1}^{N} ([x_i + 0.5])^2. \tag{5.20}$$

Its minimum is $f(\mathbf{x}^*) = 0$ at $\mathbf{x}^* = \mathbf{0}$. The recommended search space is $\mathbf{x} \in [-500, 500]^N$.

5.4.3 Future Expansion

Numerical results from parametric study on differential evolution over the present benchmark test bed will be presented in Chapters 6 and 7. Interesting results have been observed from completed simulations.

The size of the present benchmark test bed is small. In addition, there is no application problem in the present benchmark test bed. To make the conclusions based on completed simulations more convincing and attractive to application engineers, more qualified test problems, especially qualified application problems from the latest collection of toy functions and application problems, will be admitted into the benchmark test bed.

At the same time, searching for new toy functions and application problems is continuing. The benchmark test bed will be further expanded if any new toy function or application problem is qualified.

5.5 An Overview of Numerical Simulation

Numerical simulation began in 2004 and is continuing.

5.5.1 Full Simulations

At the time of writing, simulation on 9 of the 13 toy functions (Ackley function, hyper-ellipsoid function, Qing function, Rastrigin function, Schwefel function 1.2, Schwefel function 2.22, Schwefel function 2.26, sphere function, and step function 2) in the tentative benchmark test bed has been completed. Simulation on the remaining four toy functions in the tentative benchmark test bed is ongoing.

5.5.1.1 Sphere function

As a benchmark toy function, the sphere function has been fully simulated. Five dimensions have been simulated: 8, 16, 24, 50, and 100. The first three simulated dimensions were dictated by demand from the application problem: the benchmark electromagnetic inverse scattering problem. Dimension 8 can be regarded as a low dimension, while dimensions 16 and 24 can be regarded as intermediate. The other two dimensions, 50 and 100, were simulated to investigate the effect of high dimension.

5.5.1.2 Other Fully Simulated Toy functions

Initially, dimension 100 was included to study the effect of problem dimension. However, simulation of the 100-dimensional sphere function takes approximately half a year on available computational resources. It is therefore impracticable to simulate other 100-dimensional toy functions in the tentative benchmark test bed. More importantly, available simulation results show that the effect of dimension is already very clear from studying dimensions, 8, 16, 24, and 50. Hence, it is not absolutely necessary to simulate 100-dimensional toy functions.

5.5.1.3 Hardness of Fully Simulated Toy functions

The hardness of a toy function depends on its features. In our practice, with default settings, it is quantitatively indicated by the highest number of successful trials and the corresponding lowest average number of objective function evaluations. The highest number of successful trials is the primary indicator, while the corresponding lowest average number of objective function evaluations is the secondary indicator since success is of greater concern.

When the former indicators are not available with the default settings, a shrunken search space and/or raised limit of number of objective function evaluations can serve as an additional hardness indicator.

The hardness of all simulated toy functions in the tentative benchmark test bed is shown in Table 5.2.

Table 5.2 Hardness of toy functions in the tentative benchmark test bed

Toy function	HNST				LANOFE			
	8D	16D	24D	50D	8D	16D	24D	50D
Step 2	1486	1370	1336	1284	586	6079.35	8847.5	23 763.06
Sphere	1477	1350	1287	1205	798.2	1881.86	4912.56	23 549.43
Hyper-ellipsoid	1460	1287	1201	1006	828.40	3393.54	8782.26	31 148.69
Qing	1398	1242	1177	1088	864	3206.64	7894.43	24 607.77
Schwefel 2.22	1388	1263	1245	1181	2110	7149.47	11 812.16	31 432.48
Schwefel 2.26	1309	1266	1208	1088	4112.06	8867.92	15 706.24	37 712.5
Schwefel 1.2	557	245	129	24	1845.32	6338.91	14 391.15	56 602.46
Rastrigin	1013	885	797	598	5024	13 769.32	23 701.60	53 663.14
Ackley	1302	1281	1242	1181	3480.19	8272.8	12 647.52	26 610.49

5.5.2 Partial Simulations

Besides the aforementioned full simulation, the following toy functions and application problems are partially simulated. The simulation is partial from the point of view of participating strategies of differential evolution. Most often, only DDE/best/1/bin and CDE/best/1/bin are simulated.

5.5.2.1 Partially Simulated Toy functions

The partially simulated toy functions include the following:

(1) Branin RCOS function
(2) Camel-back function (six-hump)
(3) Easom function
(4) Goldstein Price function
(5) 3- and 6-dimensional Hartman function
(6) 9-dimensional Hilbert function
(7) 8-dimensional Katsuura function ($m = 32$)
(8) 5- and 8-dimensional Langerman function
(9) 8-dimensional Levy function 7 penalized
(10) 8-dimensional Levy function 13 penalized

(11) Michalewicz function 2
(12) 8-dimensional Neumaier function 3
(13) Odd square function
(14) 8-dimensional Rana function
(15) Rotated Ellipse function
(16) Rotated Ellipse function 2
(17) 8-dimensional sine envelope sine wave function
(18) 8-dimensional Schwefel function 2.21
(19) Shekel function ($m = 5, 7, 10$)
(20) 5- and 10-dimensional Shekel foxholes function
(21) Shekel foxholes function 2
(22) 8-, 16-, 24-, 50-, and 100-dimensional translated sphere function
(23) 9-dimensional Storn–Chebyshev function
(24) 8-dimensional Weierstrass function ($a = 1$, $b = 1$).

5.5.2.2 Partially Simulated Application Problems

The partially simulated application problems are the following:

(1) 8-, 16-, and 24-dimensional Benchmark Electromagnetic Inverse Scattering Problem
(2) Kowalik Enzyme Reaction Problem
(3) 6- and 9-dimensional Lennard-Jones Potential Problem.

References

[1] Branin, F.H. Jr. (1972) Widely convergent method for finding multiple solutions of simultaneous nonlinear equations. *IBM Journal of Research and Development*, **16**(5), 504–522.

[2] Suganthan, P.N., Hansen, N., Liang, J.J. *et al.* (2005) Problem Definitions and Evaluation Criteria for the CEC 2005 Special Session on Real-Parameter Optimization. *Technical Report*, Nanyang Technological University, Singapore, also KanGAL Report 2005005, IIT, Kanpur, India.

[3] Chung, C.J. and Reynolds, R.G. (1998) CAEP: an evolution-based tool for real-valued function optimization using cultural algorithms. *International Journal on Artificial Intelligence Tools*, **7**(3), 239–291.

[4] De Jong, K. (1975) An analysis of the behavior of a class of genetic adaptive systems. PhD thesis, Department of Computer and Communication Sciences, University of Michigan, Ann Arbor, MI.

[5] Dekkers, A. and Aarts, E. (1991) Global optimization and simulated annealing. *Mathematical Programming*, **50** (1–3), 367–393.

[6] Dixon, L.C.W. and Szegö, G.P. (eds) (1978) *Towards Global Optimization 2*, North-Holland, Amsterdam.

[7] Aluffi-Pentini, F., Parisi, V. and Zirilli, F. (1985) Global optimization and stochastic differential equations. *Journal of Optimization Theory and Applications*, **47**(1), 1–16.

[8] Dixon, L.C.W. and Szegö, G. (eds) (1975) *Towards Global Optimization*. North-Holland, Amsterdam.

[9] Hennart, J.P. (ed.) (1982) *Numerical Analysis: Proceedings of the 3rd IIMAS Workshop*, Lecture Notes in Mathematics, **909**, Springer, Berlin.

[10] Levy, A.V. and Montalvo, A. (1985) The tunneling algorithm for the global minimization of functions. *SIAM Journal on Scientific Computing*, **6**(1), 15–29.

[11] Averick, B.M., Carter, R.G. and Moré, J.J.(May 1991) The MINPACK-2 Test Problem Collection (Preliminary Version), Technical Memorandum No 150, Argonne National Laboratory, Mathematics and Computer Science Division.

[12] Moré, J.J., Garbow, B.S. and Hillstrom, K.E. (1981) Testing unconstrained optimization. *ACM Transactions on Mathematical Software*, **7**(1), 17–41.

[13] Powell, M.J.D. (1962) An iterative method for finding stationary values of a function of several variables. *Computer Journal*, **5**(2), 147–151.

[14] Fletcher, R. and Powell, M.J.D. (1963) An rapidly convergent descent method for minimization. *Computer Journal*, **6**(2), 163–168.

[15] Powell, M.J.D. (1964) An efficient method for finding the minimum of a function of several variables without calculating derivatives. *Computer Journal*, **7**(2), 155–162.

[16] Price, K.V., Storn, R.M. and Lampinen, J.A. (2005) *Differential Evolution: A Practical Approach to Global Optimization*, Springer, Berlin.

[17] Salomon, R. (1996) Re-evaluating genetic algorithm performance under coordinate rotation of benchmark functions: a survey of some theoretical and practical aspects of genetic algorithms. *BioSystems*, **39**(3), 263–278.

[18] Schwefel, H.P. (1981) *Numerical Optimization for Computer Models*, John Wiley & Sons, Ltd, Chichester.

[19] Schwefel, H.P. (1995) *Evolution and Optimum Seeking*, John Wiley & Sons, Inc., New York.

[20] Whitley, D., Rana, S., Dzubera, J. and Mathias, K.E. (1996) Evaluating evolutionary algorithms. *Artificial Intelligence*, **85**(1–2), 245–276.

[21] Yao, X., Liu, Y. and Lin, G. (1999) Evolutionary programming made faster. *IEEE Transactions Evolutionary Computation*, **3**(2), 82–102.

[22] Bongartz, I., Conn, A.R., Gould, N. and Toint, P.L. (1995) CUTE: constrained and unconstrained testing environment. *ACM Transactions on Mathematical Software*, **21**(1), 123–160.

[23] Bäck, T. and Schwefel, H.P. (1993) An overview of evolutionary algorithms for parameter optimization. *Evolutionary Computation*, **1**(1), 1–23.

[24] Ackley, D.H. (1987) *A Connectionist Machine for Genetic Hillclimbing*, Kluwer Academic Publishers, Boston.

[25] Adjiman, C.S., Sallwig, S., Floudas, C.A. and Neumaier, A. (1998) A global optimization method, aBB, for general twice-differentiable NLPs – I. Theoretical advances. *Computers Chemical Engineering*, **22**(9), 1137–1158.

[26] Rahnamayan, S., Tizhoosh, H.R. and Salama, M.M.A. (2007) A novel population initialization method for accelerating evolutionary algorithms. *Computers and Mathematics with Applications*, **53**(10), 1605–1614.

[27] Clerc, M. (1999) The swarm and the queen: towards a deterministic and adaptive particle swarm optimization. 1999 IEEE Congress Evolutionary Computation, Washington DC, July 6–9, 3, pp. 1951–1957.

[28] Bard, Y. (1970) Comparison of gradient methods for the solution of nonlinear parametric estimation problems. *SIAM Journal on Numerical Analysis*, **7**(1), 157–186.

[29] Rosen, J.B., Mangasarian, O.L. and Ritter, K. (eds) (1970) *Nonlinear Programming*. Academic Press, New York.

[30] Shah, R.V., Buehler, R.J. and Kempthorne, O. (1964) Some algorithms for minimizing a function of several variables. *Journal of the Society for Industrial and Applied Mathematics*, **12**(1), 74–92.

[31] Biggs, M.C. (1971) A new variable metric technique taking account of non-quadratic behaviour of the objective function. *IMA Journal of Applied Mathematics*, **8**(3), 315–327.

[32] Mishra, S. (2007) Global Optimization by Differential Evolution and Particle Swarm Methods: Evaluation on Some Benchmark functions. Munich Personal RePEc Archive, MPRA paper no. 1005 (http://mpra.ub.uni-muenchen.de/1005/1/MPRA_paper_1005.pdf), April.

[33] Bohachevsky, I.O., Johnson, M.E. and Stein, M.L. (1986) Generalized simulated annealing for function optimization. *Technometrics*, **28**(3), 209–217.

[34] Booth, A.D. (1949) An application of the method of steepest descents to the solution of systems of non-linear simultaneous equations. *Quarterly Journal of Mechanics and Applied Mathematics*, **2**(4), 460–468.

[35] Rosenbrock, H.H. (1960) An automatic method for finding the greatest or least value of a function. *Computer Journal*, **3**(3), 175–184.

[36] Box, M.J. (1966) A comparison of several current optimization methods and the use of transformations in constrained problems. *Computer Journal*, **9**(1), 67–77.

[37] Munteanu, C. and Lazarescu, V. (1998) Global search using a new evolutionary framework: the adaptive reservoir genetic algorithm. *Complexity International*, **5**.

[38] Brown, K.M. (1969) A quadratically convergent Newton-like method based upon Gaussian elimination. *SIAM Journal on Numerical Analysis*, **6**(4), 560–569.

[39] Ferraris, G.B. and Tronconi, E. (1986) BUNLSI – a FORTRAN program for solution of systems of nonlinear algebraic equations. *Computers and Chemical Engineering*, **10**(2), 129–141.

[40] Brown, K.M. and Gearhardt, W.B. (1971) Deflation techniques for the calculation of further solutions of a nonlinear system. *Numerische Mathematik*, **16**(4), 334–342.

[41] Broyden, C.G. (1971) The convergence of an algorithm for solving sparse nonlinear systems. *Mathematics of Computation*, **25**(115), 285–294.

[42] Broyden, C.G. (1965) A class of methods for solving nonlinear simultaneous equations. *Mathematics of Computation*, **19**(92), 577–593.

[43] Silagadze, Z.K. (2007) Finding two-dimensional peaks. *Physics of Particles and Nuclei Letters*, **4**(1), 73–80.

[44] Fletcher, R. (1965) function minimization without evaluating derivatives – a review. *Computer Journal*, **8**(1), 33–41.

[45] Chen, Y. (August 2003) Computer Simulation of Electron Positron Annihilation Processes. *Technical Report SLAC-Report-646*, Stanford Linear Accelerator Center, Stanford University.

[46] Chichinadze, V. (1969) The ψ-transform for solving linear and nonlinear programming problems. *Automatica*, **5**(3), 347–355.

[47] Adorio, E.P. and Diliman, U.P. (2005) MVF – Multivariate Test functions Library in C for Unconstrained Global Optimization. Unpublished.

[48] Corana, A., Marchesi, M., Martini, C. and Ridella, S. (1987) Minimizing multimodal functions of continuous variables with the simulated annealing algorithm. *ACM Transactions on Mathematical Software*, **13**(3), 262–280.

[49] Breiman, L. and Cutler, A. (1993) A deterministic algorithm for global optimization. *Mathematical Programming*, **58**, 179–199.

[50] Lavi, A. and Vogl, T.P. (eds) (1966) *Recent Advances in Optimization Techniques*, John Wiley & Sons, Inc., New York.

[51] Jensen, M.T. (Oct. 2001) Robust and Flexible Scheduling with Evolutionary Computation, Ph.D. thesis, Basic Research in Computer Science (BRICS), Department of Computer Science University of Aarhus, Denmark.

[52] de Villers, N. and Glasser, D. (1981) A continuation method for nonlinear regression. *SIAM Journal on Numerical Analysis*, **18**(6), 1139–1154.

[53] Goldberg, D.E., Deb, K. and Clark, J.H. (1992) Genetic algorithms, noise and the sizing of populations. *Complex Systems*, **6**, 333–362.

[54] Lootsma, F.A. (ed.) (1972) *Numerical Methods for Non-linear Optimization*. Academic Press, London.

[55] Moré, J.J. and Cosnard, M.Y. (1979) Numerical solution of nonlinear equations. *ACM Transactions on Mathematical Software*, **5**(1), 64–85.

[56] Dixon, L.C.W. and Price, R.C. (1989) The truncated Newton method for sparse unconstrained optimisation using automatic differentiation. *Journal of Optimization Theory and Applications*, **60**(2), 261–275.

[57] El-Attar, R.A., Vidyasagar, M. and Dutta, S.R.K. (1979) An algorithm for l1-norm minimization with application to nonlinear l1-approximation. *SIAM Journal on Numerical Analysis*, **16**(1), 70–86.

[58] Elwakeil, O.A. and Arora, J.S. (1996) Two algorithms for global optimization of general NLP problems. *International Journal for Numerical Methods in Engineering*, **39**(19), 3305–3325.

[59] Csendes, T. and Ratz, D. (1997) Subdivision direction selection in interval methods for global optimization. *SIAM Journal on Numerical Analysis*, **34**(3), 922–938.

[60] Rahnamayan, S., Tizhoosh, H.R. and Salama, N.M.M. (2007) Opposition-based differential evolution (ODE) with variable jumping rate. 2007 IEEE Symp. Foundations Computation Intelligence, Honolulu, HI, April 1–5, pp. 81–88.

[61] Dobmann, M., Liepelt, M. and Schittkowski, K. (1995) Algorithm 746: PCOMP: A Fortran code for automatic differentiation. *ACM Transactions on Mathematical Software*, **21**(3), 233–266.

[62] Floudas, C.A. and Pardalos, P.M. (eds) (1992) *Recent Advances in Global Optimization*, Princeton Series in Computer Science. Princeton University Press, Princeton, NJ.

[63] Bäck, T. (1996) *Evolutionary Algorithms in Theory and Practice*. Oxford University Press, Oxford.

[64] Freudenstein, F. and Roth, B. (1963) Numerical solutions of systems of nonlinear equations. *Journal of the Association for Computing Machinery*, **10**(4), 550–556.

[65] Gheri, G. and Mancino, O. (1971) A significant example to test methods for solving systems of nonlinear equations. *Calcolo*, **8**(1–2), 107–114.

[66] Goldstein, A.A. and Price, J.F. (1971) On descent from local minima. *Mathematics Computation*, **25**(115), 569–574.

[67] Griewank, A.O. (1981) Generalized Descent for global optimization. *Journal of Optimization Theory and Applications*, **34**(1), 11–39.

[68] Shanno, D.F. (1970) Conditioning of quasi-Newton methods for function minimization. *Mathematics of Computation*, **24**(111), 647–656.

[69] Fraley, C. (1989) Software Performances on Nonlinear Least-Squares Problems. *Technical Report no. STAN-CS-89-1244*, Department of Computer Science, Stanford University.

[70] Salane, D.E. (1987) A continuation approach for solving large residual nonlinear least squares problems. *SIAM Journal on Scientific Statistical Computing*, **8**(4), 655–671.

[71] Hartman, J.K. (1972) Some experiments in global optimization. *Naval Research Logistics Quarterly*, **20**(3), 569–576.

[72] Gao, F. and Tong, H. (2006) UEAS: a novel united evolutionary algorithm scheme, in *Neural Information Processing*, Part III (eds I. King, J. Wang, L. Chan and D. Wang), Lecture Notes in Computer Science, **4234**. Springer, Berlin, pp. 772–780.

[73] Törn, A. and Zilinskas, A. (1989) *Global Optimization*, Springer-Verlag, Berlin.

[74] Hiebert, K.L. (1982) An evaluation of mathematical software that solves systems of nonlinear equations. *ACM Transactions on Mathematical Software*, **8**(1), 5–20.

[75] Himmelblau, D.M. (1972) *Applied Nonlinear Programming*, McGraw-Hill, New York.

[76] Floudas, C.A., Pardalos, P.M., Adjiman, C.S. *et al.* (1999) *Handbook of Test Problems in Local and Global Optimization*, Kluwer Academic Publishers, Dordrecht.

[77] Bekey, G.A. and Ung, M.T. (1974) A comparative evaluation of two global search algorithms. *IEEE Transactions on Systems, Man and Cybernetics*, **4**, 112–116.

[78] Voigt, H.M. and Anheyer, T. (1994) Modal mutations in evolutionary algorithms. 1994 IEEE World Congress Computational Intelligence, also 1st IEEE Conf. Evolutionary Computation, Orlando, FL, June 27–29, 1, pp. 88–92.

[79] Jennrich, R.I. and Sampson, P.F. (1968) Application of stepwise regression to non-linear estimation. *Technometrics*, **10**(1), 63–72.

[80] Katsuura, H. (1991) Continuous nowhere-differentiable functions – an application of contraction mappings. *American Mathematical Monthly*, **98**(4), 411–416.

[81] Kjellström, G. (1969) Network optimization by random variation of component values. *Ericsson Technics*, **25** (3), 133–151.

[82] McDonnell, J.R., Reynolds, R.G. and Fogel, D.B. (eds) (1995) *Evolutionary Programming IV: Proceedings of the fourth Annual Conference on Evolutionary Programming*, MIT Press, Cambridge, MA.

[83] Bersini, H., Dorigo, M., Langerman, S. *et al.* (1996) Results of the first international contest on evolutionary optimization. IEEE International Conference on Evolutionary Computation, May 20–22, pp. 611–615.

[84] Luenberger, D.G. (1973) *Introduction to Linear and Nonlinear Programming*, Addison-Wesley, Reading, MA.

[85] Nataray, P.S.V. and Kotecha, K. (2002) An algorithm for global optimization using the Taylor-Bernstein form as inclusion function. *Journal of Global Optimization*, **24**(4), 417–436.

[86] Matyas, J. (1965) Random optimization. *Automation and Remote Control*, **26**, 244–251.

[87] McCormick, G.P. (1976) Computability of global solutions to factorable nonconvex programs I. Convex underestimating problems. *Mathematical Programming*, **10**(2), 147–175.

[88] McKeown, J.J. (1975) Specialized versus general-purpose algorithms for functions that are sums of squared terms. *Mathematical Programming*, **9**(1), 57–68.

[89] Michalewicz, Z. (1994) *Genetic Algorithms + Data Structures = Evolution Programs*, 2nd extended edn. Springer-Verlag, Berlin.

[90] Cragg, E.E. and Levy, A.V. (1969) Study on a supermemory gradient method for the minimization of functions. *Journal of Optimization Theory and Applications*, **4**(3), 191–205.

[91] Mishra, S. (2007) Performance of differential evolution and particle swarm methods on some relatively harder multi-modal benchmark functions. Munich Personal RePEc Archive, MPRA paper no. 449 (http://mpra.ub.uni-muenchen.de/449/1/MPRA_paper_449.pdf), April.

[92] Moore, R.E. (1988) *Reliability in Computing*. Academic Press, San Diego, CA.

[93] Ali, M.M., Khompatraporn, C. and Zabinsky, Z.B. (2005) A numerical evaluation of several stochastic algorithms on selected continuous global optimization test problems. *Journal of Global Optimization*, **31** (4), 635–672.

[94] Osborne, M.R. and Watson, G.A. (1971) On an algorithm for discrete nonlinear L1 approximation. *Computer Journal*, **14**(2), 184–188.

[95] Pahner, U. and Hameyer, K. (2000) Adaptive coupling of differential evolution and multiquadrics approximation for the tuning of the optimization process. *IEEE Transactions on Magnetics*, **36**(4), 1047–1051.

[96] Rahnamayan, S., Tizhoosh, H.R. and Salama, M.M.A. (2008) Opposition-based differential evolution. *IEEE Transactions on Evolutionary Computation*, **12**(1), 64–79.

[97] Pintér, J.D. (1996) *Global Optimization in Action: Continuous and Lipschitz Optimization Algorithms, Implementations, and Applications*, Kluwer Academic, Dordrecht.

[98] Rabinowitz, P. (ed.) (1970) *Numerical Methods for Nonlinear Algebraic Equations*, Gordon and Breach, London.

[99] Fu, M.C., Hu, J. and Marcus, S.I. (2006) Model-based randomized methods for global optimization. 17th Int. Symp. Mathematical Theory Networks Systems, Kyoto, Japan, July 24–28, pp. 355–363.

[100] Price, W.L. (1977) A controlled random search procedure for global optimization. *Computer Journal*, **20**(4), 367–370.

[101] Qing, A. (2006) Dynamic differential evolution strategy and applications in electromagnetic inverse scattering problems. *IEEE Transactions on Geoscience and Remote Sensing*, **44**(1), 116–125.

[102] Schumer, M.A. and Steiglitz, K. (1968) Adaptive step size random search. *IEEE Transactions on Automatic Control*, **13**(3), 270–276.

[103] Hoffmeister, F. and Bäck, T. (1991) Genetic algorithms and evolution strategies: similarities and differences, in *Parallel Problem Solving from Nature* (eds. H.P. Schwefel and R. Männer), Lecture Notes in Computer Science, **496**, Springer, Berlin, pp. 455–469.

[104] Liang, J.J., Qin, A.K., Suganthan, P.N. and Baskar, S. (2006) Comprehensive learning particle swarm optimizer for global optimization of multimodal functions. *IEEE Transactions on Evolutionary Computation*, **10**(3), 281–295.

[105] Mishra, S. (2007) Some new test functions for global optimization and performance of repulsive particle swarm method. *Munich Personal RePEc Archive*, MPRA paper no. 2718 (http://mpra.ub.uni-muenchen.de/2718/), April.

[106] Sekaj, I. (2004) Robust parallel genetic algorithms with re-initialisation, in *Parallel Problem Solving from Nature – PPSN VIII* (eds X. Yao, E. Burke and J.A. Lozano *et al.*), Lecture Notes in Computer Science, **3242**, Springer, Berlin, pp. 411–419.

[107] Opačić, J. (1973) A heuristic method for finding most extrema of a nonlinear functional. *IEEE Transactions on Systems, Man and Cybernetics*, **3**(1), 102–107.

[108] Shekel, J. (Mar. 1971) Test functions for multimodal search techniques. 5th Princeton Conf. Information Sciences Systems.

[109] Shubert, B.O. (1972) A. sequential method seeking the global maximum of a function. *SIAM Journal on Numerical Analysis*, **9**, 379–388.

[110] Schaffer, J.D., Caruana, R.A., Eshelman, L.J. and Das, R. (1989) A study of control parameters affecting online performance of genetic algorithms for function optimization. 3rd International Conference on Genetic Algorithms, George Mason University, June 4–7, pp. 51–60.

[111] Kowalik, J. and Osborne, M.R. (1968) *Methods for Unconstrained Optimization Problems*, American Elsevier, New York.

[112] Watson, L.T. (1981) Engineering applications of the Chow-Yorke algorithm. *Applied Mathematics and Computation*, **9**(2), 111–133.

[113] Courrieu, P. (1997) The hyperbell algorithm for global optimization: a random walk using Cauchy densities. *Journal of Global Optimization*, **10**(1), 37–55.

[114] Wayburn, T.L. and Seader, J.D. (1987) Homotopy continuation methods for computer-aided process design. *Computers and Chemical Engineering*, **11**(1), 7–25.

[115] Andresen, P.R., Nielsen, M. and Kreiborg, S. (1998) 4D shape-preserving modelling of bone growth, in *Medical Image Computing and Computer-Assisted Intervention – MICCAI '98* (eds W.M. Wells, A. Colchester and S. Delp), Lecture Notes in Computer Science, **1496**, Springer, Berlin, pp. 710–719.

[116] Qing, A., Lee, C.K. and Jen, L. (2001) Electromagnetic inverse scattering of two-dimensional perfectly conducting objects by real-coded genetic algorithm. *IEEE Transactions on Geoscience and Remote. Sensing*, **39**(3), 665–676.

[117] Hantos, Z., Suki, B., Csendes, T. *et al.* (1990) Modeling of low-frequency pulmonary impedance in dogs. *Journal of Applied Physiology*, **68**(3), 849–860.

[118] Peták, F., Hantos, Z., Adamicza, Á. and Daróczy, B. (1993) Partitioning of pulmonary impedance: modeling vs. alveolar capsule approach. *Journal of Applied Physiology*, **75**(3), 513–521.

[119] Floudas, C. and Pardalos, P.M. (eds) (1995) *State of the Art in Global Optimization: Computational Methods and Applications*, Kluwer, Dordrecht.

[120] Corne, D., Dorigo, M. and Glover, F. (eds) (1999) *New Ideas in Optimization*, McGraw-Hill, London.

[121] Meintjes, K. and Morgan, A.P. (1990) Chemical equilibrium systems as numerical test problems. *ACM Transactions on Mathematical Software*, **16**(2), 143–151.

[122] Bullard, L.G. and Biegler, L.T. (1991) Iterative linear-programming strategies for constrained simulation. *Computers and Chemical Engineering*, **15**(4), 239–254.

[123] Hougen, O.A. and Watson, K.M. (1947) *Chemical Reaction Principles*, John Wiley & Sons, Inc., New York.

[124] Bates, D. and Watts, D. (1988) *Nonlinear Regression Analysis and Its Applications*, John Wiley & Sons, Inc., New York.

[125] Staelens, N., Reyniers, M.F. and Marin, G.B. (2002) Langmuir–Hinshelwood–Hougen–Watson rate equations for the transalkylation of methylamines. *Chemical Engineering Journal*, **90**(1), 185–193.

[126] Nelson, C.V. and Hodgkin, B.C. (1981) Determination of magnitudes, directions and locations of two independent dipoles in a circular conducting region from boundary potential measurements. *IEEE Transactions on Biomedical Engineering*, **28**(12), 817–823.

[127] Dennis, J.E., Gay, D.M. and Vu, P.A. (1985) A New Nonlinear Equations Test Problem. *Technical Report 83-16*, Mathematical Sciences Department, Rice University, December.

[128] Kowalik, J.S. (1967) A note on nonlinear regression analysis. *Australian Computer Journal*, **1**, 51–53.

[129] Kowalik, J.S. and Morrison, J.F. (1968) Analysis of kinetic data for allosteric enzyme reactions as a nonlinear regression problem. *Mathematical Biosciences*, **2**, 57–66.

[130] Jacobs, R.A. (1988) Increased rates of convergence through learning rate adaptation. *Neural Networks*, **1**(4), 295–307.

[131] Magoulas, G.D., Vrahatis, M.N. and Androulakis, G.S. (1997) Effective backpropagation training with variable stepsize. *Neural Networks*, **10**(1), 69–82.

[132] Vrahatis, M.N., Androulakis, G.S., Lambrinos, J.N. and Magoulas, G.D. (2000) A class of gradient unconstrained minimization algorithms with adaptive stepsize. *Journal of Computational and Applied Mathematics*, **114**(2), 367–386.

[133] Hadjisavvas, N. and Pardalos, P. (eds) (2001) *Advances in Convex Analysis and Global Optimization*. Kluwer Academic, Dordrecht.

[134] Parsopoulos, K.E., Plagianakos, V.P., Magoulas, G.D. and Vrahatis, M.N. (2001) Stretching technique for obtaining global minimizers through particle swarm optimization. Particle Swarm Optimization Workshop, Indianapolis, USA, pp. 22–29.

[135] Babu, B.V. and Chaturvedi, G. (2000) Evolutionary computation strategy for optimization of an alkylation reaction. Int. Symp. 53rd Annual Session of IIChE (CHEMCON-2000), Science City, Calcutta, December 18–21.

[136] Andreoni, J. and Miller, J.H. (1995) Auctions with artificial adaptive agents. *Games and Economic Behavior*, **10** (1), 39–64.

[137] Anthony, P. and Jennings, N.R. (2002) Evolving bidding strategies for multiple auctions. 15th European Conf. Artificial Intelligence, Lyon, France, pp. 178–182.

[138] Ausubel, L.M. (2004) An efficient ascending-bid auction for multiple objects. *American Economic Review*, **94** (5), 1452–1475.

[139] Paprzycki, M., Dent, D. and Kucaba-Pietal, A. (2002) Solvers for nonlinear algebraic equations; where are we today? in *Parallel Processing and Applied Mathematics* (eds. R. Wyrzykowski, J. Dongarra, M. Paprzycki and J. Wasniewski), Lecture Notes in Computer Science, **2328**, Springer, Berlin, pp. 719–728.

[140] Babu, B.V. and Chaurasia, A.S. (2003) Optimization of pyrolysis of biomass using differential evolution approach. 2nd Int. Conf. Computational Intelligence Robotics Autonomous Systems (CIRAS-2003), Singapore, December 15–18, 2003, PS0504, pp. 01–06.

[141] Babu, B.V. and Jehan, M.M.L. (2003) Differential evolution for multi-objective optimization. 2003 Congress Evolutionary Computation, Canberra, Australia, December 8–12, pp. 2696–2703.

[142] Angira, R. and Babu, B.V. (2005) Non-dominated sorting differential evolution (NSDE): an extension of differential evolution for multi-objective optimization. 2nd Indian Int. Conf Artificial Intelligence, Pune, India, December 20–22, pp. 1428–1443.

[143] Babu, B.V., Syed Mubeen, J.H. and Chakole, P.G. (2005) Multiobjective optimization using differential evolution. *Tech Genesis – The Journal of Information Technology*, **2**(2), 4–12.

[144] Bracken, J. and McCormick, G.P. (1968) *Selected Applications of Nonlinear Programming*, John Wiley & Sons, Inc., New York.

[145] Carnahan, B., Luther, H.A. and Wilkes, J.O. (1969) *Applied Numerical Methods*, John Wiley & Sons, Inc., New York.

[146] Zhang, D. and Huang, W. (2004) A simulated annealing algorithm for the circles packing problem, in *Computational Science – ICCS 2004, Part I* (eds. M. Bubak, G.D. van Albada, P.M.A. Sloot and J.J. Dongarra), Lecture Notes in Computer Science, **3036**, Springer, Berlin, pp. 206–214.

[147] Liu, J. and Huang, W.Q. (2006) A modified differential evolution algorithm for solving circles packing problem with constraints of equilibrium. *Information Control*, **35**(1), 103–107, 113.

[148] Ebers, J.J. and Moll, J.L. (1954) Large-scale behavior of junction transistors. *Proceedings of IRE*, **42**(12), 1761–1772 Dec.

[149] Cutteridge, O.P.D. (1974) Powerful 2-part program for solution of nonlinear simultaneous equations. *Electronics Letters*, **10**(10), 182–184.

[150] Babu, B.V. and Gautam, K. (2001) Evolutionary computation for scenario-integrated optimization of dynamic systems. Int. Symp. 54th Annual Session IIChE (CHEMCON-2001), CLRI, Chennai, India, December 19–22.

[151] Babu, B.V. and Angira, R. (2003) Optimization of water pumping system using differential evolution strategies. 2nd Int. Conf. Computational Intelligence Robotics Autonomous Systems (CIRAS-2003), Singapore, December 15–18, 2003, PS0205, pp. 25–30.

[152] Reklaitis, G.V., Ravindran, A. and Ragsdell, K.M. (2006) *Engineering Optimization*, 2nd edn, John Wiley & Sons, Inc., Hoboken, NJ.

[153] Booker, A.J., Dennis, J.E. Jr, Frank, P.D. *et al.* (1998) A Rigorous Framework for Optimization of Expensive functions by Surrogates. *Technical Report 98-47*, Institute for Computer Applications in Science and, Engineering.

[154] Storn, R. (1996) On the usage of differential evolution for function optimisation. 1996 Biennial Conf. North American Fuzzy Information Processing Society, Berkeley, CA, 19–22 June, pp. 519–523.

[155] Judge, G.G., Hill, R.C., Griffiths, W.E. *et al.* (1982) *Introduction to the Theory and Practice of Econometrics*, John Wiley & Sons, Inc., New York.

[156] Judge, G.G., Griffiths, W.E., Hill, R.C. *et al.* (1985) *The Theory and Practice of Econometrics*, 2nd edn, John Wiley & Sons, Inc., New York.

[157] Kubíček, M., Hofmann, H., Hlaváček, V. and Sinkule, J. (1980) Multiplicity and stability in a sequence of two nonadiabatic nonisothermal CSTRs. *Chemical Engineering Science*, **35**(4), 987–996.

[158] Cogotti, E., Fanni, A. and Pilo, F. (2000) A comparison of optimization techniques for Loney's solenoids design: an alternative tabu search algorithm. *IEEE Transactions on Magnetics*, **36**(4), 1153–1157.

[159] Coelho, L.S. and Mariani, V.C. (2006) Chaotic differential evolution applied to electromagnetics optimization. 12th Biennial IEEE Conf. Electromagnetic Field Computation, Miami, April 30–May 3, p. 230.

[160] Coelho, L.S. and Mariani, V.C. (2006) Combining of differential evolution and implicit filtering algorithm applied to electromagnetic design optimization. 11th Online World Conf. Soft Computing Industrial Applications, September 18–October 6.

[161] Rashid, K., Farina, M., Ramirez, J.A. *et al.* (2001) A comparison of two generalized response surface methods for optimisation in electromagnetics. *International Journal for Computation and Mathematics in Electrical and Electronic Engineering*, **20**(3), 740–752.

[162] Francken, K., Vancorenland, P. and Gielen, G. (2000) DAISY: a simulation-based high-level synthesis tool for Delta Sigma modulators. IEEE/ACM Int. Conf. Computer Aided Design, San Jose, CA, November 5–9, pp. 188–192.

[163] Rae, A. and Parameswaran, S. (1998) Application-specific heterogeneous multiprocessor synthesis using differential-evolution. 11th Int. Symp. System Synthesis, Hsinchu, Taiwan, December 2–4, pp. 83–88.

[164] Joshi, R. and Sanderson, A.C. (1996) Multisensor fusion and model selection using a minimal representation size framework. 1996 IEEE/SICE/RSJ Int. Conf. Multisensor Fusion Integration Intelligent Systems, Washington, DC, December 8–11, 1996, pp. 25–32.

[165] Joshi, R. and Sanderson, A.C. (1997) Experimental studies on minimal representation multisensor fusion. 8th Int. Conf. Advanced Robotics, Monterey, CA, July 7–9, pp. 603–610.

[166] Joshi, R. and Sanderson, A.C. (1997) Minimal representation multisensor fusion using differential evolution. 1997 IEEE Int. Symp. Computational Intelligence Robotics Automation, Monterey, CA, July 10–11, pp. 266–273.

[167] Joshi, R. and Sanderson, A.C. (1997) Multisensor fusion of touch and vision using minimal representation size. 1997 IEEE/RSJ Int. Conf. Intelligent Robots Systems, Grenoble, September 7–11, 3, pp. v4–v5.

[168] Joshi, R. and Sanderson, A.C. (1999) Minimal representation multisensor fusion using differential evolution. *IEEE Transactions on Systems, Man and Cybernetics*, **29**(1), 63–76.

[169] Bates, D. (1983) The derivative of |X#X| and its uses. *Technometrics*, **25**(4), 373–376.

[170] Babu, B.V. and Angira, R. (2002) Optimization of non-linear chemical processes using evolutionary algorithm. Int. Symp. 55th Annual Session IIChE (CHEMCON-2002), OU, Hyderabad, December 19–22.

[171] Priem Mendes, S., Gómez Pulido, J.A., Vega Rodríguez, M.A. *et al.* (2006) A differential evolution based algorithm to optimize the radio network design problem. 2nd IEEE Int. Conf. e-Science Grid Computing, Amsterdam, The Netherlands, Dec, pp. 119–125.

[172] Priem Mendes, S., Gómez Pulido, J.A., Vega Rodríguez, M.A. *et al.* (2007) Fast wide area network design optimisation using differential evolution. 2007 Int. Conf. Advanced Engineering Computing Applications Sciences, Papeete, France, Nov. 4–9, pp. 3–10.

[173] Vega-Rodriguez, M.A., Gomez-Pulido, J.A., Alba, E. *et al.* (2007) Evaluation of different metaheuristics solving the RND problem, in *Applications of Evolutionary Computing* (eds. M. Giacobini *et al.*), Lecture Notes in Computer Science, **4448**, Springer, Berlin, pp. 101–110.

[174] Vega-Rodriguez, M.A., Gomez-Pulido, J.A., Alba, E. *et al.* (2007) Using omnidirectional BTS and different evolutionary approaches to solve the RND problem, in *Computer Aided Systems Theory – EUROCAST 2007* (eds. R. Moreno-Diaz, F. Pichler and A. Quesada Arencibia), Lecture Notes in Computer Science, **4739**, Springer, Berlin, pp. 853–860.

[175] Wu, X.G. and Zhu, Y.P. (2001) A global optimization method for three-dimensional conformal radiotherapy treatment planning. *Physics Medicine Biology*, **46**(1), 107–119.

[176] Parsopoulos, K.E., Papageorgiou, E.I., Groumpos, P.P. and Vrahatis, M.N. (2004) Evolutionary computation techniques for optimizing fuzzy cognitive maps in radiation therapy systems, in *Genetic and Evolutionary Computation – GECCO 2004, Part I* (eds. K. Deb, R. Poli, W. Banzhaf *et al.*), Lecture Notes in Computer Science, **3102**, Springer, Berlin, pp. 402–413.

[177] Li, Y., Yao, D., Chen, W. *et al.* (2005) Ant colony system for the beam angle optimization problem in radiotherapy planning: a preliminary study. 2005 IEEE Congress Evolutionary Computation, Edinburgh, UK, September 2–5, 2, pp. 1532–1538.

[178] Dent, D., Paprzycki, M. and Kucaba-Pietal, A. (1998) Testing convergence of nonlinear system solvers. Southern Symp. Computing, The University of Southern Mississippi, Dec. 4–5.

[179] Shacham, M. (1985) Comparing software for the solution of systems of nonlinear algebraic equations arising in chemical engineering. *Computers and Chemical Engineering*, **9**(2), 103–112.

[180] Babu, B.V. and Munawar, S.A. (2001) Optimal design of shell-and-tube heat exchangers using different strategies of differential evolution. *PreJournal*, Article No. 003873, March 03.

[181] Battiti, R. and Tecchiolli, G. (1996) The continuous reactive tabu search: blending combinatorial optimization and stochastic search for global optimization. *Annals of Operations Research*, **63**(2), 151–188.

[182] Alotto, P., Kuntsevitch, A.V., Magele, C. *et al.* (1996) Multiobjective optimization in magnetostatics: a proposal for benchmark problems. *IEEE Transactions on Magnetics*, **32**(3), 1238–1241.

[183] Alotto, P., Girdinio, P., Molinari, G. and Nervi, M. (2000) Hybrid deterministic stochastic fuzzy methods for the optimization of electromagnetic devices. *International Journal for Computation and Mathematics in Electrical and Electronic Engineering*, **19**(1), 30–38.

[184] Coelho, L.S. and Alotto, P. (2006) Electromagnetic device optimization using improved differential evolution methods. 12th Biennial IEEE Conf. Electromagnetic Field Computation, Miami, April 30–May 3, p. 295.

[185] Babu, B.V. and Angira, R. (2001) Optimization of thermal cracker operation using differential evolution. Int. Symp. 54th Annual Session IIChE (CHEMCON-2001), CLRI, Chennai, December 19–22.

[186] Krink, T., Filipič, B., Fogel, G.B. and Thomsen, R. (2004) Noisy optimization problems – a particular challenge for differential evolution? 2004 IEEE Congress Evolutionary Computation, Portland, OR, June 19–23, 1, pp. 332–339.

[187] Vesterstrøm, J. and Thomsen, R. (2004) A comparative study of differential evolution, particle swarm optimization, and evolutionary algorithms on numerical benchmark problems. 2004 IEEE Congress Evolutionary Computation, Portland, OR, June 19–23, 2004, 2, pp. 1980–1987.

[188] Das, S., Konar, A. and Chakraborty, U.K. (2005) Improved differential evolution algorithms for handling noisy optimization problems. 2005 IEEE Congress Evolutionary Computation, Edinburgh, UK, September 2–5, 2, pp. 1691–1698.

[189] Brest, J., Greiner, S., Bošković, B. *et al.* (2006) Self-adapting control parameters in differential evolution: a comparative study on numerical benchmark problems. *IEEE Transactions Evolutionary Computation*, **10**(6), 646–657.

6

Differential Evolution Strategies

This chapter aims to answer the first question facing the parametric study. In particular, the intention is to find which strategy is more suitable for a particular problem with specific features. Although there are many differential evolution strategies, as discussed in Chapters 2 and 3, we are only concerned with the eight strategies mentioned in Chapter 4 since far too much time would be required to study all available differential evolution strategies.

This chapter also aims to answer the third question facing the parametric study from the point of view of problem features. As mentioned in Chapter 5, nine of the 13 toy functions in the tentative benchmark test bed have been fully simulated. Each toy function has distinctive features. Examining the simulation results together with problem features may give us insight into the choice of the most appropriate strategy for future application problems with certain features.

Although partial simulation on other toy functions and application problems has also been conducted, these simulation results are not presented and analyzed here, one of our major concerns being to avoid committing evolutionary crimes. It is also our concern here not to drown readers in a sea of simulation results, or lull them to sleep with endless simulation data. Of course, space limitations are another important issue.

6.1 Sphere Function

We begin with the sphere fuction. As mentioned in Chapter 5, this is a benchmark for studying problem features.

6.1.1 8-Dimensional Sphere Function

6.1.1.1 Robustness

The number of successful trials of differential evolution is shown in Figure 6.1. From Figure 6.1, we note the following

(a) All curves are bell-shaped. The position, height and width of the bell for each strategy are different.
(b) Dynamic differential evolution is more robust than classic differential evolution.

Differential Evolution Anyong Qing

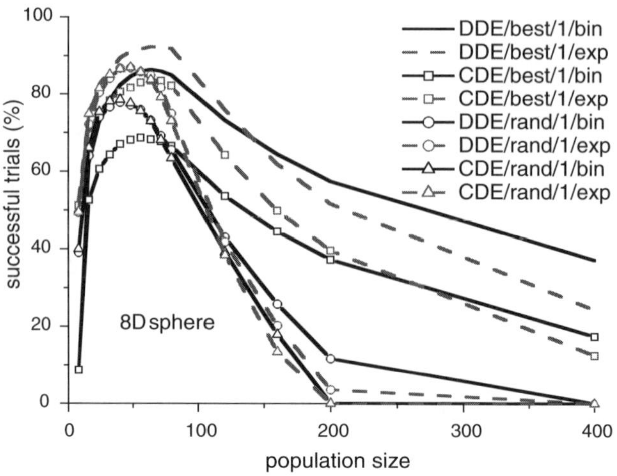

Figure 6.1 Robustness of differential evolution for the 8-dimensional sphere function

(c) DDE/best/1/* is much more robust than DDE/rand/1/*, when population size is not very small.
(d) At smaller population size, DDE/*/1/exp is more robustness than DDE/*/1/bin. However, DDE/*/1/exp is less robust than DDE/*/1/bin at larger population size.
(e) When population size is small, CDE/best/1/* is less robust than classic CDE/rand/1/*. However, CDE/best/1/* outperforms CDE/rand/1/* after population size passes a certain threshold.
(f) CDE/*/1/bin behaves less robustly than CDE/*/1/exp when population size is small. On the other hand, CDE/*/1/bin slightly outperforms CDE/*/1/exp when population size is above a certain threshold.

6.1.1.2 Efficiency

The minimal average number of objective function evaluations is shown in Figure 6.2. It is readily apparent from Figure 6.2 that:

(a) The minimal average number of objective function evaluations of all strategies grows almost linearly with population size.
(b) The growth rate strongly depends on the evolution mechanism and differential mutation base.
(c) The growth rate is much more weakly dependent on crossover.
(d) Dynamic differential evolution is more efficient than classic differential evolution.
(e) DE/best/1/* is much more efficient than DE/rand/1/*.
(f) Exponential crossover contributes more efficiency to DE/best/1/exp. On the other hand, it makes DE/rand/1/exp slightly less efficient.
(g) Interestingly, the minimal average number of objective function evaluations of DDE/best/1/bin and DDE/best/1/exp remains almost flat within quite a wide range of population size. From the point of view of efficiency, this gives us much freedom to choose population size.

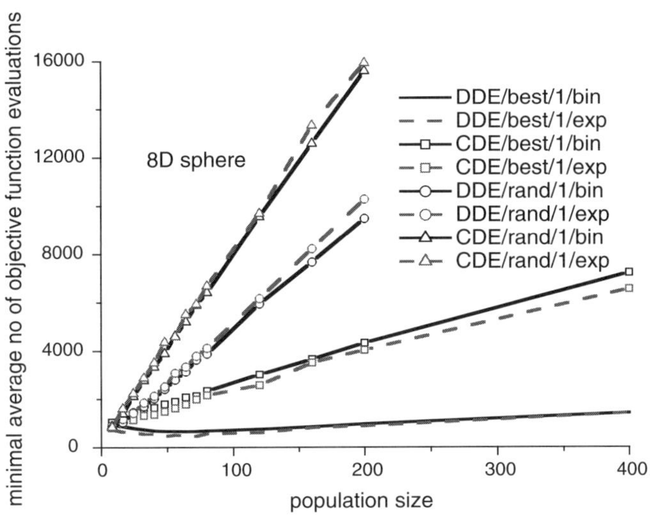

Figure 6.2 Efficiency of differential evolution for the 8-dimensional sphere function

6.1.1.3 Analysis

Balancing population diversity is the fundamental mechanism behind all these observations. Qualitatively, classic differential evolution and best differential mutation base limit population diversity, while dynamic differential evolution and random base result in more diverse population. In general, a larger population is more diverse. The role of crossover is a little more complicated. However, for this case, the optimal crossover probability, as shown later in Figures 7.1–7.4, is very close to 1. In this situation, the binomial crossover leads to a less diverse population than the exponential crossover does.

At smaller population size, dynamic differential evolution, random base, and exponential crossover are more favorable since they can increase population diversity. As population size grows, the population diversity may be over-compensated by dynamic differential evolution, random differential mutation base, and exponential crossover. Differential evolution performs best when these forces are in harmony.

6.1.1.4 Recommendations

Apparently, for this case, DDE/best/1/exp is the best strategy in terms of both efficiency and robustness.

It is mentioned in Chapter 3 that dynamic differential evolution is not parallelizable. From the point of view of parallelization, classic differential evolution is also of interest. In terms of efficiency, CDE/best/1/exp outperforms the other three classic differential evolution strategies investigated, as shown in Figure 6.2. In contrast, in terms of robustness, CDE/rand/1/exp exhibits the largest highest number of successful trials. However, CDE/best/1/exp seems also acceptable since the bell is much wider but only slightly shorter.

6.1.1.5 Misconceptions on Differential Evolution Revisited

Let us now revisit the misconceptions on differential evolution mentioned in Chapter 1.

A Do dynamic differential evolution and classic differential evolution perform equally well?
Absolutely not. Dynamic differential evolution always outperforms classic differential evolution in terms of both efficiency and robustness. This is understandable. Its faster response to population status change makes dynamic differential evolution more efficient. Virtually doubling population size provides more population diversity and consequently makes dynamic differential evolution more robust.

B Is differential evolution with random base more robust?
No. From Figure 6.1, it can be seen that DDE/best/1/* is almost always more robust than its counterpart DDE/rand/1/*. However, below a certain population size, CDE/rand/1/* is more robust than CDE/best/1/*. The highest number of successful trials of CDE/rand/1/* is also larger than that of CDE/best/1/*. This therefore does provide limited support to this claim.
Increasing population size was recommended in Chapters 2 and 3 in order to assure success. However, differential evolution will not benefit from such a recommendation unless the original population size is below the threshold. Even worse, Figure 6.2 shows that increasing population size makes differential evolution less efficient.

C Is crossover unimportant for differential evolution?
No. Crossover is very important for differential evolution. Figure 6.2 shows that the efficiency of DE/*/1/bin differs very little from that of DE/*/1/exp. However, in general, DE/*/1/exp is seen much more robust than DE/*/1/bin, as demonstrated in Figure 6.1.

D Is binomial crossover never worse than exponential?
No. DE/*/1/exp is seen to be more robust than DE/*/1/bin. In terms of efficiency, DE/best/1/exp is more efficient than DE/best/1/bin.

According to the aforementioned analysis, the above misconceptions are even more ridiculous from the point of view of population diversity. The first misconception does not distinguish between dynamic differential evolution and classic differential evolution. The third misconception on crossover does not distinguish binomial and exponential crossover. These two misconceptions imply that differences in population diversity originated from evolution mechanism and that crossover does not affect differential evolution. The second misconception favors random base instead of best base. This implies that a diverse population is better. However, the third misconception implies that less diverse population is more favorable.

6.1.2 16-Dimensional Sphere Function

Henceforth, observations similar to the foregoing will not be repeated.

6.1.2.1 Robustness

The number of successful trials of differential evolution is shown in Figure 6.3. From Figure 6.3, we note the following:

(a) All bell-shaped curves are shifted to the left.
(b) All bell-shaped curves are reduced in height. The reduction in height is sharpest for DE/*/1/bin.

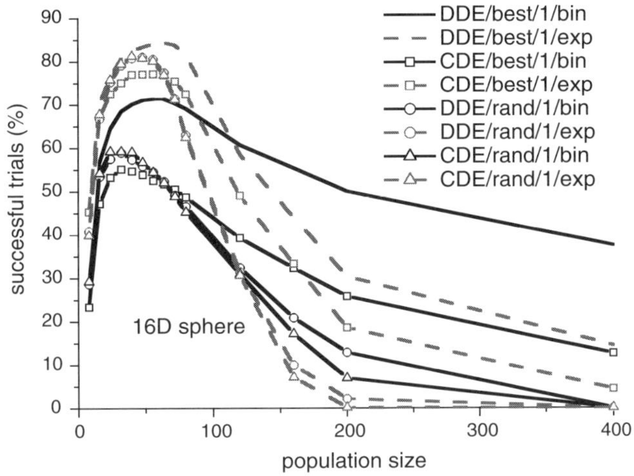

Figure 6.3 Robustness of differential evolution for the 16-dimensional sphere function

(c) Bells of DE/*/1/exp are getting narrower while bells of DE/*/1/bin are getting wider (compare Figure 6.1).

6.1.2.2 Efficiency

The minimal average number of objective function evaluations is shown in Figure 6.4. In general, Figure 6.4 is similar in appearance to Figure 6.2. However, it should be pointed out that DE/*/1/exp is now almost always less efficient than DE/*/1/bin.

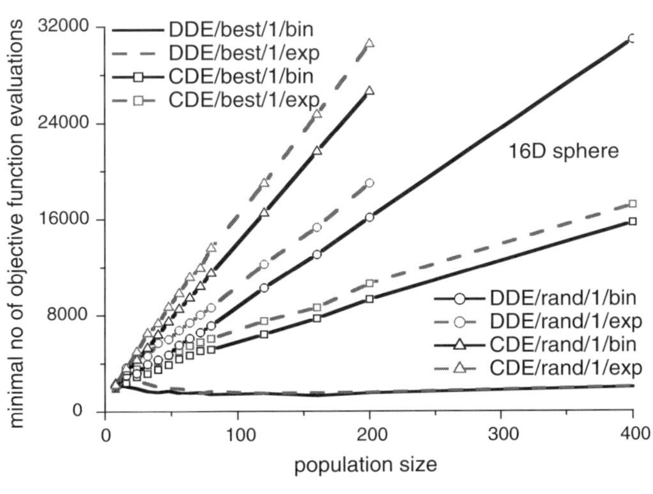

Figure 6.4 Efficiency of differential evolution for the 16-dimensional sphere function

6.1.2.3 Analysis

The difference in population diversity arising from different evolution mechanism and differential mutation base is independent of problem dimension. However, that arising from different crossover strongly depends on problem dimension. As problem dimension increases, exponential crossover makes the population more diverse. Therefore, additional diversity from a larger population may be potentially harmful.

6.1.2.4 Recommendations

In terms of efficiency, DDE/best/1/bin is slightly better than DDE/best/1/exp. However, DDE/best/1/exp appears to be much more robust than DDE/best/1/bin. DDE/best/1/exp is regarded as the best strategy because success is our first priority.

Determining the best classic differential evolution becomes much more difficult. None of the four investigated strategies of classic differential evolution dominates in terms of both efficiency and robustness. CDE/best/1/bin is the most efficient classic differential evolution, while CDE/rand/1/exp is the most robust one.

6.1.3 24-Dimensional Sphere Function

6.1.3.1 Observations

The performance of differential evolution for the 24-dimensional sphere function is shown in Figures 6.5 and 6.6:

(a) The bell-shaped curves are shifted further to the left and getting shorter.
(b) The height difference between DE/*/1/exp and DE/*/1/bin is growing.
(c) DDE/rand/1/exp is now slightly less efficient than CDE/rand/1/bin.

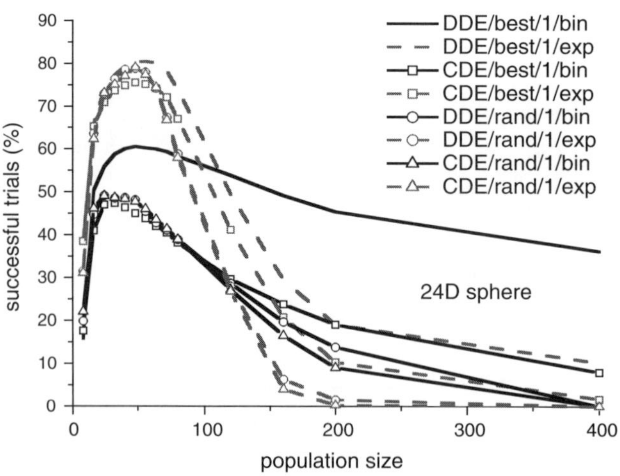

Figure 6.5 Robustness of differential evolution for the 24-dimensional sphere function

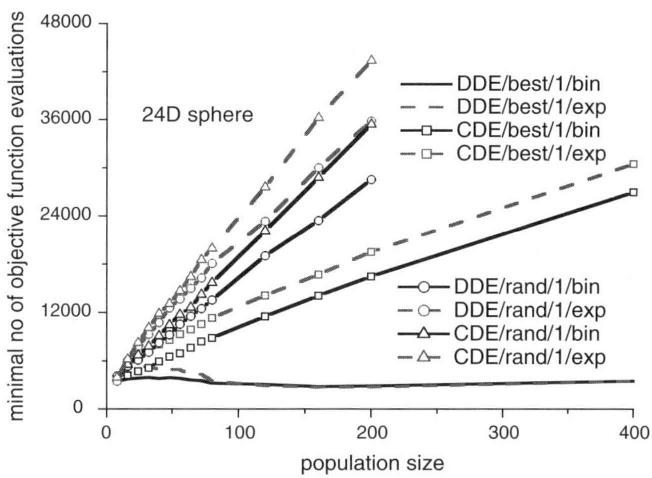

Figure 6.6 Efficiency of differential evolution for the 24-dimensional sphere function

6.1.3.2 Analysis

Similarly, problem dimension is the cause behind all the observed differences. Exponential crossover results in more and more population diversity as problem dimension increases. This makes differential evolution more robust but less efficient.

The above observations also imply that crossover is playing a more and more crucial role. Its effect even exceeds the effect of evolution mechanism and differential mutation base.

6.1.3.3 Recommendations

In terms of efficiency, DDE/best/1/bin is slightly better than DDE/best/1/exp. However, DDE/best/1/exp is apparently much more robust than DDE/best/1/bin. DDE/best/1/exp is therefore regarded as the best strategy again.

It seems that CDE/best/1/exp provides a good tradeoff between efficiency and robustness although it is dominated by CDE/best/1/bin in terms of efficiency while outperformed by CDE/rand/1/exp in terms of robustness.

6.1.4 50-Dimensional Sphere Function

6.1.4.1 Observations

The performance of differential evolution for the 50-dimensional sphere function is shown in Figures 6.7 and 6.8.

It is noted from Figure 6.7 that the eight strategies form three clusters. The four DE/*/1/exp strategies form the first cluster. CDE/best/1/bin, DDE/rand/1/bin, and CDE/rand/1/bin are in the second cluster, while DDE/best/1/bin is the only member of the last cluster.

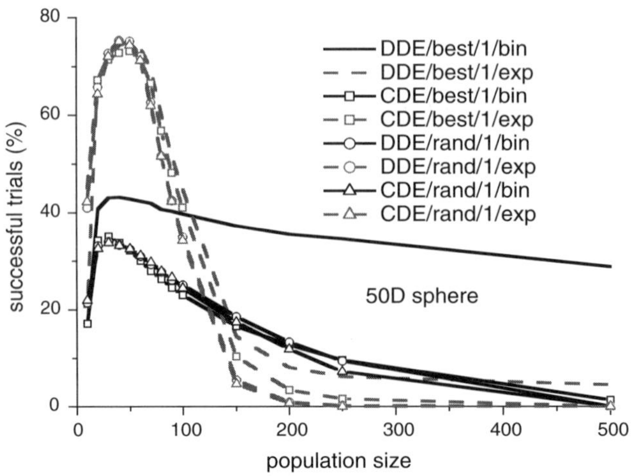

Figure 6.7 Robustness of differential evolution for the 50-dimensional sphere function

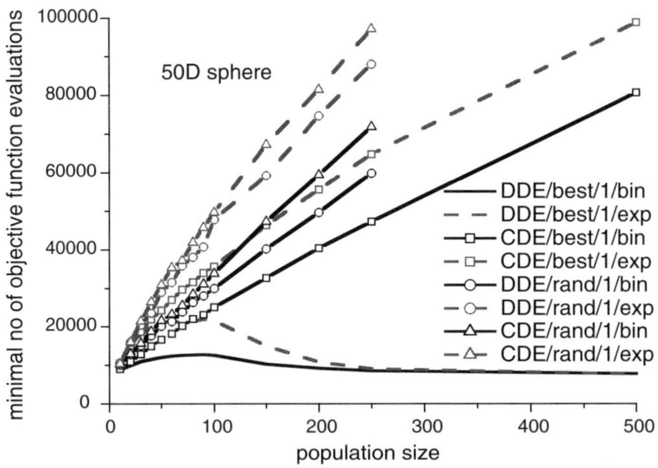

Figure 6.8 Efficiency of differential evolution for the 50-dimensional sphere function

It is interesting to note from Figure 6.8 that CDE/best/1/exp is not more efficient than DDE/rand/1/bin any more. It is also interesting to note that the robustness and efficiency of DDE/best/1/bin do not change much when the population size is not too small.

6.1.4.2 Recommendations

It is now necessary to seek a tradeoff between efficiency and robustness when choosing dynamic differential evolution. In terms of efficiency, DDE/best/1/bin is better than

DDE/best/1/exp. However, DDE/best/1/exp is seen to be much more robust than DDE/best/1/bin. The flatness of curves in Figures 6.7 and 6.8 corresponding to DDE/best/1/bin makes the choice more difficult.

6.1.5 100-Dimensional Sphere Function

6.1.5.1 Observations

The performance of differential evolution for the 100-dimensional sphere function is shown in Figures 6.9 and 6.10. Clustering is now even more evident.

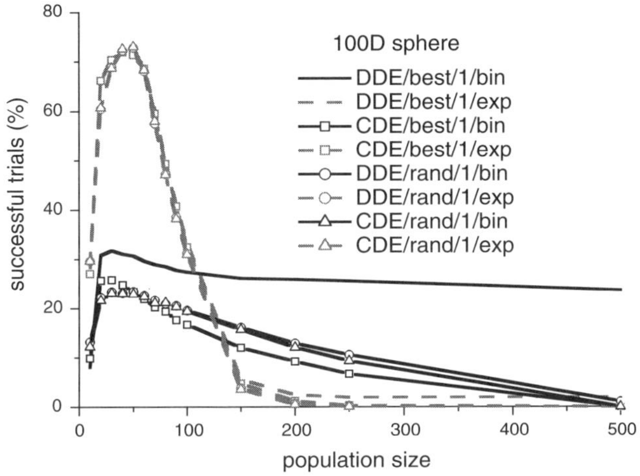

Figure 6.9 Robustness of differential evolution for the 100-dimensional sphere function

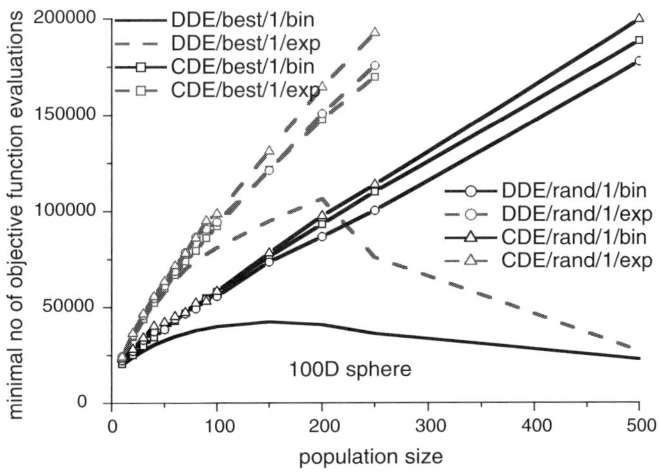

Figure 6.10 Efficiency of differential evolution for the 100-dimensional sphere function

Although DDE/best/1/exp is still more efficient than the other three strategies of differential evolution with exponential crossover, DE/*/1/exp is in general less efficient than DE/*/1/bin. In contrast, DE/*/1/exp is much more robust than DE/*/1/bin.

Inspection of the corresponding optimal crossover probability tells us that population diversity is still the cause behind the odd behavior of DDE/best/1/exp when population size is very large. The optimal crossover probability for DDE/best/1/exp is exactly 1 when population size is 250 or 500 while that for DDE/best/1/bin is slightly smaller than 1.

6.1.5.2 Recommendations

Users now have to think about the conflict between efficiency and robustness more seriously in choosing the most appropriate dynamic differential evolution strategy. It seems that DDE/best/1/bin has gained more and more support.

As to classic differential evolution, it seems that differential mutation base is now not very important. Crossover becomes the dominant factor for both efficiency and robustness.

6.1.6 Effect of Dimension

Apparently, problem dimension affects differential evolution through crossover. Higher dimension makes it more difficult for a mutant to donate optimization parameters to a child through exponential crossover. It makes the population more diverse. Such diversity is in general helpful for robustness. However, it may result in loss of efficiency.

6.1.6.1 Robustness

The effect of dimension on robustness of differential evolution for sphere function is shown in Figure 6.11:

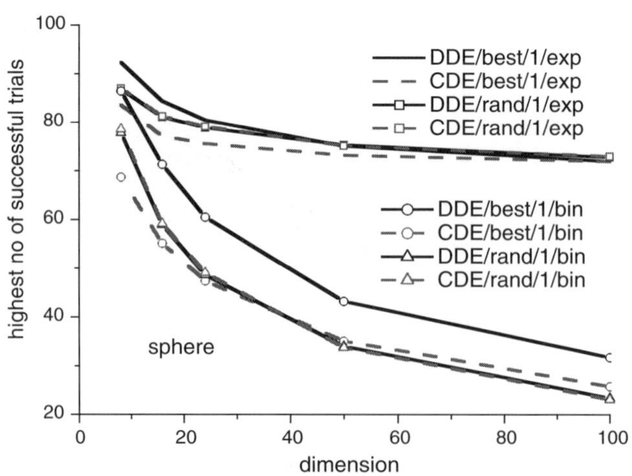

Figure 6.11 Effect of dimension on robustness of differential evolution for the sphere function

(a) When dimension is not very high, dynamic evolution makes differential evolution with best base more robust. Evolution mechanism plays a weaker and weaker role as dimension increases.
(b) The effect of evolution mechanism on robustness of DE/rand/1/* appears to be negligible.
(c) DDE/best/1/* is more robust than DDE/rand/1/*. However, best base makes classic differential evolution less robust.
(d) In general, DE/*/1/exp is more robust than DE/*/1/bin.
(e) DDE/best/1/bin is seen to be more robust than the other three differential evolution strategies with binomial crossover.
(f) CDE/best/1/bin becomes more robust than DE/rand/1/bin when dimension is high.

6.1.6.2 Efficiency

The effect of dimension on efficiency of differential evolution for the sphere function is shown in Figure 6.12. DDE/best/1/bin is always more efficient within the problem dimension range investigated. It seems that the minimal average number of objective function evaluations grows exponentially with problem dimension.

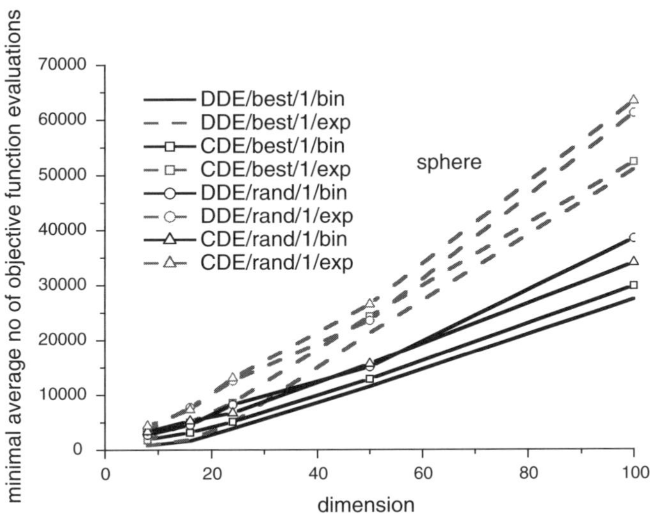

Figure 6.12 Effect of dimension on efficiency of differential evolution for the sphere function

It must be pointed out that the minimal average number of objective function evaluations shown in Figure 6.12 is not the lowest average number of objective function evaluations of all successful trials.

6.1.7 General Recommendations

According to Figures 6.1-6.12, DDE/best/1/exp seems to be a good choice. DDE/best/1/bin is a good alternative for users who are more concerned about efficiency. DDE/rand/1/bin and DDE/

rand/1/exp are not recommended for problems having features the same as or similar to those of the sphere function.

For users who are keen to parallelize differential evolution, a tradeoff between efficiency and robustness is always necessary. Problem dimension is also an important issue to account for. None of the four classic differential evolution strategies is found to be dominant.

6.2 Step Function 2

Step function 2 is designed for studying the effect of continuity.

6.2.1 8-Dimensional Step Function 2

6.2.1.1 Robustness

The number of successful trials of differential evolution is shown in Figure 6.13. A comparison with Figure 6.1 shows that the bell-shaped curves in Figure 6.13 are taller, fatter and further to the right. It is observed that DDE/best/1/* is less robust than DDE/rand/1/* below certain population size. The population size threshold in Figure 6.16 is much higher.

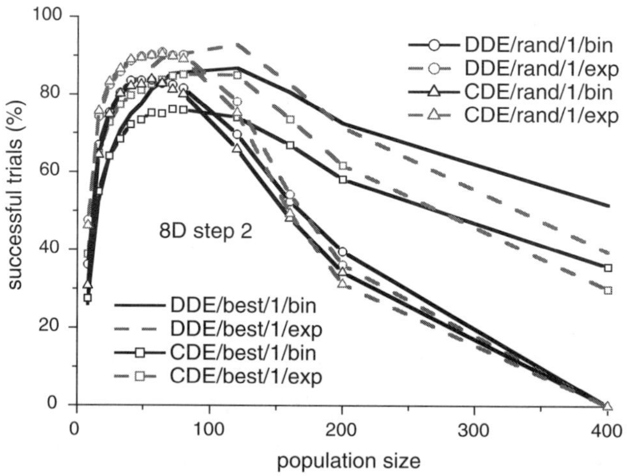

Figure 6.13 Robustness of differential evolution for the 8-dimensional step function 2

6.2.1.2 Efficiency

The minimal average number of objective function evaluations for the 8-dimensional step function 2 is shown in Figure 6.14. No significant difference compared to the 8-dimensional sphere function is observed.

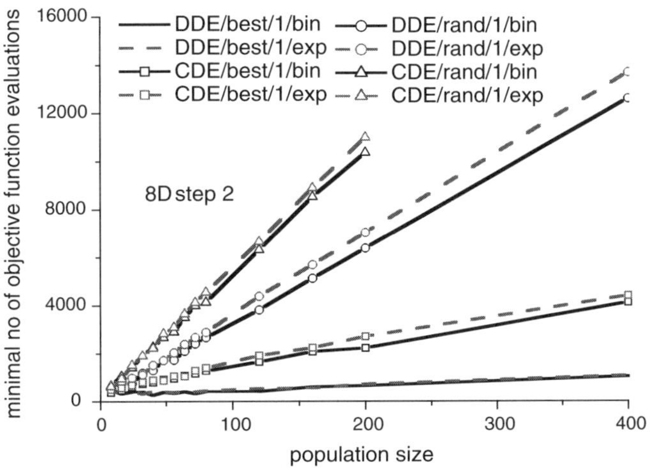

Figure 6.14 Efficiency of differential evolution for the 8-dimensional step function 2

6.2.1.3 Analysis

The discontinuity of step function 2 introduces further loss of population diversity. Such loss has to be compensated by dynamic evolution mechanism, random differential mutation base, exponential crossover, and larger population size.

6.2.1.4 Recommendations

DDE/best/1/exp is still the best differential evolution. Among the four classic differential evolution strategies, CDE/rand/1/exp is the most robust, while CDE/best/1/bin is the most efficient.

6.2.2 16-Dimensional Step Function 2

6.2.2.1 Robustness

The number of successful trials is shown in Figure 6.15. DE/rand/1/exp is seen to be more robust. It seems that robustness is now becoming insensitive to evolution mechanism.

6.2.2.2 Efficiency

The minimal average number of objective function evaluations for the 16-dimensional step function 2 is shown in Figure 6.16. Again, no significant difference compared to the 16-dimensional sphere function is observed.

6.2.2.3 Analysis

Loss of population diversity due to discontinuity of step function 2 now appears to be more serious.

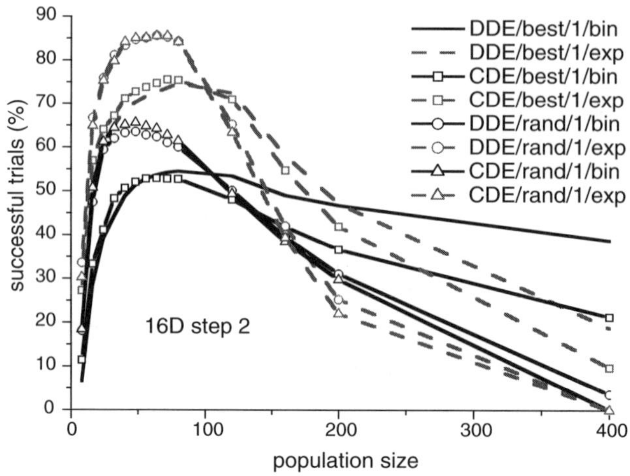

Figure 6.15 Robustness of differential evolution for the 16-dimensional step function 2

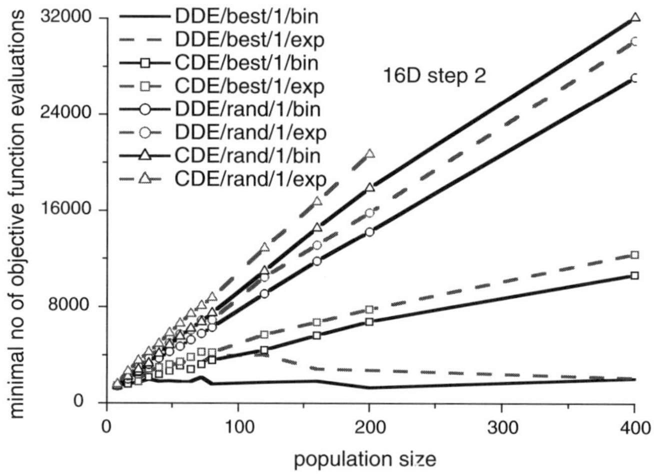

Figure 6.16 Efficiency of differential evolution for the 16-dimensional step function 2

6.2.2.4 Recommendations

It is now necessary to seek a tradeoff between efficiency and robustness even for dynamic differential evolution. DE/rand/1/exp is more robust, while DE/best/1/bin is more efficient.

6.2.3 24-Dimensional Step Function 2

6.2.3.1 Observations

The performance of differential evolution for the 24-dimensional step function 2 is shown in Figures 6.17 and 6.18. The trends mentioned in Sections 6.1.3.1, 6.2.2.1 and 6.2.2.2 are also observed.

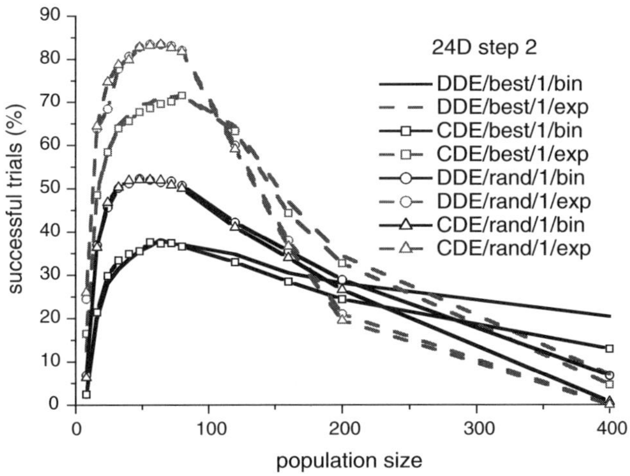

Figure 6.17 Robustness of differential evolution for the 24-dimensional step function 2

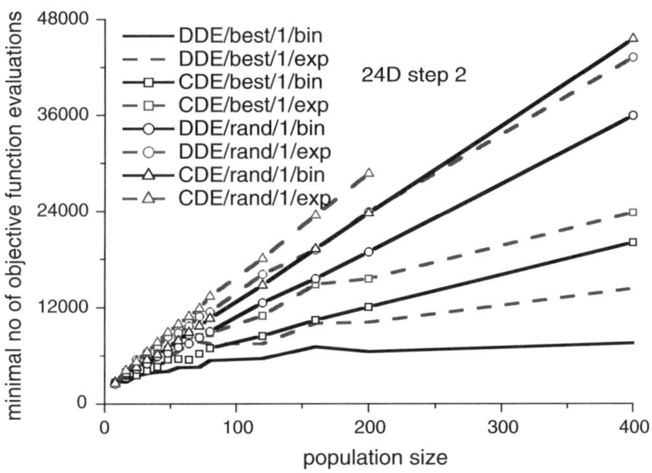

Figure 6.18 Efficiency of differential evolution for the 24-dimensional step function 2

6.2.3.2 Recommendations

The conflict between efficiency and robustness becomes much more serious. The user faces a tough choice between efficiency and robustness.

6.2.4 50-Dimensional Step Function 2

6.2.4.1 Observations

The performance of differential evolution for the 50-dimensional step function 2 is shown in Figures 6.19 and 6.20.

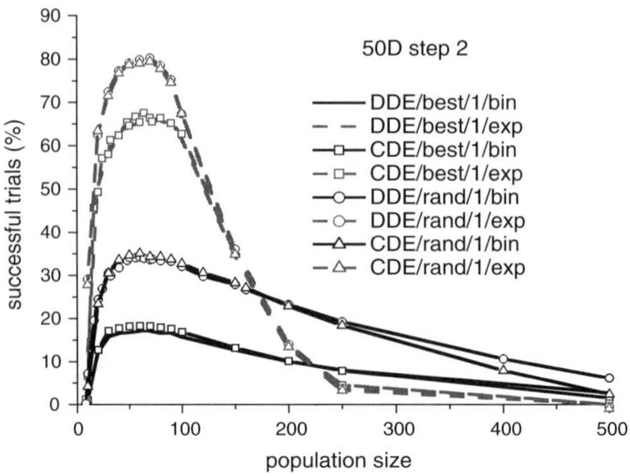

Figure 6.19 Robustness of differential evolution for the 50-dimensional step function 2

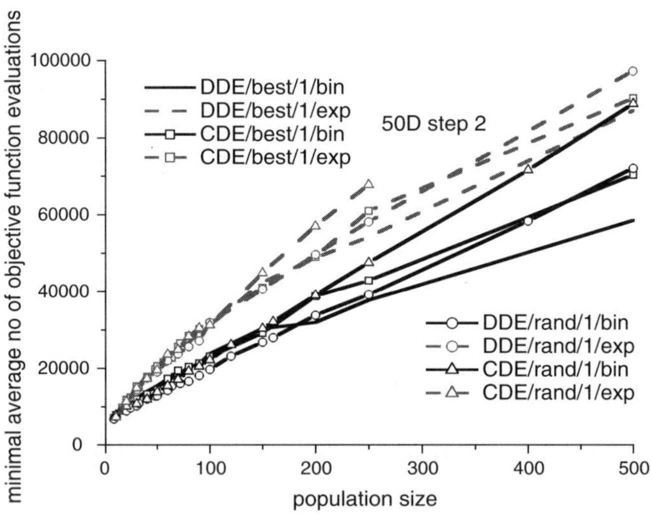

Figure 6.20 Efficiency of differential evolution for the 50-dimensional step function 2

It is apparent that DE/*/1/exp is much more robust than DE/*/1/bin. It is also seen that best base is harmful to robustness.

DE/*/1/exp is much less efficient than DE/*/1/bin. In addition, the minimal average number of objective function evaluations of DDE/best/1/bin is not flat any more. It grows almost linearly with population size at a notable rate, although the growth rate is still the smallest among all differential evolution strategies investigated.

6.2.4.2 Recommendations

Dynamic differential evolution is seen to be slightly better than classic differential evolution. However, users have to play the tradeoff game very seriously to determine base and crossover.

6.2.5 *Effect of Dimension*

6.2.5.1 Robustness

The effect of dimension on robustness of differential evolution for step function 2 is shown in Figure 6.21. Obviously, robustness is insensitive to evolution mechanism although dynamic differential evolution looks favorable at small population size. Random differential mutation base and exponential crossover make differential evolution more robust.

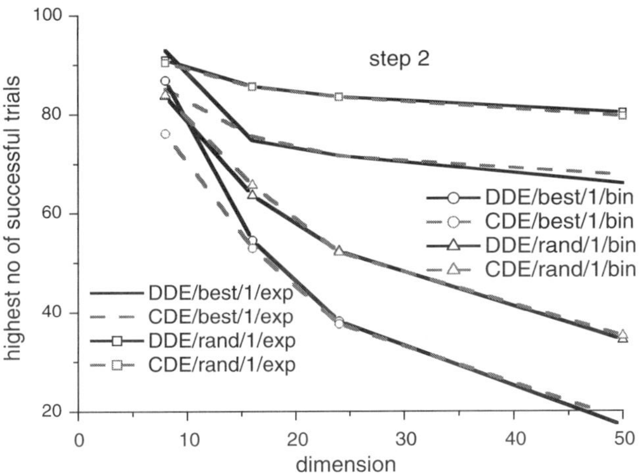

Figure 6.21 Effect of dimension on robustness of differential evolution for step function 2

6.2.5.2 Efficiency

The effect of dimension on efficiency of differential evolution for step function 2 is shown in Figure 6.22. At first sight, the results seem strange as problem dimension increases. DDE/best/1/bin and CDE/best/1/bin are less efficient than DDE/rand/1/bin and CDE/rand/1/bin. However, as pointed out in Section 6.1.6.2, the minimal average number of objective function evaluations of all successful trials at population size when differential evolution behaves most robustly is not the lowest average number of objective functions of all successful trials. Such an observation clearly shows the conflict between efficiency and robustness.

6.2.6 *Effect of Discontinuity of Objective Function*

Objective function discontinuity negatively affects differential evolution in two ways.

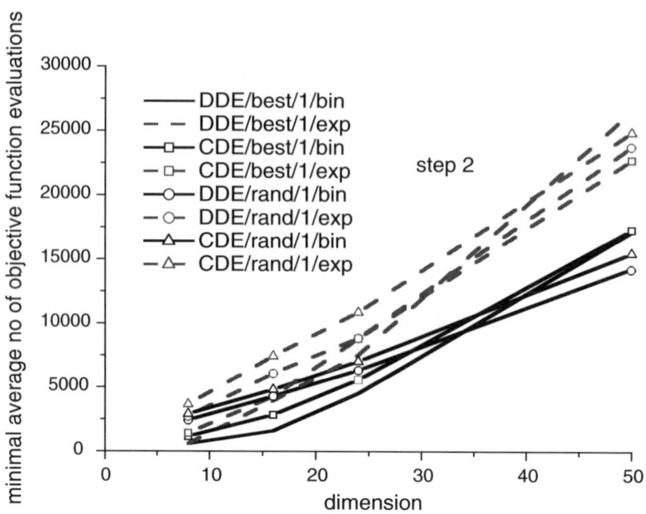

Figure 6.22 Effect of dimension on efficiency of differential evolution for step function 2

(a) *Reduction of population diversity*. It is understandable that population is more compact because two individuals with completely differential optimization parameters may have identical objective function value
(b) *Trapping*. The landscape of step function 2 looks like a basin with many staircases. Guidance is lost within such staircases. Children closer to the solution may be falsely rejected.

6.3 Hyper-ellipsoid Function

The hyper-ellipsoid function is implemented to study the effect of objective function symmetry.

6.3.1 8-Dimensional Hyper-ellipsoid Function

The performance of differential evolution for the 8-dimensional hyper-ellipsoid function is shown in Figures 6.23 and 6.24. The robustness of differential evolution seems intact. However, the growth rates in Figure 6.24 are considerably higher than those in Figure 6.1.

6.3.2 16-Dimensional Hyper-ellipsoid Function

The performance of differential evolution for the 16-dimensional hyper-ellipsoid function is shown in Figures 6.25 and 6.26. A loss of efficiency as well as a slight loss of robustness are observed.

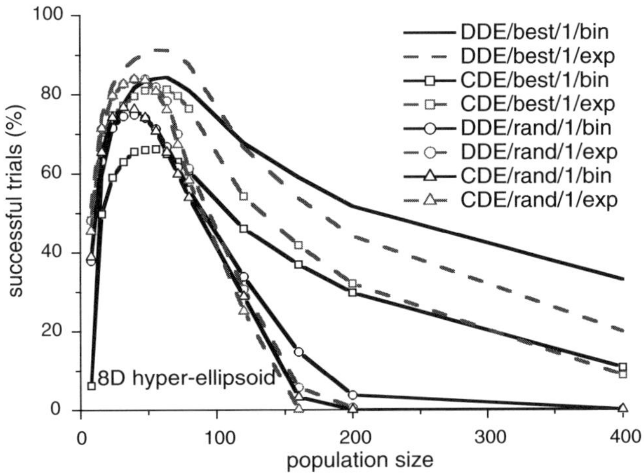

Figure 6.23 Robustness of differential evolution for the 8-dimensional hyper-ellipsoid function

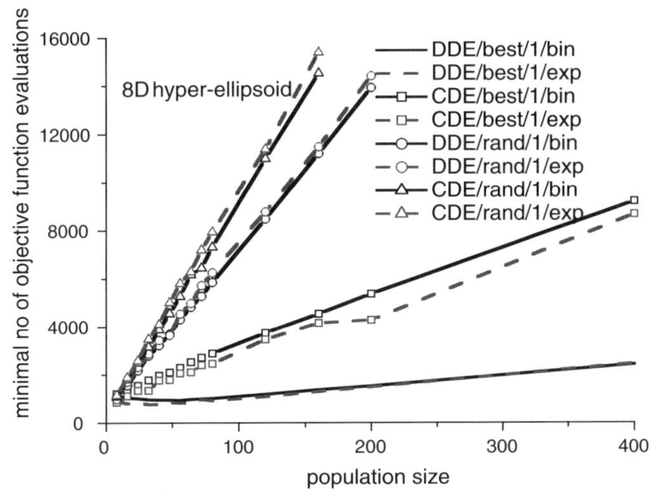

Figure 6.24 Efficiency of differential evolution for the 8-dimensional hyper-ellipsoid function

6.3.3 24-Dimensional Hyper-ellipsoid Function

The performance of differential evolution for the 24-dimensional hyper-ellipsoid function is shown in Figures 6.27 and 6.28. Efficiency and robustness continue to fall.

6.3.4 50-Dimensional Hyper-ellipsoid Function

The performance of differential evolution for the 50-dimensional hyper-ellipsoid function is shown in Figures 6.29 and 6.30.

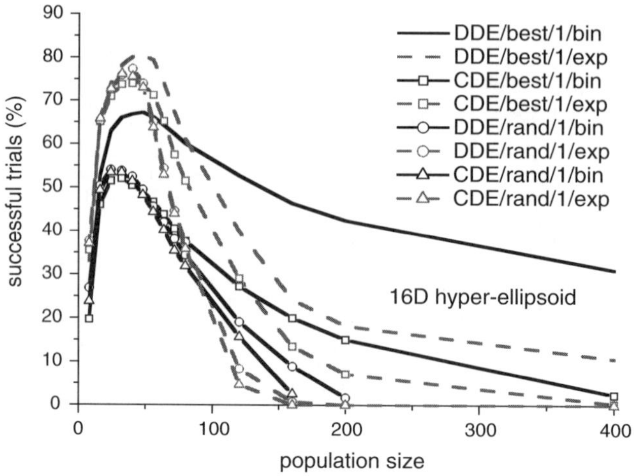

Figure 6.25 Robustness of differential evolution for the 16-dimensional hyper-ellipsoid function

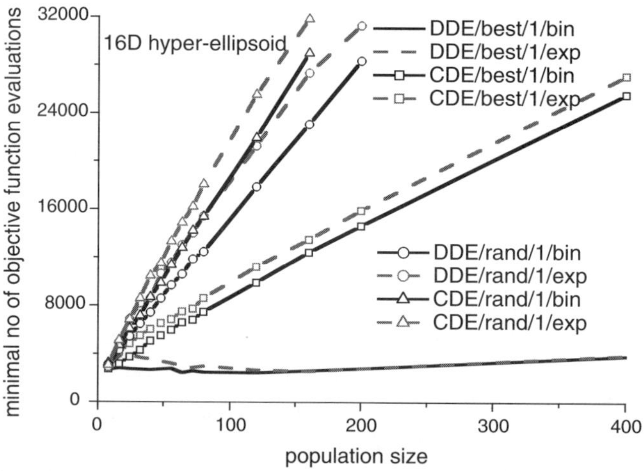

Figure 6.26 Efficiency of differential evolution for the 16-dimensional hyper-ellipsoid function

6.3.5 Effect of Dimension

6.3.5.1 Robustness

The effect of dimension on robustness of differential evolution for the hyper-ellipsoid function is shown in Figure 6.31. It is interesting to note that DE/rand/1/exp is less robust that DE/best/exp when the problem dimension is not very small.

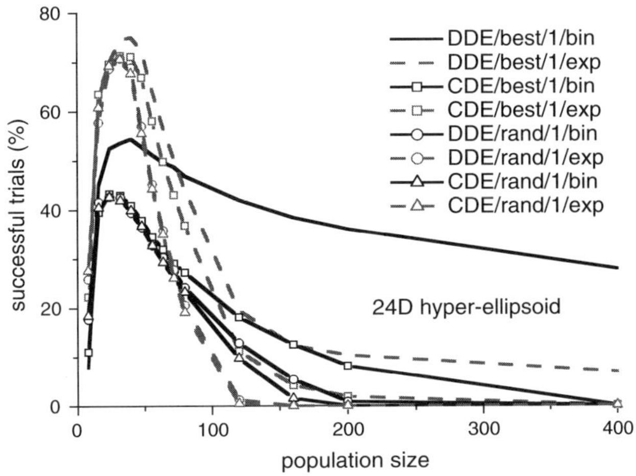

Figure 6.27 Robustness of differential evolution for the 24-dimensional hyper-ellipsoid function

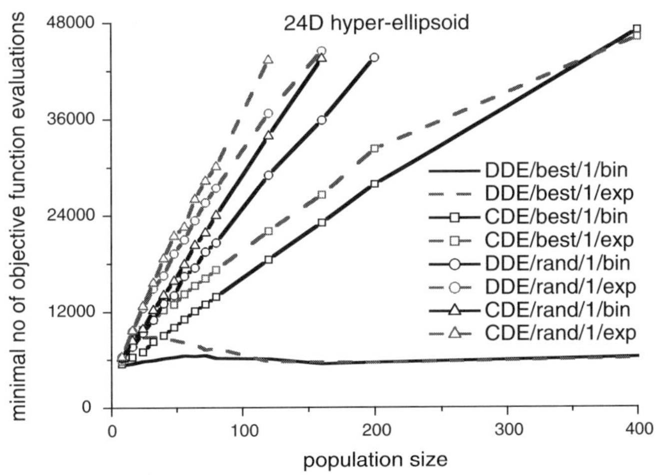

Figure 6.28 Efficiency of differential evolution for the 24-dimensional hyper-ellipsoid function

6.3.5.2 Efficiency

The effect of dimension on efficiency of differential evolution for the hyper-ellipsoid function is shown in Figure 6.32. Apart from loss of efficiency, there is no notable difference between Figures 6.12 and 6.32.

6.4 Qing Function

The Qing function is implemented to study the effect of non-uniqueness of solution.

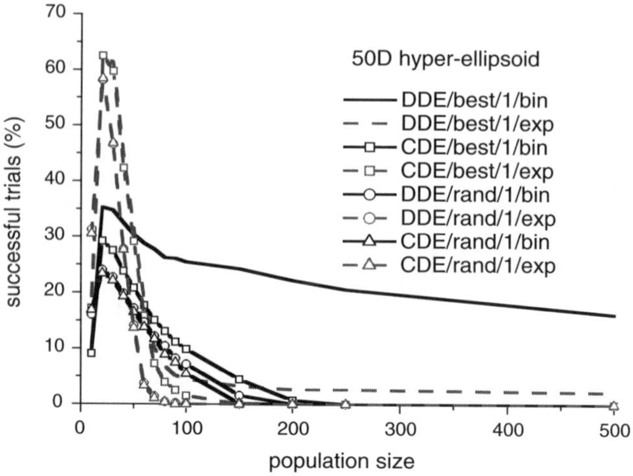

Figure 6.29 Robustness of differential evolution for the 50-dimensional hyper-ellipsoid function

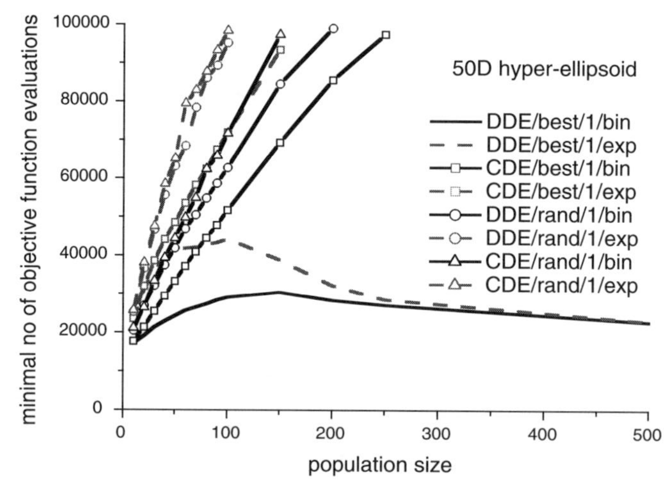

Figure 6.30 Efficiency of differential evolution for the 50-dimensional hyper-ellipsoid function

6.4.1 8-Dimensional Qing Function

6.4.1.1 Robustness

The number of successful trials for the 8-dimensional Qing function is shown in Figure 6.33. The bell-shaped curves are now shorter, narrower and moving toward the left. Those of DE/rand/1/* are now very narrow. It is also interesting to note that DE/rand/1/* is very insensitive to evolution mechanism.

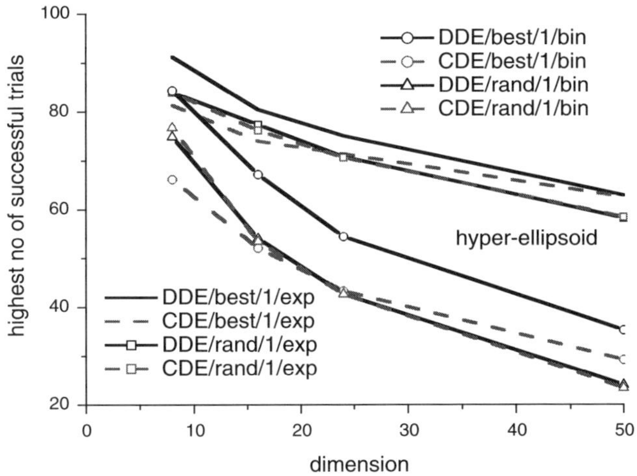

Figure 6.31 Effect of dimension on robustness of differential evolution for the hyper-ellipsoid function

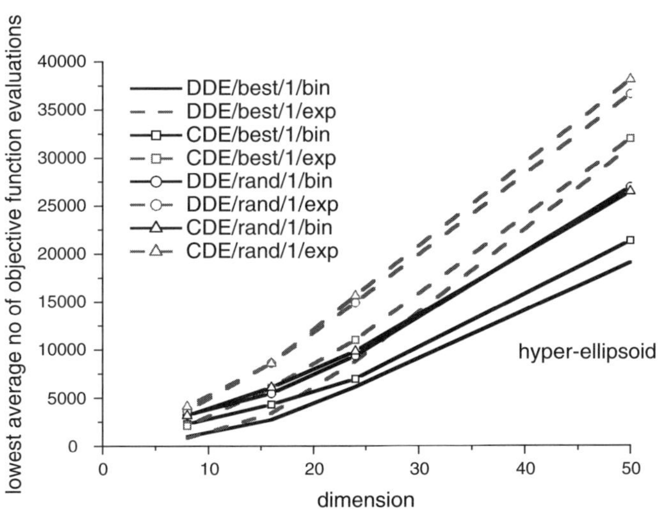

Figure 6.32 Effect of dimension on efficiency of differential evolution for the hyper-ellipsoid function

6.4.1.2 Efficiency

The minimal average number of objective function evaluations for the 8-dimensional Qing function is shown in Figure 6.34. The odd improvement of efficiency of DDE/rand/bin at population size 200 is misleading since no trial is fully successful. The efficiency of DE/best/1/* remains almost unchanged. However, DE/rand/1/* is much less efficient. It is also noticed that the DE/rand/1/* strategies are now indistinguishable from each other.

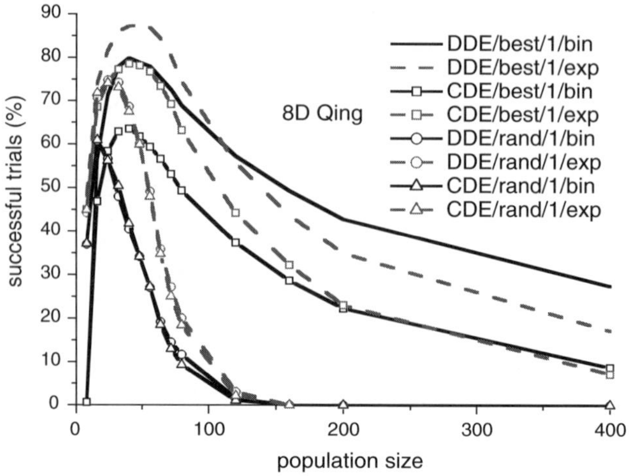

Figure 6.33 Robustness of differential evolution for the 8-dimensional Qing function

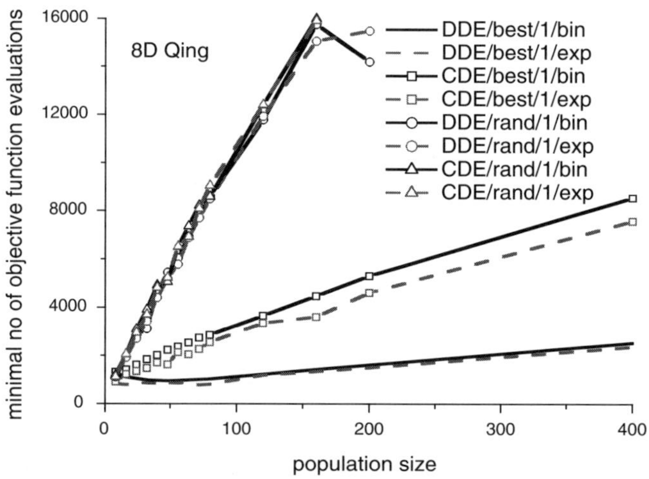

Figure 6.34 Efficiency of differential evolution for the 8-dimensional Qing function

6.4.1.3 Analysis

The Qing function solutions divide the search space into subdomains. Each subdomain contains one solution. DDE/best/1/* provides the best local search ability.

6.4.1.4 Recommendations

In terms of both robustness and efficiency, DDE/best/1/exp is obviously the best strategy, while CDE/best/1/exp is the best classic differential evolution strategy.

6.4.2 16-Dimensional Qing Function

6.4.2.1 Robustness

The number of successful trials for the 16-dimensional Qing function is shown in Figure 6.35. The bell-shaped curves are quickly losing height and width and are moving further to the left. Classic differential evolution is catching up with dynamic differential evolution.

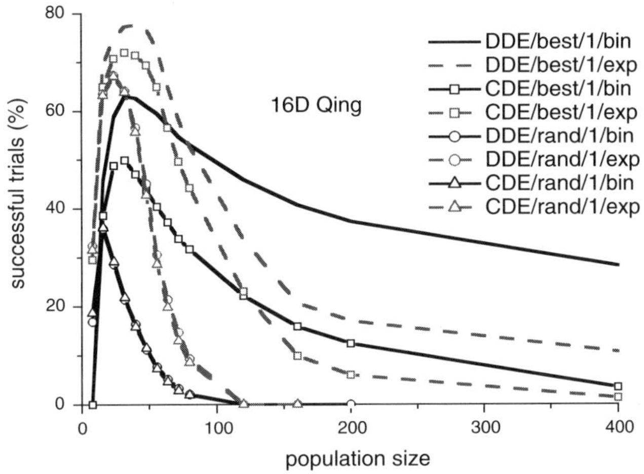

Figure 6.35 Robustness of differential evolution for the 16-dimensional Qing function

6.4.2.2 Efficiency

The minimal average number of objective function evaluations for the 16-dimensional Qing function is shown in Figure 6.36. No significant difference is observed.

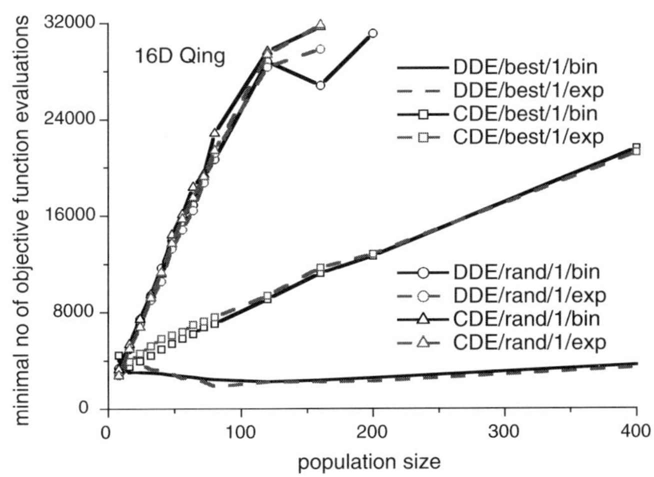

Figure 6.36 Efficiency of differential evolution for the 16-dimensional Qing function

6.4.3 24-Dimensional Qing Function

The performance of differential evolution for the 24-dimensional Qing function is shown in Figures 6.37 and 6.38. The aforementioned general trend remains.

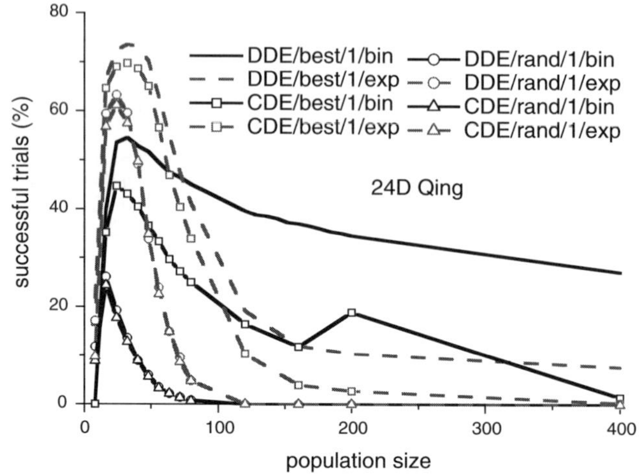

Figure 6.37 Robustness of differential evolution for the 24-dimensional Qing function

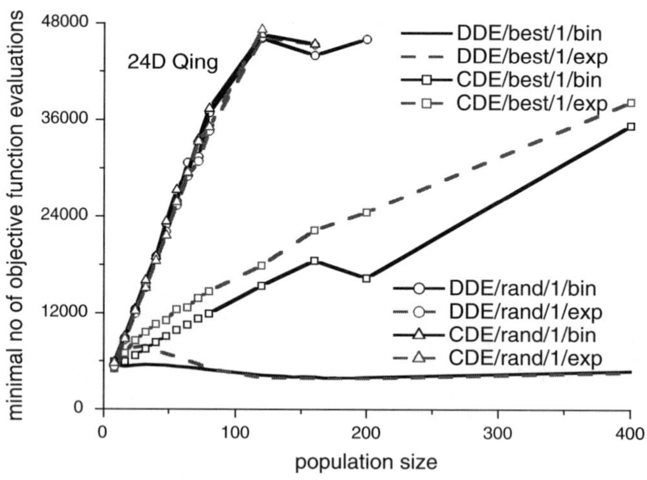

Figure 6.38 Efficiency of differential evolution for the 24-dimensional Qing function

6.4.4 50-Dimensional Qing Function

The performance of differential evolution for the 50-dimensional Qing function is shown in Figures 6.39 and 6.40. DDE/best/1/exp is now almost indistinguishable from CDE/best/1/exp in terms of robustness.

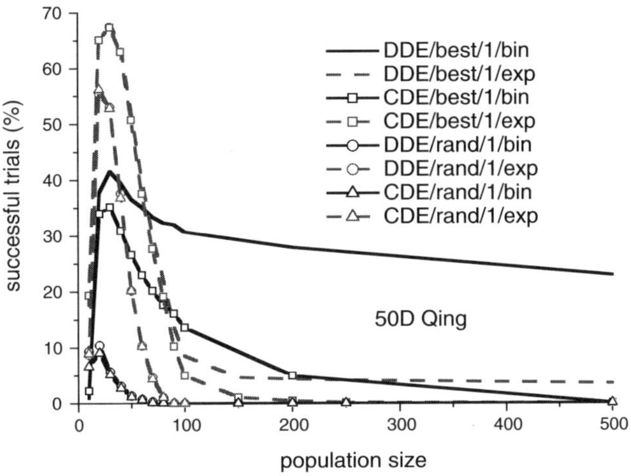

Figure 6.39 Robustness of differential evolution for the 50-dimensional Qing function

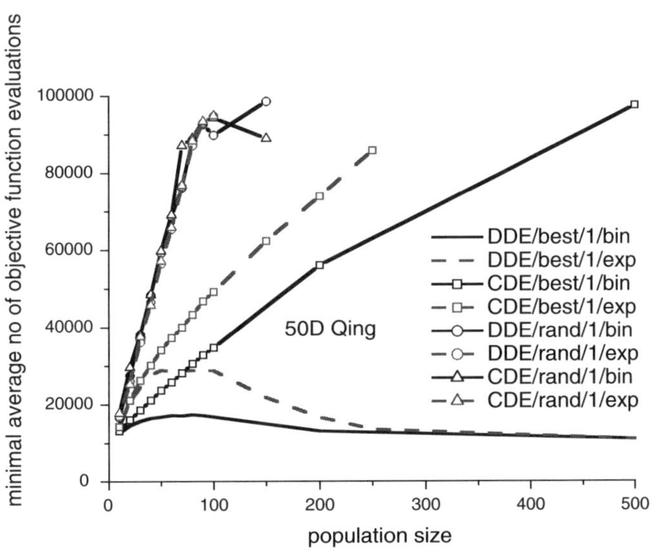

Figure 6.40 Efficiency of differential evolution for the 50-dimensional Qing function

6.4.5 Effect of Dimension

6.4.5.1 Robustness

The effect of dimension on robustness of differential evolution for the Qing function is shown in Figure 6.41. Within the problem dimension range investigated, DDE/best/1/exp is the most robust differential evolution strategy. CDE/best/1/exp is approaching DDE/best/1/exp as problem dimension increases.

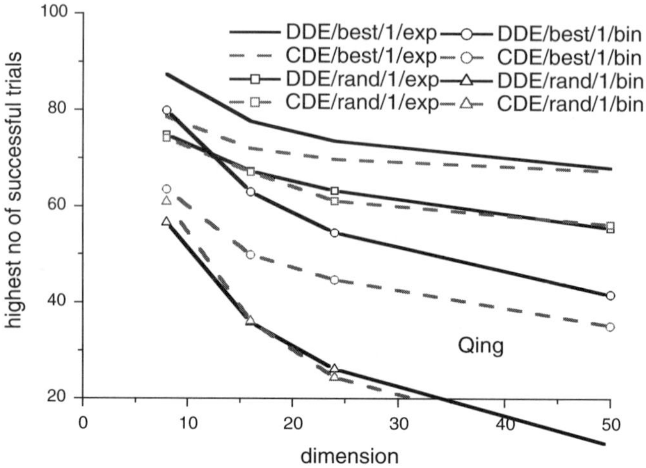

Figure 6.41 Effect of dimension on robustness of differential evolution for the Qing function

6.4.5.2 Efficiency

The effect of dimension on efficiency of differential evolution for the Qing function is shown in Figure 6.42. DE/rand/1/bin is now not always more efficient than DE/*/1/exp.

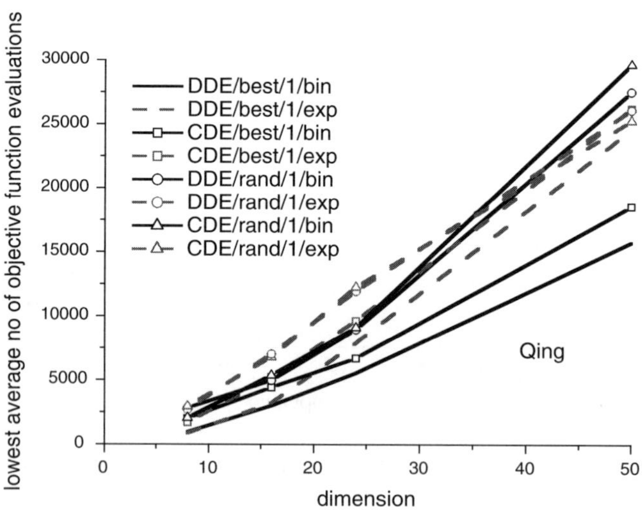

Figure 6.42 Effect of dimension on efficiency of differential evolution for the Qing function

6.5 Schwefel Function 2.22

The Schwefel function 2.22 is good for studying the effect of differentiability.

6.5.1 8-Dimensional Schwefel Function 2.22

6.5.1.1 Robustness

The number of successful trials for the 8-dimensional Schwefel function 2.22 is shown in Figure 6.43. DE/rand/1/* is catching up with DE/best/1/*.

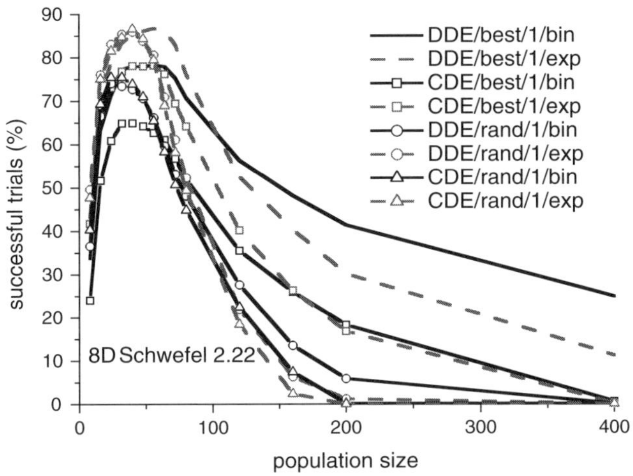

Figure 6.43 Robustness of differential evolution for the 8-dimensional Schwefel function 2.22

6.5.1.2 Efficiency

The minimal average number of objective function evaluations for the 8-dimensional Schwefel function 2.22 is shown in Figure 6.44. DE/*/1/bin is more efficient than DE/*/1/exp. Note also

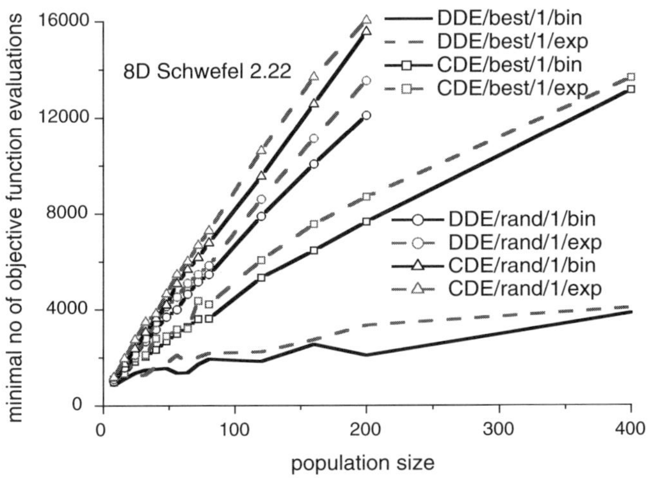

Figure 6.44 Efficiency of differential evolution for the 8-dimensional Schwefel function 2.22

that the minimal average number of objective function evaluations of DE/best/1/* grows almost linearly with population size at a notable rate, although the growth rate is still the smallest among all differential evolution strategies investigated.

6.5.2 16-Dimensional Schwefel Function 2.22

6.5.2.1 Robustness

The number of successful trials for the 16-dimensional Schwefel function 2.22 is shown in Figure 6.45. Crossover is evidently now playing a more essential role.

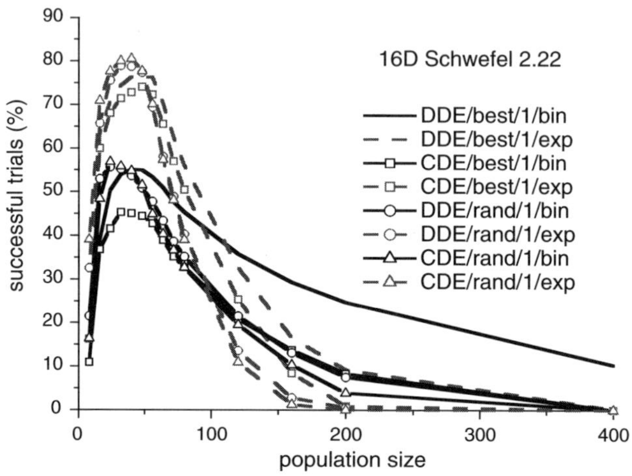

Figure 6.45 Robustness of differential evolution for the 16-dimensional Schwefel function 2.22

6.5.2.2 Efficiency

The minimal average number of objective function evaluations for the 16-dimensional Schwefel function 2.22 is shown in Figure 6.46. There is no significant difference compared to the 8-dimensional Schwefel function 2.22.

6.5.3 24-Dimensional Schwefel Function 2.22

6.5.3.1 Robustness

The number of successful trials for the 24-dimensional Schwefel function 2.22 is shown in Figure 6.47. DE/rand/1/exp is now more robust than DE/best/1/exp. In addition, CDE/best/1/bin is much less robust than other differential evolution strategies investigated.

6.5.3.2 Efficiency

The minimal average number of objective function evaluations for the 24-dimensional Schwefel function 2.22 is shown in Figure 6.48.

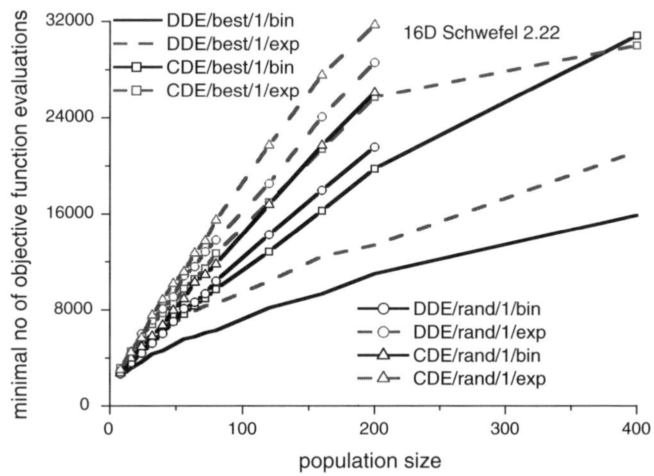

Figure 6.46 Efficiency of differential evolution for the 16-dimensional Schwefel function 2.22

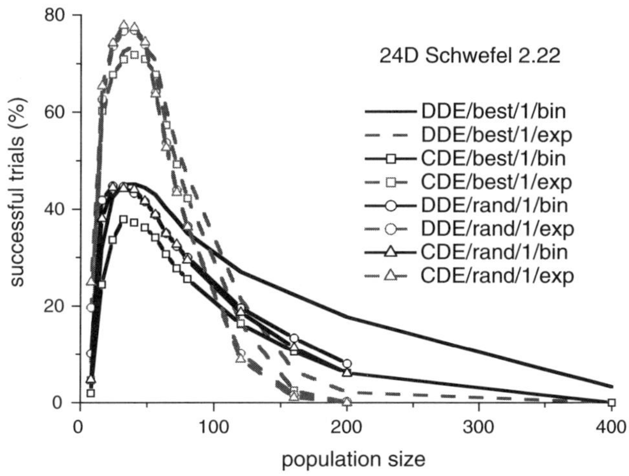

Figure 6.47 Robustness of differential evolution for the 24-dimensional Schwefel function 2.22

6.5.4 50-Dimensional Schwefel Function 2.22

The performance of differential evolution for the 50-dimensional Schwefel function 2.22 is shown in Figures 6.49 and 6.50. The aforementioned general trend remains evident.

6.5.5 Effect of Dimension

6.5.5.1 Robustness

The effect of dimension on robustness of differential evolution for the Schwefel function 2.22 is shown in Figure 6.51. It is seen that random base is more helpful for robustness.

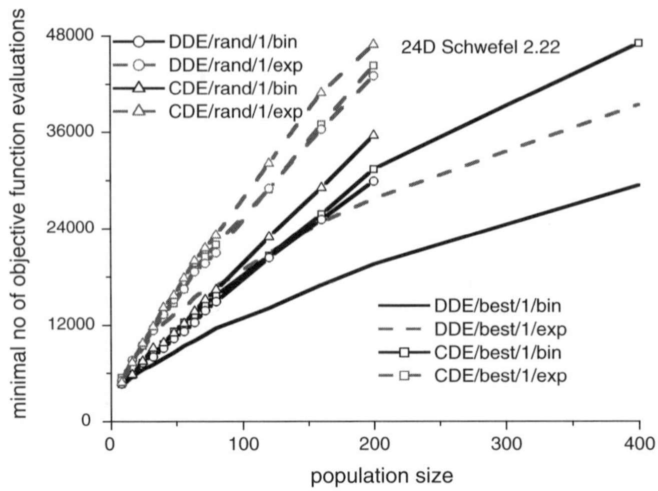

Figure 6.48 Efficiency of differential evolution for the 24-dimensional Schwefel function 2.22

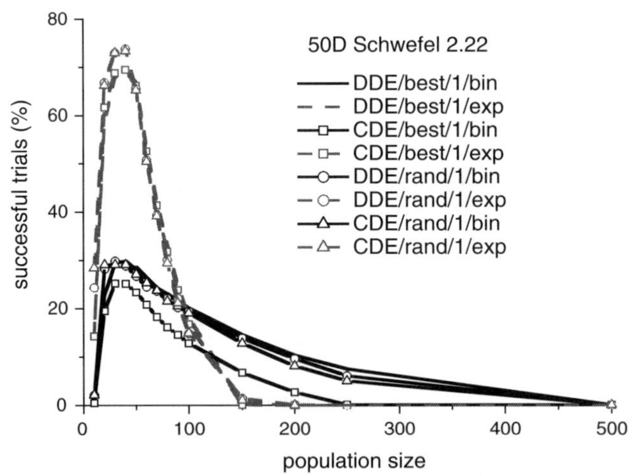

Figure 6.49 Robustness of differential evolution for the 50-dimensional Schwefel function 2.22

6.5.5.2 Efficiency

The effect of dimension on efficiency of differential evolution for the Schwefel function 2.22 is shown in Figure 6.52. Binomial crossover makes differential evolution more efficient.

6.6 Schwefel Function 2.26

The Schwefel function 2.26 is good for studying the effect of differentiability and modality.

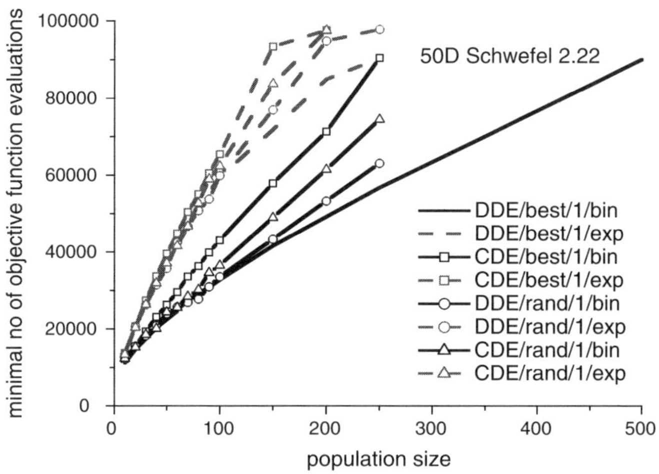

Figure 6.50 Efficiency of differential evolution for the 50-dimensional Schwefel function 2.22

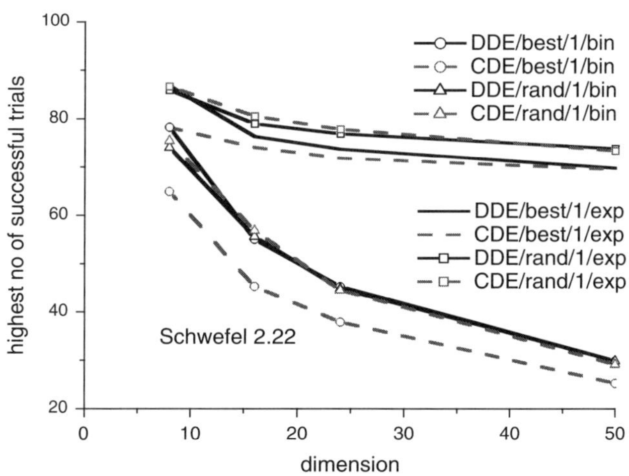

Figure 6.51 Effect of dimension on robustness of differential evolution for the Schwefel function 2.22

6.6.1 8-Dimensional Schwefel Function 2.26

6.6.1.1 Robustness

The number of successful trials for the 8-dimensional Schwefel function 2.26 is shown in Figure 6.53. The effect of evolution mechanism is now very weak. Random base and exponential crossover make differential evolution more robust.

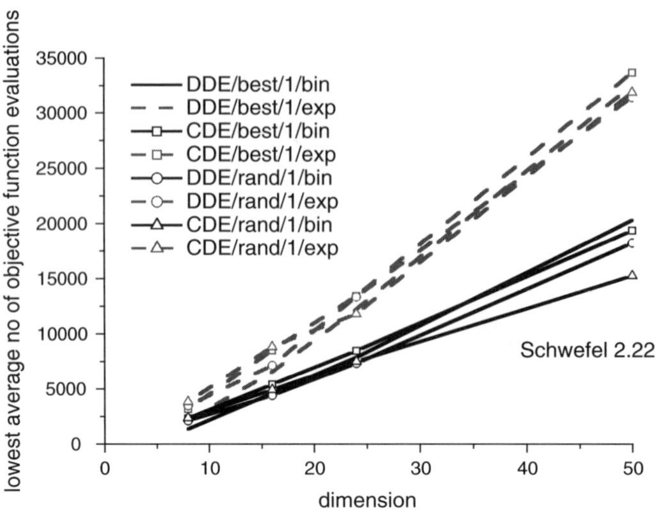

Figure 6.52 Effect of dimension on efficiency of differential evolution for the Schwefel function 2.22

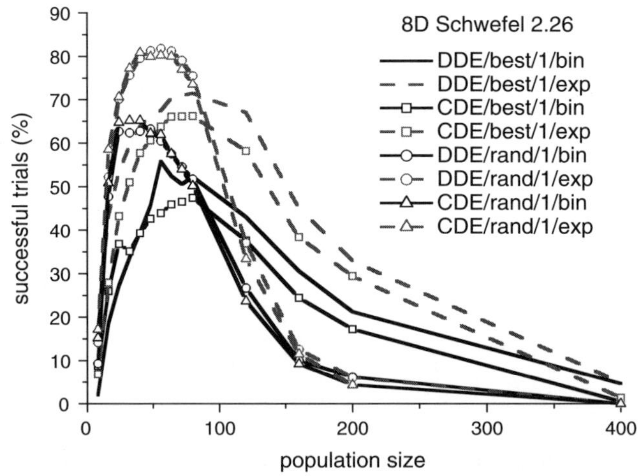

Figure 6.53 Robustness of differential evolution for the 8-dimensional Schwefel function 2.26

6.6.1.2 Efficiency

The minimal average number of objective function evaluations for the 8-dimensional Schwefel function 2.26 is shown in Figure 6.54. An increase in minimal average number of objective function evaluations of DDE/best/1/* with population size is seen when population size is not very large.

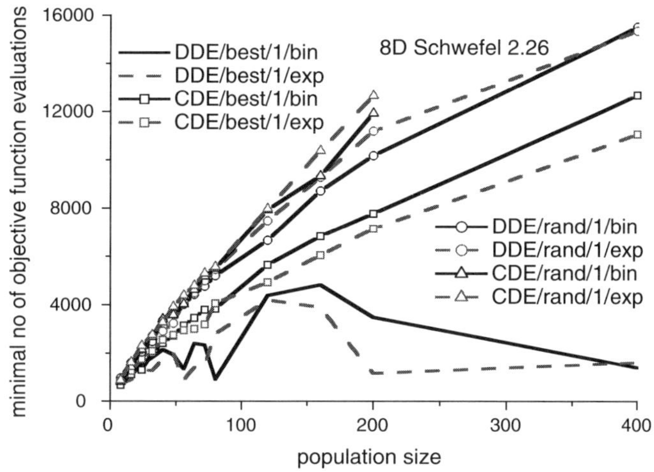

Figure 6.54 Efficiency of differential evolution for the 8-dimensional Schwefel function 2.26

6.6.2 16-Dimensional Schwefel Function 2.26

6.6.2.1 Robustness

The number of successful trials for the 16-dimensional Schwefel function 2.26 is shown in Figure 6.55. The effect of evolution mechanism on robustness is now almost negligible.

6.6.2.2 Efficiency

The minimal average number of objective function evaluations for the 16-dimensional Schwefel function 2.26 is shown in Figure 6.56. DE/rand/1/bin is now outperforming the other differential evolution strategies investigated.

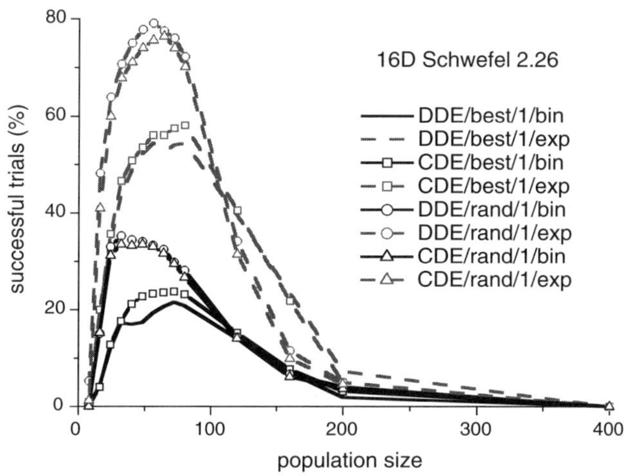

Figure 6.55 Robustness of differential evolution for the 16-dimensional Schwefel function 2.26

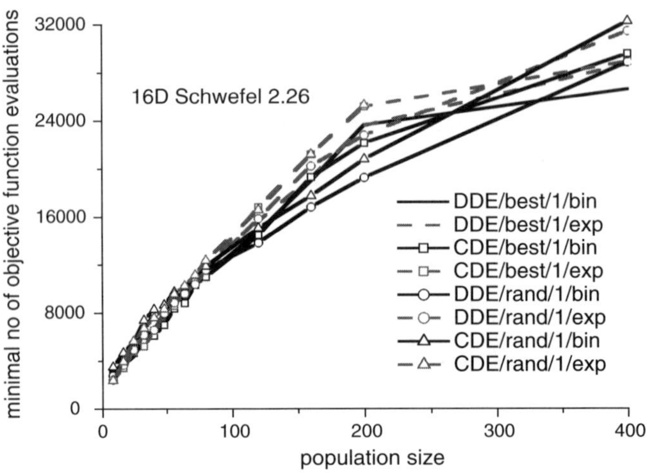

Figure 6.56 Efficiency of differential evolution for the 16-dimensional Schwefel function 2.26

6.6.3 *24-Dimensional Schwefel Function 2.26*

6.6.3.1 Robustness

The number of successful trials for the 24-dimensional Schwefel function 2.26 is shown in Figure 6.57. Dynamic differential evolution and classic differential evolution are now performing equally well.

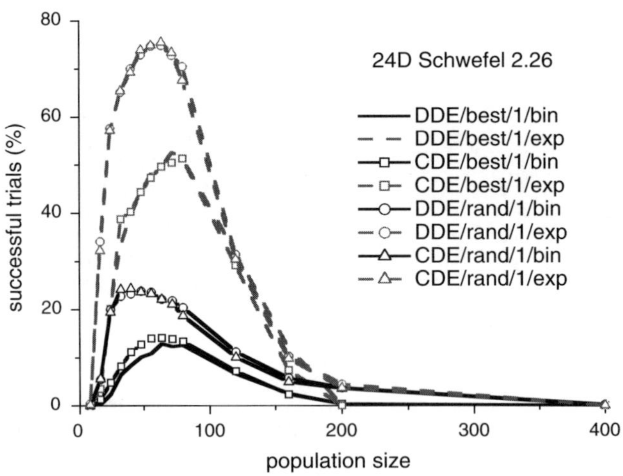

Figure 6.57 Robustness of differential evolution for the 24-dimensional Schwefel function 2.26

6.6.3.2 Efficiency

The minimal average number of objective function evaluations for the 24-dimensional Schwefel function 2.26 is shown in Figure 6.58. DE/rand/1/exp is now more efficient than DE/best/1/*.

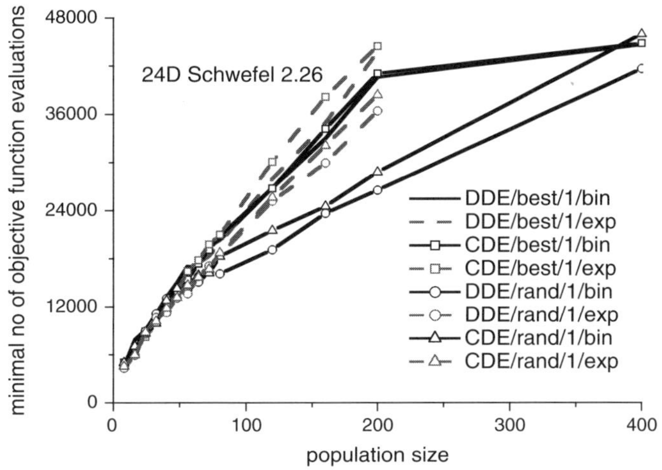

Figure 6.58 Efficiency of differential evolution for the 24-dimensional Schwefel function 2.26

6.6.4 50-Dimensional Schwefel Function 2.26

The performance of differential evolution for the 50-dimensional Schwefel function 2.26 is shown in Figures 6.59 and 6.60.

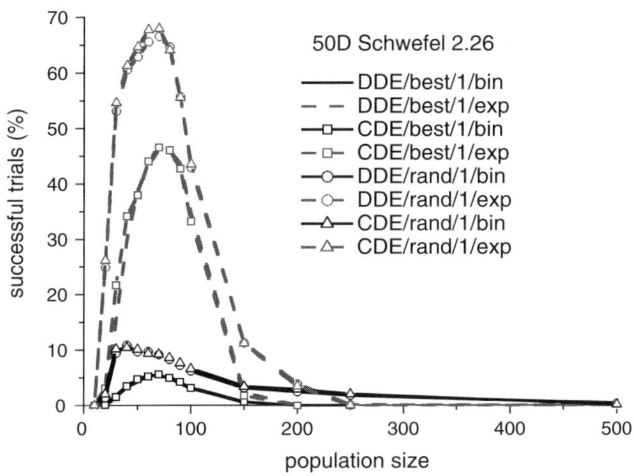

Figure 6.59 Robustness of differential evolution for the 50-dimensional Schwefel function 2.26

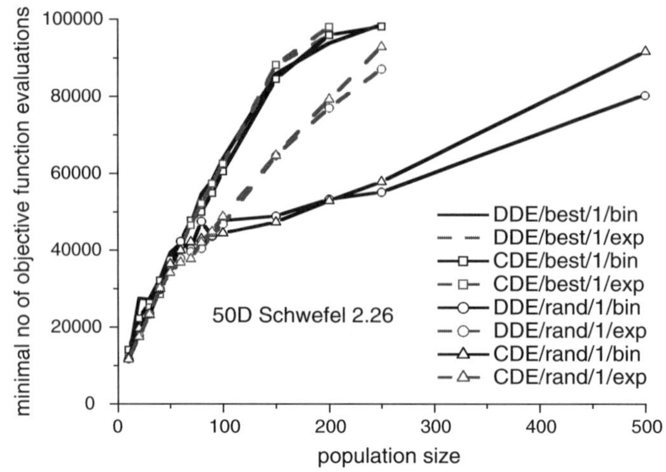

Figure 6.60 Efficiency of differential evolution for the 50-dimensional Schwefel function 2.26

6.6.5 Effect of Dimension

6.6.5.1 Robustness

The effect of dimension on robustness of differential evolution for the Schwefel function 2.26 is shown in Figure 6.61. It is very clear that the effect of evolution mechanism is very weak if the problem dimension is not very low. DE/rand/1/exp behaves most robustly.

6.6.5.2 Efficiency

The effect of dimension on efficiency of differential evolution for the Schwefel function 2.26 is shown in Figure 6.62. DE/rand/bin is more efficient if problem dimension is not very low.

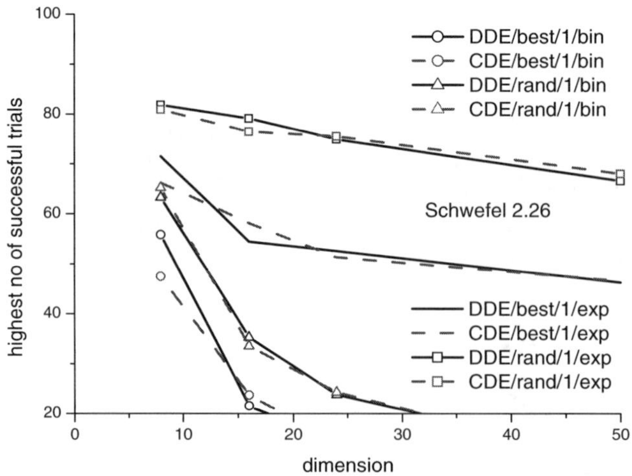

Figure 6.61 Effect of dimension on robustness of differential evolution for the Schwefel function 2.26

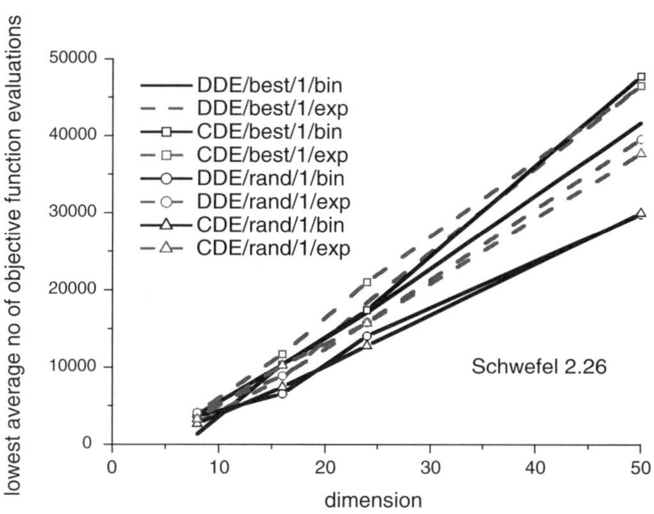

Figure 6.62 Effect of dimension on efficiency of differential evolution for the Schwefel function 2.26

6.7 Schwefel Function 1.2

The Schwefel function 1.2 is good for studying the effect of separability of optimization parameters. The optimization parameters become non-separable through coordinate rotation.

6.7.1 8-Dimensional Schwefel Function 1.2

6.7.1.1 Robustness

The number of successful trials for the 8-dimensional Schwefel function 1.2 is shown in Figure 6.63. The bell-shaped curves are significantly shorter and moving far to the left.

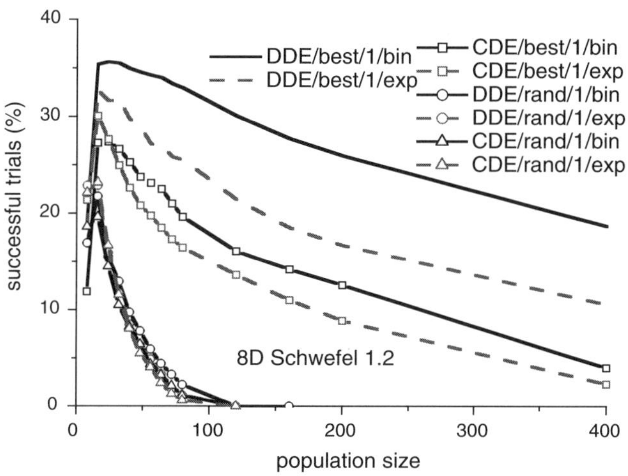

Figure 6.63 Robustness of differential evolution for the 8-dimensional Schwefel function 1.2

6.7.1.2 Efficiency

The minimal average number of objective function evaluations for the 8-dimensional Schwefel function 1.2 is shown in Figure 6.64. Significant loss of efficiency is observed.

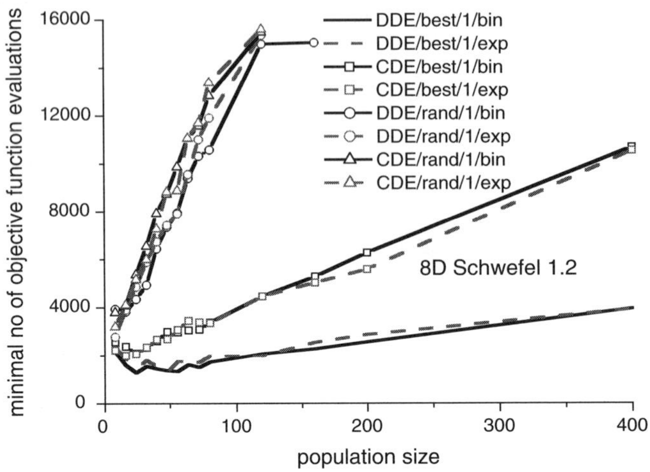

Figure 6.64 Efficiency of differential evolution for the 8-dimensional Schwefel function 1.2

6.7.2 16-Dimensional Schwefel Function 1.2

6.7.2.1 Robustness

The number of successful trials for the 16-dimensional Schwefel function 1.2 is shown in Figure 6.65. DDE/best/1/bin is much more robust than other differential evolution strategies investigated.

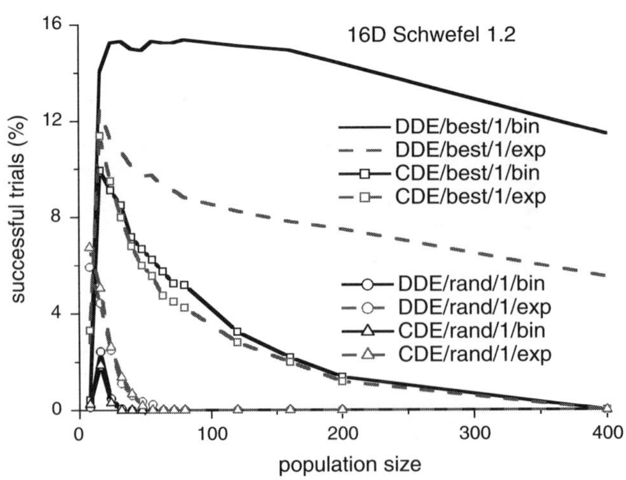

Figure 6.65 Robustness of differential evolution for the 16-dimensional Schwefel function 1.2

6.7.2.2 Efficiency

The minimal average number of objective function evaluations for the 16-dimensional Schwefel function 1.2 is shown in Figure 6.66.

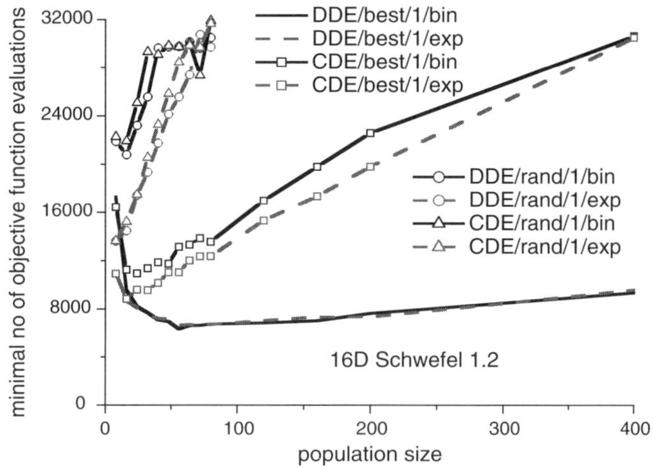

Figure 6.66 Efficiency of differential evolution for the 16-dimensional Schwefel function 1.2

6.7.3 24-Dimensional Schwefel Function 1.2

The performance of differential evolution for the 24-dimensional Schwefel function 1.2 is shown in Figures 6.67 and 6.68.

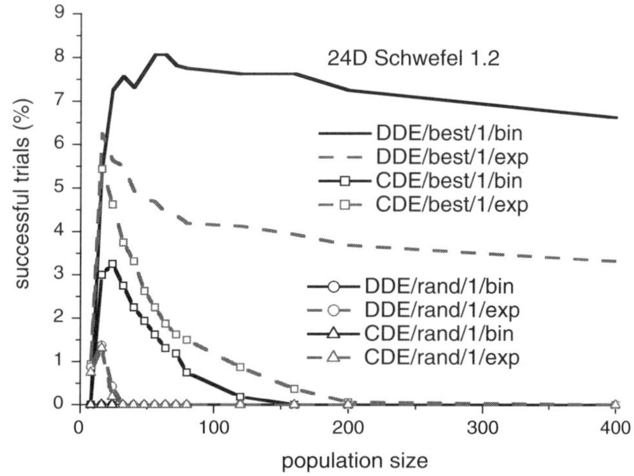

Figure 6.67 Robustness of differential evolution for the 24-dimensional Schwefel function 1.2

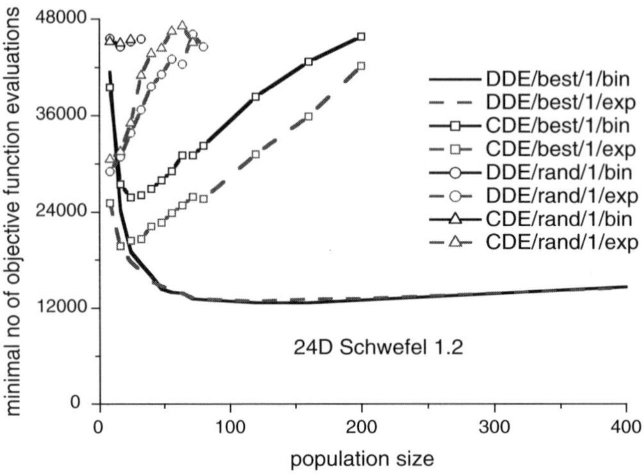

Figure 6.68 Efficiency of differential evolution for the 24-dimensional Schwefel function 1.2

6.7.4 50-Dimensional Schwefel Function 1.2

The performance of differential evolution for the 50-dimensional Schwefel function 1.2 is shown in Figures 6.69 and 6.70. No successful trial is observed for classic differential evolution.

6.7.5 Effect of Dimension

6.7.5.1 Robustness

The effect of dimension on robustness of differential evolution for Schwefel function 1.2 is shown in Figure 6.71. Dynamic differential evolution, best differential mutation base, and exponential crossover consistently make differential evolution more robust.

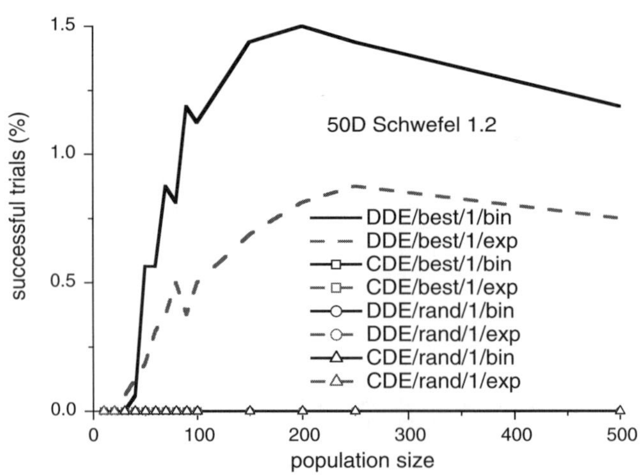

Figure 6.69 Robustness of differential evolution for the 50-dimensional Schwefel function 1.2

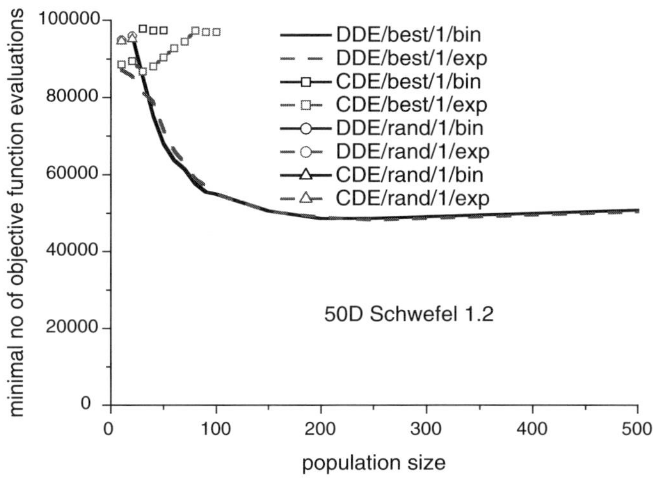

Figure 6.70 Efficiency of differential evolution for the 50-dimensional Schwefel function 1.2

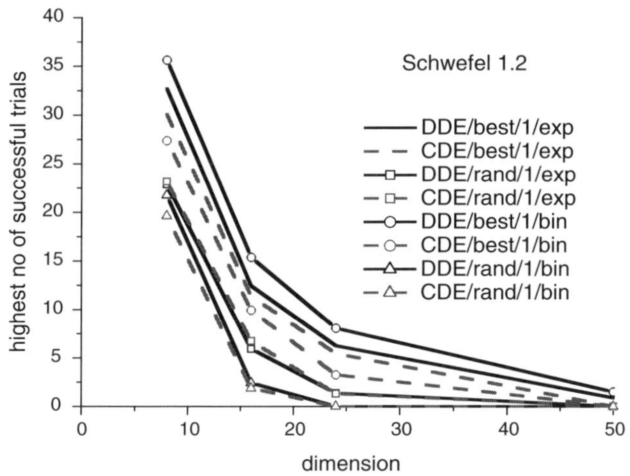

Figure 6.71 Effect of dimension on robustness of differential evolution for the Schwefel function 1.2

6.7.5.2 Efficiency

The effect of dimension on efficiency of differential evolution for Schwefel function 1.2 is shown in Figure 6.72. Consistency with robustness is observed.

6.7.5.3 Analysis

The Schwefel function 1.2 is the rotated sphere function. Non-separability introduced through rotation makes the Schwefel function 1.2 very hard to optimize. However, it does not change the

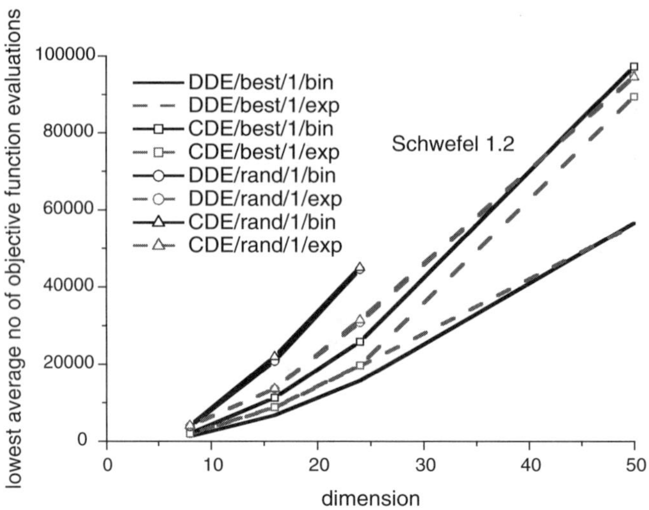

Figure 6.72 Effect of dimension on efficiency of differential evolution for the Schwefel function 1.2

basic behavior of differential evolution. The consistency between robustness and efficiency may be due to insufficient limit of number of objective function evaluations.

6.8 Rastrigin Function

The Rastrigin function is suitable for studying the effect of modality.

6.8.1 8-Dimensional Rastrigin Function

6.8.1.1 Robustness

The number of successful trials for the 8-dimensional Rastrigin function is shown in Figure 6.73. The effect of evolution mechanism is almost invisible. This observation agrees well with that for the Schwefel function 2.26.

6.8.1.2 Efficiency

The minimal average number of objective function evaluations for the 8-dimensional Rastrigin function is shown in Figure 6.74. Although best differential mutation base still looks beneficial to efficiency, its advantage is very marginal.

6.8.2 16-Dimensional Rastrigin Function

6.8.2.1 Robustness

The number of successful trials for the 16-dimensional Rastrigin function is shown in Figure 6.75. Random differential mutation base makes differential evolution more robust now.

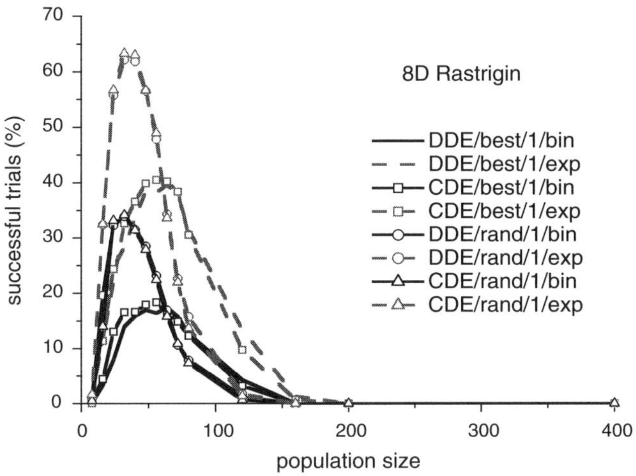

Figure 6.73 Robustness of differential evolution for the 8-dimensional Rastrigin function

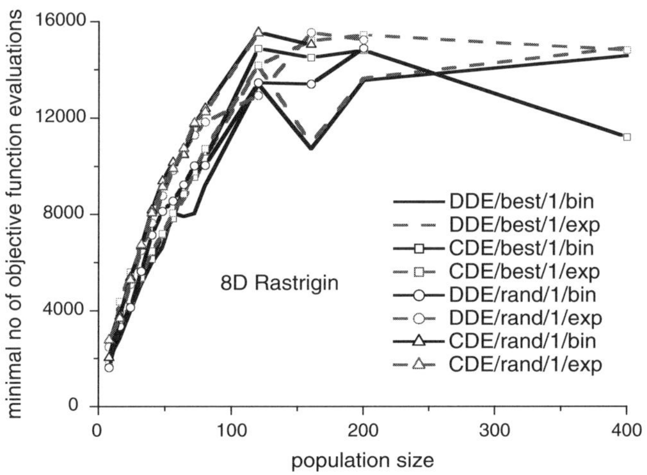

Figure 6.74 Efficiency of differential evolution for the 8-dimensional Rastrigin function

6.8.2.2 Efficiency

The minimal average number of objective function evaluations for the 16-dimensional Rastrigin function is shown in Figure 6.76. DE/rand/1/exp is now more efficient.

6.8.3 24-Dimensional Rastrigin Function

The performance of differential evolution for the 24-dimensional Rastrigin function is shown in Figures 6.77 and 6.78. The above general trend is again observed.

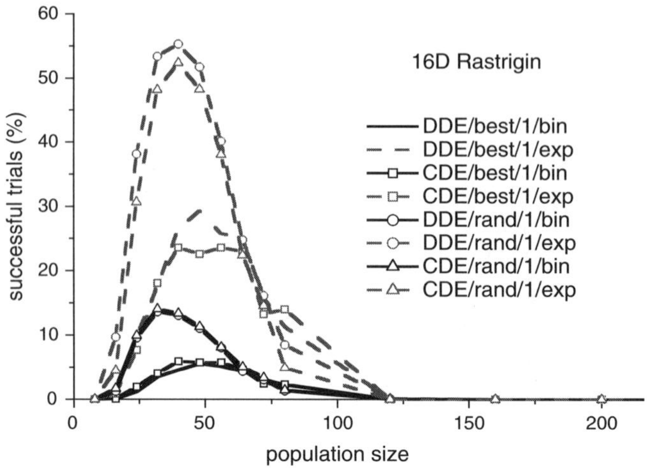

Figure 6.75 Robustness of differential evolution for the 16-dimensional Rastrigin function

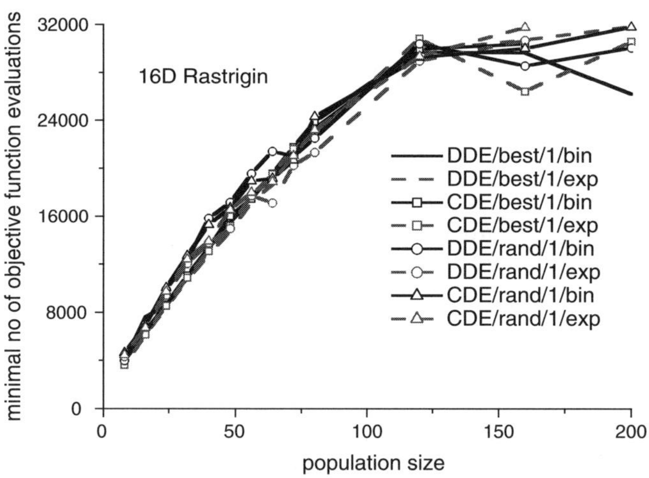

Figure 6.76 Efficiency of differential evolution for the 16-dimensional Rastrigin function

6.8.4 50-Dimensional Rastrigin Function

The performance of differential evolution for the 50-dimensional Rastrigin function is shown in Figures 6.79 and 6.80.

6.8.5 Effect of Dimension

6.8.5.1 Robustness

The effect of dimension on robustness of differential evolution for the Rastrigin function is shown in Figure 6.81. It looks very similar to Figure 6.61. Evolution mechanism has little effect

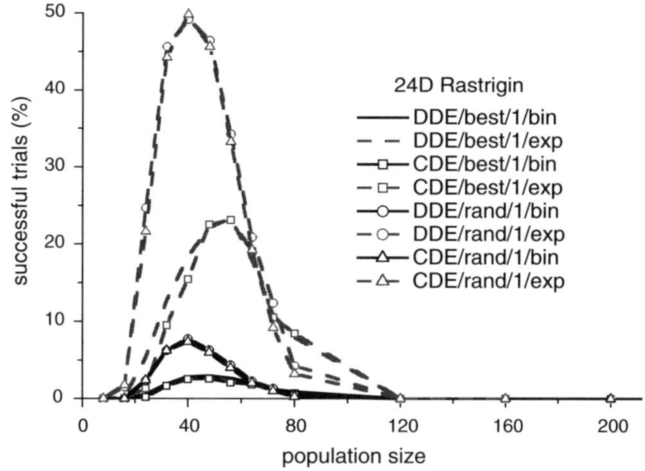

Figure 6.77 Robustness of differential evolution for the 24-dimensional Rastrigin function

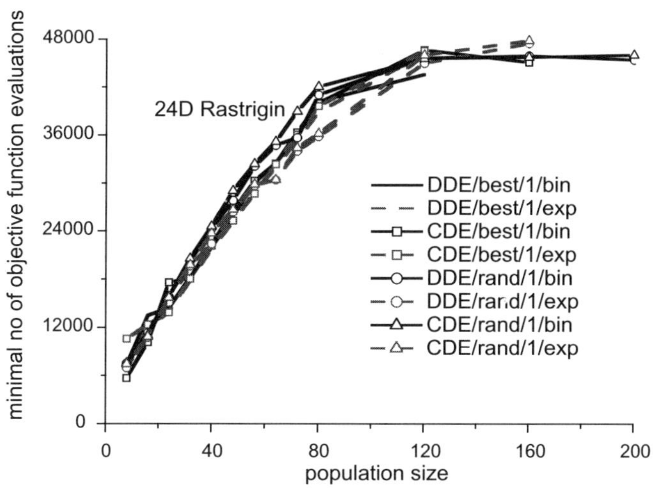

Figure 6.78 Efficiency of differential evolution for the 24-dimensional Rastrigin function

on differential evolution's robustness, while random differential mutation base and exponential crossover make differential evolution more robust.

6.8.5.2 Efficiency

The effect of dimension on efficiency of differential evolution for the Rastrigin function is shown in Figure 6.82. Likewise, evolution mechanism has little effect on differential evolution's efficiency, while random base and exponential crossover make differential evolution more efficient.

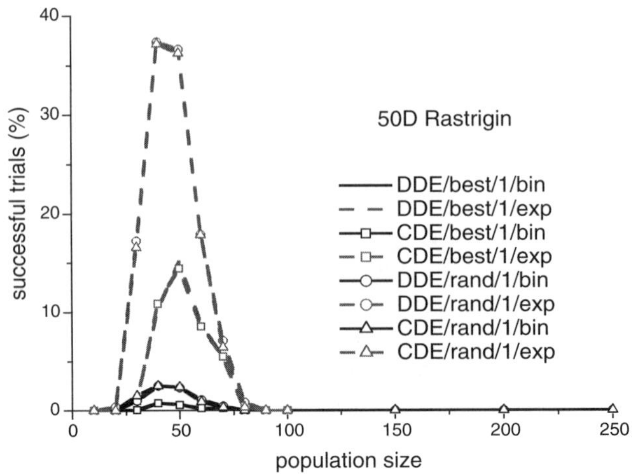

Figure 6.79 Robustness of differential evolution for the 50-dimensional Rastrigin function

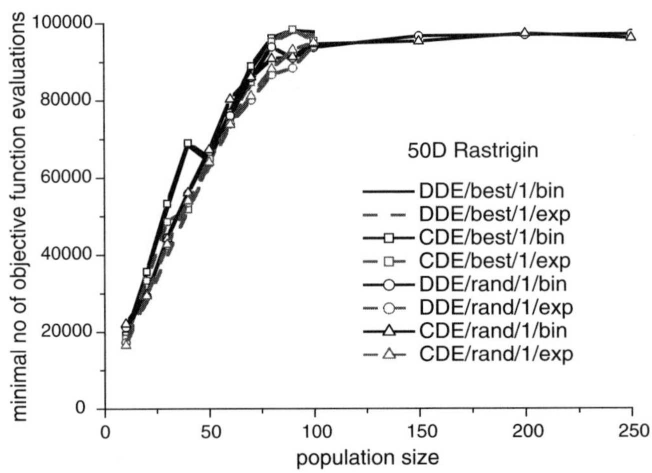

Figure 6.80 Efficiency of differential evolution for the 50-dimensional Rastrigin function

6.9 Ackley Function

The Ackley function is good for studying the effect of modality and separability.

6.9.1 8-Dimensional Ackley Function

6.9.1.1 Robustness

The number of successful trials for the 8-dimensional Ackley function is shown in Figure 6.83. There is no major difference between the Ackley function and Schwefel function 2.26.

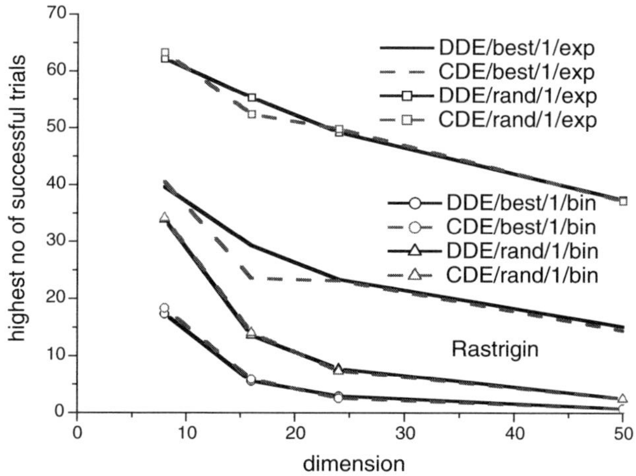

Figure 6.81 Effect of dimension on robustness of differential evolution for the Rastrigin function

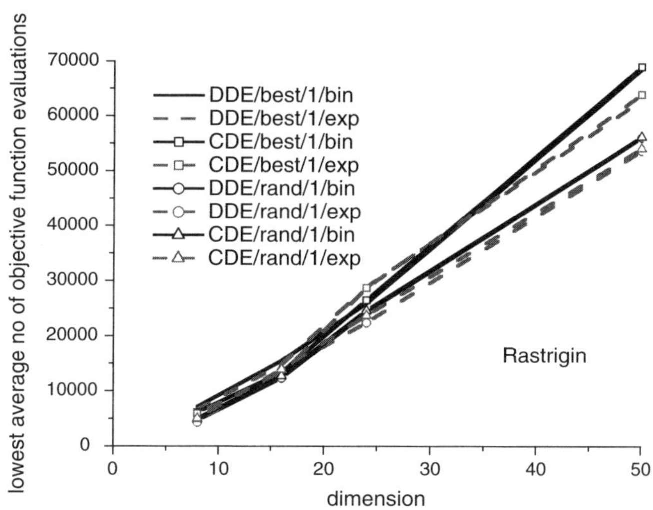

Figure 6.82 Effect of dimension on efficiency of differential evolution for the Rastrigin function

6.9.1.2 Efficiency

The minimal average number of objective function evaluations for the 8-dimensional Ackley function is shown in Figure 6.84. It looks quite similar to the sphere function.

6.9.2 16-Dimensional Ackley Function

The performance of differential evolution for the 16-dimensional Ackley function is shown in Figures 6.85 and 6.86. Similar observations apply.

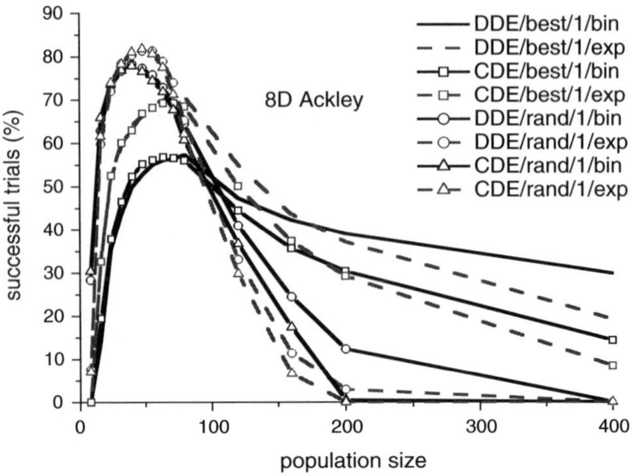

Figure 6.83 Robustness of differential evolution for the 8-dimensional Ackley function

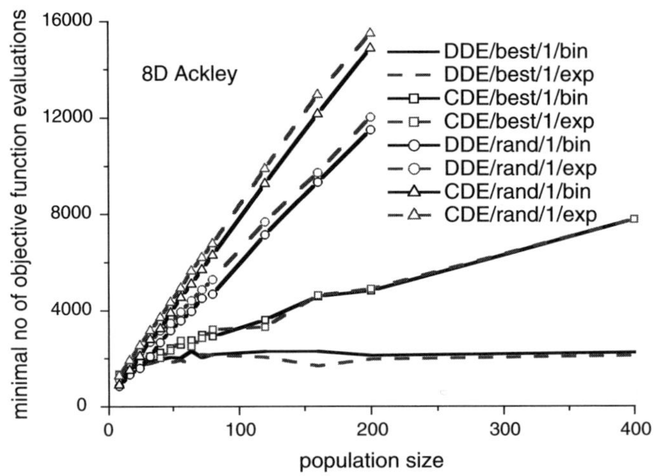

Figure 6.84 Efficiency of differential evolution for the 8-dimensional Ackley function

6.9.3 24-Dimensional Ackley Function

The performance of differential evolution for the 24-dimensional Ackley function is shown in Figures 6.87 and 6.88.

6.9.4 50-Dimensional Ackley Function

The performance of differential evolution for the 50-dimensional Ackley function is shown in Figures 6.89 and 6.90.

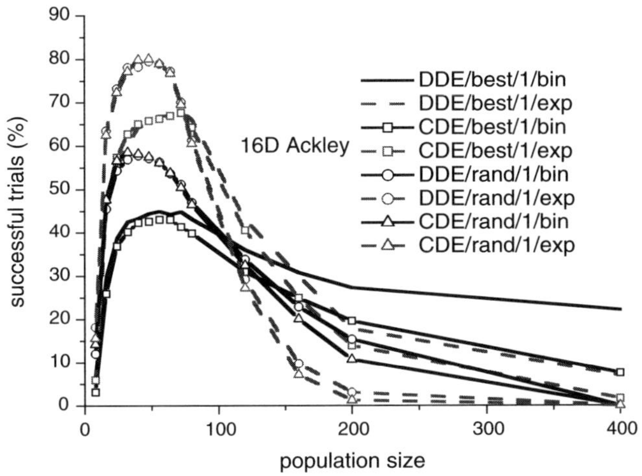

Figure 6.85 Robustness of differential evolution for the 16-dimensional Ackley function

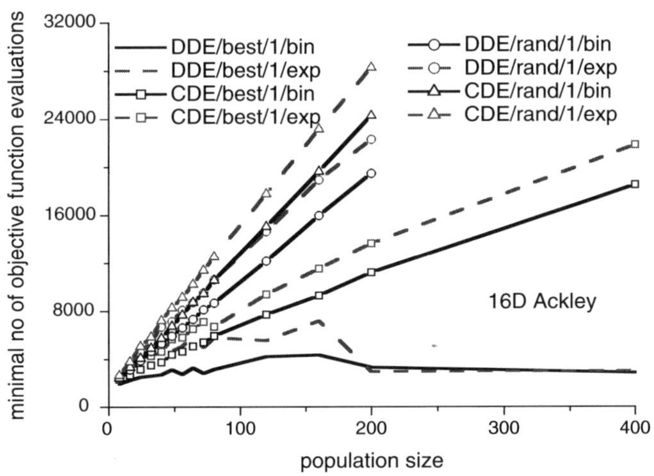

Figure 6.86 Efficiency of differential evolution for the 16-dimensional Ackley function

6.9.5 Effect of Dimension

6.9.5.1 Robustness

The effect of dimension on robustness of differential evolution for the Ackley function is shown in Figure 6.91. The general trend is identical with that of the Schwefel function 2.26.

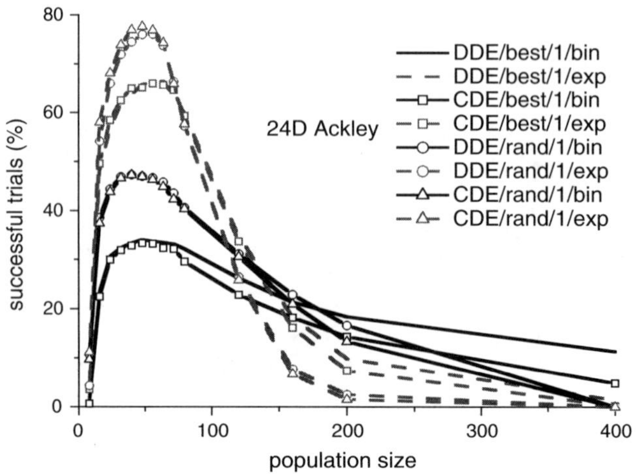

Figure 6.87 Robustness of differential evolution for the 24-dimensional Ackley function

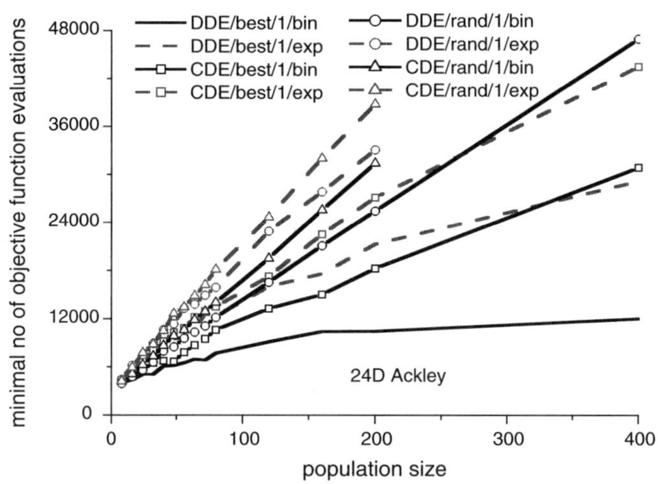

Figure 6.88 Efficiency of differential evolution for the 24-dimensional Ackley function

6.9.5.2 Efficiency

The effect of dimension on efficiency of differential evolution for Ackley function is shown in Figure 6.92. It looks very similar to that of the sphere function.

6.9.5.3 Analysis

The Schwefel function 2.26, Rastrigin function, and Ackley function are all multimodal. It is encouraging to note that the robustness of differential evolution

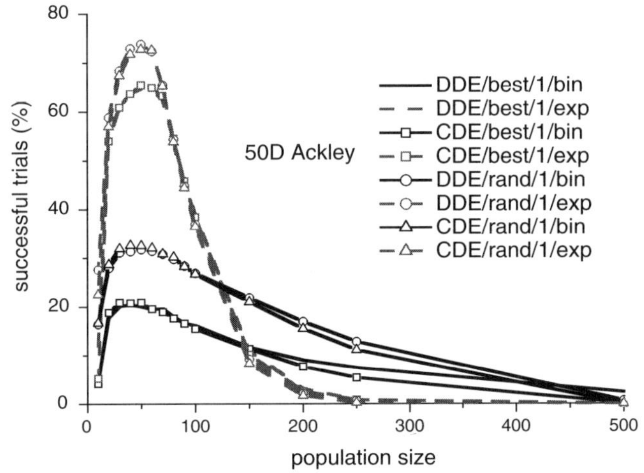

Figure 6.89 Robustness of differential evolution for the 50-dimensional Ackley function

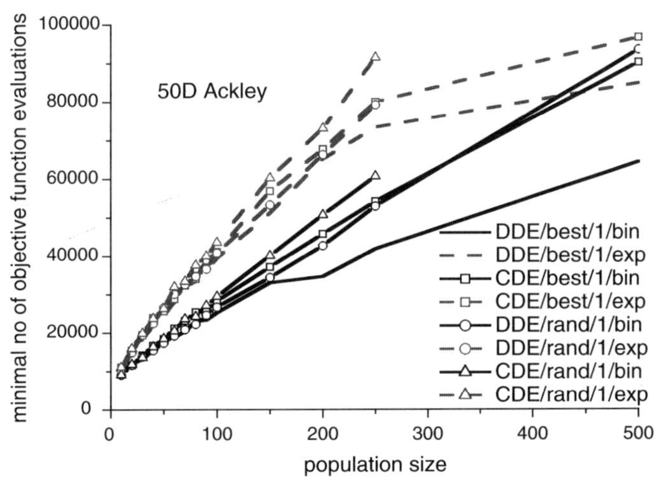

Figure 6.90 Efficiency of differential evolution for the 50-dimensional Ackley function

over these toy functions is observed to be similar. This implies that problem features have a profound impact on the robustness of differential evolution. On the other hand, the efficiency of differential evolution over these functions is observed to be inconsistent.

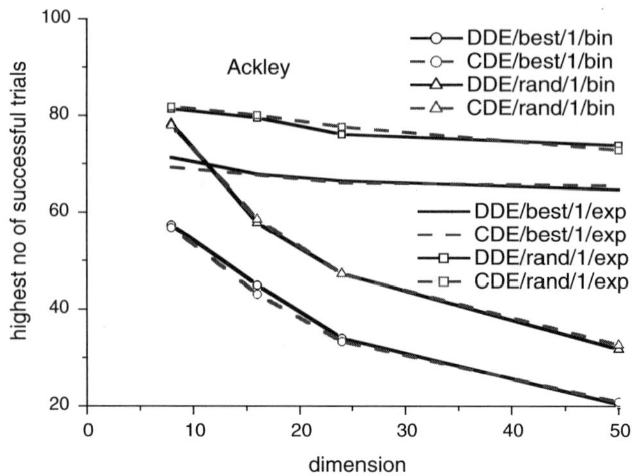

Figure 6.91 Effect of dimension on robustness of differential evolution for the Ackley function

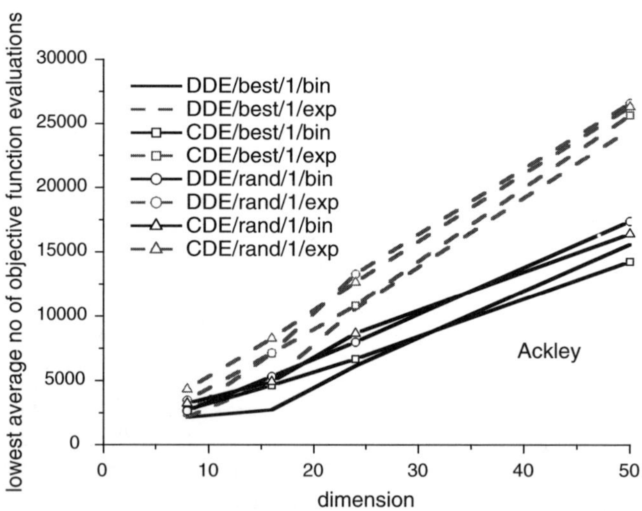

Figure 6.92 Effect of dimension on efficiency of differential evolution for the Ackley function

7

Optimal Intrinsic Control Parameters

This chapter aims to answer the second question facing the parametric study. In particular, the intention is to find out the optimal values of intrinsic control parameters by which differential evolution performs best. We are interested in the relationship between optimal intrinsic control parameters, evolutionary operators, and problem features. The relationship will partially answer the third question facing the parametric study too.

7.1 Sphere Function

7.1.1 Optimal Population Size

7.1.1.1 Observations

The optimal population size and the safeguard zone of population size of all our strategies of differential evolution for the sphere function are summarized in Table 7.1. The safeguard

Table 7.1 Optimal population size and safeguard zone of population size for the sphere function

Strategy	Optimal population size					Safeguard zone				
	8D	16D	24D	50D	100D	8D	16D	24D	50D	100D
DDE/best/1/bin	64	56	48	40	30	[40,80]	[32,80]	[24,80]	[20,150]	[20,250]
CDE/best/1/bin	56	32	32	30	30	[32,80]	[24,80]	[16,64]	[20,60]	[20,80]
DDE/rand/1/bin	40	32	40	30	50	[24,64]	[16,64]	[16,64]]	[20,80]	[20,100]
CDE/rand/1/bin	40	32	24	30	30	[24,64]	[16,64]	[16,64]	[20,80]	[20,150]
DDE/best/1/exp	64	64	56	50	40	[32,80]	[32,80]	[24,80]	[30,70]	[20,60]
CDE/best/1/exp	64	56	48	50	40	[32,80]	[24,80]	[24,72]	[20,70]	[20,60]
DDE/rand/1/exp	48	48	48	40	50	[32,72]	[24,64]	[24,64]	[20,60]	[30,70]
CDE/rand/1/exp	48	40	48	40	50	[24,72]	[24,64]	[24,64]	[30,60]	[30,60]

zone of population size is the zone in which the number of successful trials is not 100 less than the highest number of successful trials.

From Table 7.1, it can be seen that

(a) A smaller population size appears to be more beneficial.
(b) The optimal population size for DE/best/1/* decreases as problem dimension increases.
(c) Higher dimension leads to a wider safeguard zone for DE/*/1/bin.
(d) The safeguard zone for DE/*/1/exp contracts as problem dimension increases.

7.1.1.2 Recommendations

A population size of 40 seems a good starting point. It is quite safe to adjust it between 20 and 60. Population size beyond this region will benefit neither efficiency nor robustness, regardless of the differential evolution strategy implemented.

It may also be worthwhile to point out that N_p/N does not seems a good criterion for choosing population size.

7.1.2 Optimal Mutation Intensity and Crossover Probability

Let us now have a look at the optimal mutation intensity and crossover probability in terms of robustness.

7.1.2.1 8-Dimensional Sphere Function

The optimal mutation intensity and crossover probability of differential evolution for the 8-dimensional sphere function are shown in Figures 7.1–7.4. In general, from Figures 7.1–7.4 we note the following:

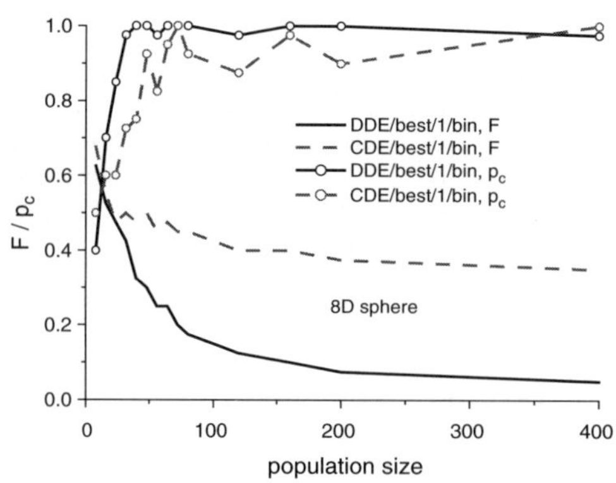

Figure 7.1 Mutation intensity and crossover probability of DE/best/1/bin for the 8-dimensional sphere function

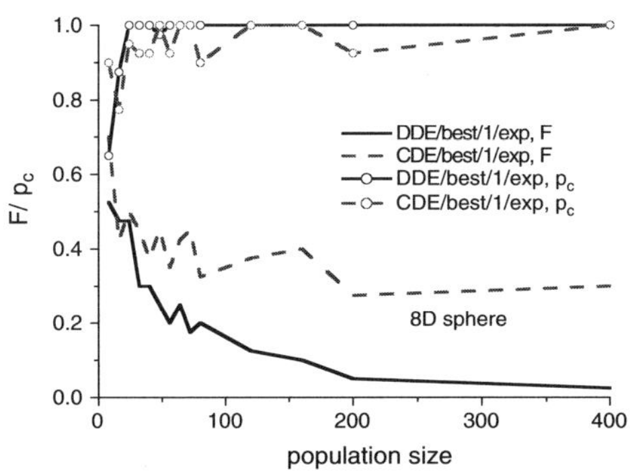

Figure 7.2 Mutation intensity and crossover probability of DE/best/1/exp for the 8-dimensional sphere function

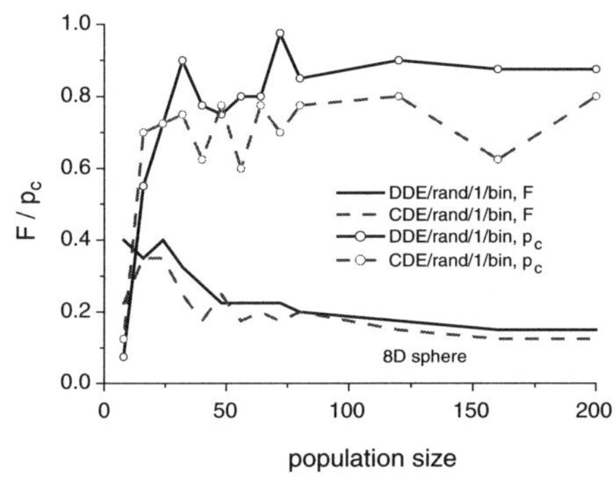

Figure 7.3 Mutation intensity and crossover probability of DE/rand/1/bin for the 8-dimensional sphere function

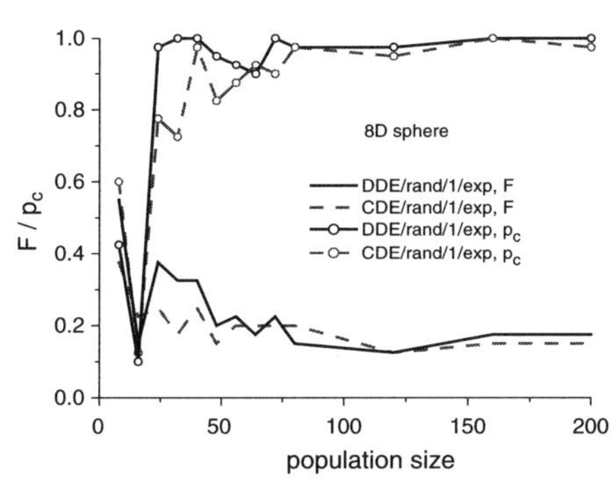

Figure 7.4 Mutation intensity and crossover probability of DE/rand/1/exp for the 8-dimensional sphere function

(a) Mutation intensity drops while crossover probability grows as population size increases.
(b) Mutation intensity for DDE/best/1/* is smaller than that for CDE/best/1/*. The difference in mutation intensity between DDE/rand/1/* and CDE/rand/1/* is marginal.
(c) Differential mutation base has very little effect on mutation intensity for dynamic differential evolution. However, its effect on mutation intensity for classic differential evolution is much more remarkable. Mutation intensity for CDE/rand/1/* is observed to be much smaller than that for CDE/best/1/*.
(d) The effect of crossover on mutation intensity is negligible.
(e) Crossover probability for dynamic differential evolution is observed to be bigger than that for classic differential evolution.
(f) Crossover probability for DE/best/1/* is bigger than that for DE/rand/1/*.
(g) Exponential crossover prefers a higher crossover probability.

7.1.2.2 16-Dimensional Sphere Function

The optimal mutation intensity and crossover probability of differential evolution for the 16-dimensional sphere function are shown in Figures 7.5–7.8.

7.1.2.3 24-Dimensional Sphere Function

The optimal mutation intensity and crossover probability of differential evolution for the 24-dimensional sphere function are shown in Figures 7.9–7.12.

7.1.2.4 50-Dimensional Sphere Function

The optimal mutation intensity and crossover probability of differential evolution for the 50-dimensional sphere function are shown in Figures 7.13–7.16.

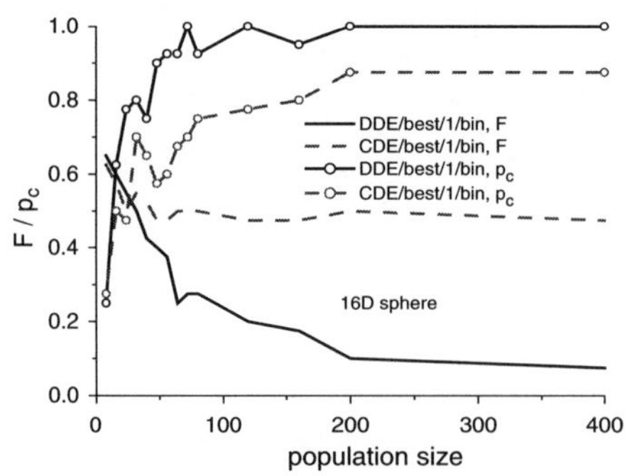

Figure 7.5 Mutation intensity and crossover probability of DE/best/1/bin for the 16-dimensional sphere function

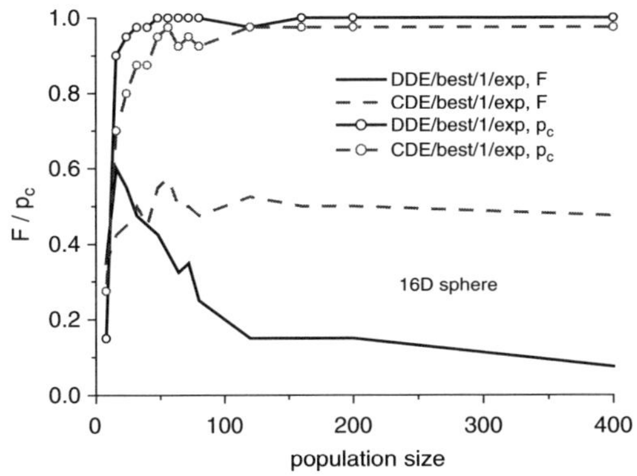

Figure 7.6 Mutation intensity and crossover probability of DE/best/1/exp for the 16-dimensional sphere function

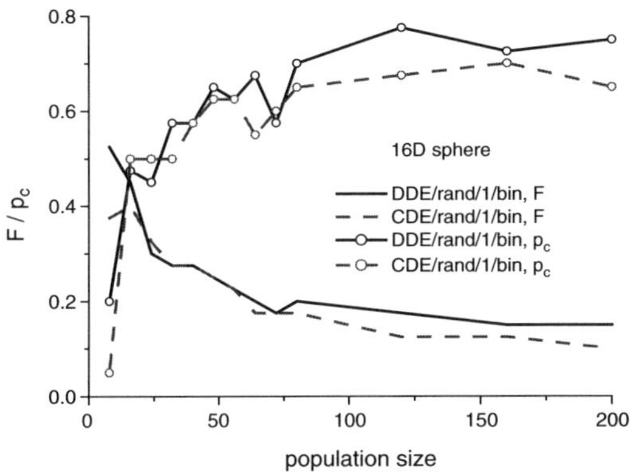

Figure 7.7 Mutation intensity and crossover probability of DE/rand/1/bin for the 16-dimensional sphere function

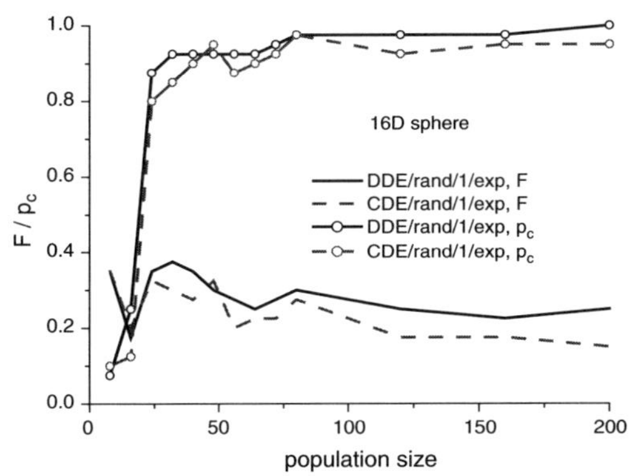

Figure 7.8 Mutation intensity and crossover probability of DE/rand/1/exp for the 16-dimensional sphere function

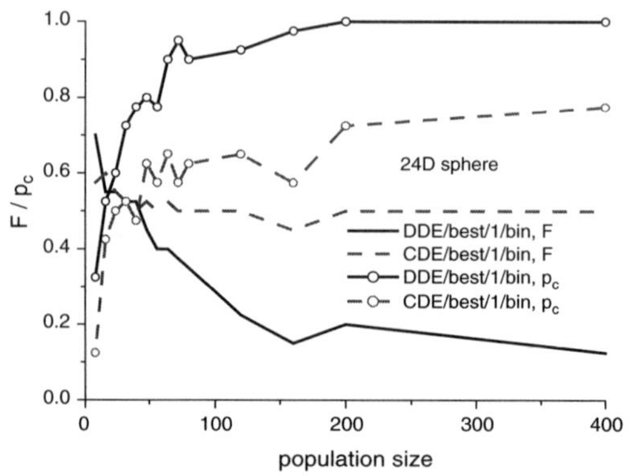

Figure 7.9 Mutation intensity and crossover probability of DE/best/1/bin for the 24-dimensional sphere function

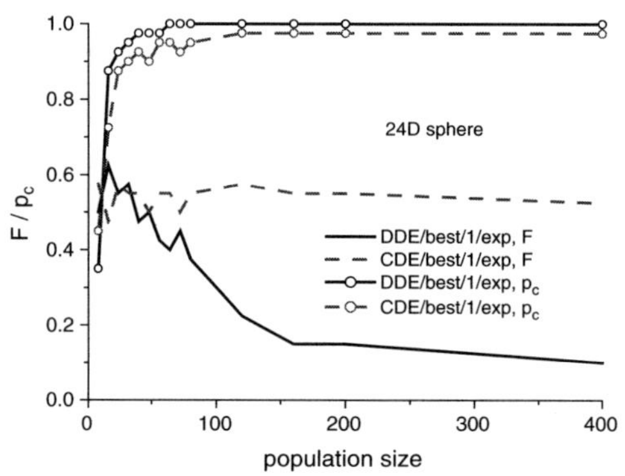

Figure 7.10 Mutation intensity and crossover probability of DE/best/1/exp for the 24-dimensional sphere function

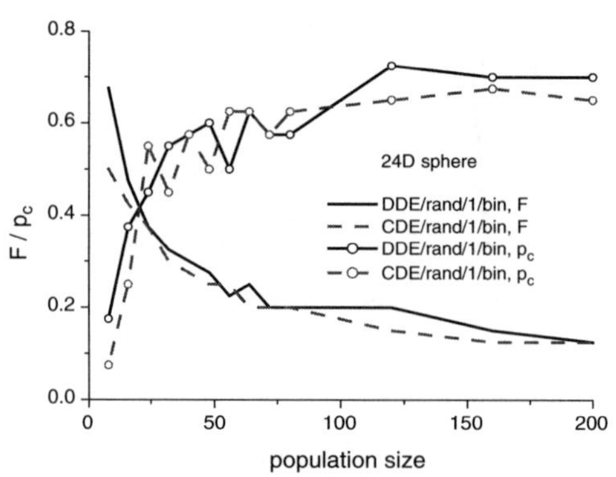

Figure 7.11 Mutation intensity and crossover probability of DE/rand/1/bin for the 24-dimensional sphere function

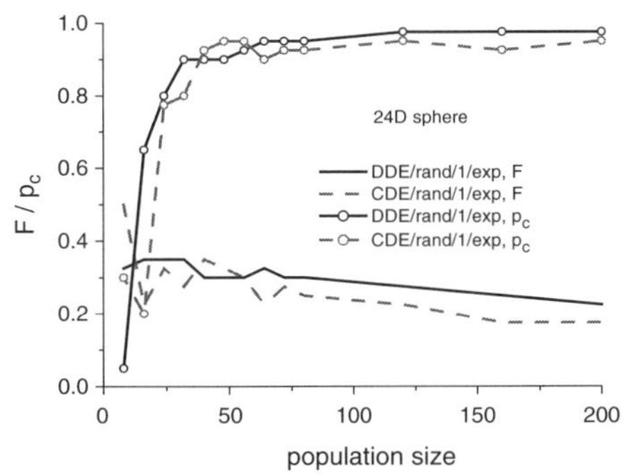

Figure 7.12 Mutation intensity and crossover probability of DE/rand/1/exp for the 24-dimensional sphere function

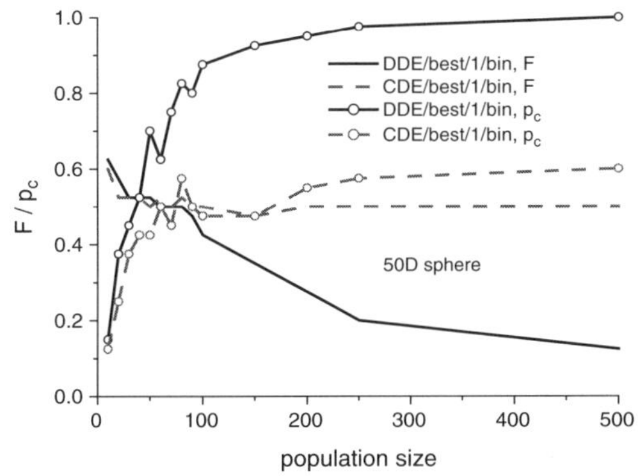

Figure 7.13 Mutation intensity and crossover probability of DE/best/1/bin for the 50-dimensional sphere function

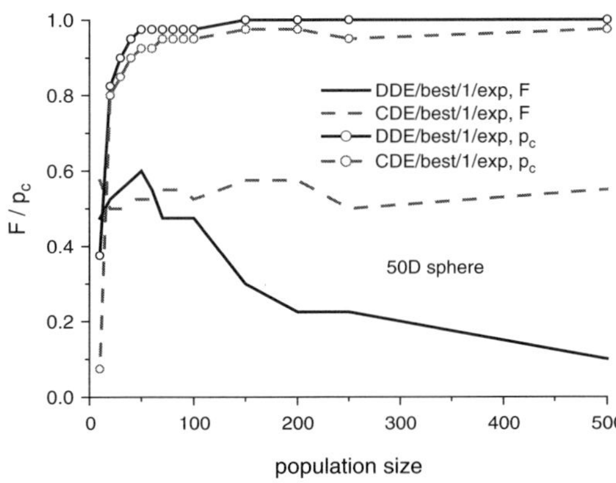

Figure 7.14 Mutation intensity and crossover probability of DE/best/1/exp for the 50-dimensional sphere function

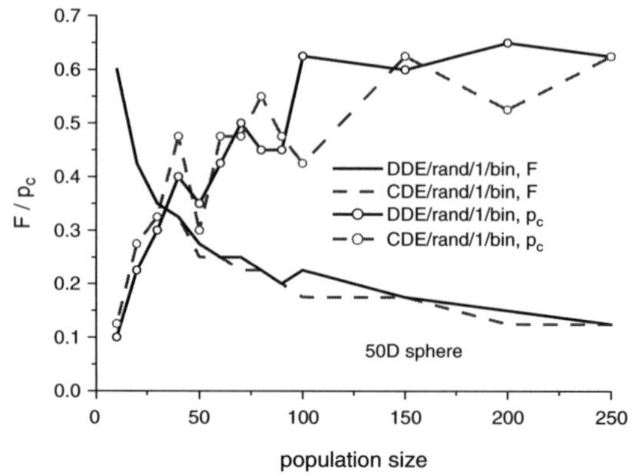

Figure 7.15 Mutation intensity and crossover probability of DE/rand/1/bin for the 50-dimensional sphere function

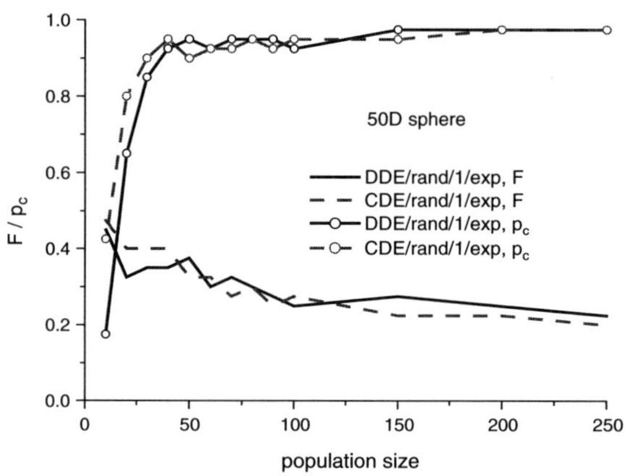

Figure 7.16 Mutation intensity and crossover probability of DE/rand/1/exp for the 50-dimensional sphere function

7.1.2.5 100-Dimensional Sphere Function

The optimal mutation intensity and crossover probability of differential evolution for the 100-dimensional sphere function are shown in Figures 7.17–7.20.

7.1.2.6 Effect of Dimension

From Figures 7.1–7.20, we note the following:

(a) Mutation intensity for DE/*/1/bin is insensitive to problem dimension. However, it may be good practice to increase mutation intensity slightly for higher dimension.

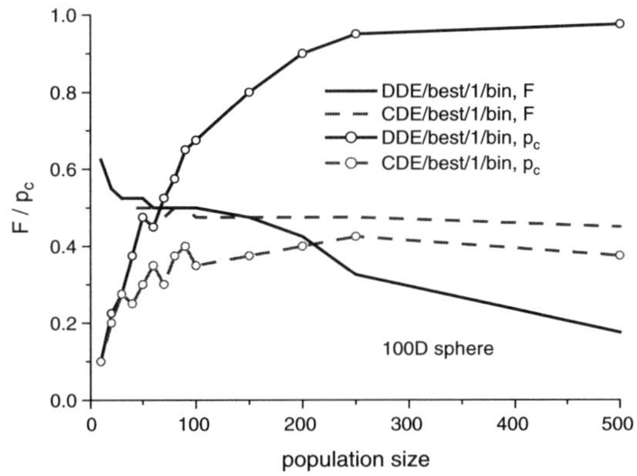

Figure 7.17 Mutation intensity and crossover probability of DE/best/1/bin for the 100-dimensional sphere function

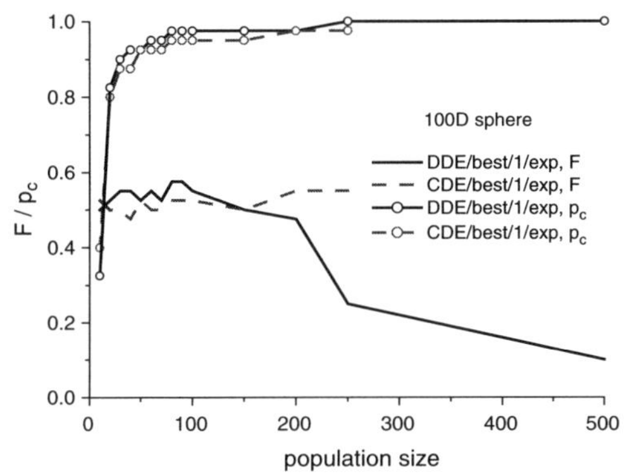

Figure 7.18 Mutation intensity and crossover probability of DE/best/1/exp for the 100-dimensional sphere function

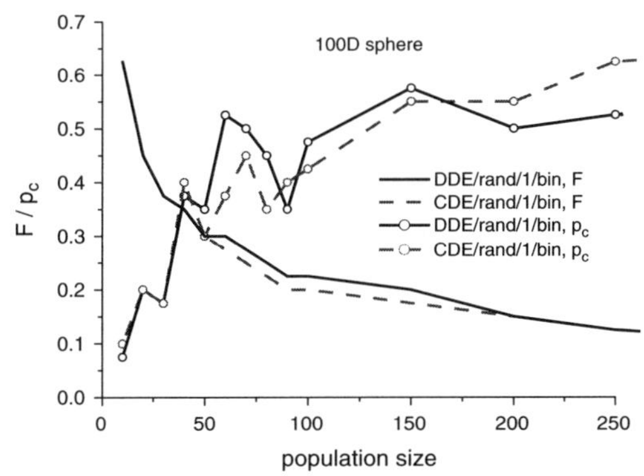

Figure 7.19 Mutation intensity and crossover probability of DE/rand/1/bin for the 100-dimensional sphere function

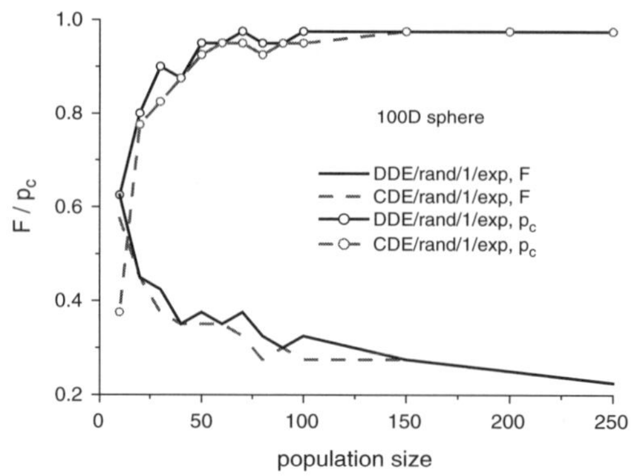

Figure 7.20 Mutation intensity and crossover probability of DE/rand/1/exp for the 100-dimensional sphere function

(b) Mutation intensity for DE/*/1/exp is sensitive to problem dimension. In general, higher dimension prefers higher mutation intensity.
(c) Crossover probability for DE/*/1/exp is also insensitive to problem dimension. It appears to be quite safe to use a crossover probability very close to 1.
(d) Crossover probability for DDE/best/1/bin is moderately sensitive to problem dimension.
(e) Crossover probability for CDE/best/1/bin, DDE/rand/1/bin, and CDE/rand/1/bin is very sensitive to problem dimension. Crossover probability falls very quickly with problem dimension.

7.2 Step Function 2

7.2.1 Optimal Population Size

The optimal population size and the safeguard zone of population size for step function 2 are summarized in Table 7.2. Both optimal population size and safeguard zone are observed to be shifting upward. The safeguard zone is also wider.

Table 7.2 Optimal population size and safeguard zone of population size for step function 2

Strategy	Optimal population size				Safeguard zone			
	8D	16D	24D	50D	8D	16D	24D	50D
DDE/best/1/bin	120	80	64	60	[56,160]	[40,160]	[40,120]	[20,150]
CDE/best/1/bin	72	64	56	60	[40,120]	[32,120]	[32,120]	[20,150]
DDE/rand/1/bin	56	48	48	60	[32,80]	[24,80]	[32,80]	[30,150]
CDE/rand/1/bin	56	48	48	60	[32,80]	[24,80]	[24,80]	[30,120]
DDE/best/1/exp	120	80	80	80	[48,160]	[40,120]	[40,120]	[40,100]
CDE/best/1/exp	80	72	80	70	[40,120]	[40,120]	[40,120]	[40,100]
DDE/rand/1/exp	64	64	56	70	[32,80]	[32,80]	[32,80]	[40,90]
CDE/rand/1/exp	64	64	64	70	[32,80]	[32,80]	[32,80]	[40,90]

7.2.2 Optimal Mutation Intensity and Crossover Probability

7.2.2.1 8-Dimensional Step Function 2

The optimal mutation intensity and crossover probability of differential evolution for the 8-dimensional step function 2 are shown in Figures 7.21–7.24.

7.2.2.2 16-Dimension Step Function 2

The optimal mutation intensity and crossover probability of differential evolution for the 16-dimensional step function 2 are shown in Figures 7.25–7.28.

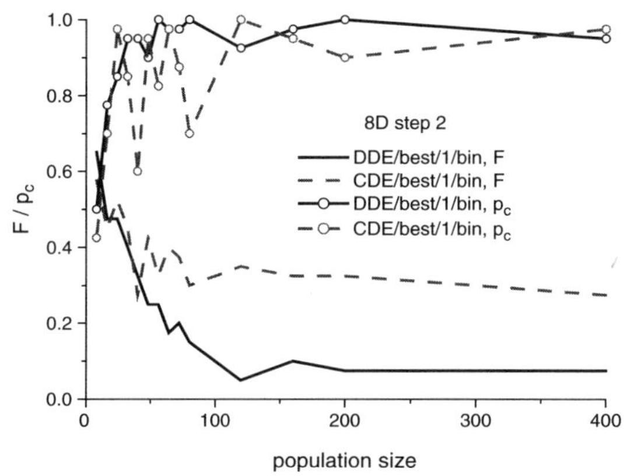

Figure 7.21 Mutation intensity and crossover probability of DE/best/1/bin for the 8-dimensional step function 2

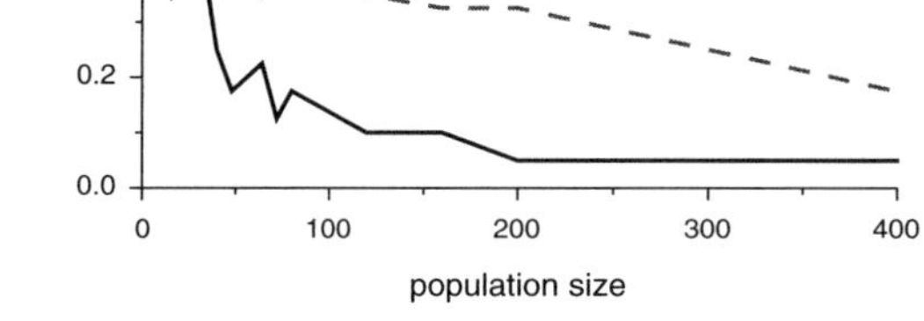

Figure 7.22 Mutation intensity and crossover probability of DE/best/1/exp for the 8-dimensional step function 2

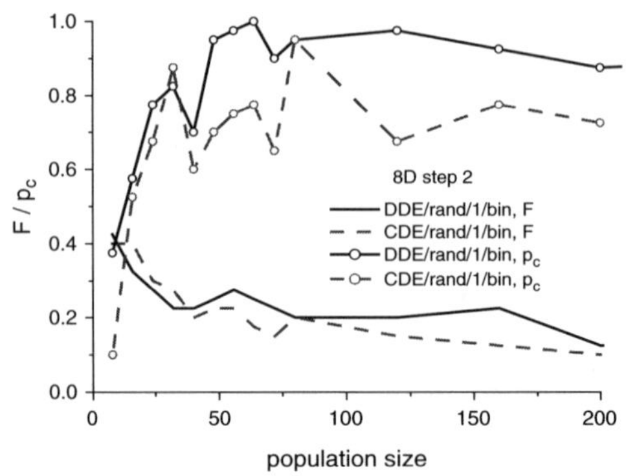

Figure 7.23 Mutation intensity and crossover probability of DE/rand/1/bin for the 8-dimensional step function 2

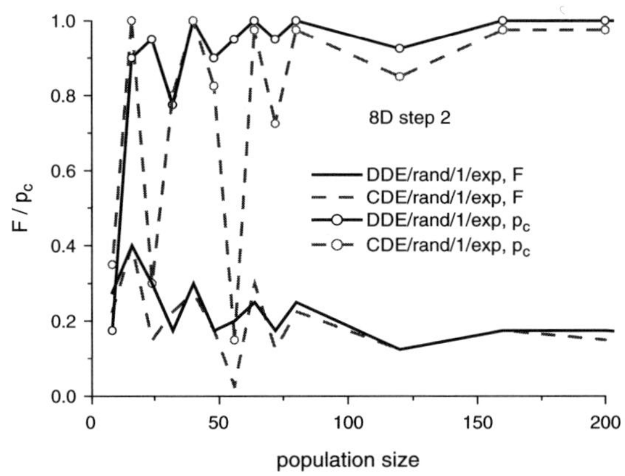

Figure 7.24 Mutation intensity and crossover probability of DE/rand/1/exp for the 8-dimensional step function 2

7.2.2.3 24-Dimensional Step Function 2

The optimal mutation intensity and crossover probability of differential evolution for the 24-dimensional step function 2 are shown in Figures 7.29–7.32.

7.2.2.4 50-Dimensional Step Function 2

The optimal mutation intensity and crossover probability of differential evolution for the 50-dimensional step function 2 are shown in Figures 7.33–7.36.

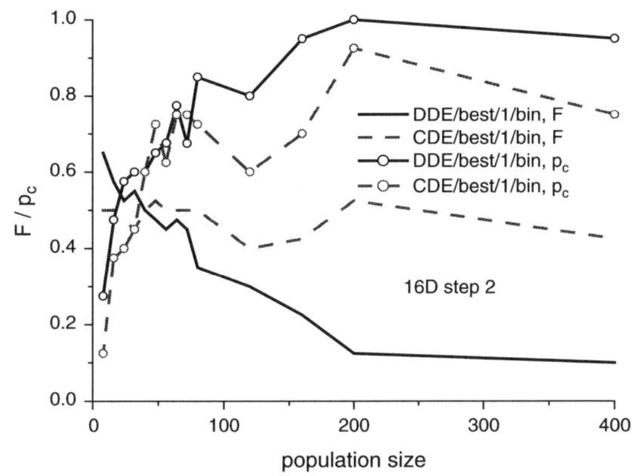

Figure 7.25 Mutation intensity and crossover probability of DE/best/1/bin for the 16-dimensional step function 2

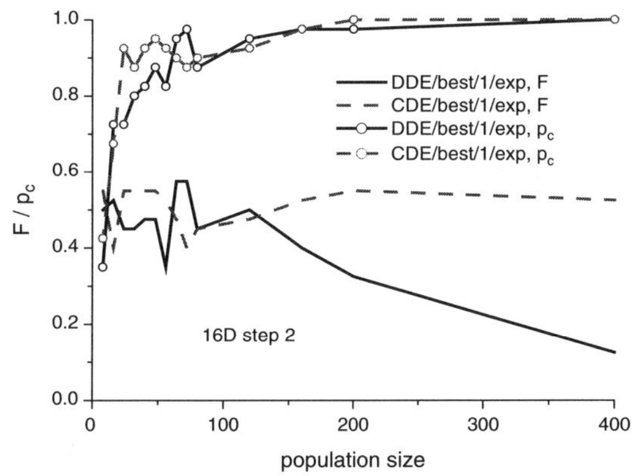

Figure 7.26 Mutation intensity and crossover probability of DE/best/1/exp for the 16-dimensional step function 2

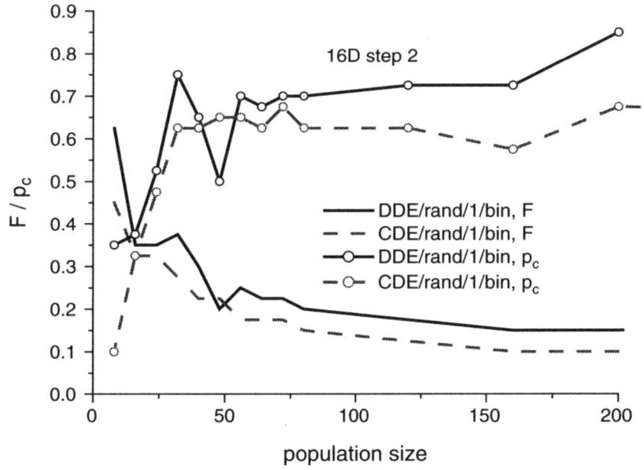

Figure 7.27 Mutation intensity and crossover probability of DE/rand/1/bin for the 16-dimensional step function 2

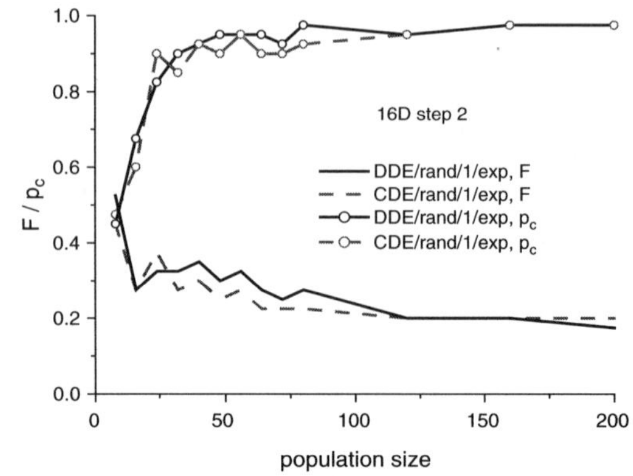

Figure 7.28 Mutation intensity and crossover probability of DE/rand/1/exp for the 16-dimensional step function 2

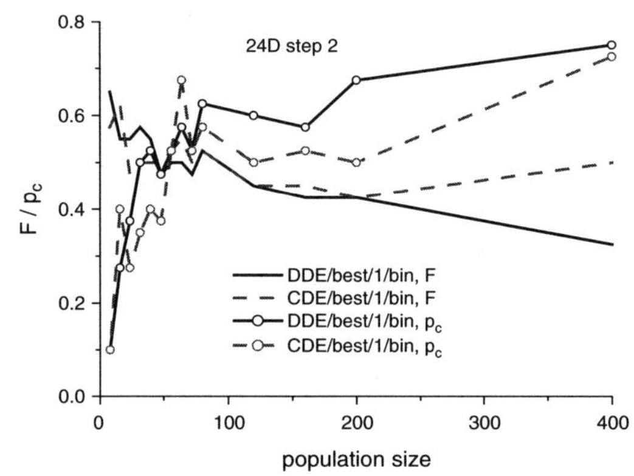

Figure 7.29 Mutation intensity and crossover probability of DE/best/1/bin for the 24-dimensional step function 2

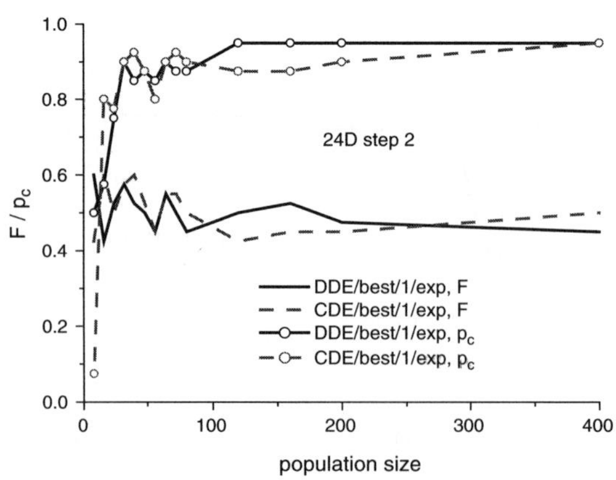

Figure 7.30 Mutation intensity and crossover probability of DE/best/1/exp for the 24-dimensional step function 2

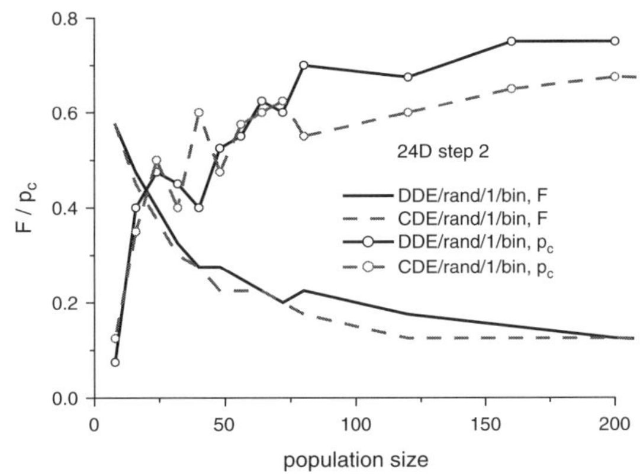

Figure 7.31 Mutation intensity and crossover probability of DE/rand/1/bin for the 24-dimensional step function 2

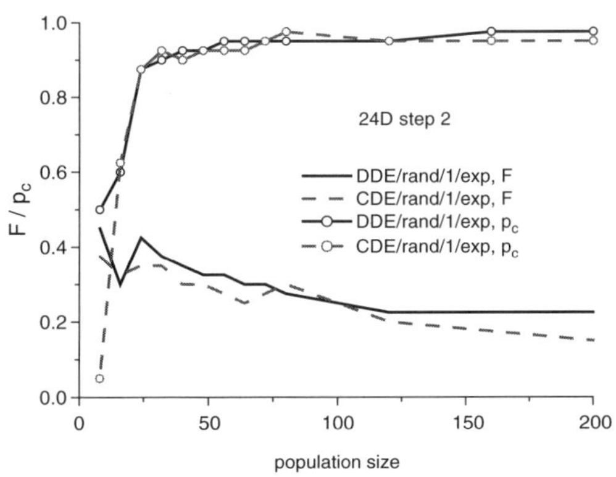

Figure 7.32 Mutation intensity and crossover probability of DE/rand/1/exp for the 24-dimensional step function 2

7.2.2.5 Comparison with Sphere Function

From Figures 7.21–7.36 and Figures 7.1–7.20, we note the following:

(a) There is no notable difference between mutation intensity and crossover for DE/rand/1/* for the sphere function and those for step function 2.
(b) Mutation intensity and crossover probability for CDE/best/1/* for step function 2 are consistent with those for the sphere function.
(c) Mutation intensity and crossover probability for DDE/best/1/bin are now very sensitive to problem dimension. They rapidly approach those for CDE/best/1/bin.
(d) Mutation intensity for DDE/best/1/exp is also very sensitive to problem now. It is also rapidly approaching that for CDE/best/1/exp.

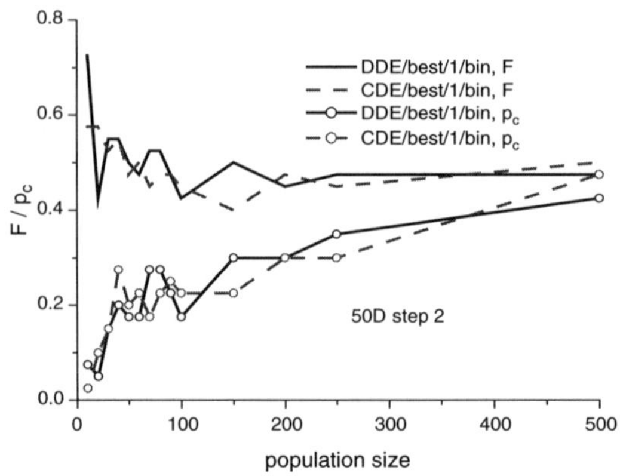

Figure 7.33 Mutation intensity and crossover probability of DE/best/1/bin for the 50-dimensional step function 2

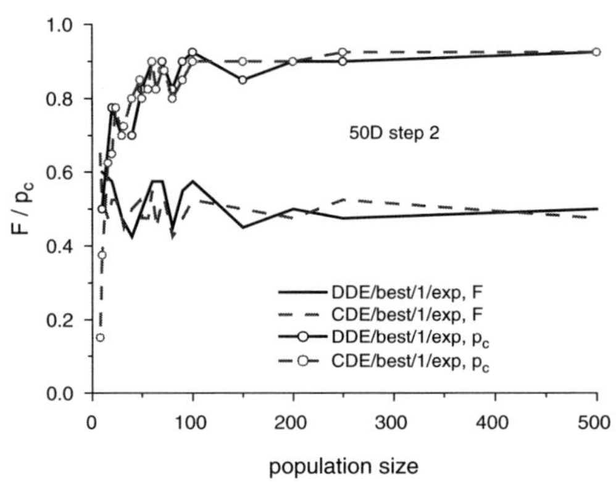

Figure 7.34 Mutation intensity and crossover probability of DE/best/1/exp for the 50-dimensional step function 2

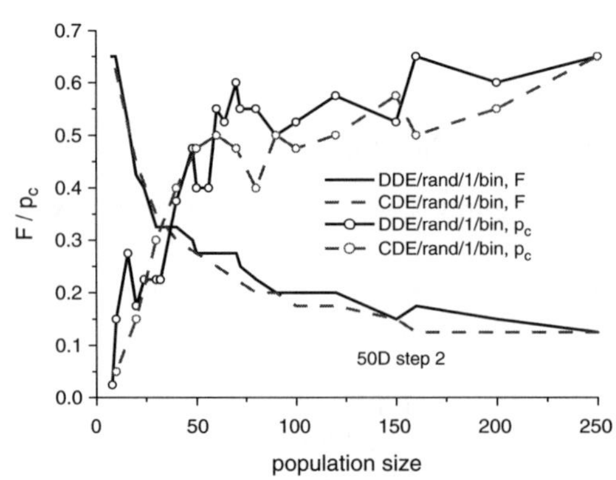

Figure 7.35 Mutation intensity and crossover probability of DE/rand/1/bin for the 50-dimensional step function 2

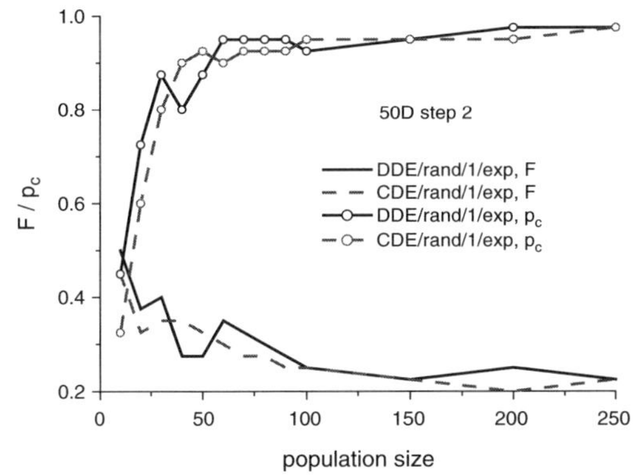

Figure 7.36 Mutation intensity and crossover Probability of DE/rand/1/exp for the 50-dimensional step function 2

7.2.2.6 Effect of Discontinuity of Objective Function

Discontinuity of step function 2 introduces an addition limitation on population diversity. Such limitation has to be compensated by dynamic evolution, larger population size, stronger mutation, and lower crossover probability.

7.3 Hyper-ellipsoid Function

7.3.1 Optimal Population Size

The optimal population size and the safeguard zone of population size for the hyper-ellipsoid function are summarized in Table 7.3. Optimal population size decreases and safeguard zone contracts as problem dimension increases.

Table 7.3 Optimal population size and safeguard zone of population size for the hyper-ellipsoid function

Strategy	Optimal population size				Safeguard zone			
	8D	16D	24D	50D	8D	16D	24D	50D
DDE/best/1/bin	64	48	40	20	[40,80]	[24,72]	[24,72]	[20,50]
CDE/best/1/bin	56	32	24	20	[32,80]	[16,56]	[16,48]	[20,40]
DDE/rand/1/bin	40	24	24	20	[24,56]	[16,48]	[16,48]	[20,40]
CDE/rand/1/bin	32	24	24	20	[24,56]	[16,48]	[16,48]	[20,40]
DDE/best/1/exp	56	48	40	20	[32,80]	[24,64]	[24,48]	[20,30]
CDE/best/1/exp	64	40	32	20	[32,80]	[24,56]	[24,48]	[20,30]
DDE/rand/1/exp	48	40	32	20	[24,64]	[24,48]	[24,40]	[20,20]
CDE/rand/1/exp	40	32	32	20	[24,56]	[24,48]	[24,40]	[20,20]

7.3.2 *Optimal Mutation Intensity and Crossover Probability*

7.3.2.1 8-Dimensional Hyper-ellipsoid Function

The optimal mutation intensity and crossover probability of differential evolution for the 8-dimensional hyper-ellipsoid function are shown in Figures 7.37–7.40.

7.3.2.2 16-Dimensional Hyper-ellipsoid Function

The optimal mutation intensity and crossover probability of differential evolution for the 16-dimensional hyper-ellipsoid function are shown in Figures 7.41–7.44.

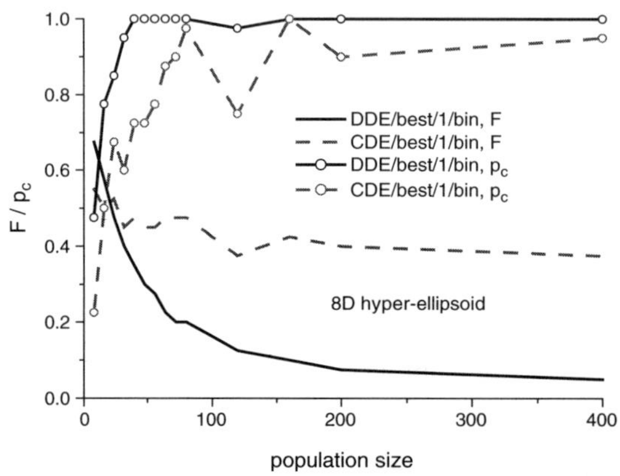

Figure 7.37 Mutation intensity and crossover probability of DE/best/1/bin for the 8-dimensional hyper-ellipsoid function

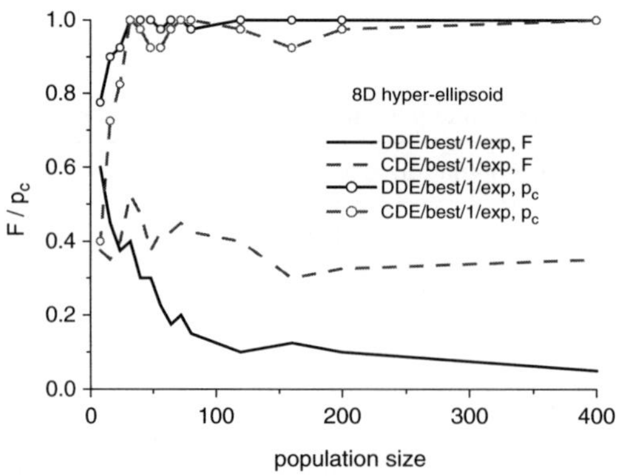

Figure 7.38 Mutation intensity and crossover probability of DE/best/1/exp for the 8-dimensional hyper-ellipsoid function

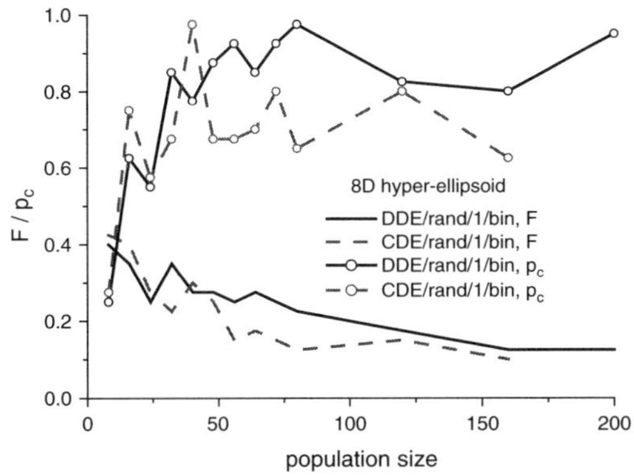

Figure 7.39 Mutation intensity and crossover probability of DE/rand/1/bin for the 8-dimensional hyper-ellipsoid function

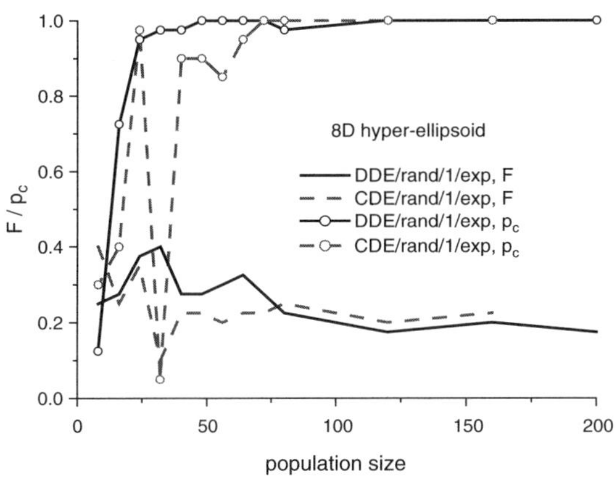

Figure 7.40 Mutation intensity and crossover probability of DE/rand/1/exp for the 8-dimensional hyper-ellipsoid function

7.3.2.3 24-Dimensional Hyper-ellipsoid Function

The optimal mutation intensity and crossover probability of differential evolution for the 24-dimensional hyper-ellipsoid function are shown in Figures 7.45–7.48.

7.3.2.4 50-Dimensional Hyper-ellipsoid Function

The optimal mutation intensity and crossover probability of differential evolution for the 50-dimensional hyper-ellipsoid function are shown in Figures 7.49–7.52.

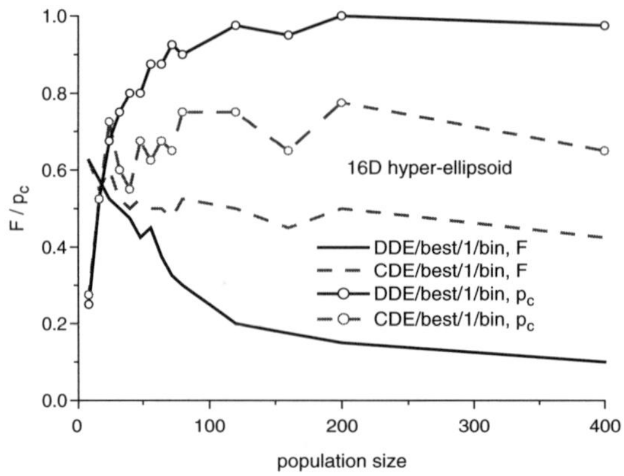

Figure 7.41 Mutation intensity and crossover probability of DE/best/1/bin for the 16-dimensional hyper-ellipsoid function

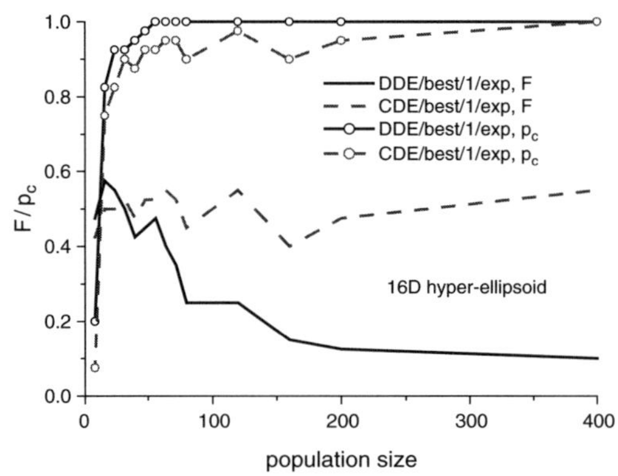

Figure 7.42 Mutation intensity and crossover probability of DE/best/1/exp for the 16-dimensional hyper-ellipsoid function

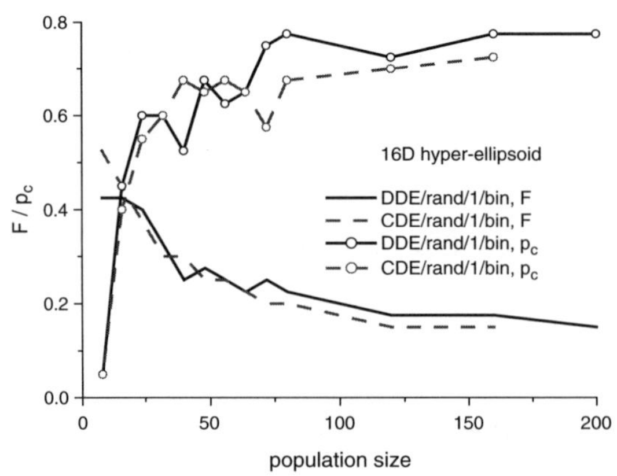

Figure 7.43 Mutation intensity and crossover probability of DE/rand/1/bin for the 16-dimensional hyper-ellipsoid function

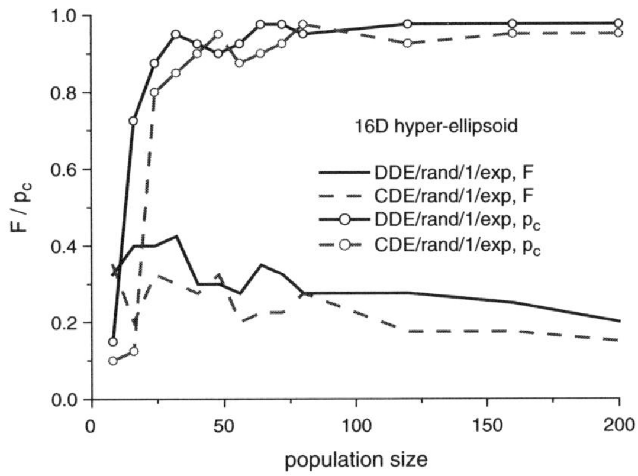

Figure 7.44 Mutation intensity and crossover probability of DE/rand/1/exp for the 16-dimensional hyper-ellipsoid function

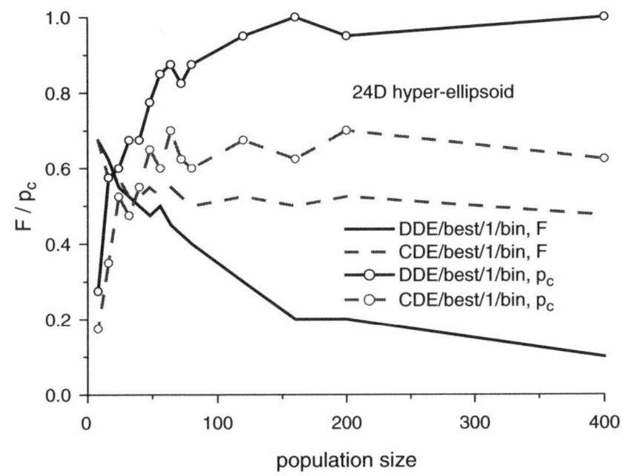

Figure 7.45 Mutation intensity and crossover probability of DE/best/1/bin for the 24-dimensional hyper-ellipsoid function

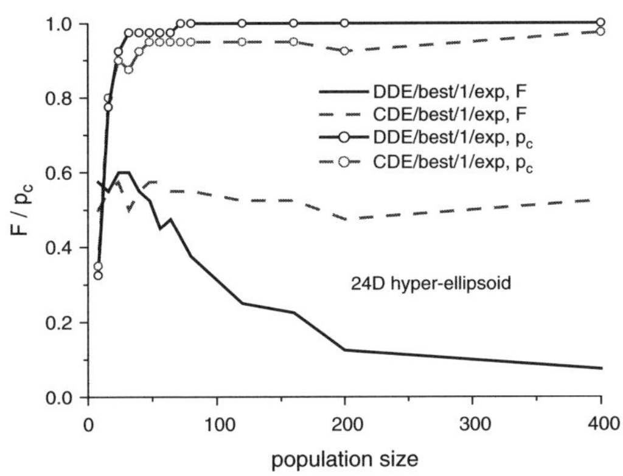

Figure 7.46 Mutation intensity and crossover probability of DE/best/1/exp for the 24-dimensional hyper-ellipsoid function

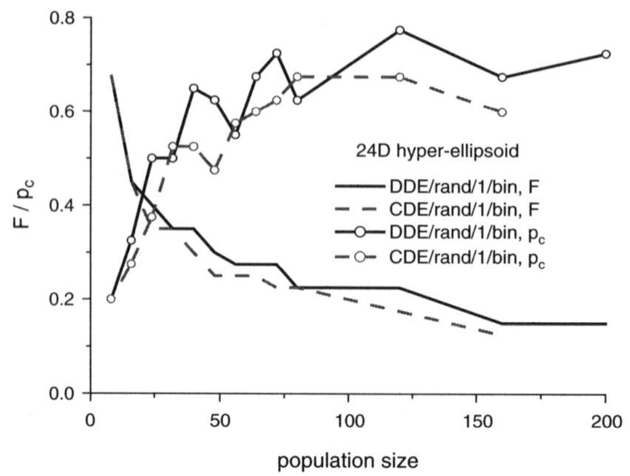

Figure 7.47 Mutation intensity and crossover probability of DE/rand/1/bin for the 24-dimensional hyper-ellipsoid function

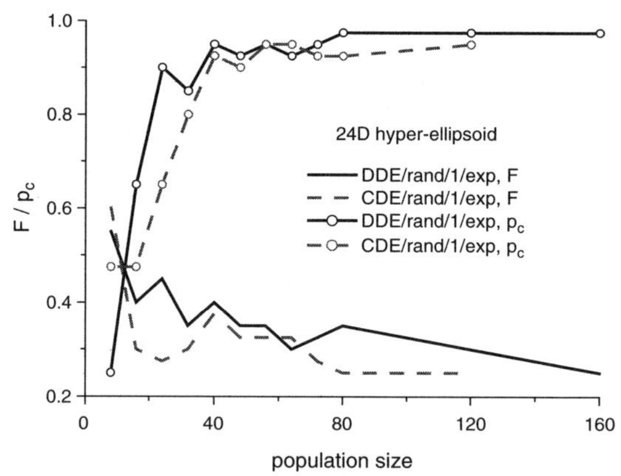

Figure 7.48 Mutation intensity and crossover probability of DE/rand/1/exp for the 24-dimensional hyper-ellipsoid function

7.3.2.5 Comparison with Sphere Function

From Figures 7.37–7.52 and Figures 7.1–7.20, we note that the general trends mentioned in Sections 7.1.2.1 and 7.1.2.6 remain present.

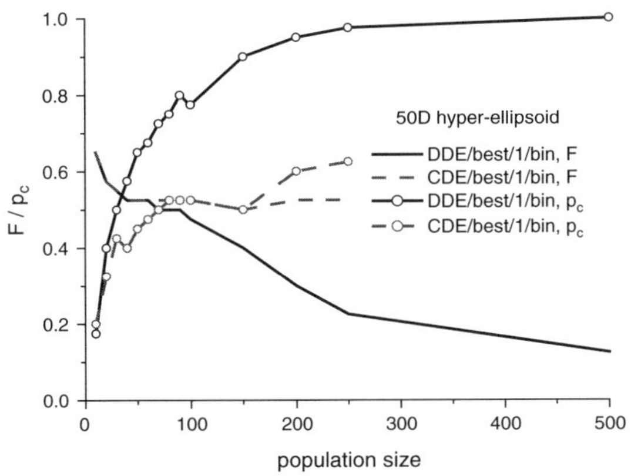

Figure 7.49 Mutation intensity and crossover probability of DE/best/1/bin for the 50-dimensional hyper-ellipsoid function

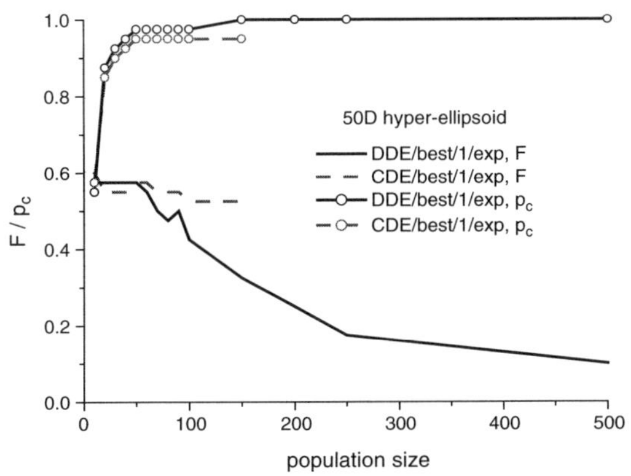

Figure 7.50 Mutation intensity and crossover probability of DE/best/1/exp for the 50-dimensional hyper-ellipsoid function

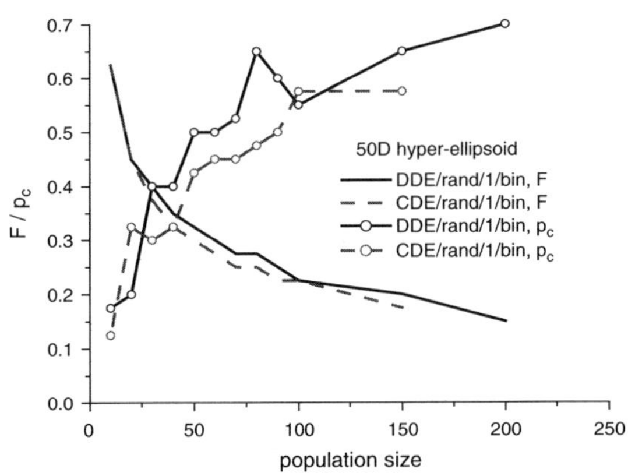

Figure 7.51 Mutation intensity and crossover probability of DE/rand/1/bin for the 50-dimensional hyper-ellipsoid function

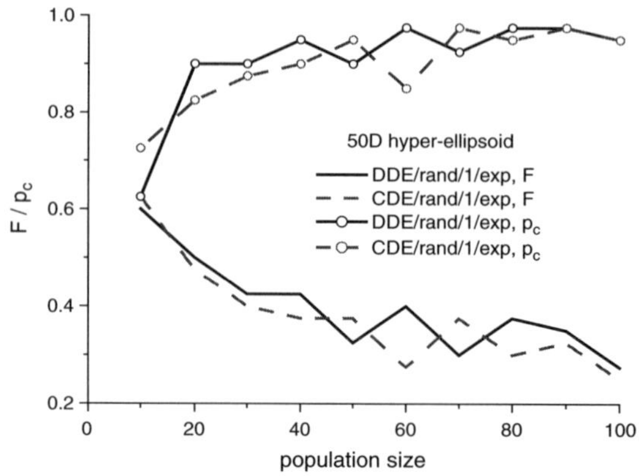

Figure 7.52 Mutation intensity and crossover probability of DE/rand/1/exp for the 50-dimensional hyper-ellipsoid function

7.4 Qing Function

7.4.1 Optimal Population Size

The optimal population size and the safeguard zone of population size for Qing function are summarized in Table 7.4. Optimal population size decreases and safeguard zone contracts as problem dimension increases. It is worthwhile to point out that the optimal population size here is significantly smaller.

Table 7.4 Optimal population size and safeguard zone of population size for the Qing function

Strategy	Optimal population size				Safeguard zone			
	8D	16D	24D	50D	8D	16D	24D	50D
DDE/best/1/bin	40	32	32	30	[32,64]	[24,64]	[24,56]	[20,50]
CDE/best/1/bin	40	32	24	30	[24,56]	[24,48]	[24,40]	[20,40]
DDE/rand/1/bin	16	16	16	20	[16,24]	[16,16]	[16,16]	[10,30]
CDE/rand/1/bin	16	16	16	20	[16,24]	[16,16]	[16,16]	[10,30]
DDE/best/1/exp	48	40	32	30	[24,64]	[24,56]	[24,48]	[20,40]
CDE/best/1/exp	40	32	32	30	[24,64]	[24,48]	[16,48]	[20,40]
DDE/rand/1/exp	24	24	24	20	[16,40]	[16,32]	[16,32]	[20,30]
CDE/rand/1/exp	24	24	24	20	[16,32]	[16,32]	[16,32]	[20,30]

7.4.2 Optimal Mutation Intensity and Crossover Probability

7.4.2.1 8-Dimensional Qing Function

The optimal mutation intensity and crossover probability of differential evolution for the 8-dimensional Qing function are shown in Figures 7.53–7.56. Note that mutation intensity and crossover probability for DE/rand/* are approaching 0.

7.4.2.2 16-Dimensional Qing Function

The optimal mutation intensity and crossover probability of differential evolution for the 16-dimensional Qing function are shown in Figures 7.57–7.60.

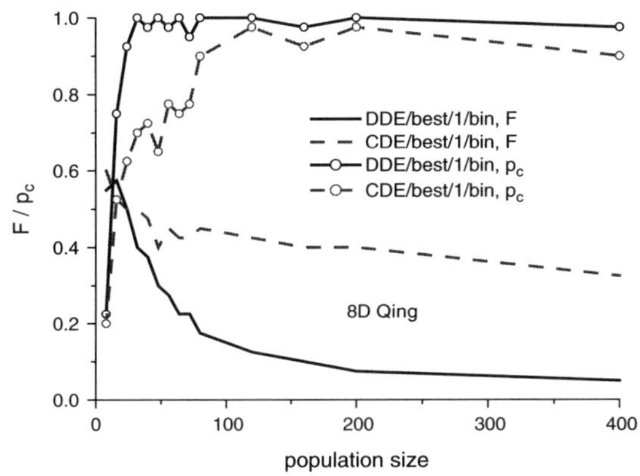

Figure 7.53 Mutation intensity and crossover probability of DE/best/1/bin for the 8-dimensional Qing function

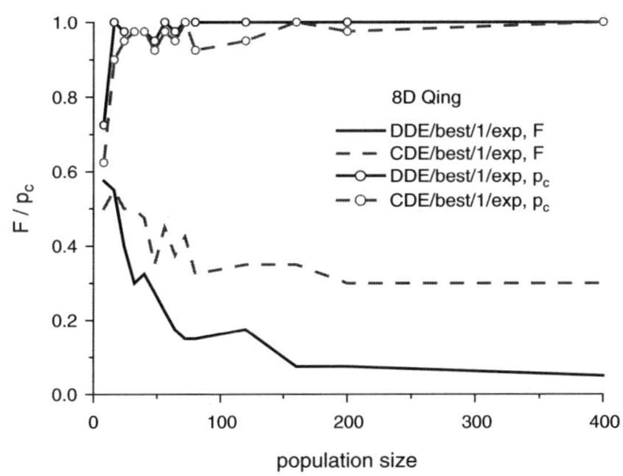

Figure 7.54 Mutation intensity and crossover probability of DE/best/1/exp for the 8-dimensional Qing function

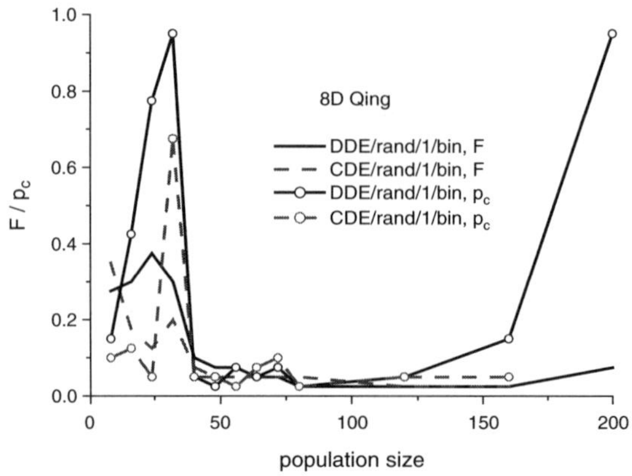

Figure 7.55 Mutation intensity and crossover probability of DE/rand/1/bin for the 8-dimensional Qing function

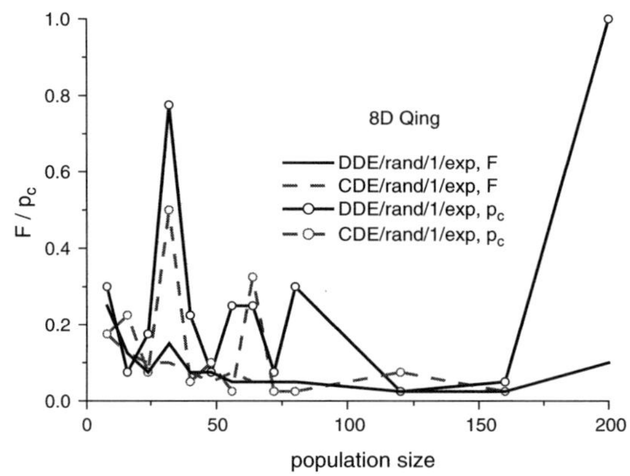

Figure 7.56 Mutation intensity and crossover probability of DE/rand/1/exp for the 8-dimensional Qing function

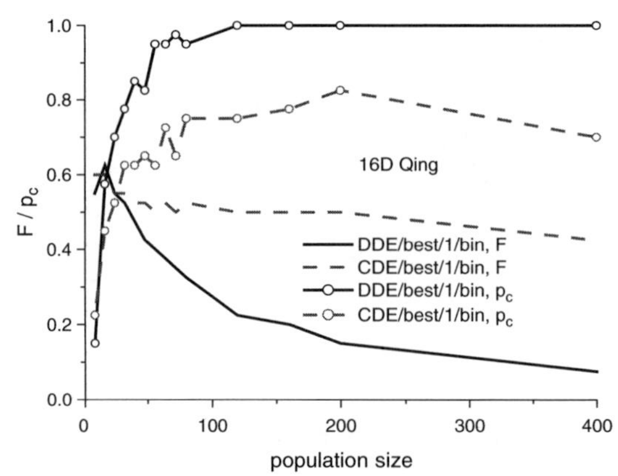

Figure 7.57 Mutation intensity and crossover probability of DE/best/1/bin for the 16-dimensional Qing function

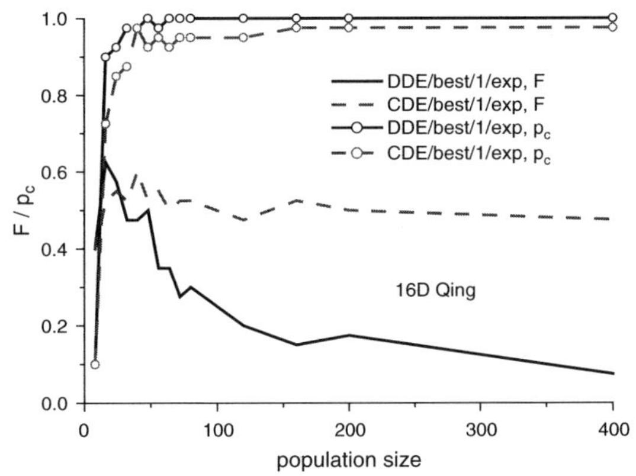

Figure 7.58 Mutation intensity and crossover probability of DE/best/1/exp for the 16-dimensional Qing function

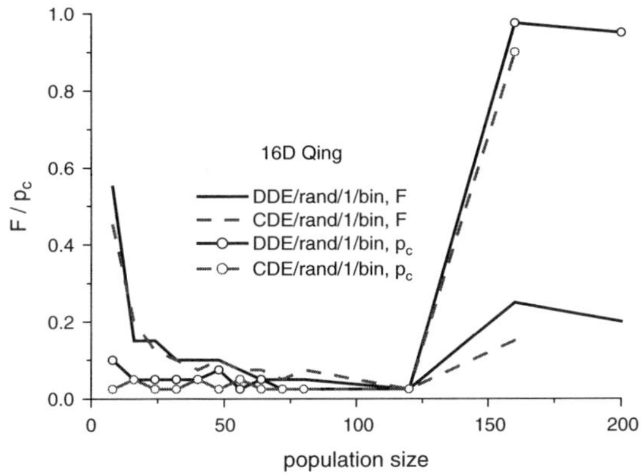

Figure 7.59 Mutation intensity and crossover probability of DE/rand/1/bin for the 16-dimensional Qing function

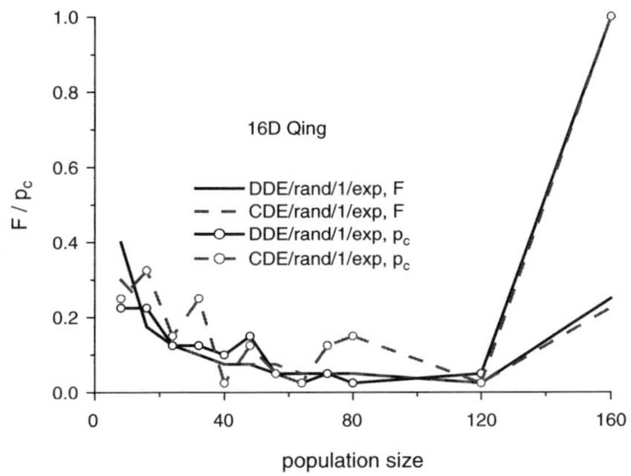

Figure 7.60 Mutation intensity and crossover probability of DE/rand/1/exp for the 16-dimensional Qing function

7.4.2.3 24-Dimensional Qing Function

The optimal mutation intensity and crossover probability of differential evolution for the 24-dimensional Qing function are shown in Figures 7.61–7.64.

7.4.2.4 50-Dimensional Qing Function

The optimal mutation intensity and crossover probability of differential evolution for the 50-dimensional Qing function are shown in Figures 7.65–7.68.

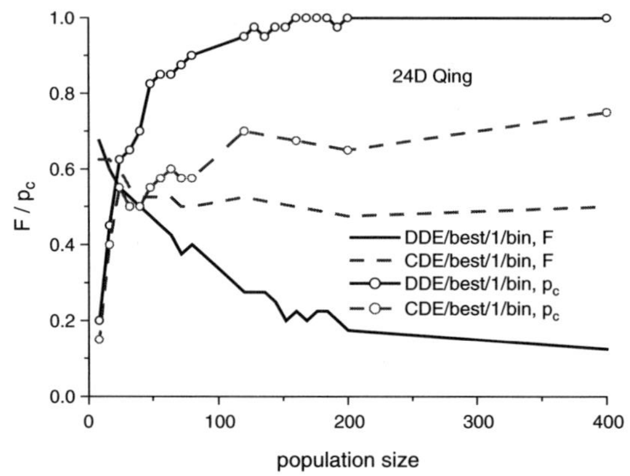

Figure 7.61 Mutation intensity and crossover probability of DE/best/1/bin for the 24-dimensional Qing function

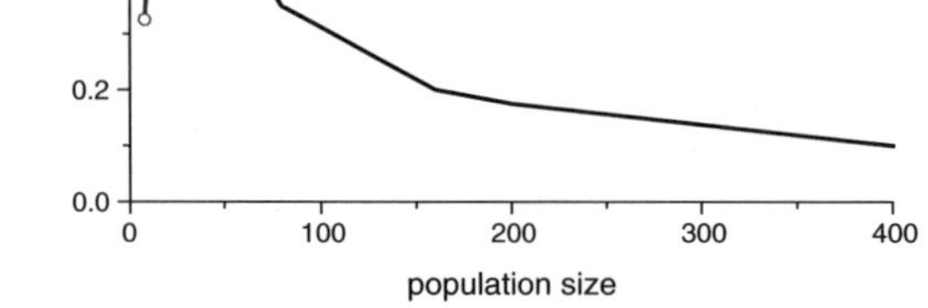

Figure 7.62 Mutation intensity and crossover probability of DE/best/1/exp for the 24-dimensional Qing function

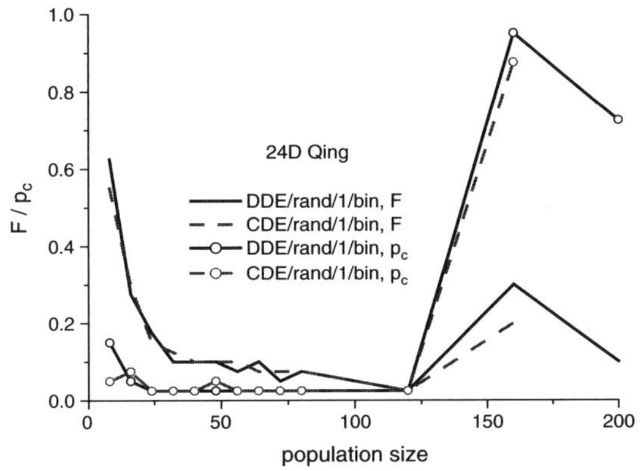

Figure 7.63 Mutation intensity and crossover probability of DE/rand/1/bin for the 24-dimensional Qing function

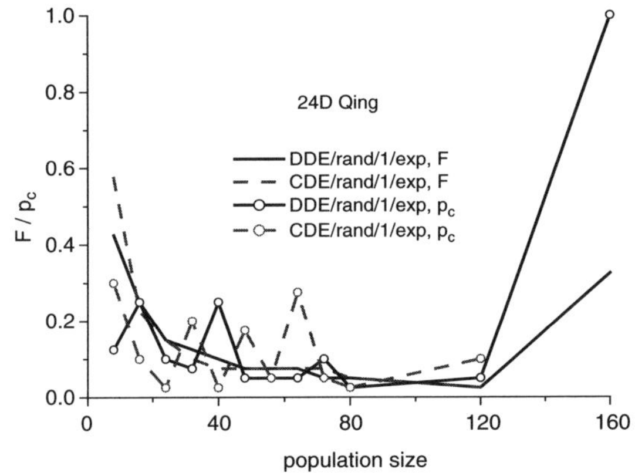

Figure 7.64 Mutation intensity and crossover probability of DE/rand/1/exp for the 24-dimensional Qing function

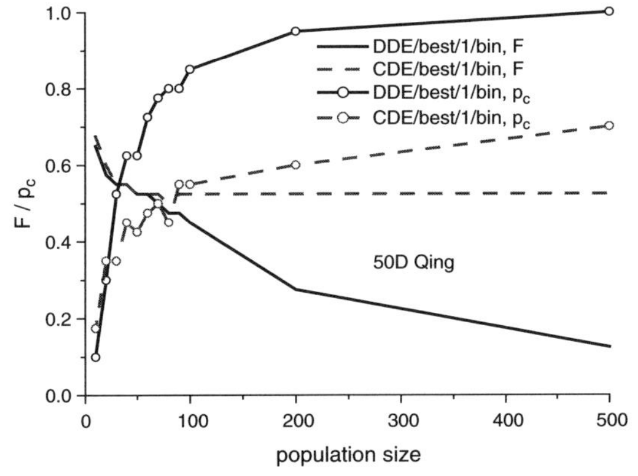

Figure 7.65 Mutation intensity and crossover probability of DE/best/1/bin for the 50-dimensional Qing function

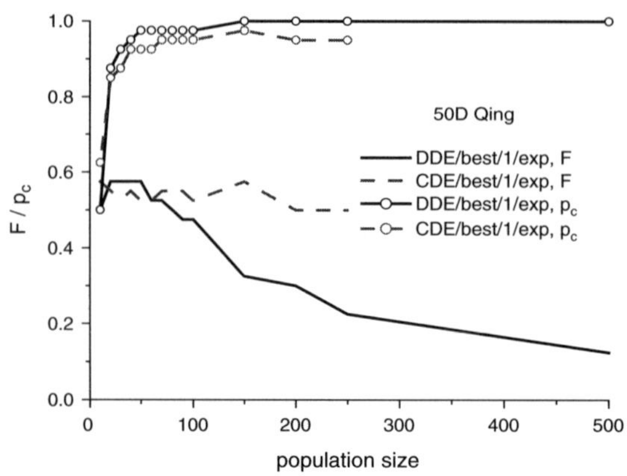

Figure 7.66 Mutation intensity and crossover probability of DE/best/1/exp for the 50-dimensional Qing function

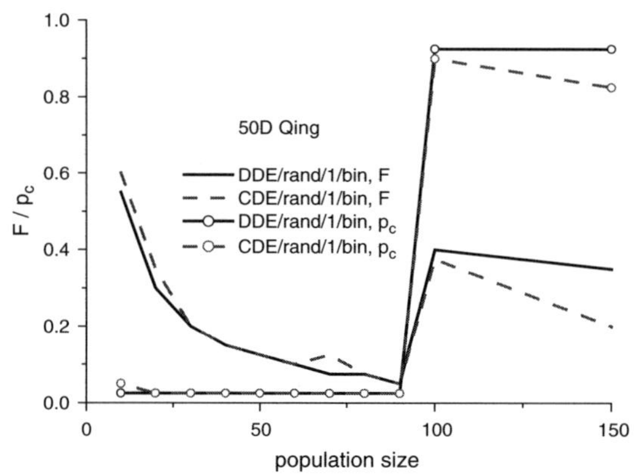

Figure 7.67 Mutation intensity and crossover probability of DE/rand/1/bin for the 50-dimensional Qing function

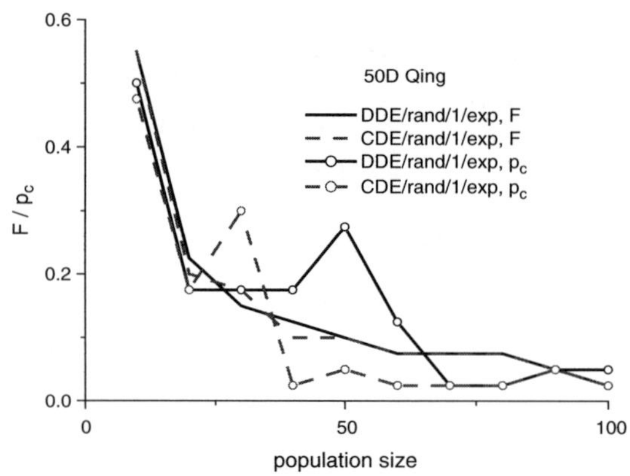

Figure 7.68 Mutation intensity and crossover probability of DE/rand/1/exp for the 50-dimensional Qing function

7.4.2.5 Effect of Solution Uniqueness

From Figures 7.53–7.68 and Figures 7.1–7.20, it is noted that the mutation intensity and crossover probability for DE/best/1/* are consistent with those for the sphere function. However, the mutation intensity and crossover probability for DE/rand/1/* are very close to 0.

7.5 Schwefel Function 2.22

7.5.1 Optimal Population Size

The optimal population size and the safeguard zone of population size for the Schwefel function 2.22 are summarized in Table 7.5. It seems that the optimal population size is independent of problem dimension. The safeguard zone of DE/*/1/bin also changes little with problem dimension. However, the safeguard zone of DE/*/1/exp decreases slowly as problem dimension increases.

Table 7.5 Optimal population size and safeguard zone of population size for the Schwefel function 2.22

Strategy	Optimal population size				Safeguard zone			
	8D	16D	24D	50D	8D	16D	24D	50D
DDE/best/1/bin	48	40	40	40	[32,72]	[24,64]	[24,64]	[30,70]
CDE/best/1/bin	40	32	32	30	[24,64]	[24,56]	[24,56]	[20,60]
DDE/rand/1/bin	24	24	24	30	[24,48]	[16,48]	[16,56]	[20,70]
CDE/rand/1/bin	24	24	24	20	[16,56]	[24,48]	[24,56]	[20,70]
DDE/best/1/exp	56	40	40	40	[32,72]	[24,64]	[24,56]	[30,50]
CDE/best/1/exp	56	48	40	40	[24,64]	[24,56]	[24,56]	[30,50]
DDE/rand/1/exp	40	32	40	40	[24,56]	[24,48]	[24,48]	[30,40]
CDE/rand/1/exp	40	40	32	40	[24,48]	[24,48]	[24,48]	[30,40]

7.5.2 Optimal Mutation Intensity and Crossover Probability

7.5.2.1 8-Dimensional Schwefel Function 2.22

The optimal mutation intensity and crossover probability of differential evolution for the 8-dimensional Schwefel function 2.22 are shown in Figures 7.69–7.72:

(a) Best differential mutation base prefers higher mutation intensity.
(b) Crossover probability for DDE/best/1/bin is higher than that for DDE/rand/1/bin, while crossover probability for CDE/best/1/bin is smaller than that for CDE/rand/1/bin.

7.5.2.2 16-Dimensional Schwefel Function 2.22

The optimal mutation intensity and crossover probability of differential evolution for the 16-dimensional Schwefel function 2.22 are shown in Figures 7.73–7.76.

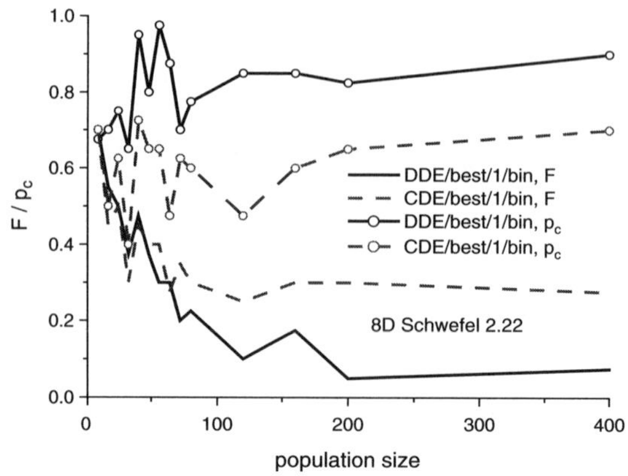

Figure 7.69 Mutation intensity and crossover probability of DE/best/1/bin for the 8-dimensional Schwefel function 2.22

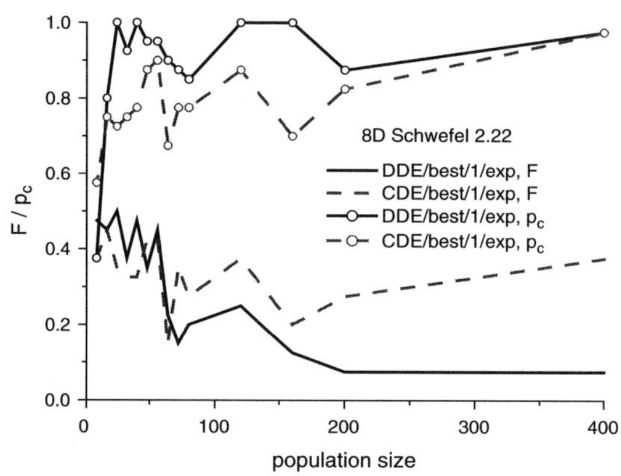

Figure 7.70 Mutation intensity and crossover probability of DE/best/1/exp for the 8-dimensional Schwefel function 2.22

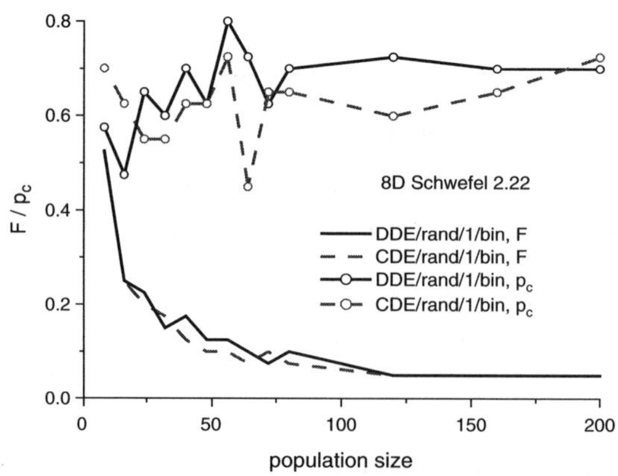

Figure 7.71 Mutation intensity and crossover probability of DE/rand/1/bin for the 8-dimensional Schwefel function 2.22

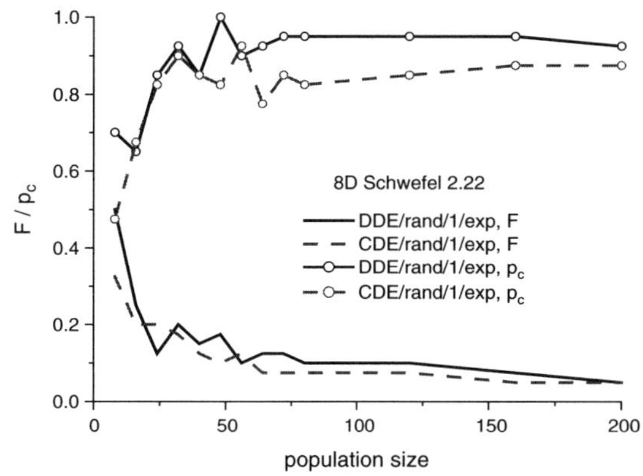

Figure 7.72 Mutation intensity and crossover probability of DE/rand/1/exp for the 8-dimensional Schwefel function 2.22

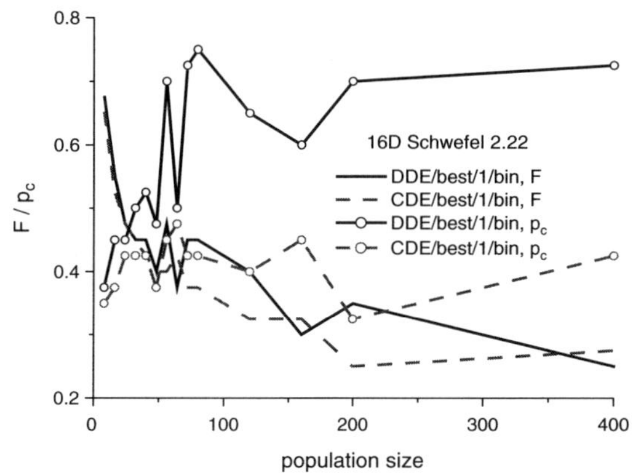

Figure 7.73 Mutation intensity and crossover probability of DE/best/1/bin for the 16-dimensional Schwefel function 2.22

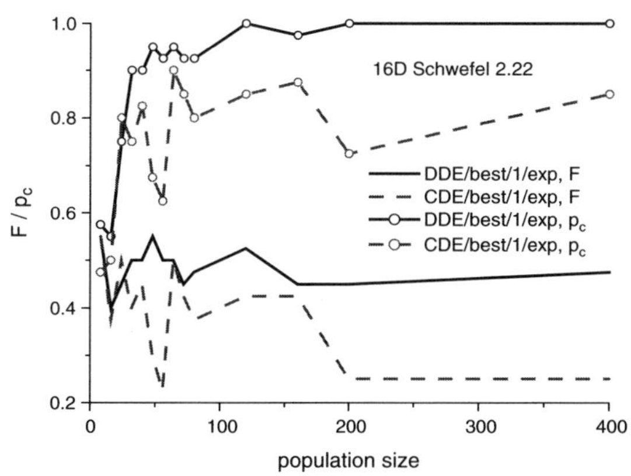

Figure 7.74 Mutation intensity and crossover probability of DE/best/1/exp for the 16-dimensional Schwefel function 2.22

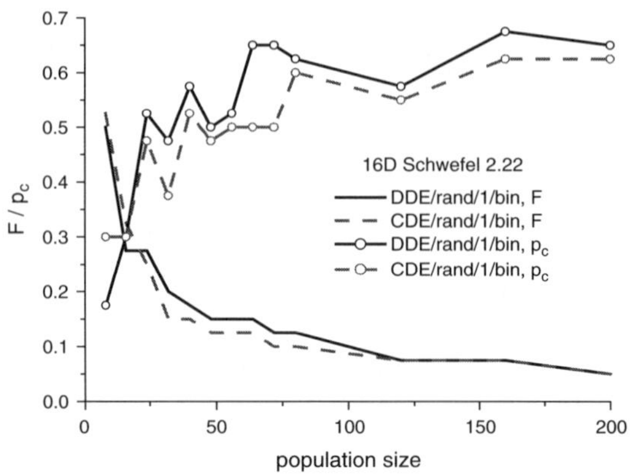

Figure 7.75 Mutation intensity and crossover probability of DE/rand/1/bin for the 16-dimensional Schwefel function 2.22

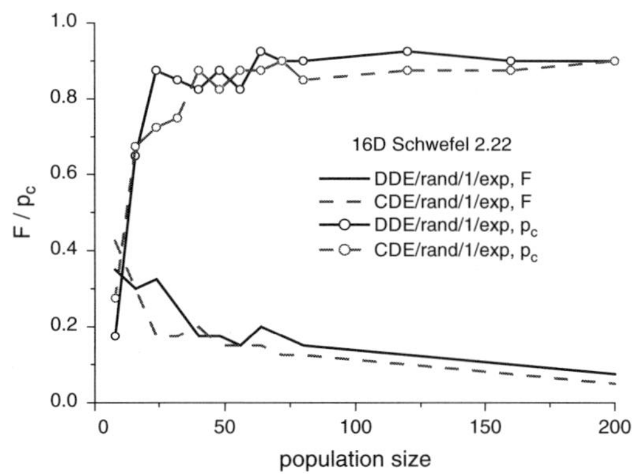

Figure 7.76 Mutation intensity and crossover probability of DE/rand/1/exp for the 16-dimensional Schwefel function 2.22

7.5.2.3 24-Dimensional Schwefel Function 2.22

The optimal mutation intensity and crossover probability of differential evolution for the 24-dimensional Schwefel 2.22 are shown in Figures 7.77–7.80.

7.5.2.4 50-Dimensional Schwefel Function 2.22

The optimal mutation intensity and crossover probability of differential evolution for the 50-dimensional Schwefel function 2.22 are shown in Figures 7.81–7.84.

7.5.2.5 Effect of Convexity

From Figures 7.69–7.84 and Figures 7.1–7.20, it is observed that the mutation intensity and crossover probability for DE/rand/1/* and CDE/best/1/* and the crossover probability for

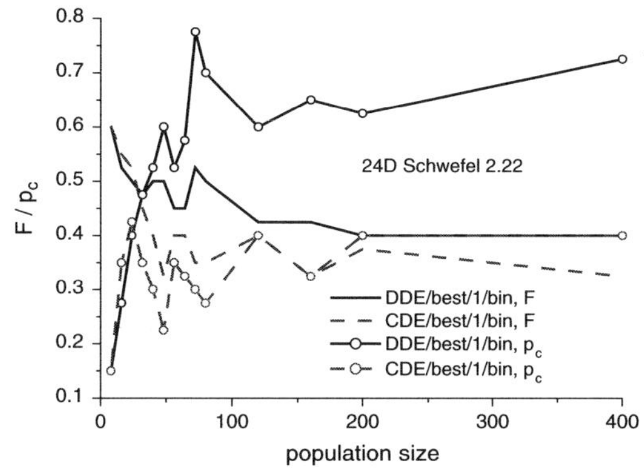

Figure 7.77 Mutation intensity and crossover probability of DE/best/1/bin for the 24-dimensional Schwefel function 2.22

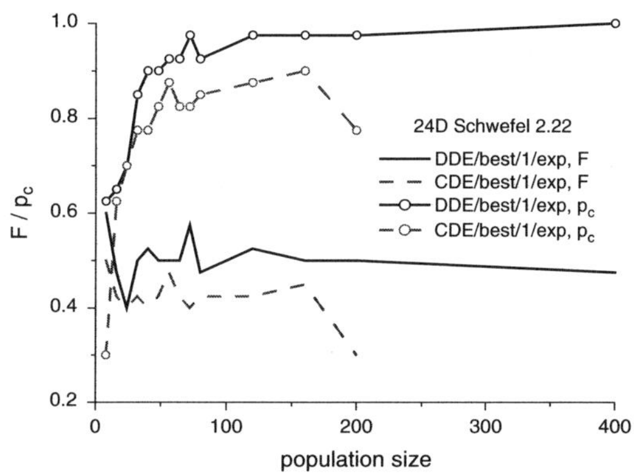

Figure 7.78 Mutation intensity and crossover probability of DE/best/1/exp for the 24-dimensional Schwefel function 2.22

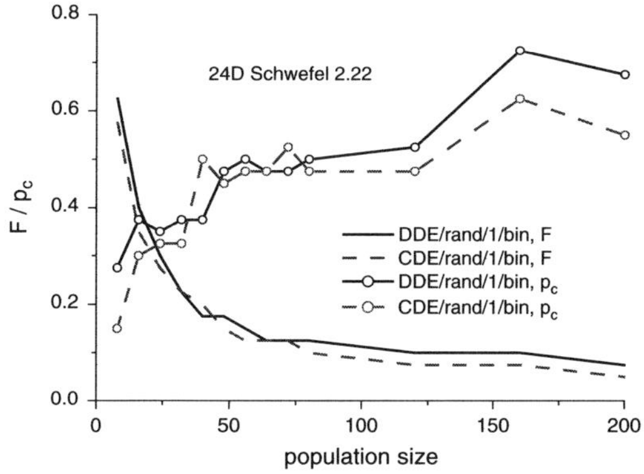

Figure 7.79 Mutation intensity and crossover probability of DE/rand/1/bin for the 24-dimensional Schwefel function 2.22

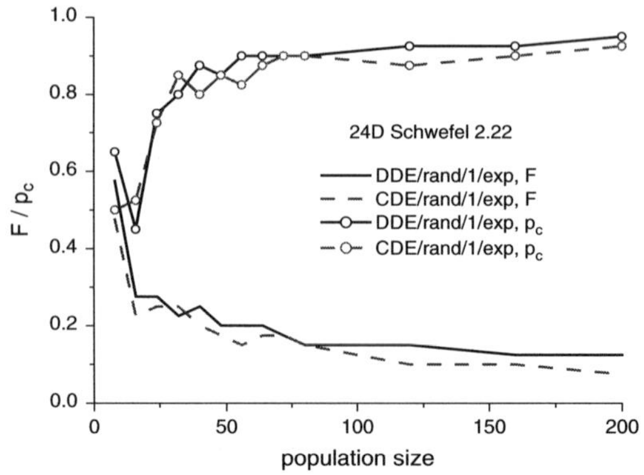

Figure 7.80 Mutation intensity and crossover probability of DE/rand/1/exp for the 24-dimensional Schwefel function 2.22

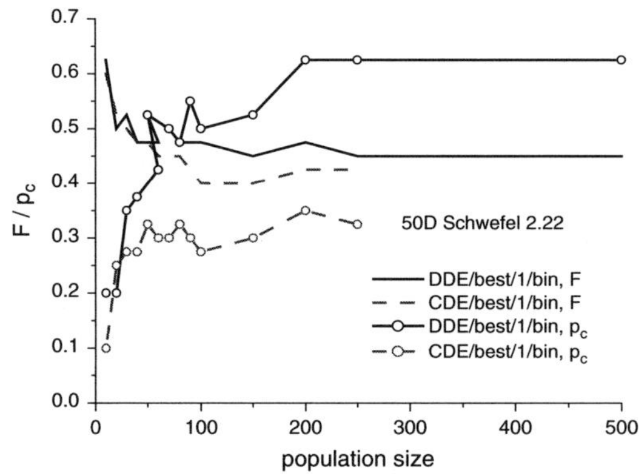

Figure 7.81 Mutation intensity and crossover probability of DE/best/1/bin for the 50-dimensional Schwefel function 2.22

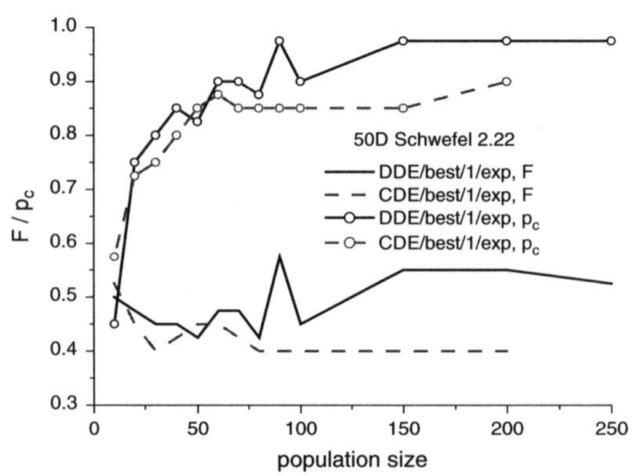

Figure 7.82 Mutation intensity and crossover probability of DE/best/1/exp for the 50-dimensional Schwefel function 2.22

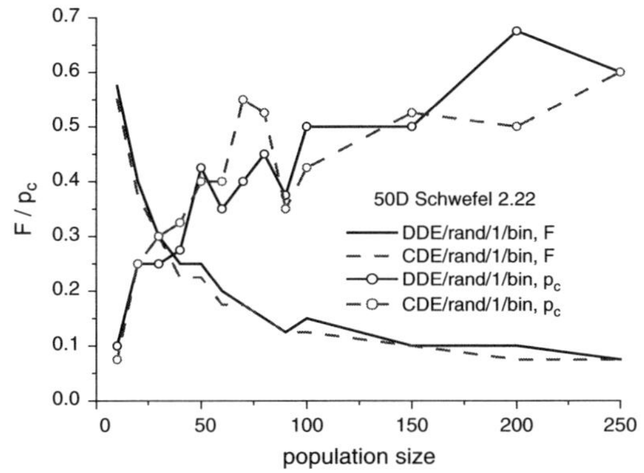

Figure 7.83 Mutation intensity and crossover probability of DE/rand/1/bin for the 50-dimensional Schwefel function 2.22

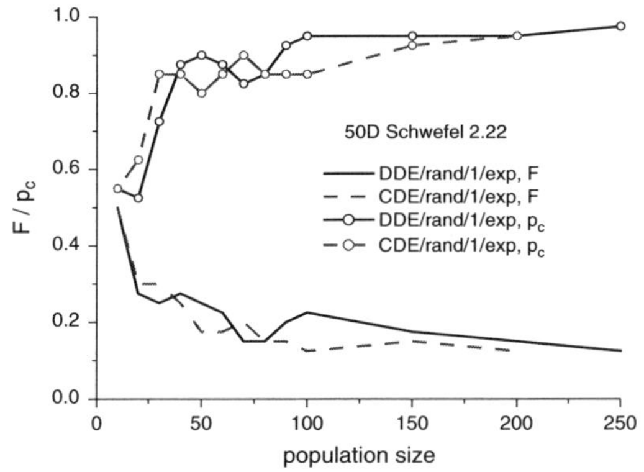

Figure 7.84 Mutation intensity and crossover probability of DE/rand/1/exp for the 50-dimensional Schwefel function 2.22

DDE/best/1/* are consistent with those for the sphere function. The mutation intensity for DDE/best/1/* here is remarkably different with that for the sphere function. It grows so quickly with problem dimension that it soon overtakes the mutation intensity for CDE/best/1/*.

7.6 Schwefel Function 2.26

7.6.1 Optimal Population Size

The optimal population size and the safeguard zone of population size for the Schwefel function 2.26 are summarized in Table 7.6. The optimal population size is moving upward.

Table 7.6 Optimal population size and safeguard zone of population size for the Schwefel function 2.26

Strategy	Optimal population size				Safeguard zone			
	8D	16D	24D	50D	8D	16D	24D	50D
DDE/best/1/bin	56	72	64	60	[56,80]	[32,80]	[40,120]	[60,80]
CDE/best/1/bin	80	72	64	70	[48,80]	[32,80]	[32,80]	[60,80]
DDE/rand/1/bin	48	32	48	40	[24,64]	[24,72]	[24,80]	[30,100]
CDE/rand/1/bin	32	32	40	40	[24,56]	[24,72]	[24,80]	[30,100]
DDE/best/1/exp	80	56	72	70	[48,120]	[40,80]	[56,80]	[60,90]
CDE/best/1/exp	80	64	80	70	[48,80]	[40,80]	[56,80]	[60,90]
DDE/rand/1/exp	56	56	64	70	[32,72]	[40,80]	[40,80]	[40,80]
CDE/rand/1/exp	56	64	64	70	[32,80]	[40,80]	[40,72]	[50,80]

The safeguard zone of population size for DE/*/1/bin seems intact, while that for DE/*/1/exp contracts as problem dimension increases.

7.6.2 Optimal Mutation Intensity and Crossover Probability

7.6.2.1 8-Dimensional Schwefel 2.26

The optimal mutation intensity and crossover probability of differential evolution for the 8-dimensional Schwefel function 2.26 are shown in Figures 7.85–7.88.

7.6.2.2 16-Dimensional Schwefel Function 2.26

The optimal mutation intensity and crossover probability of differential evolution for the 16-dimensional Schwefel function 2.26 are shown in Figures 7.89–7.92.

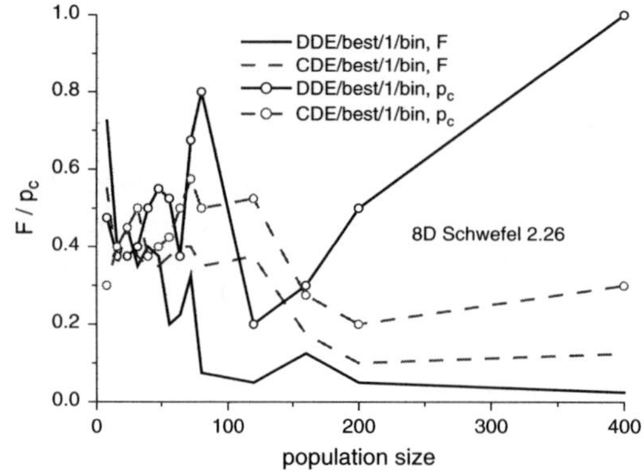

Figure 7.85 Mutation intensity and crossover probability of DE/best/1/bin for the 8-dimensional Schwefel function 2.26

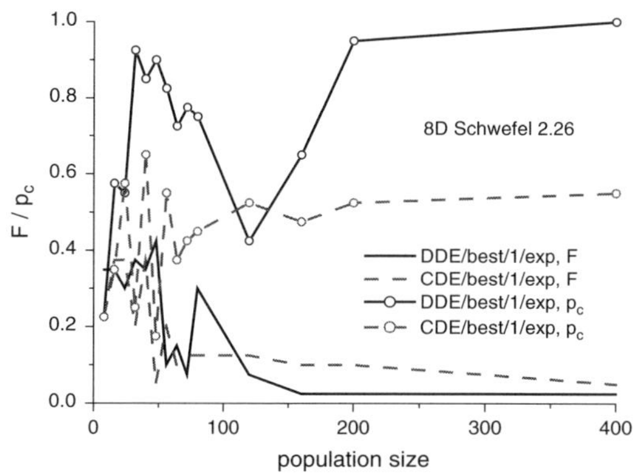

Figure 7.86 Mutation intensity and crossover probability of DE/best/1/exp for the 8-dimensional Schwefel function 2.26

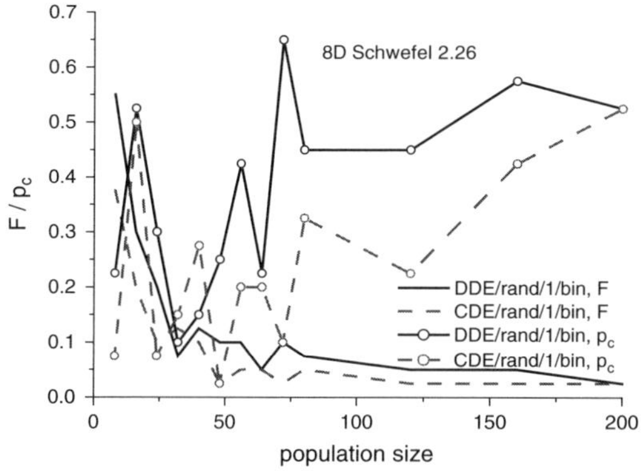

Figure 7.87 Mutation intensity and crossover probability of DE/rand/1/bin for the 8-dimensional Schwefel function 2.26

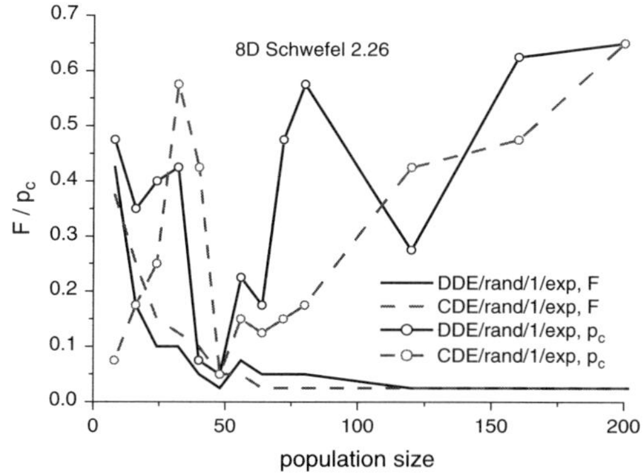

Figure 7.88 Mutation intensity and crossover probability of DE/rand/1/exp for the 8-dimensional Schwefel function 2.26

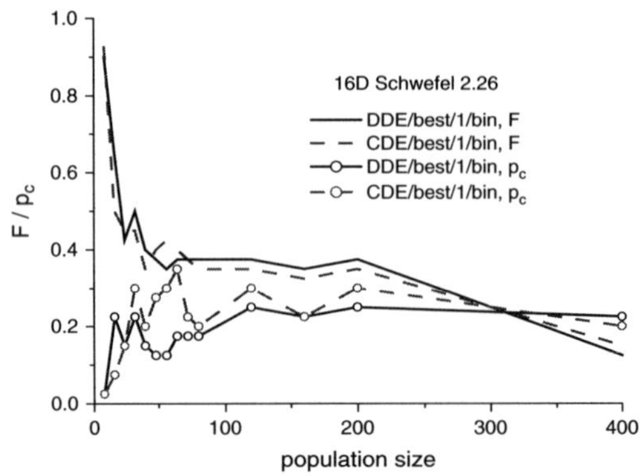

Figure 7.89 Mutation intensity and crossover probability of DE/best/1/bin for the 16-dimensional Schwefel function 2.26

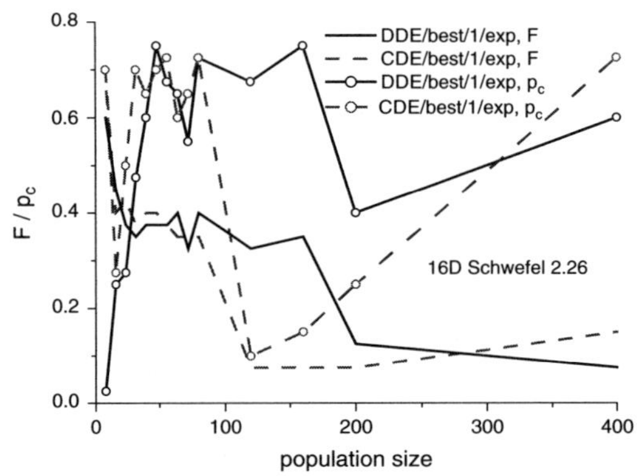

Figure 7.90 Mutation intensity and crossover probability of DE/best/1/exp for the 16-dimensional Schwefel function 2.26

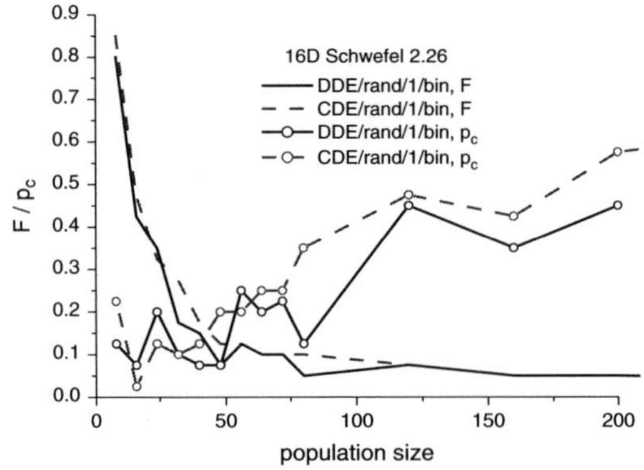

Figure 7.91 Mutation intensity and crossover probability of DE/rand/1/bin for the 16-dimensional Schwefel function 2.26

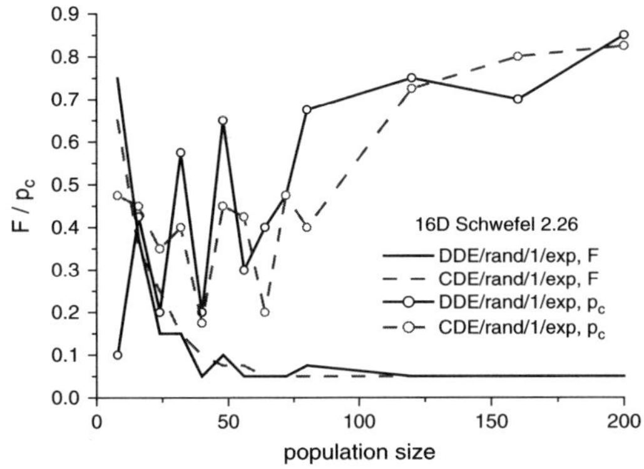

Figure 7.92 Mutation intensity and crossover probability of DE/rand/1/exp for the 16-dimensional Schwefel function 2.26

7.6.2.3 24-Dimensional Schwefel Function 2.26

The optimal mutation intensity and crossover probability of differential evolution for the 24-dimensional Schwefel function 2.26 are shown in Figures 7.93–7.96.

7.6.2.4 50-Dimensional Schwefel Function 2.26

The optimal mutation intensity and crossover probability of differential evolution for the 50-dimensional Schwefel function 2.26 are shown in Figures 7.97–7.100.

7.6.2.5 Effect of Modality

From Figures 7.85–7.100, we note that the mutation intensity and crossover probability here are quite different from those for the sphere function, step function 2, hyper-ellipsoid function,

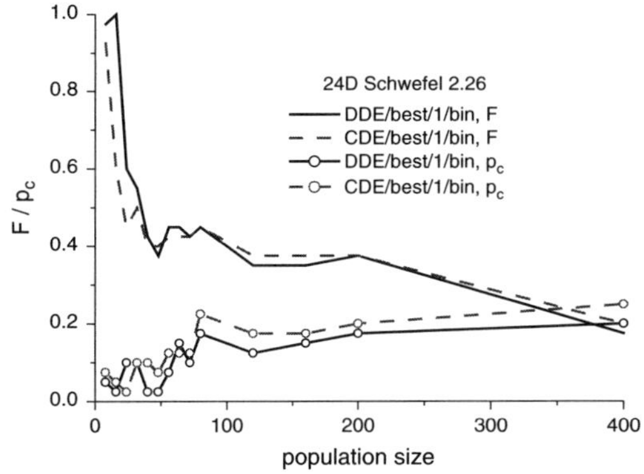

Figure 7.93 Mutation intensity and crossover probability of DE/best/1/bin for the 24-dimensional Schwefel function 2.26

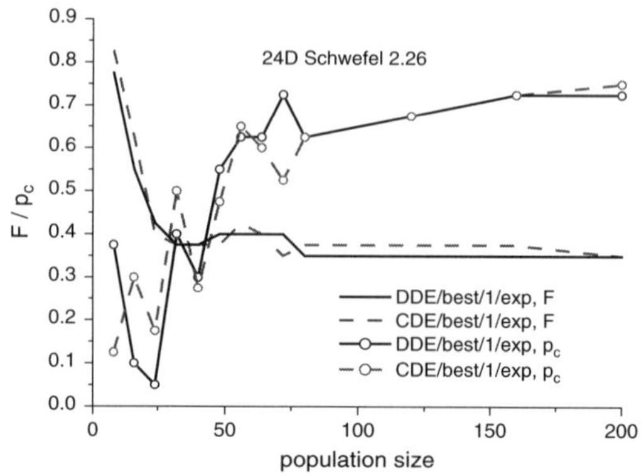

Figure 7.94 Mutation intensity and crossover probability of DE/best/1/exp for the 24-dimensional Schwefel function 2.26

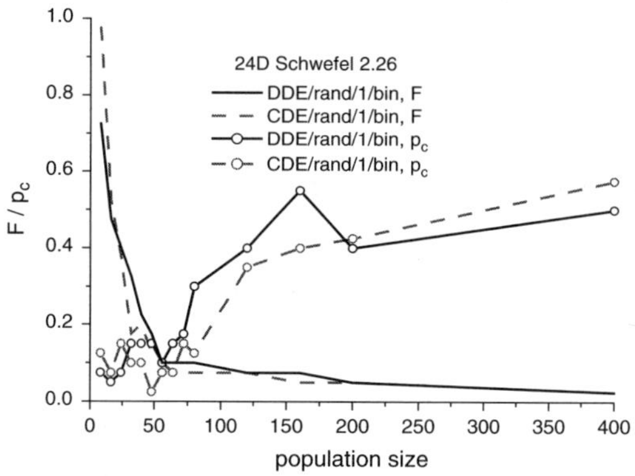

Figure 7.95 Mutation intensity and crossover probability of DE/rand/1/bin for the 24-dimensional Schwefel function 2.26

Qing function, and Schwefel function 2.22. In general:

(a) Mutation intensity decreases and crossover probability increases as population size increases.
(b) Mutation intensity and crossover probability appear to be independent of evolution mechanism.
(c) Mutation intensity for DE/best/1/* is larger than that for DE/rand/1/*.
(d) Mutation intensity for DE/best/1/* goes up slightly as problem dimension increases, while mutation intensity for DE/rand/1/* seems insensitive to problem dimension.
(e) Crossover probability has little effect on mutation intensity.
(f) Crossover probability for DE/rand/1/bin is larger than that for DE/best/1/bin. In contrast, crossover probability for DE/best/1/exp is larger than that for DE/rand/1/exp.

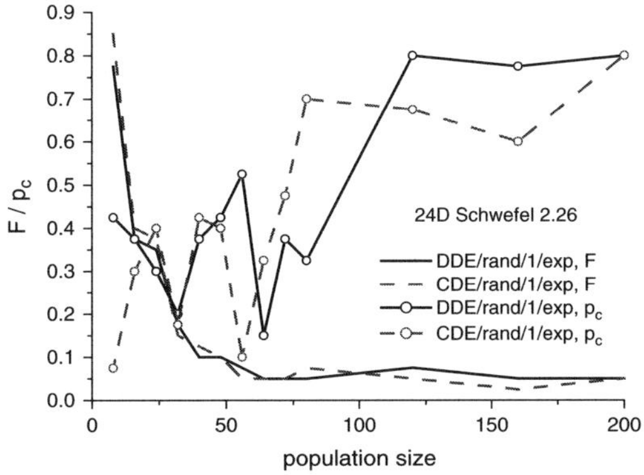

Figure 7.96 Mutation intensity and crossover probability of DE/rand/1/exp for the 24-dimensional Schwefel function 2.26

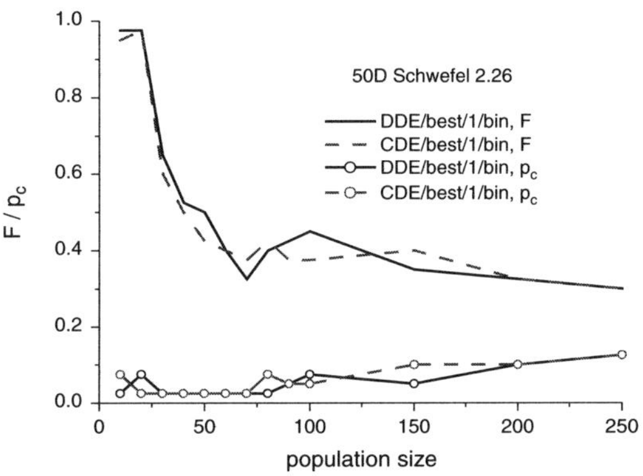

Figure 7.97 Mutation intensity and crossover probability of DE/best/1/bin for the 50-dimensional Schwefel function 2.26

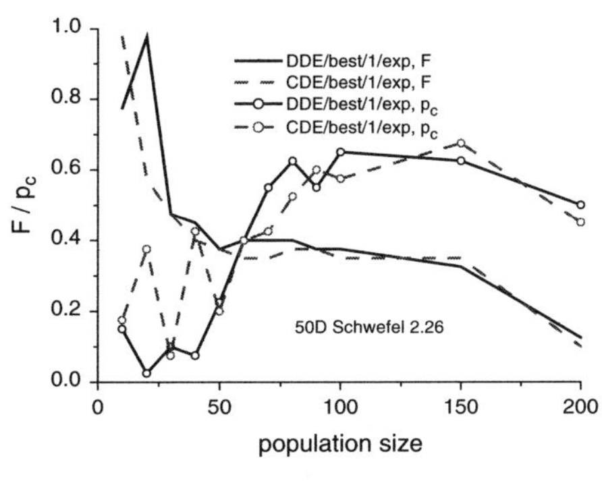

Figure 7.98 Mutation intensity and crossover probability of DE/best/1/exp for the 50-dimensional Schwefel function 2.26

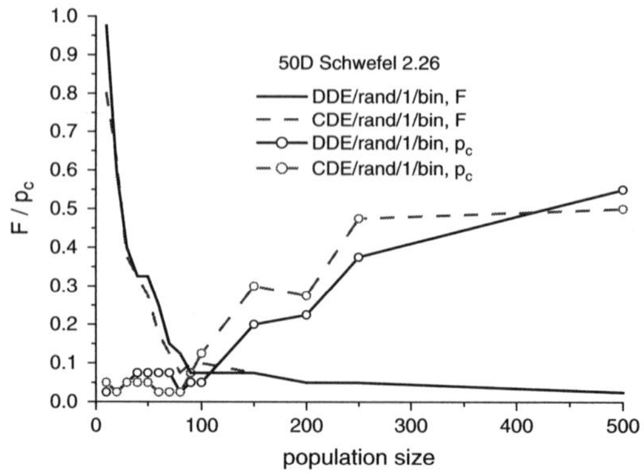

Figure 7.99 Mutation intensity and crossover probability of DE/rand/1/bin for the 50-dimensional Schwefel function 2.26

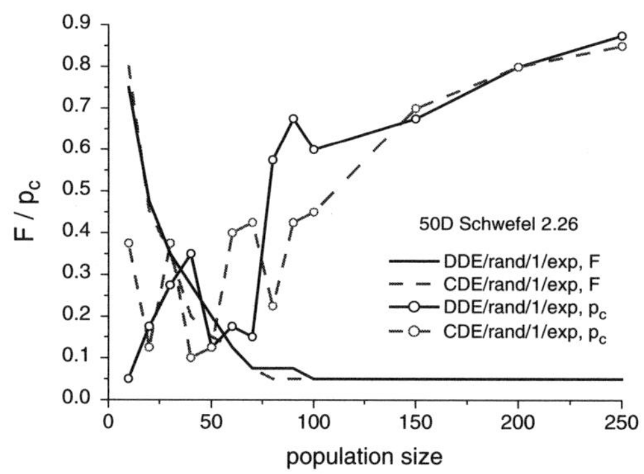

Figure 7.100 Mutation intensity and crossover probability of DE/rand/1/exp for the 50-dimensional Schwefel function 2.26

(g) Crossover probability for DE/*/1/exp is higher that that for DE/*/1/bin. Both decrease as problem dimension increases. Crossover probability for DE/*/1/bin quickly falls to almost 0.

7.7 Schwefel Function 1.2

7.7.1 Optimal Population Size

The optimal population size and the safeguard zone of the optimal population size for the Schwefel function 1.2 are summarized in Table 7.7. The optimal population size is now significantly smaller. The high optimal population size for DDE/best/1/* for the 50-dimensional Schwefel function 1.2 may be questionable since the number of successful trials is extremely small.

Table 7.7 Optimal population size and safeguard zone of optimal population size for the Schwefel function 1.2

Strategy	Optimal population size				Safeguard zone			
	8D	16D	24D	50D	8D	16D	24D	50D
DDE/best/1/bin	24	32	56	200	[16,120]	[16,400]	[16,400]	[40,500]
CDE/best/1/bin	24	16	24	—	[16,72]	[16,80]	[16,120]	—
DDE/rand/1/bin	16	16	—	—	[8,24]	[16,16]	—	—
CDE/rand/1/bin	16	16	—	—	[8,24]	[16,16]	—	—
DDE/best/1/exp	16	16	16	250	[16,40]	[16,200]	[8,400]	[30,500]
CDE/best/1/exp	16	16	16	—	[16,32]	[16,56]	[8,200]	—
DDE/rand/1/exp	16	16	16	10	[8,16]	[8,56]	[8,24]	[10,20]
CDE/rand/1/exp	16	8	16	—	[8,24]	[8,40]	[8,24]	—

7.7.2 Optimal Mutation Intensity and Crossover Probability

7.7.2.1 8-Dimensional Schwefel Function 1.2

The optimal mutation intensity and crossover probability of differential evolution for the 8-dimensional Schwefel function 1.2 are shown in Figures 7.101–7.104.

7.7.2.2 16-Dimensional Schwefel Function 1.2

The optimal mutation intensity and crossover probability of differential evolution for the 16-dimensional Schwefel function 1.2 are shown in Figures 7.105–7.108.

7.7.2.3 24-Dimensional Schwefel Function 1.2

The optimal mutation intensity and crossover probability of differential evolution for the 24-dimensional Schwefel function 1.2 are shown in Figures 7.109–7.112.

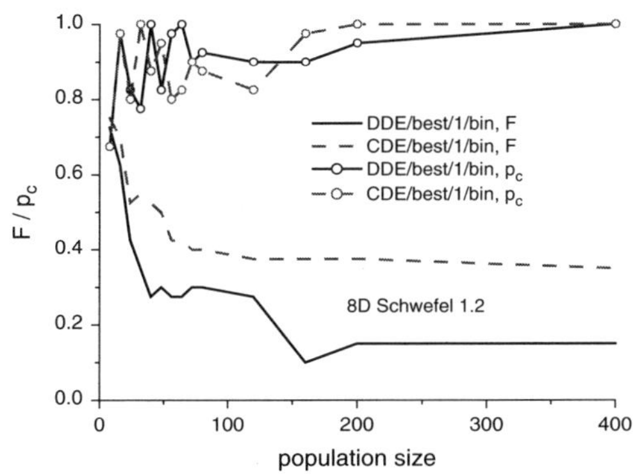

Figure 7.101 Mutation intensity and crossover probability of DE/best/1/bin for the 8-dimensional Schwefel function 1.2

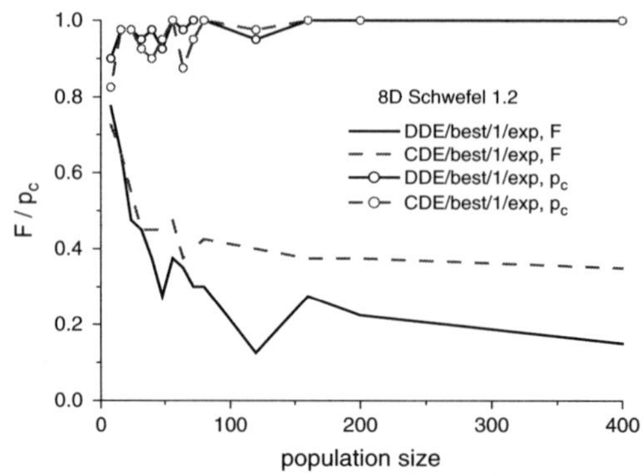

Figure 7.102 Mutation intensity and crossover probability of DE/best/1/exp for the 8-dimensional Schwefel function 1.2

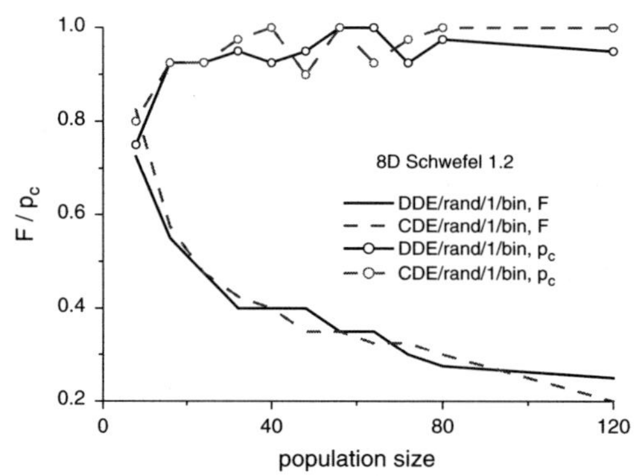

Figure 7.103 Mutation intensity and crossover probability of DE/rand/1/bin for the 8-dimensional Schwefel function 1.2

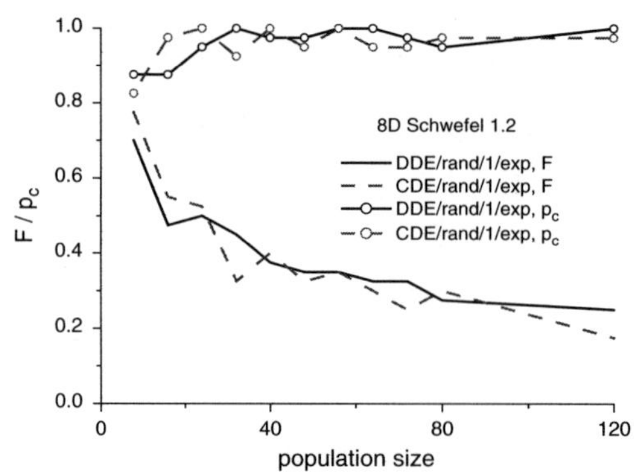

Figure 7.104 Mutation intensity and crossover probability of DE/rand/1/exp for the 8-dimensional Schwefel function 1.2

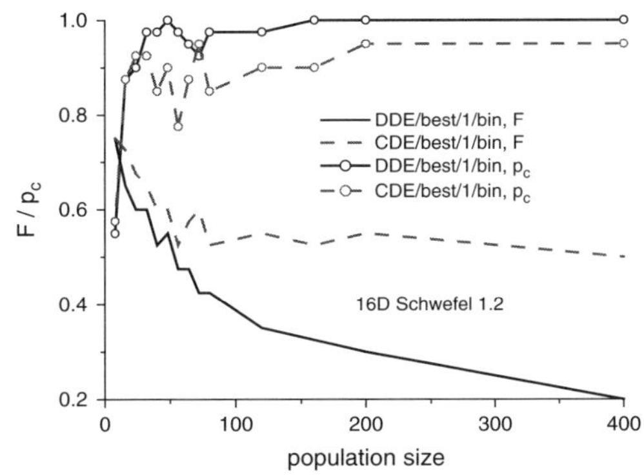

Figure 7.105 Mutation intensity and crossover probability of DE/best/1/bin for the 16-dimensional Schwefel function 1.2

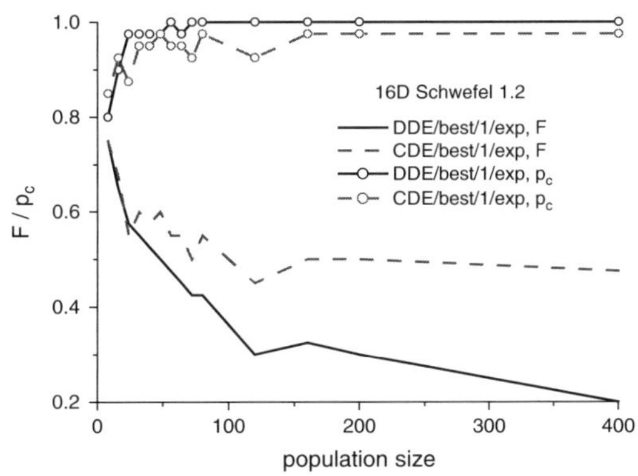

Figure 7.106 Mutation intensity and crossover probability of DE/best/1/exp for the 16-dimensional Schwefel function 1.2

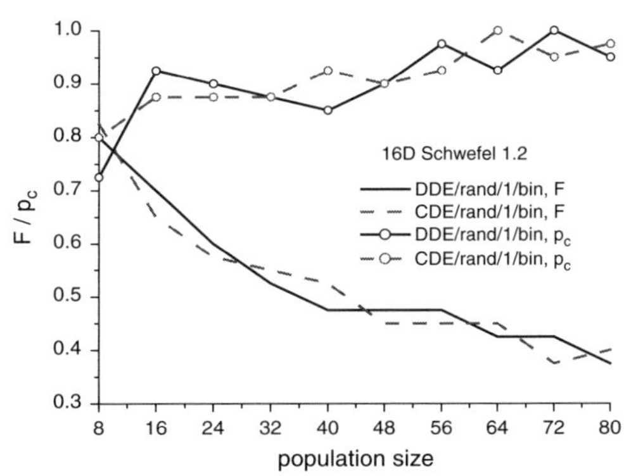

Figure 7.107 Mutation intensity and crossover probability of DE/rand/1/bin for the 16-dimensional Schwefel function 1.2

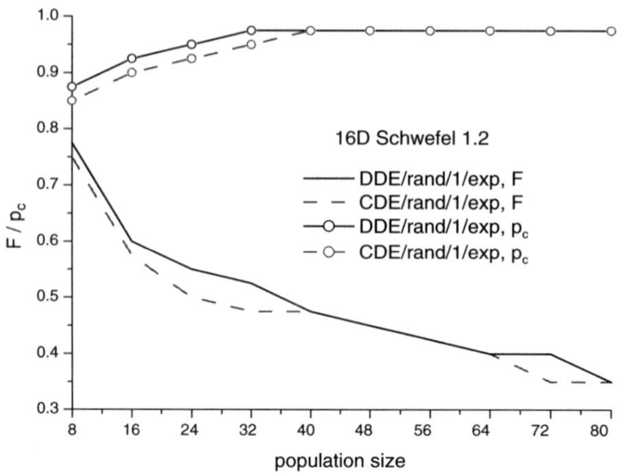

Figure 7.108 Mutation intensity and crossover probability of DE/rand/1/exp for the 16-dimensional Schwefel function 1.2

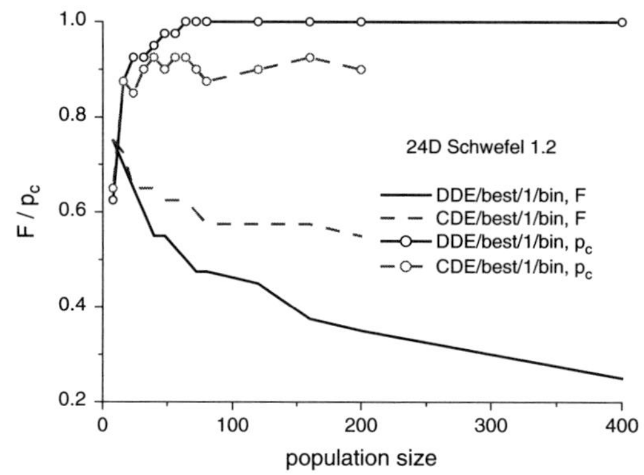

Figure 7.109 Mutation intensity and crossover probability of DE/best/1/bin for the 24-dimensional Schwefel function 1.2

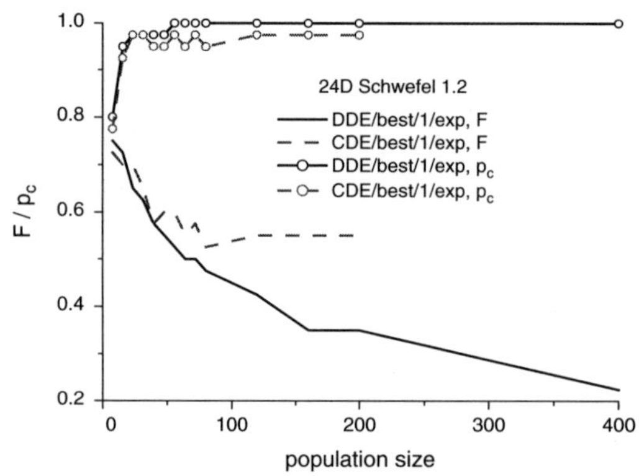

Figure 7.110 Mutation intensity and crossover probability of DE/best/1/exp for the 24-dimensional Schwefel function 1.2

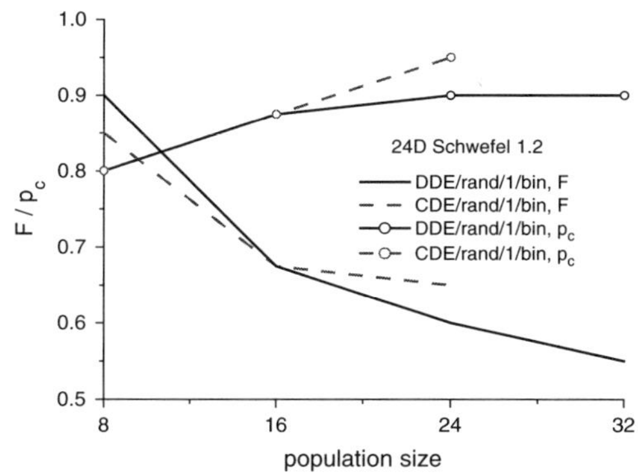

Figure 7.111 Mutation intensity and crossover probability of DE/rand/1/bin for the 24-dimensional Schwefel function 1.2

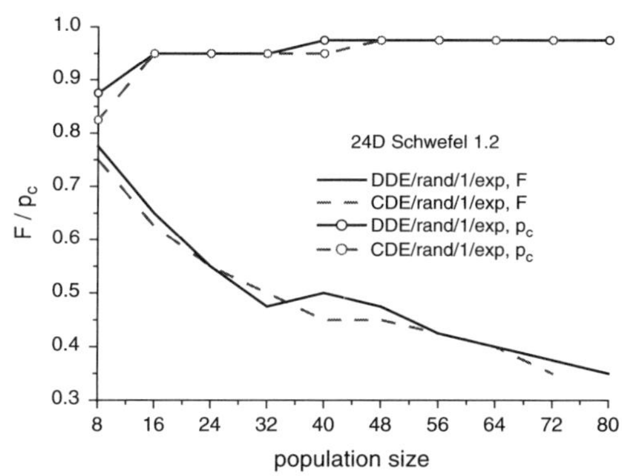

Figure 7.112 Mutation intensity and crossover probability of DE/rand/1/exp for the 24-dimensional Schwefel function 1.2

7.7.2.4 50-Dimensional Schwefel Function 1.2

The optimal mutation intensity and crossover probability of differential evolution for the 50-dimensional Schwefel function 1.2 are shown in Figures 7.113–7.116.

7.7.2.5 Effect of Rotation

The mutation intensity here is slightly larger than that of sphere function, although the general trend is consistent.

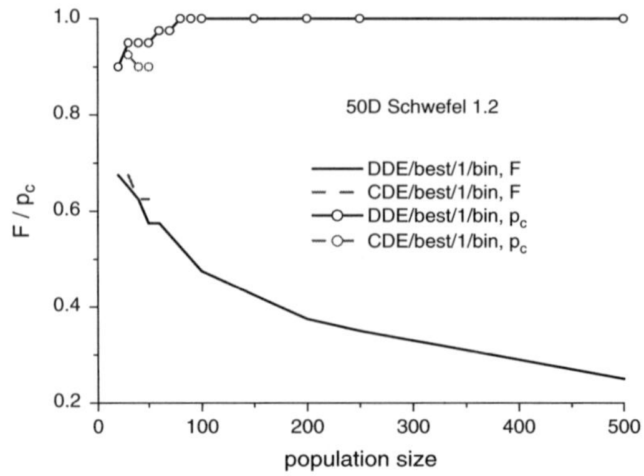

Figure 7.113 Mutation intensity and crossover probability of DE/best/1/bin for the 50-dimensional Schwefel function 1.2

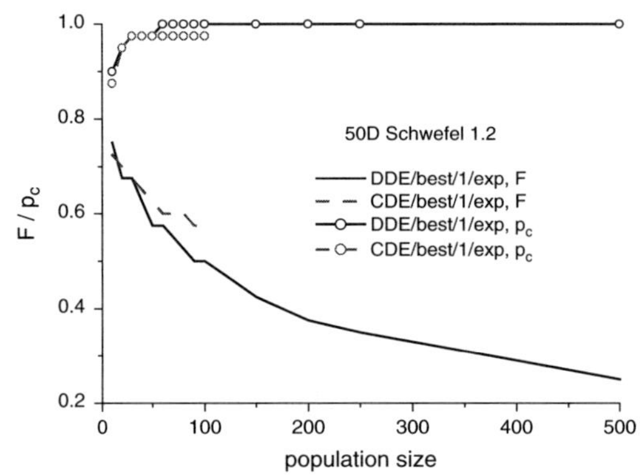

Figure 7.114 Mutation intensity and crossover probability of DE/best/1/exp for the 50-dimensional Schwefel function 1.2

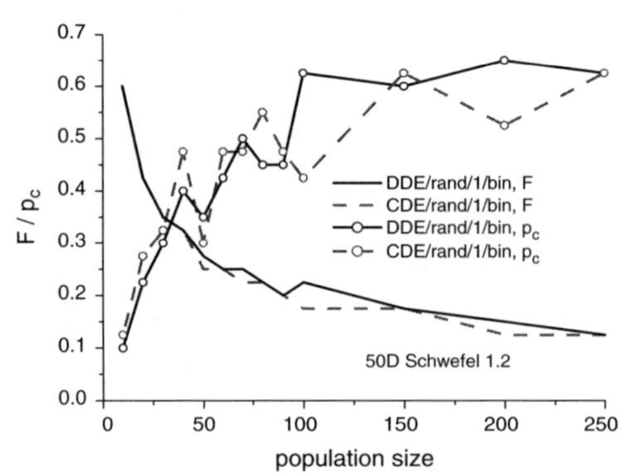

Figure 7.115 Mutation intensity and crossover probability of DE/rand/1/bin for the 50-dimensional Schwefel function 1.2

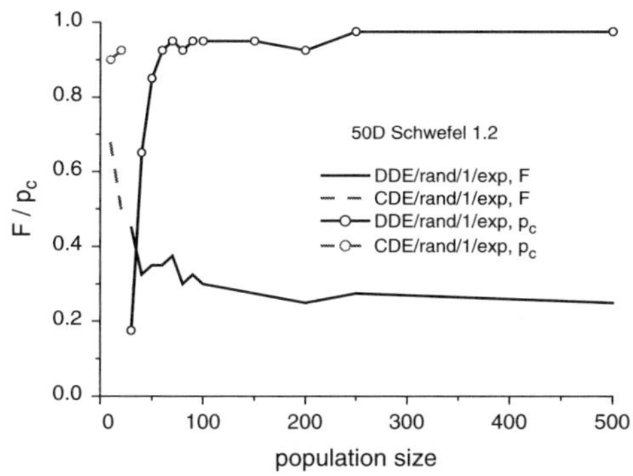

Figure 7.116 Mutation intensity and crossover probability of DE/rand/1/exp for the 50-dimensional Schwefel function 1.2

7.8 Rastrigin Function

7.8.1 Optimal Population Size

The optimal population size and the safeguard zone of the optimal population size for the Rastrigin function are summarized in Table 7.8. It seems that the optimal population size is independent of problem dimension.

Table 7.8 Optimal population size and safeguard zone of optimal population size for the Rastrigin function

Strategy	Optimal population size				Safeguard zone			
	8D	16D	24D	50D	8D	16D	24D	50D
DDE/best/1/bin	48	48	48	40	[32,80]	[16,80]	[24,80]	[40,70]
CDE/best/1/bin	56	40	48	40	[24,80]	[16,80]	[24,80]	[40,70]
DDE/rand/1/bin	32	32	40	40	[24,48]	[24,56]	[24,64]	[30,70]
CDE/rand/1/bin	32	32	40	40	[24,40]	[24,56]	[24,64]	[30,70]
DDE/best/1/exp	64	48	56	50	[40,72]	[40,64]	[40,64]	[30,40]
CDE/best/1/exp	56	40	56	50	[40,72]	[32,64]	[48,64]	[40,60]
DDE/rand/1/exp	32	40	40	40	[32,48]	[32,48]	[32,48]	[40,50]
CDE/rand/1/exp	32	40	40	40	[32,40]	[32,48]	[32,48]	[40,50]

7.8.2 Optimal Mutation Intensity and Crossover Probability

7.8.2.1 8-Dimensional Rastrigin Function

The optimal mutation intensity and crossover probability of differential evolution for the 8-dimensional Rastrigin function are shown in Figures 7.117–7.120. The general trend of mutation intensity and crossover is consistent with that for the Schwefel function 2.26.

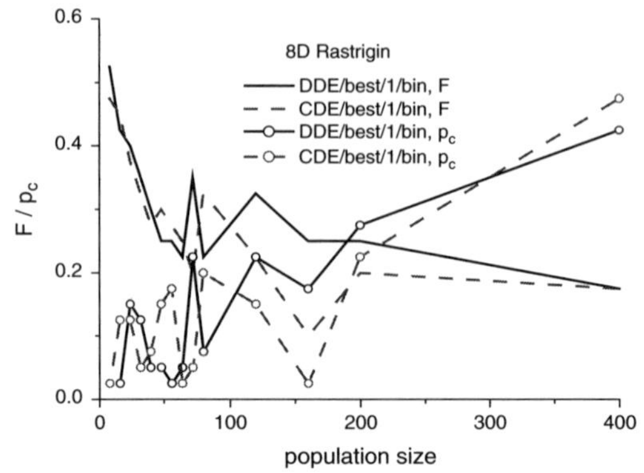

Figure 7.117 Mutation intensity and crossover probability of DE/best/1/bin for the 8-dimensional Rastrigin function

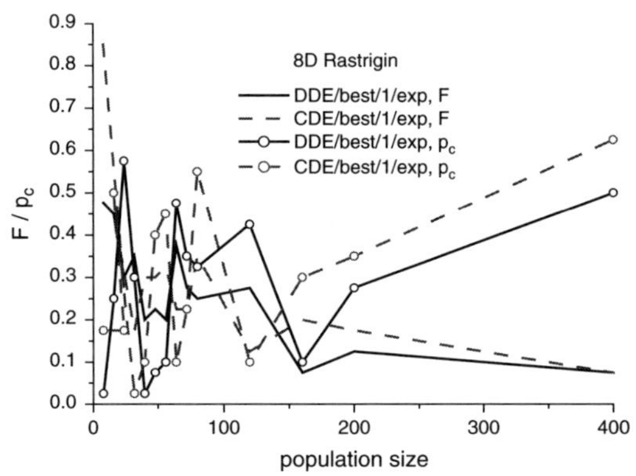

Figure 7.118 Mutation intensity and crossover probability of DE/best/1/exp for the 8-dimensional Rastrigin function

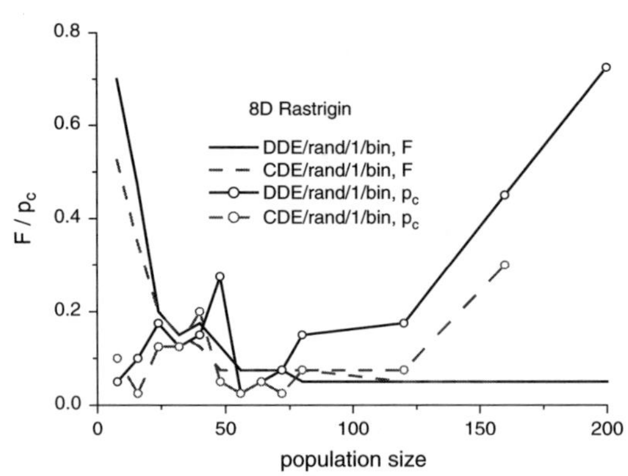

Figure 7.119 Mutation intensity and crossover probability of DE/rand/1/bin for the 8-dimensional Rastrigin function

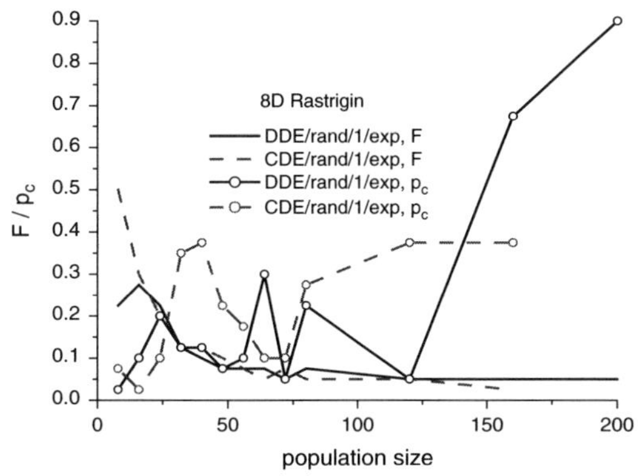

Figure 7.120 Mutation intensity and crossover probability of DE/rand/1/exp for the 8-dimensional Rastrigin function

7.8.2.2 16-Dimensional Rastrigin Function

The optimal mutation intensity and crossover probability of differential evolution for the 16-dimensional Rastrigin function are shown in Figures 7.121–7.124.

7.8.2.3 24-Dimensional Rastrigin Function

The optimal mutation intensity and crossover probability of differential evolution for the 24-dimensional Rastrigin function are shown in Figures 7.125–7.128.

7.8.2.4 50-Dimensional Rastrigin Function

The optimal mutation intensity and crossover probability of differential evolution for the 50-dimensional Rastrigin function are shown in Figures 7.129–7.132.

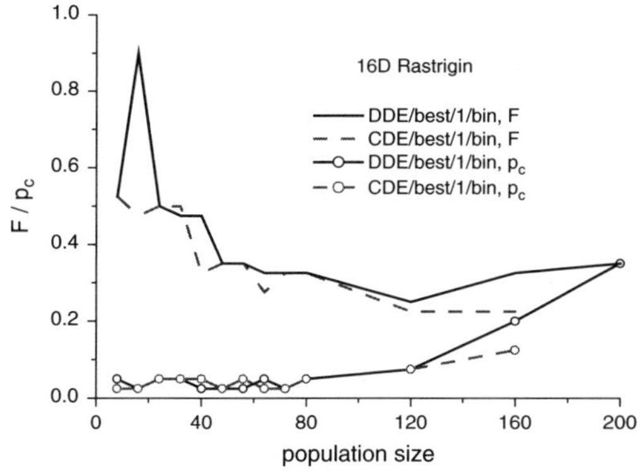

Figure 7.121 Mutation intensity and crossover probability of DE/best/1/bin for the 16-dimensional Rastrigin function

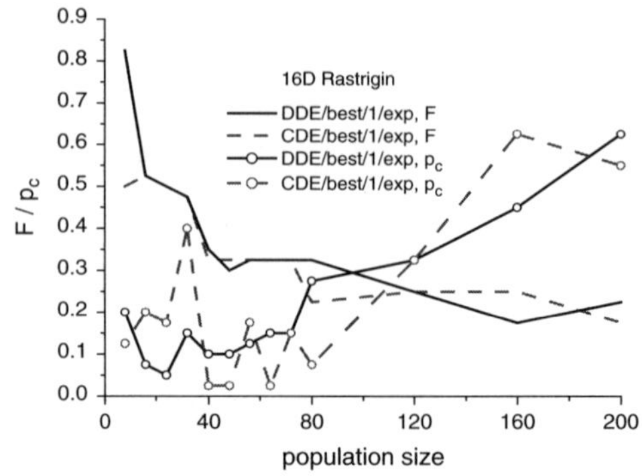

Figure 7.122 Mutation intensity and crossover probability of DE/best/1/exp for the 16-dimensional Rastrigin function

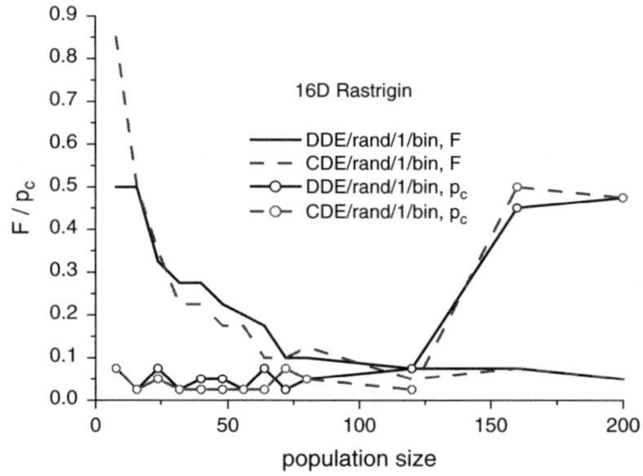

Figure 7.123 Mutation intensity and crossover probability of DE/rand/1/bin for the 16-dimensional Rastrigin function

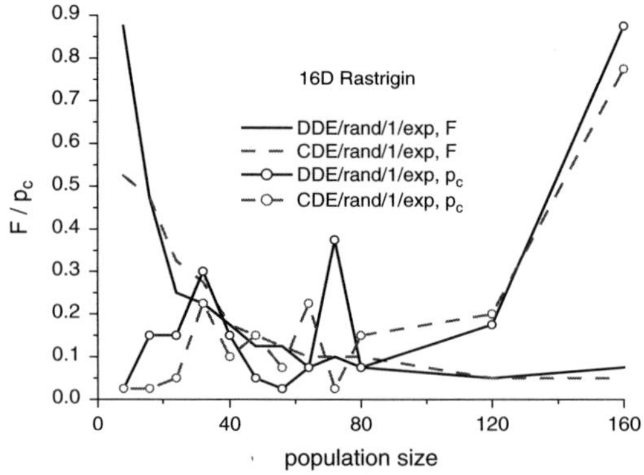

Figure 7.124 Mutation intensity and crossover probability of DE/rand/1/exp for the 16-dimensional Rastrigin function

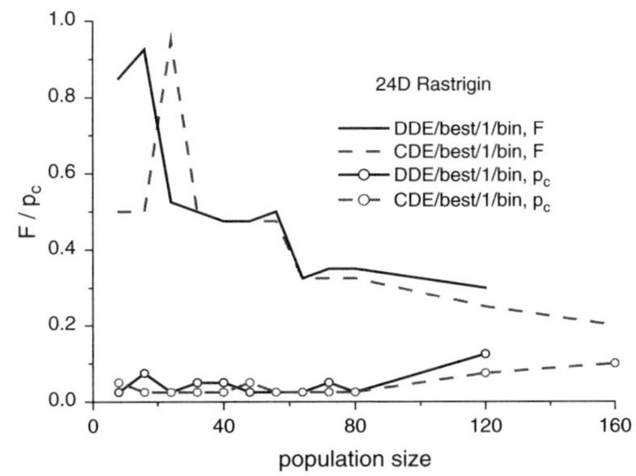

Figure 7.125 Mutation intensity and crossover probability of DE/best/1/bin for the 24-dimensional Rastrigin function

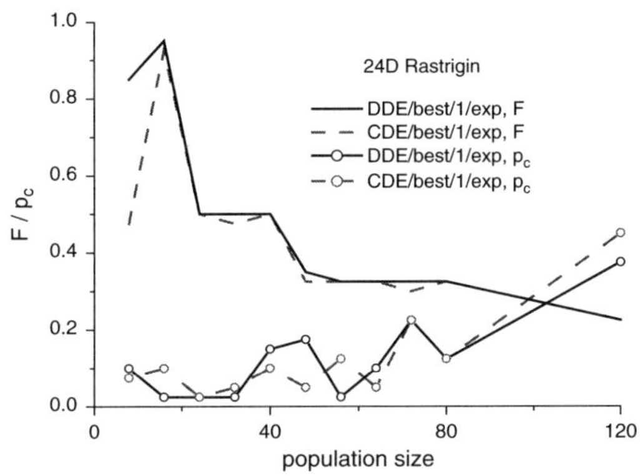

Figure 7.126 Mutation intensity and crossover probability of DE/best/1/exp for the 24-dimensional Rastrigin function

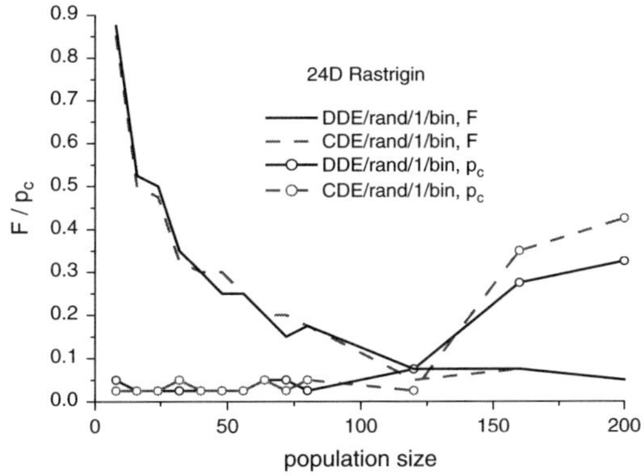

Figure 7.127 Mutation intensity and crossover probability of DE/rand/1/bin for the 24-dimensional Rastrigin function

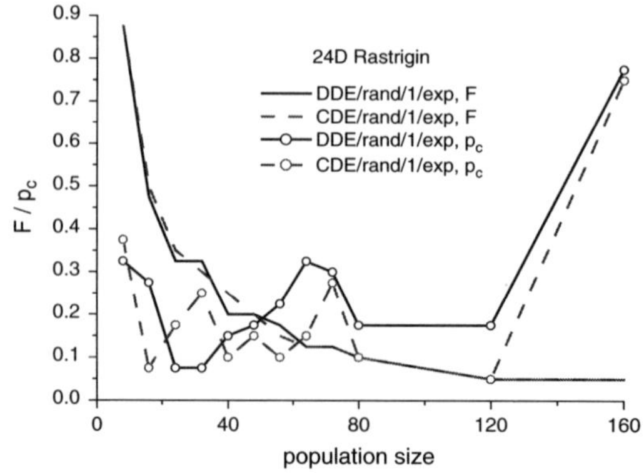

Figure 7.128 Mutation intensity and crossover probability of DE/rand/1/exp for the 24-dimensional Rastrigin function

7.8.2.5 Observations

It is very interesting to note that the mutation intensity and crossover probability for the Rastrigin function look very similar to those for the Schwefel function 2.26, although mutation intensity here is larger while crossover probability here is smaller.

7.9 Ackley Function

7.9.1 Optimal Population Size

The optimal population size and the safeguard zone of the optimal population size for the Ackley function are summarized in Table 7.9. Optimal population size for DE/rand/1/* does not

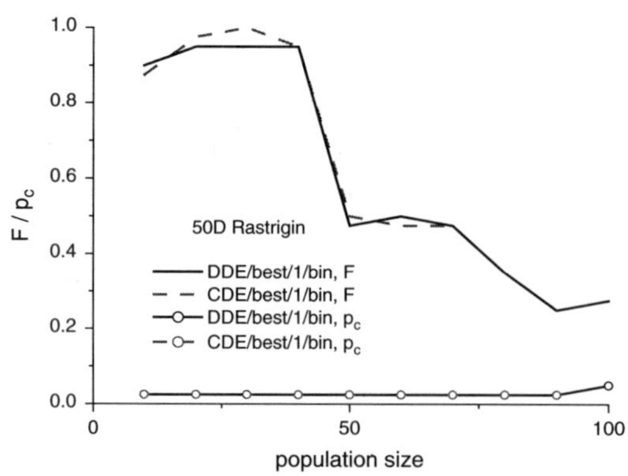

Figure 7.129 Mutation intensity and crossover probability of DE/best/1/bin for the 50-dimensional Rastrigin function

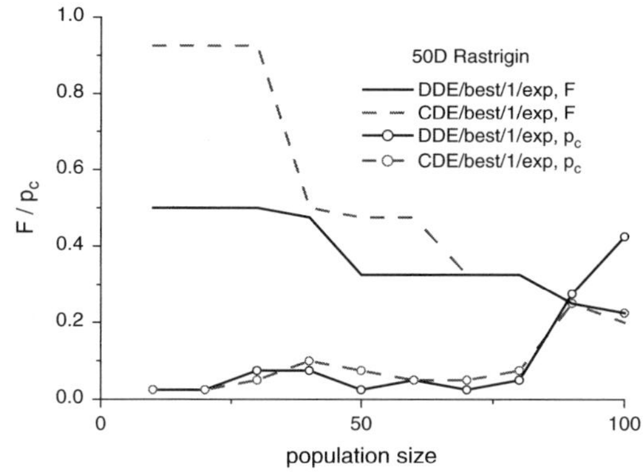

Figure 7.130 Mutation intensity and crossover probability of DE/best/1/exp for the 50-dimensional Rastrigin function

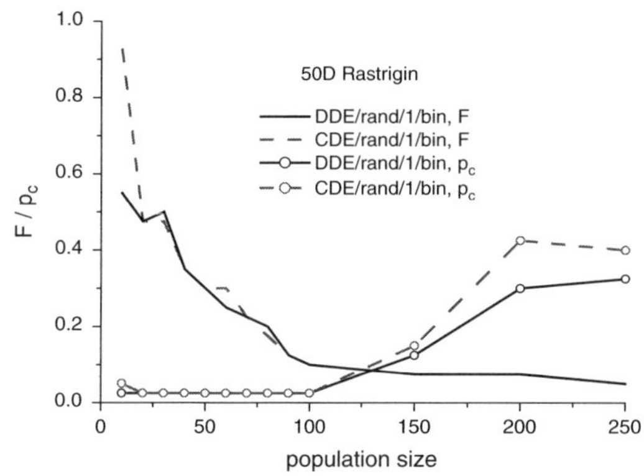

Figure 7.131 Mutation intensity and crossover probability of DE/rand/1/bin for the 50-dimensional Rastrigin function

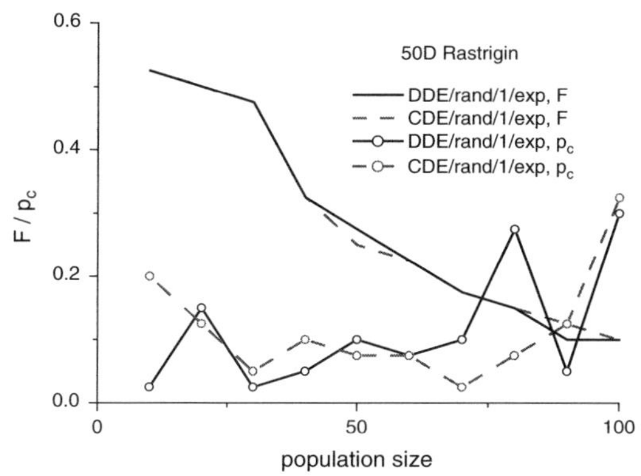

Figure 7.132 Mutation intensity and crossover probability of DE/rand/1/exp for the 50-dimensional Rastrigin function

change much as problem dimension increases, while optimal population size for DE/best/1/* decreases with problem dimension. The safeguard zone for DE/*/1/bin expands as problem dimension increases. The safeguard zone for DE/*/1/exp changes very little with problem dimension.

Table 7.9 Optimal population size and safeguard zone of optimal population size for the Ackley function

Strategy	Optimal population size				Safeguard zone			
	8D	16D	24D	50D	8D	16D	24D	50D
DDE/best/1/bin	80	56	48	40	[48,80]	[32,80]	[24,80]	[20,100]
CDE/best/1/bin	64	56	48	30	[40,80]	[24,80]	[24,80]	[20,100]
DDE/rand/1/bin	40	40	40	50	[24,64]	[24,64]	[24,72]	[20,100]
CDE/rand/1/bin	40	32	40	40	[24,64]	[24,64]	[24,72]	[20,100]
DDE/best/1/exp	72	72	64	50	[48,80]	[40,80]	[32,72]	[30,70]
CDE/best/1/exp	64	72	56	50	[40,80]	[32,80]	[32,72]	[30,70]
DDE/rand/1/exp	48	48	56	50	[32,64]	[24,64]	[32,64]	[30,60]
CDE/rand/1/exp	48	48	48	50	[32,64]	[32,64]	[32,64]	[30,60]

7.9.2 Optimal Mutation Intensity and Crossover Probability

7.9.2.1 8-Dimensional Ackley Function

The optimal mutation intensity and crossover probability of differential evolution for the 8-dimensional Ackley function are shown in Figures 7.133–7.136.

7.9.2.2 16-Dimensional Ackley Function

The optimal mutation intensity and crossover probability of differential evolution for the 16-dimensional Ackley function are shown in Figures 7.137–7.140.

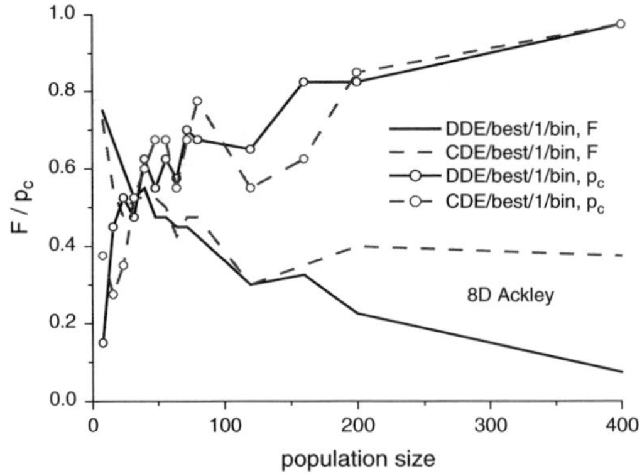

Figure 7.133 Mutation intensity and crossover probability of DE/best/1/bin for the 8-dimensional Ackley function

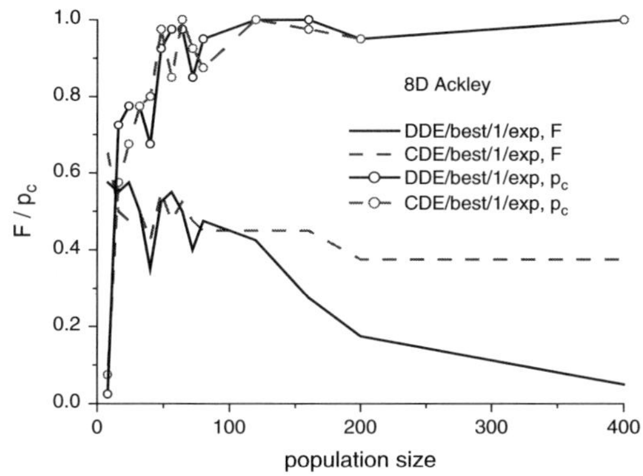

Figure 7.134 Mutation intensity and crossover probability of DE/best/1/exp for the 8-dimensional Ackley function

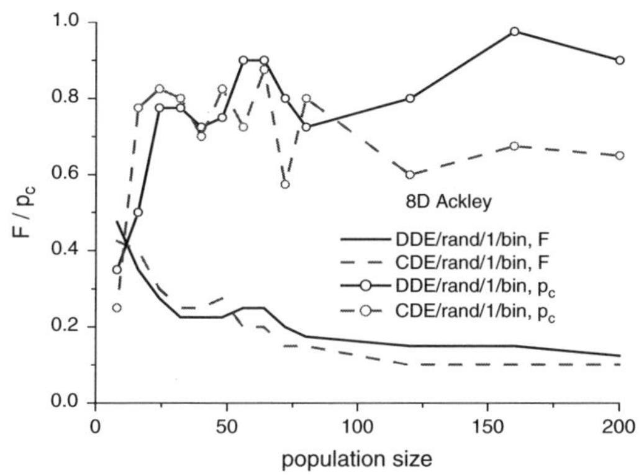

Figure 7.135 Mutation intensity and crossover probability of DE/rand/1/bin for the 8-dimensional Ackley function

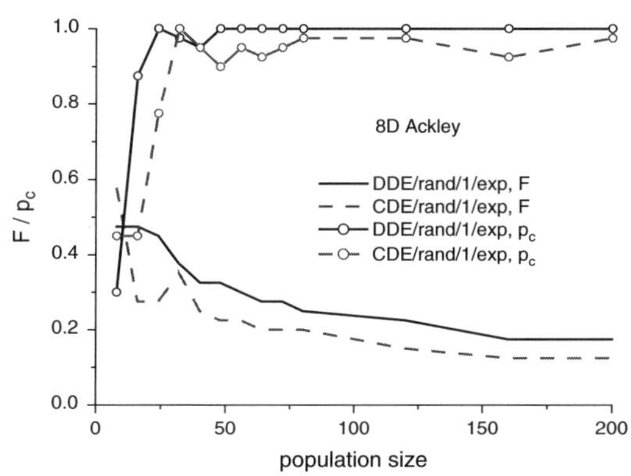

Figure 7.136 Mutation intensity and crossover probability of DE/rand/1/exp for the 8-dimensional Ackley function

7.9.2.3 24-Dimensional Ackley Function

The optimal mutation intensity and crossover probability of differential evolution for the 24-dimensional Ackley function are shown in Figures 7.141–7.144.

7.9.2.4 50-Dimensional Ackley Function

The optimal mutation intensity and crossover probability of differential evolution for the 50-dimensional Ackley function are shown in Figures 7.145–7.148.

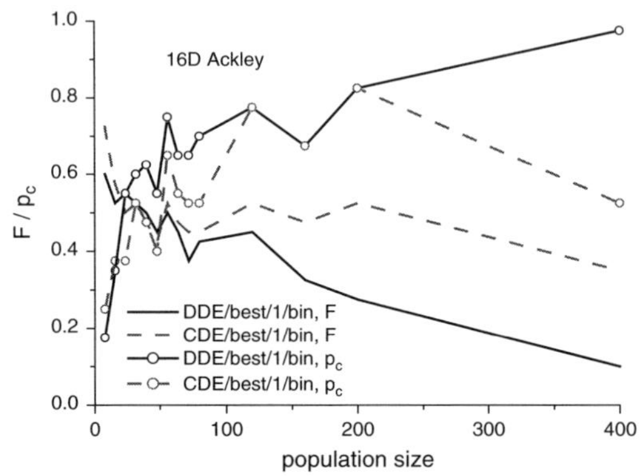

Figure 7.137 Mutation intensity and crossover probability of DE/best/1/bin for the 16-dimensional Ackley function

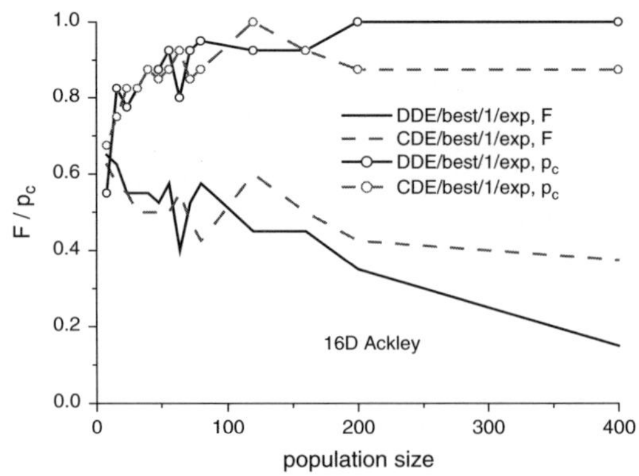

Figure 7.138 Mutation intensity and crossover probability of DE/best/1/exp for the 16-dimensional Ackley function

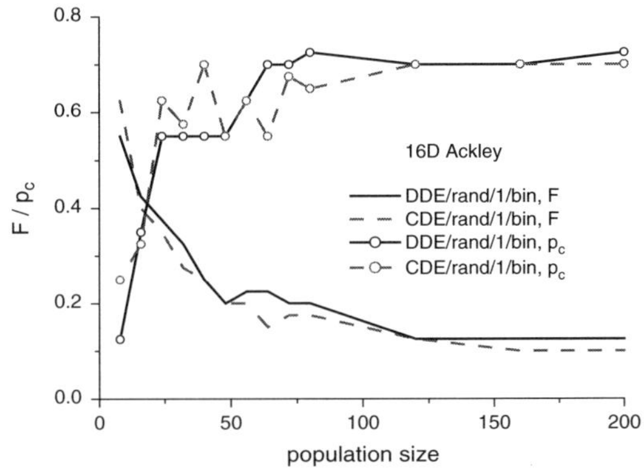

Figure 7.139 Mutation intensity and crossover probability of DE/rand/1/bin for the 16-dimensional Ackley function

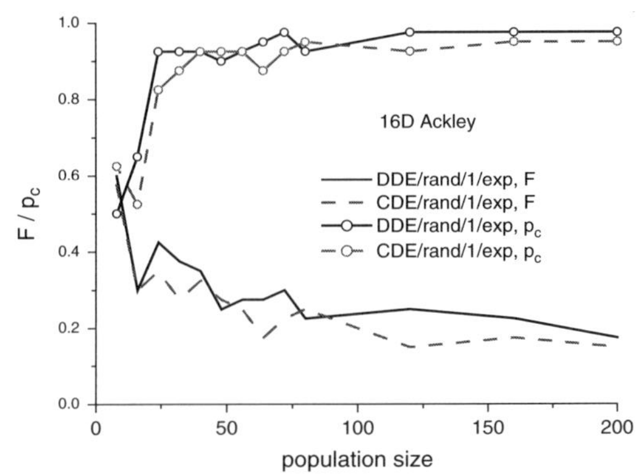

Figure 7.140 Mutation intensity and crossover probability of DE/rand/1/exp for the 16-dimensional Ackley function

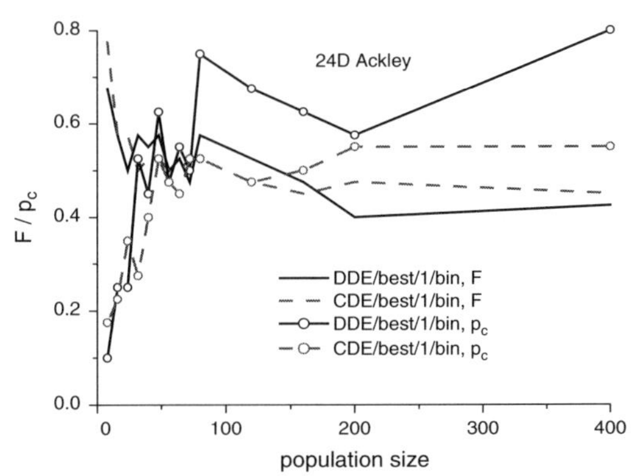

Figure 7.141 Mutation intensity and crossover probability of DE/best/1/bin for the 24-dimensional Ackley function

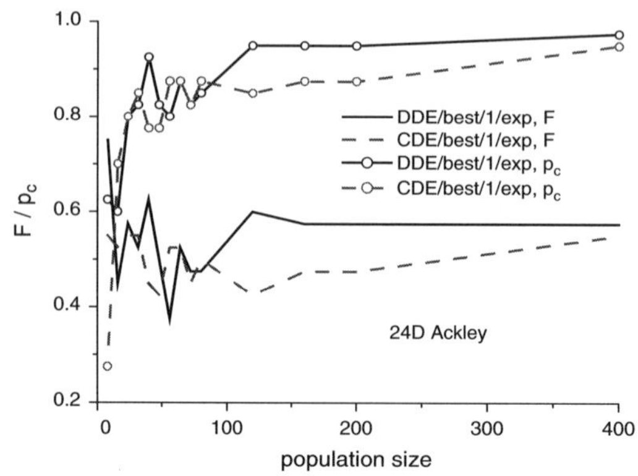

Figure 7.142 Mutation intensity and crossover probability of DE/best/1/exp for the 24-dimensional Ackley function

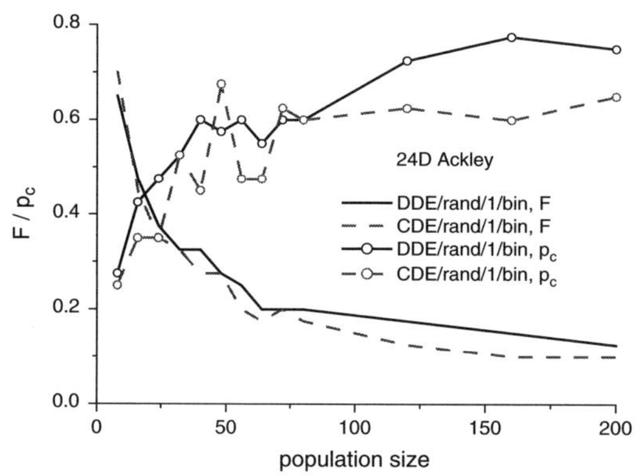

Figure 7.143 Mutation intensity and crossover probability of DE/rand/1/bin for the 24-dimensional Ackley function

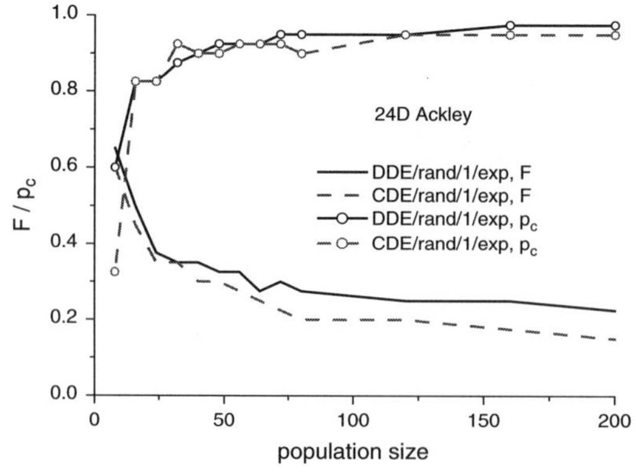

Figure 7.144 Mutation intensity and crossover probability of DE/rand/1/exp for the 24-dimensional Ackley function

7.9.2.5 Observations

The optimal mutation intensity and crossover probability of differential evolution look very similar to those for step function 2.

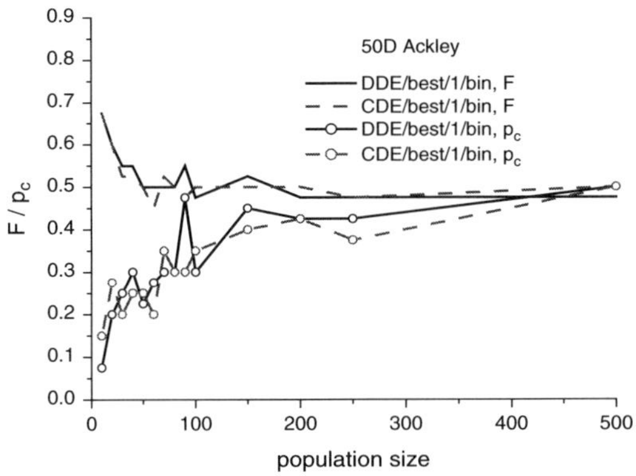

Figure 7.145 Mutation intensity and crossover probability of DE/best/1/bin for the 50-dimensional Ackley function

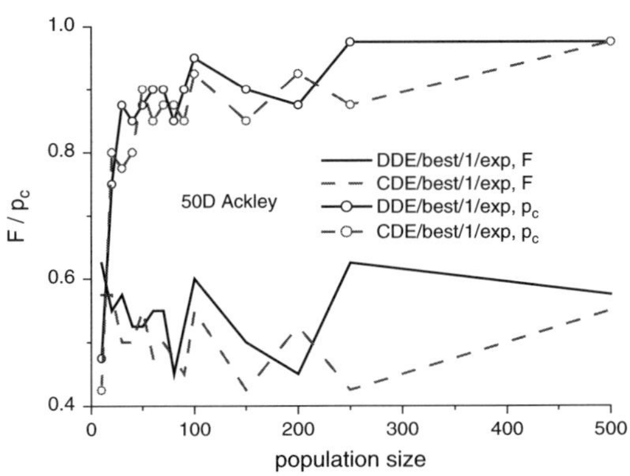

Figure 7.146 Mutation intensity and crossover probability of DE/best/1/exp for the 50-dimensional Ackley function

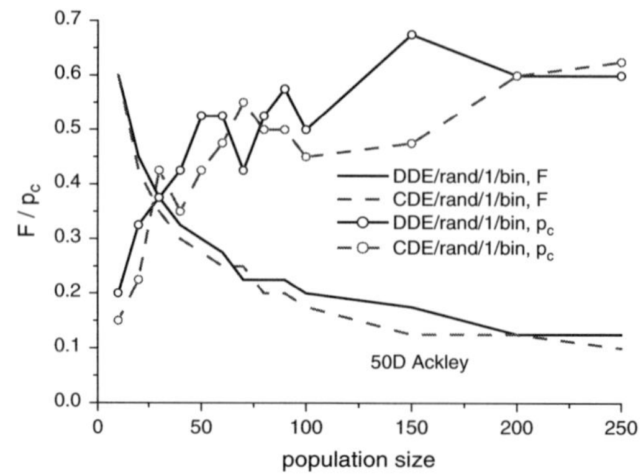

Figure 7.147 Mutation intensity and crossover probability of DE/rand/1/bin for the 50-dimensional Ackley function

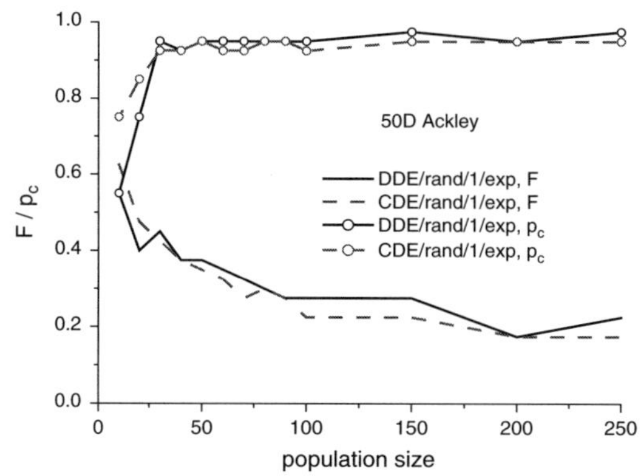

Figure 7.148 Mutation intensity and crossover probability of DE/rand/1/exp for the 50-dimensional Ackley function

8

Non-Intrinsic Control Parameters

8.1 Introduction

This chapter discusses the relationship between problem features, differential evolution strategies, intrinsic control parameters, and non-intrinsic control parameters.

Value to reach, limit of number of objective function evaluations, limit of population diversity, and search space are the four non-intrinsic control parameters for differential evolution. From the point of view of practical application, search space is the most critical non-intrinsic control parameter and is of great concern to application engineers.

Value to reach comes directly from practical application requirements. A solution that does not meet practical application requirements is absolutely unacceptable. On the other hand, it makes no sense to look for a solution that is more accurate than necessary. Therefore, little can be done about value to reach.

Both the limit of number of objective function evaluations and limit of population diversity decide when to terminate differential evolution. Neither is active during evolution. A higher limit of number of objective function evaluations and a lower limit of population diversity will not lengthen the evolution. As such, the study of these two non-intrinsic control parameters is not urgent from the point of view of practical applications.

In contrast, the search space participates in the evolution very actively. A properly chosen search space is widely believed to benefit differential evolution in terms of both robustness and efficiency. Unfortunately, choosing an appropriate search space usually requires a priori knowledge of the practical application problem concerned, which is usually not available. It is a common practice to choose a wide search space when no or little a priori knowledge is available, so that at least a quasi-optimal solution can be guaranteed. This has been one of the major attractions for application engineers to implement evolutionary algorithms for their problems of interest. It is also believed that different intrinsic control parameter values have to be applied to differential evolution if different search spaces are applied. Hence, it is very important to know how the performance of differential evolution and its optimal intrinsic control parameters change as the search space changes.

However, no systematic study has ever been carried out to examine the effect of search space on any evolutionary algorithm. Therefore, it is not known exactly how sensitive differential evolution and its optimal intrinsic control parameter values are to the search space. Consequently, most users still make a great deal of effort to choose as narrow a search space as possible. In this sense, people are still suffering the same challenge facing the deterministic optimization community instead of enjoying the strong global search ability of evolution computation. In addition, when they find it difficult to narrow the search space for their specific application problems, they usually worry about the applicability of reported intrinsic control parameter values. Such a lack of confidence may even stop people from implementing differential evolution to solve their application problems.

Thus non-intrinsic control parameters, especially search space, are an important part of the parametric study.

8.2 Alternative Search Space

So far, the parametric study on differential evolution has been carried out over the standard search space mentioned in Chapter 5. To investigate the effect of search space on differential evolution, an alternative search space has been chosen for each member toy function in the tentative benchmark test bed, as shown in Table 8.1.

Table 8.1 Standard and alternative search spaces for fully simulated member toy functions in the tentative benchmark test bed

Member toy function	Standard search space	Alternative search space
Step 2	[−500, 500]	[−2500, 2500]
Sphere	[−500, 500]	[−2500, 2500]
Hyper-ellipsoid	[−500, 500]	[−2500, 2500]
Qing	[−500, 500]	[−2500, 2500]
Schwefel 2.22	[−500, 500]	[−2500, 2500]
Schwefel 1.2	[−500, 500]	[−100, 100]
Rastrigin	[−100, 100]	[−500, 500]
Ackley	[−30, 30]	[−100, 100]

The alternative search space is significantly different from the standard search space to make sure that the effect of search space, if any, is as visible as possible. Furthermore, the alternative search space is usually much wider than the standard search space to further confirm the strong global search ability of differential evolution. The Schwefel function 2.26 is omitted since its minimal value in [−2500, 2500] is different from that in [−500, 500].

As expected, the performance of differential evolution deteriorates in the wider search space. The performance and the optimal population size of differential evolution have been observed to be very sensitive to search space for the Ackley function. Fortunately, the optimal mutation intensity and crossover probability have been observed to be very insensitive to search space, even for the Ackley function.

In this regard, numerical results are presented here in a different way. The performance of the sphere and Ackley functions is shown first. Next, the optimal population size and safeguard zone for all fully simulated member toy functions are given. Finally, the optimal mutation intensity and crossover probability of the sphere function are presented.

8.3 Performance of Differential Evolution

8.3.1 Sphere Function

8.3.1.1 8-Dimensional Sphere Function

The performance of differential evolution for the 8-dimensional sphere function is shown in Figures 8.1–8.4. It is seen that the wider search space makes differential evolution less robust and efficient.

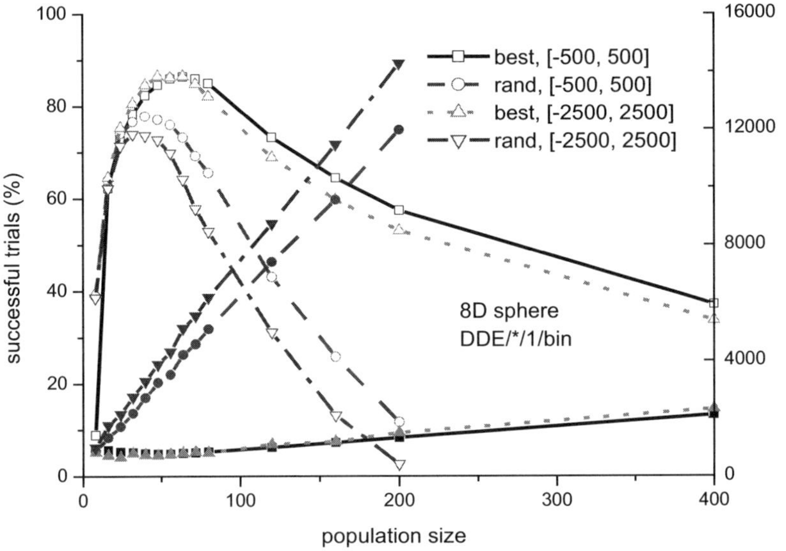

Figure 8.1 Robustness and efficiency of DDE/*/1/bin for the 8-dimensional sphere function

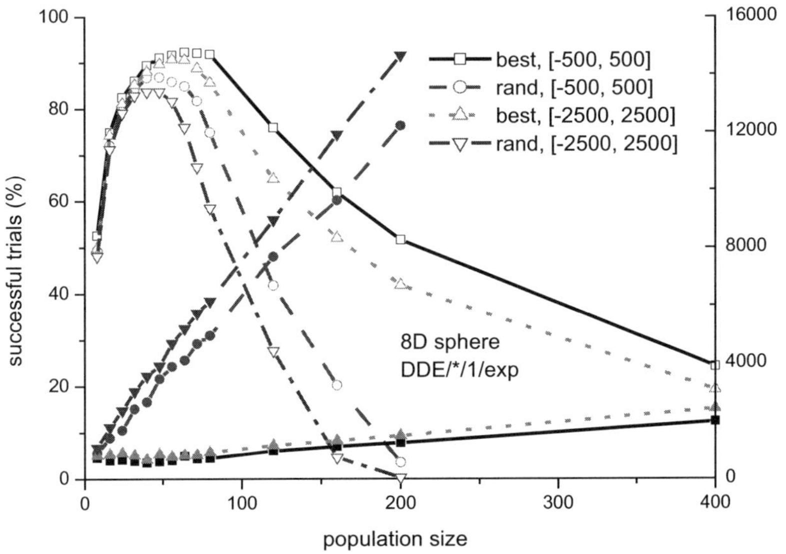

Figure 8.2 Robustness and efficiency of DDE/*/1/exp for the 8-dimensional sphere function

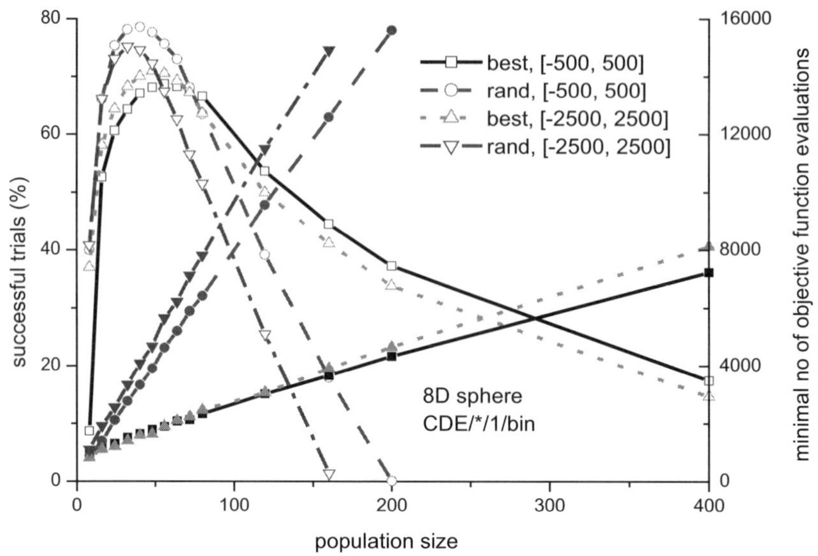

Figure 8.3 Robustness and efficiency of CDE/*/1/bin for the 8-dimensional sphere function

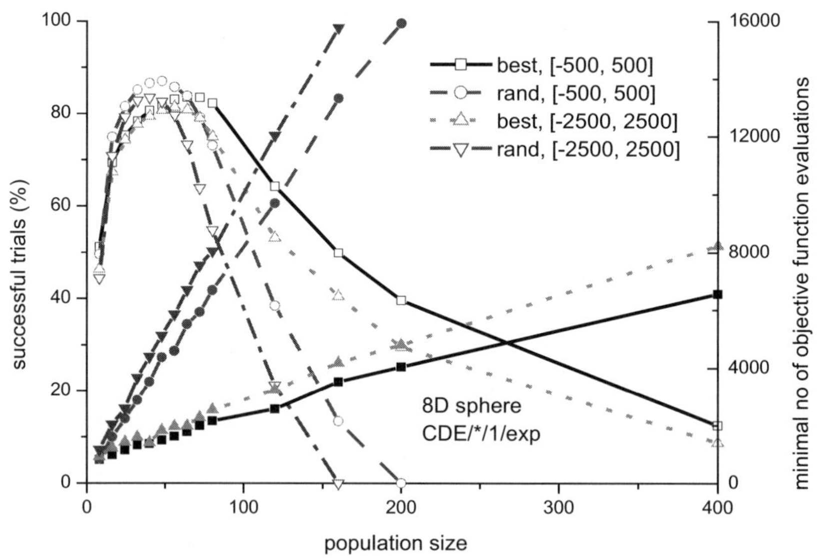

Figure 8.4 Robustness and efficiency of CDE/*/1/exp for the 8-dimensional sphere function

8.3.1.2 16-Dimensional Sphere Function

The performance of differential evolution for the 16-dimensional sphere function is shown in Figures 8.5–8.8.

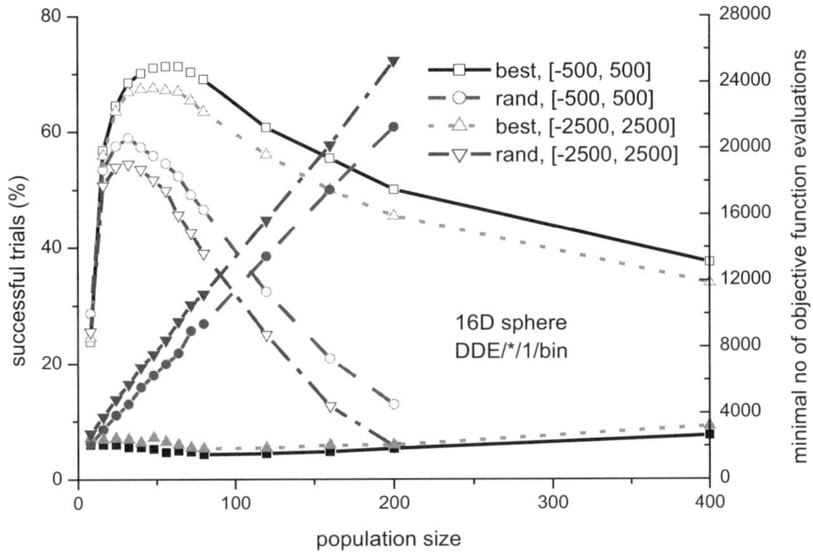

Figure 8.5 Robustness and efficiency of DDE/*/1/bin for the 16-dimensional sphere function

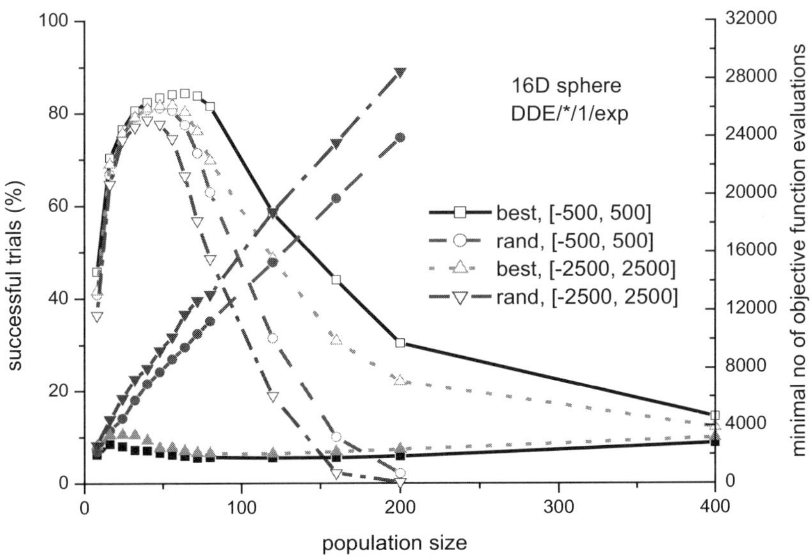

Figure 8.6 Robustness and efficiency of DDE/*/1/exp for the 16-dimensional sphere function

8.3.1.3 24-Dimensional Sphere Function

The performance of differential evolution for the 24-dimensional sphere function is shown in Figures 8.9–8.12.

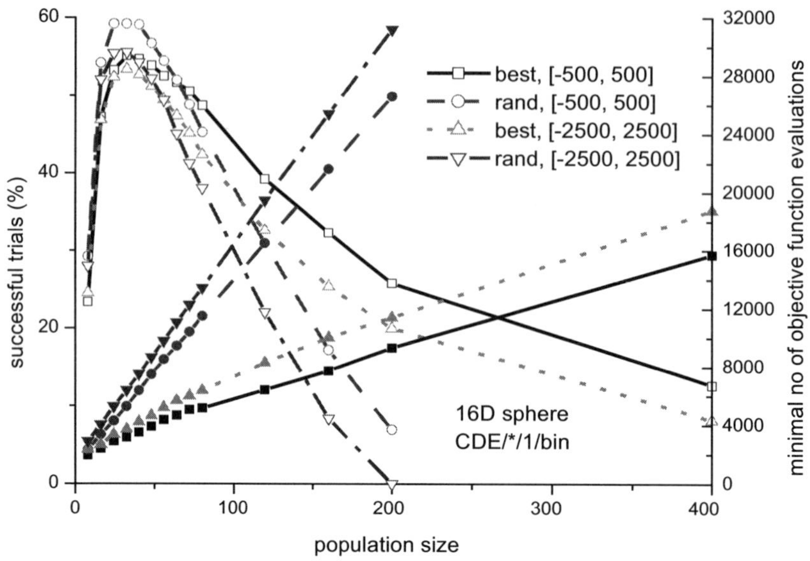

Figure 8.7 Robustness and efficiency of CDE/*/1/bin for the 16-dimensional sphere function

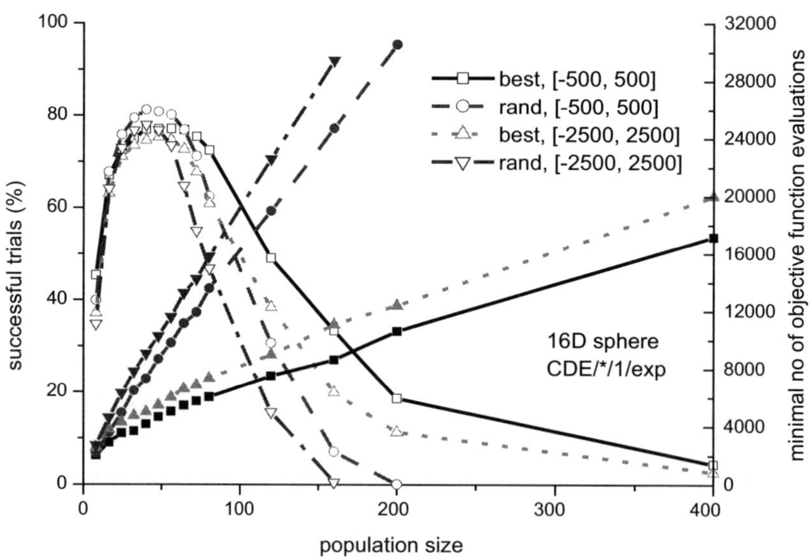

Figure 8.8 Robustness and efficiency of CDE/*/1/exp for the 16-dimensional sphere function

8.3.1.4 50-Dimensional Sphere Function

The performance of differential evolution for the 50-dimensional sphere function is shown in Figures 8.13–8.16.

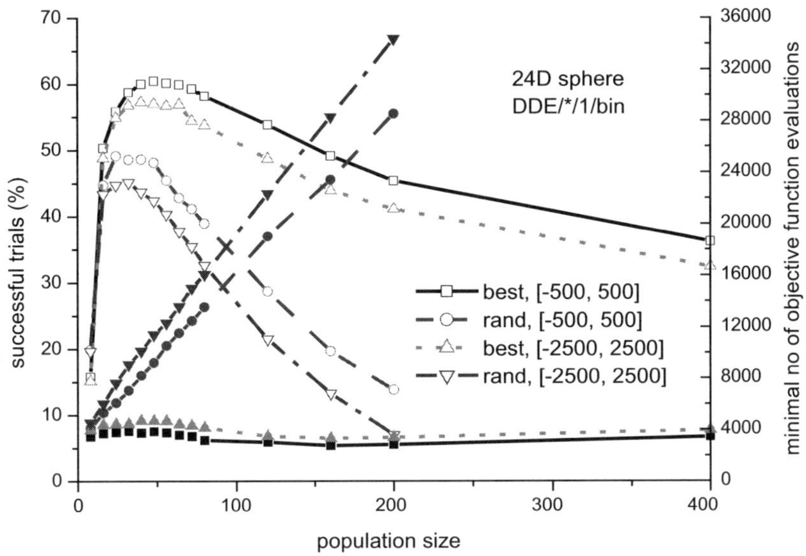

Figure 8.9 Robustness and efficiency of DDE/*/1/bin for the 24-dimensional sphere function

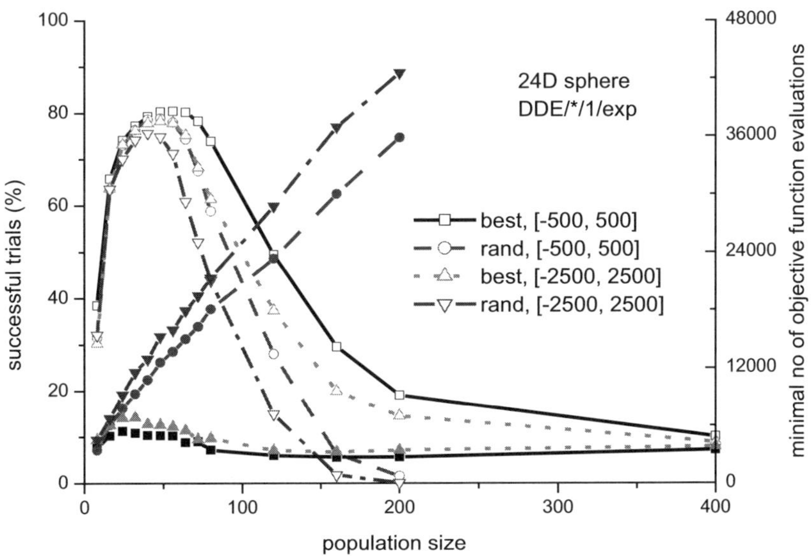

Figure 8.10 Robustness and efficiency of DDE/*/1/exp for the 24-dimensional sphere function

8.3.2 Ackley Function

8.3.2.1 8-Dimensional Ackley Function

The performance of differential evolution for the 8-dimensional Ackley function is shown in Figures 8.17–8.20. The sensitivity of differential evolution with respect to search space here is

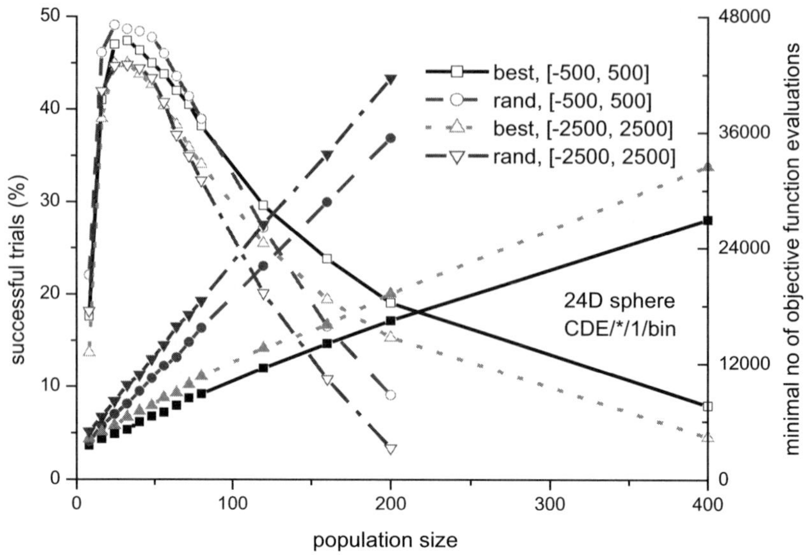

Figure 8.11 Robustness and efficiency of CDE/*/1/bin for the 24-dimensional sphere function

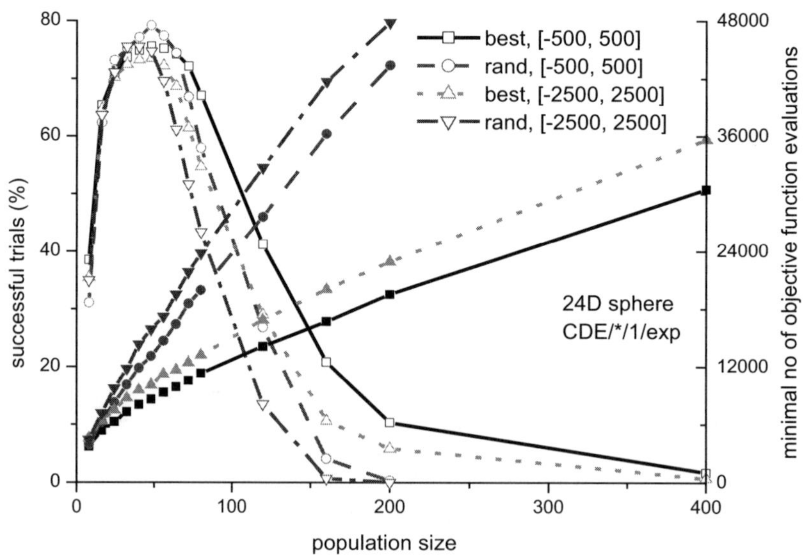

Figure 8.12 Robustness and efficiency of CDE/*/1/exp for the 24-dimensional sphere function

very clear. The successful trials fall by at least 50% when the search space is widened from $[-30, 30]$ to $[-100, 100]$, compared to at most a 10% decrease for the sphere function when the search space is widened from $[-500, 500]$ to $[-2500, 2500]$.

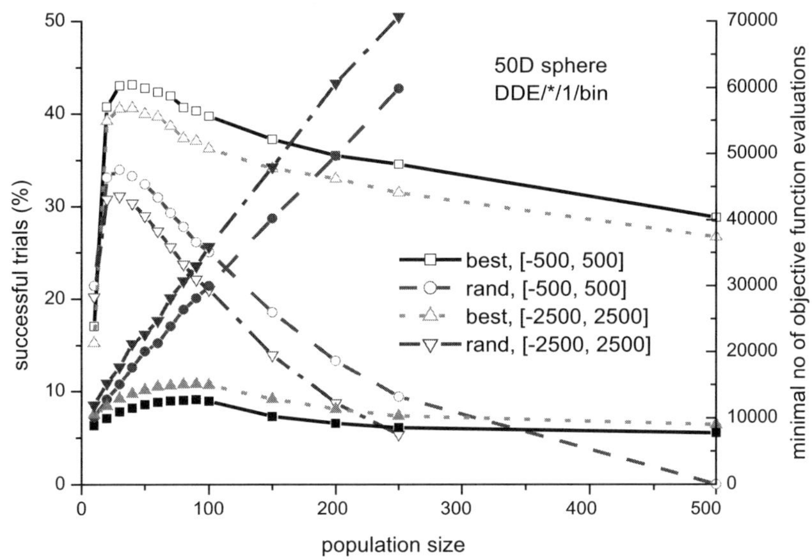

Figure 8.13 Robustness and efficiency of DDE/*/1/bin for the 50-dimensional sphere function

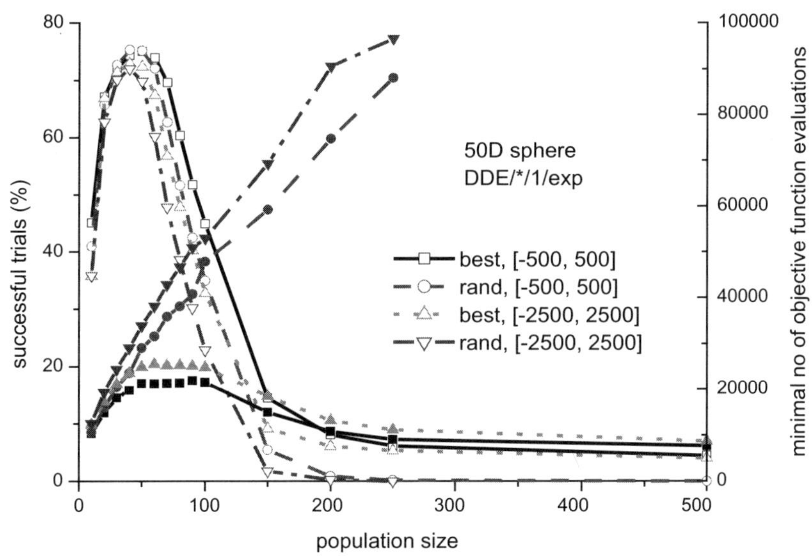

Figure 8.14 Robustness and efficiency of DDE/*/1/exp for the 50-dimensional sphere function

8.3.2.2 16-Dimensional Ackley Function

The performance of differential evolution for the 16-dimensional Ackley function is shown in Figures 8.21–8.24.

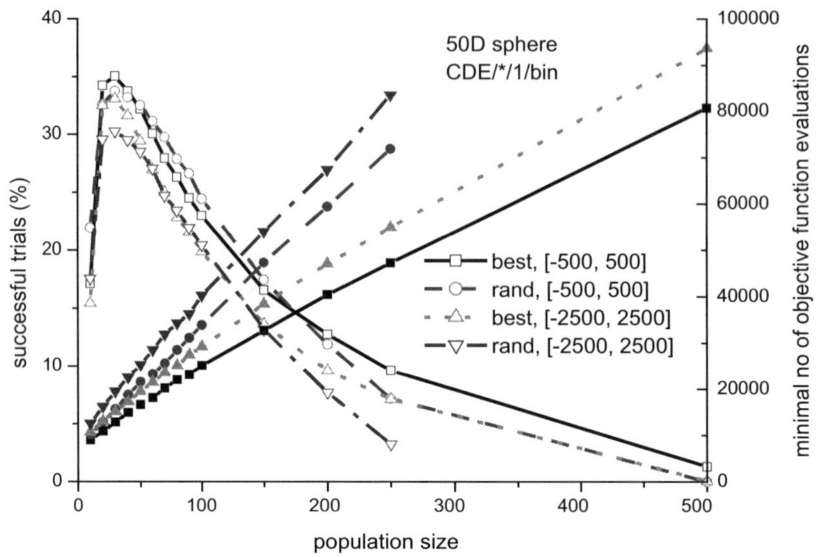

Figure 8.15 Robustness and efficiency of CDE/*/1/bin for the 50-dimensional sphere function

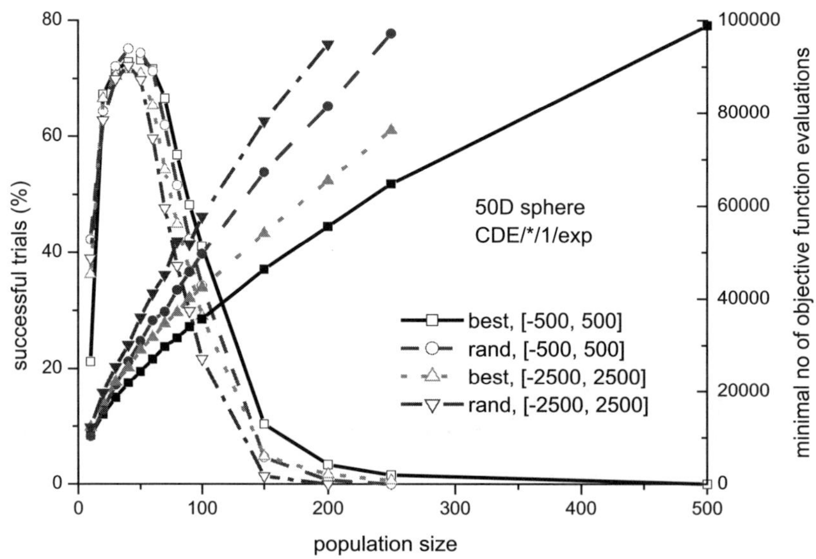

Figure 8.16 Robustness and efficiency of CDE/*/1/exp for the 50-dimensional sphere function

8.3.2.3 24-Dimensional Ackley Function

The performance of differential evolution for the 24-dimensional Ackley function is shown in Figures 8.25–8.28.

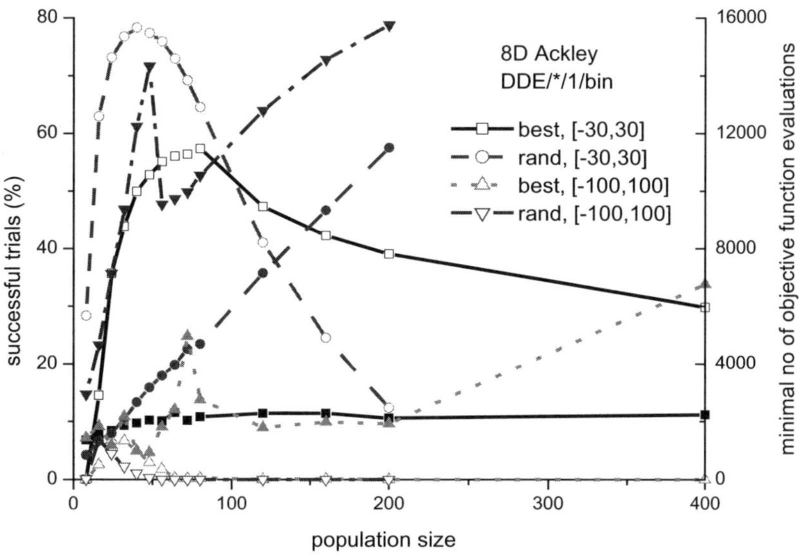

Figure 8.17 Robustness and efficiency of DDE/*/1/bin for the 8-dimensional Ackley function

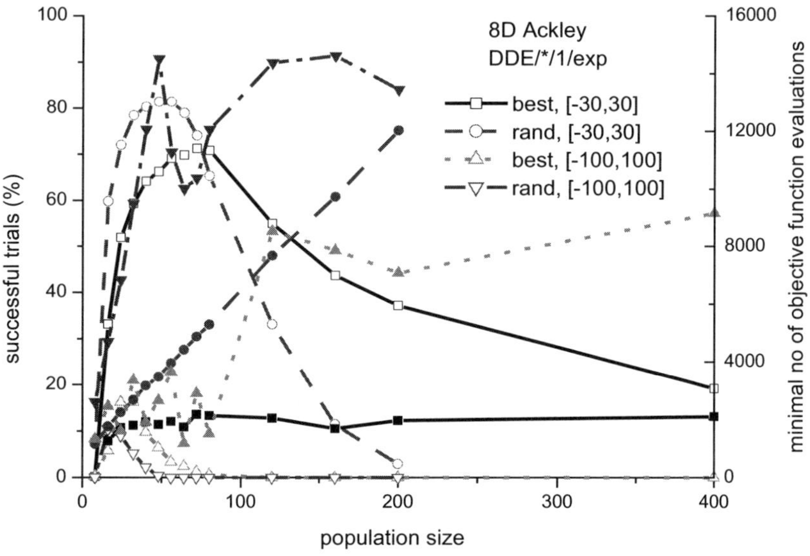

Figure 8.18 Robustness and efficiency of DDE/*/1/exp for the 8-dimensional Ackley function

8.3.2.4 50-Dimensional Ackley Function

The performance of differential evolution for the 50-dimensional Ackley function is shown in Figures 8.29–8.32.

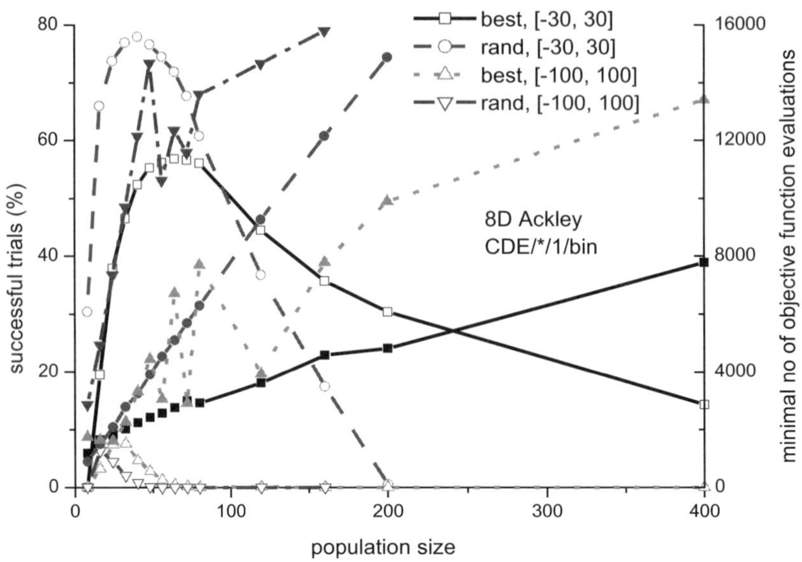

Figure 8.19 Robustness and efficiency of CDE/*/1/bin for the 8-dimensional Ackley function

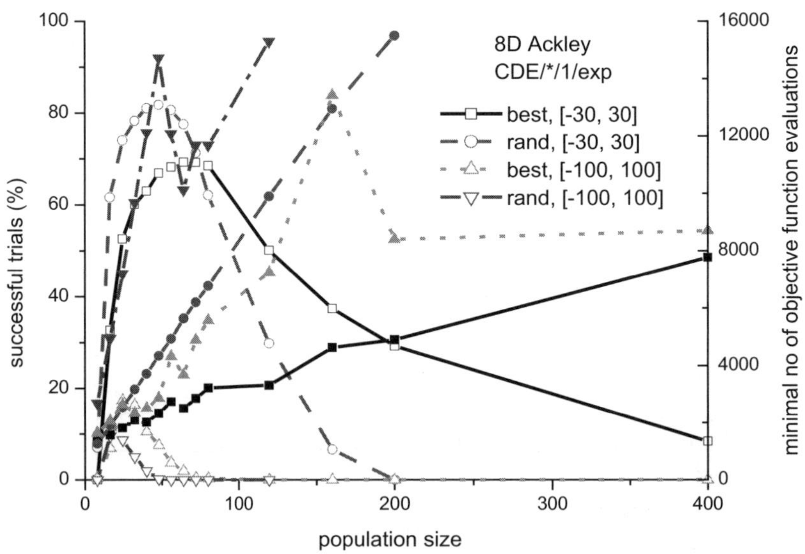

Figure 8.20 Robustness and efficiency of CDE/*/1/exp for the 8-dimensional Ackley function

8.4 Optimal Population Size and Safeguard Zone

8.4.1 Sphere Function

The optimal population size and safeguard zone for the sphere function are shown in Table 8.2, where the non-shaded rows correspond to the standard search space [−500, 500] while the

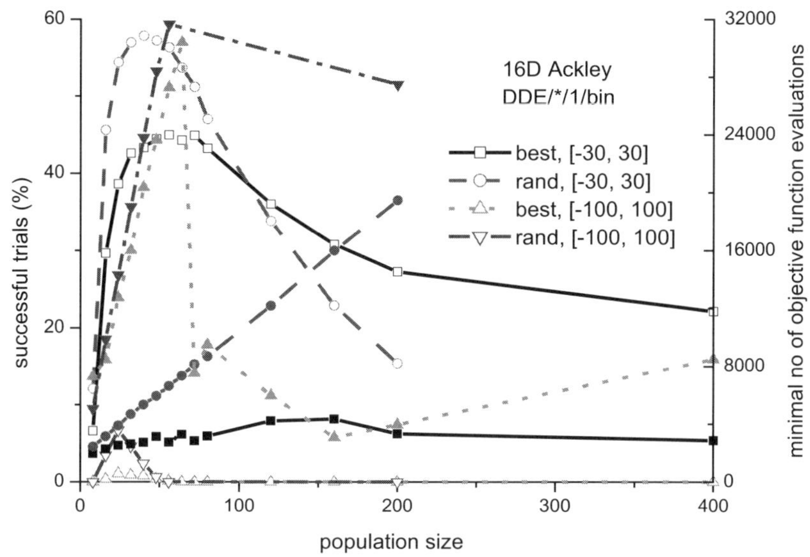

Figure 8.21 Robustness and efficiency of DDE/*/1/bin for the 16-dimensional Ackley function

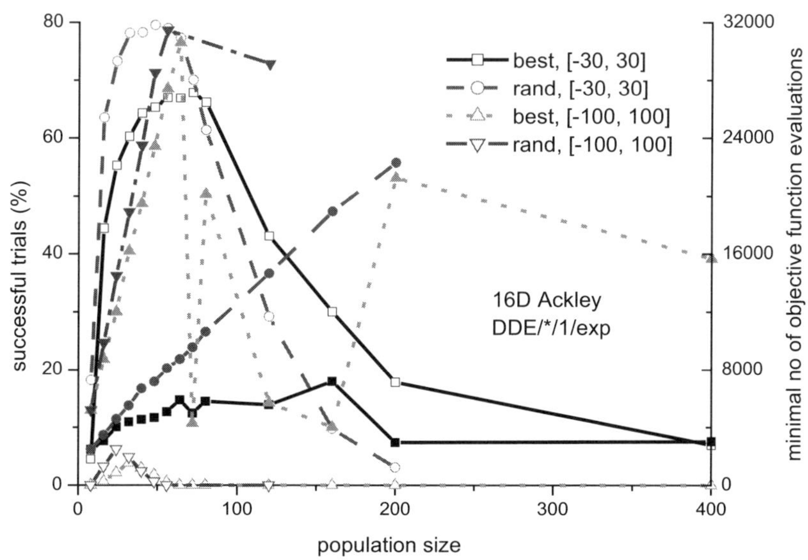

Figure 8.22 Robustness and efficiency of DDE/*/1/exp for the 16-dimensional Ackley function

shadowed rows correspond to the wider search space [−2500, 2500]. It is readily evident that the optimal population size decreases and the safeguard zone shrinks. However, the effect seems quite marginal.

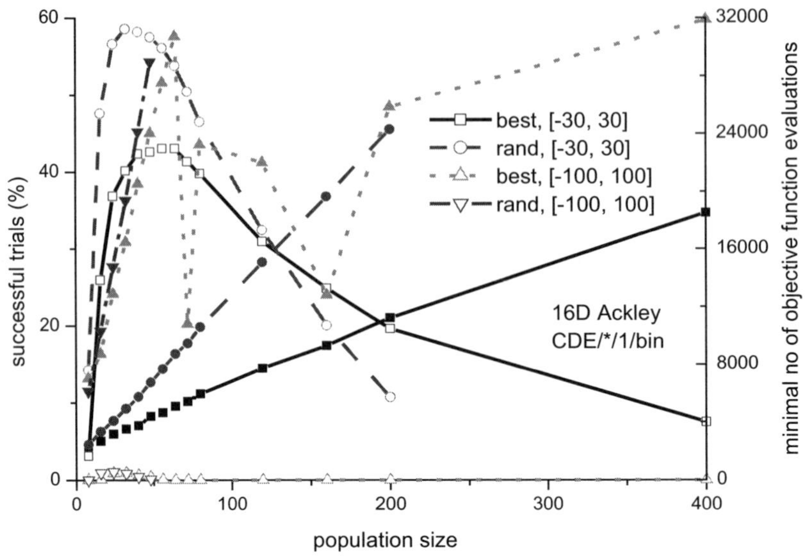

Figure 8.23 Robustness and efficiency of CDE/*/1/bin for the 16-dimensional Ackley function

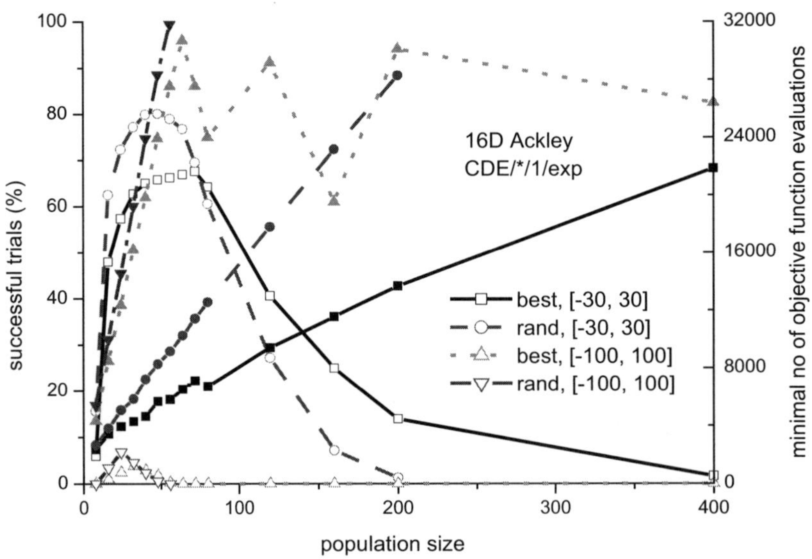

Figure 8.24 Robustness and efficiency of CDE/*/1/exp for the 16-dimensional Ackley function

8.4.2 Step Function 2

The optimal population size and the safeguard zone of population size for step function 2 are summarized in Table 8.3. The trend observed for the sphere function is repeated here.

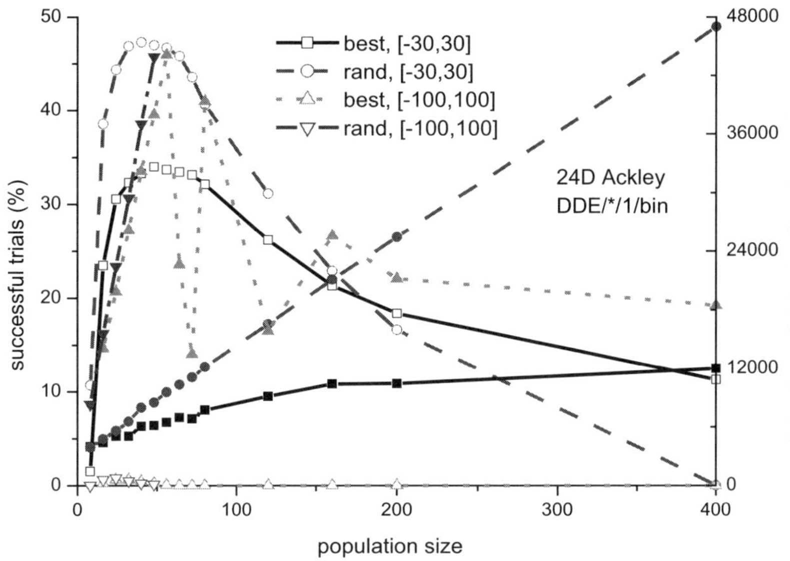

Figure 8.25 Robustness and efficiency of DDE/*/1/bin for the 24-dimensional Ackley function

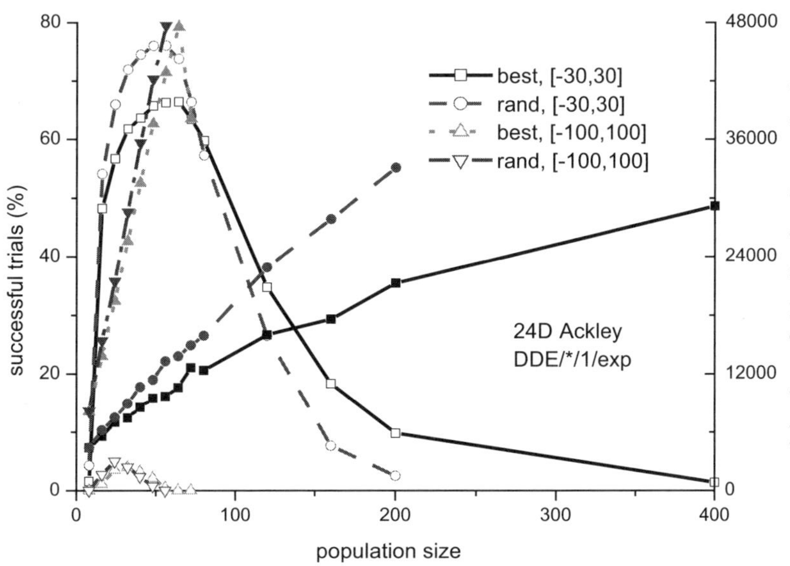

Figure 8.26 Robustness and efficiency of DDE/*/1/exp for the 24-dimensional Ackley function

8.4.3 Hyper-Ellipsoid Function

The optimal population size and the safeguard zone of population size for the hyper-ellipsoid function are summarized in Table 8.4. Without exception, the optimal population size decreases and the safeguard zone shrinks just a little.

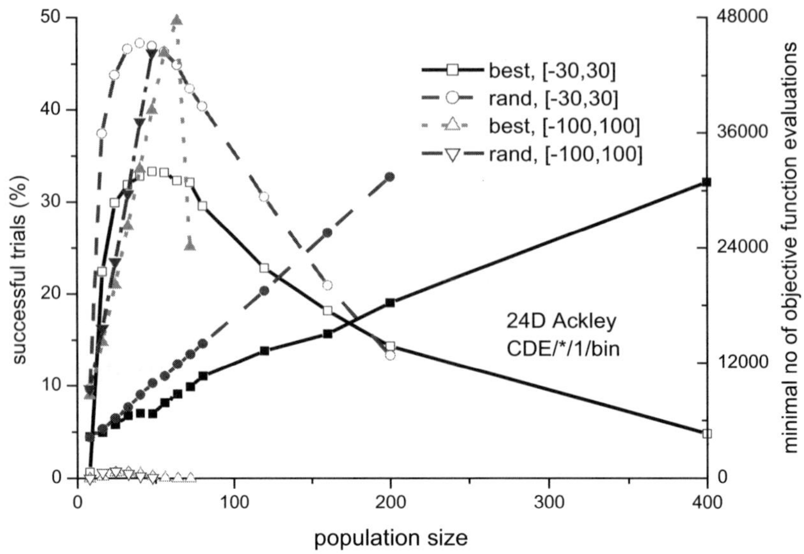

Figure 8.27 Robustness and efficiency of CDE/*/1/bin for the 24-dimensional Ackley function

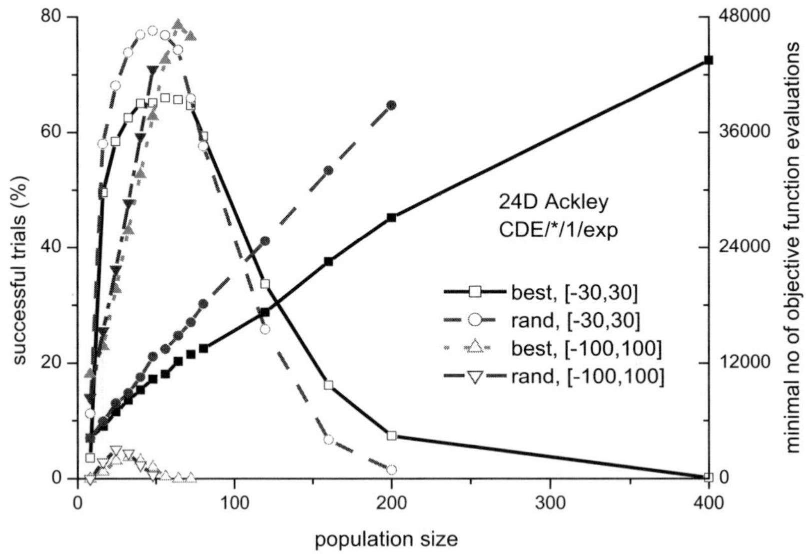

Figure 8.28 Robustness and efficiency of CDE/*/1/exp for the 24-dimensional Ackley function

8.4.4 Qing Function

The optimal population size and the safeguard zone of population size for the Qing function are summarized in Table 8.5.

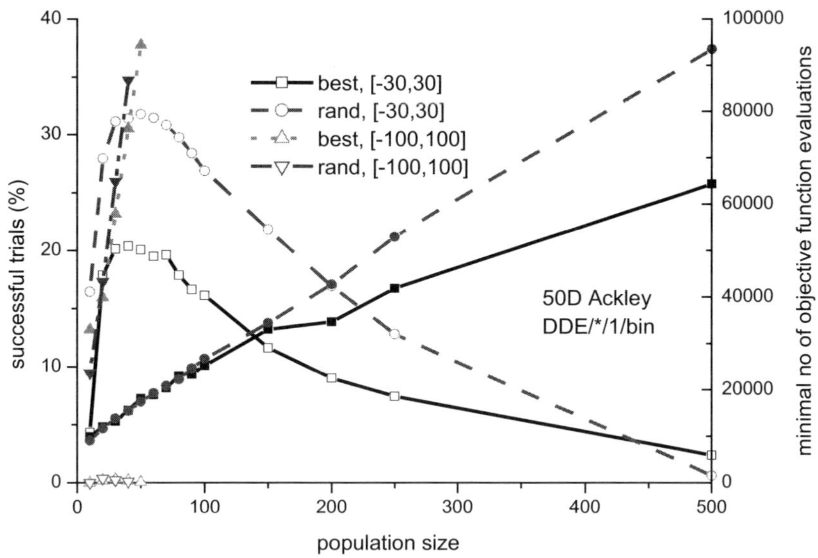

Figure 8.29 Robustness and efficiency of DDE/*/1/bin for the 50-dimensional Ackley function

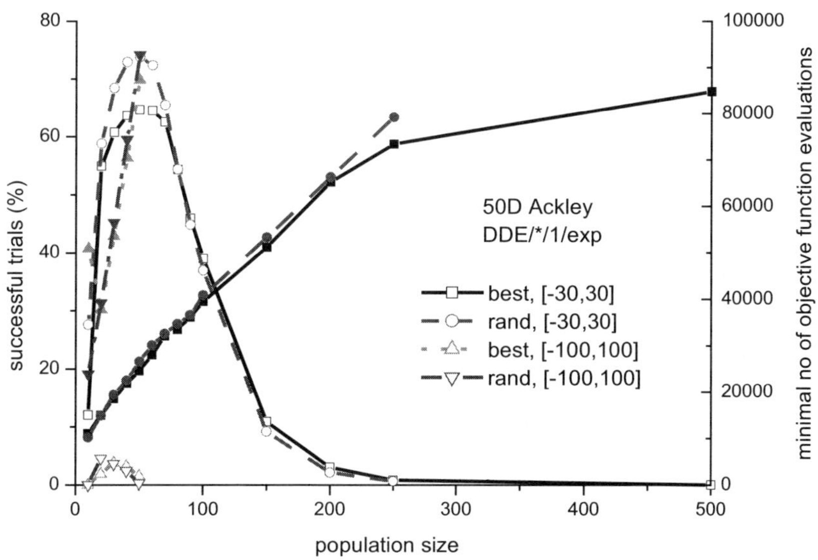

Figure 8.30 Robustness and efficiency of DDE/*/1/exp for the 50-dimensional Ackley function

8.4.5 Schwefel Function 2.22

The optimal population size and the safeguard zone of population size for the Schwefel function 2.22 are summarized in Table 8.6.

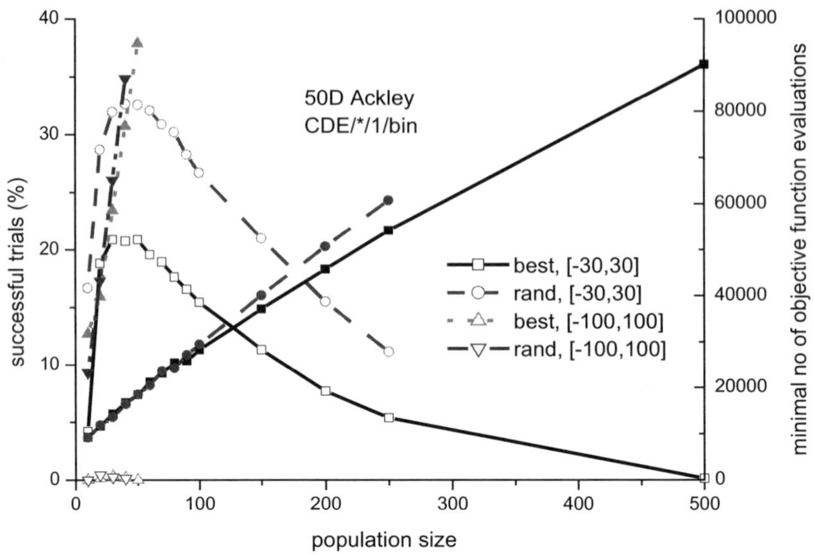

Figure 8.31 Robustness and efficiency of CDE/*/1/bin for the 50-dimensional Ackley function

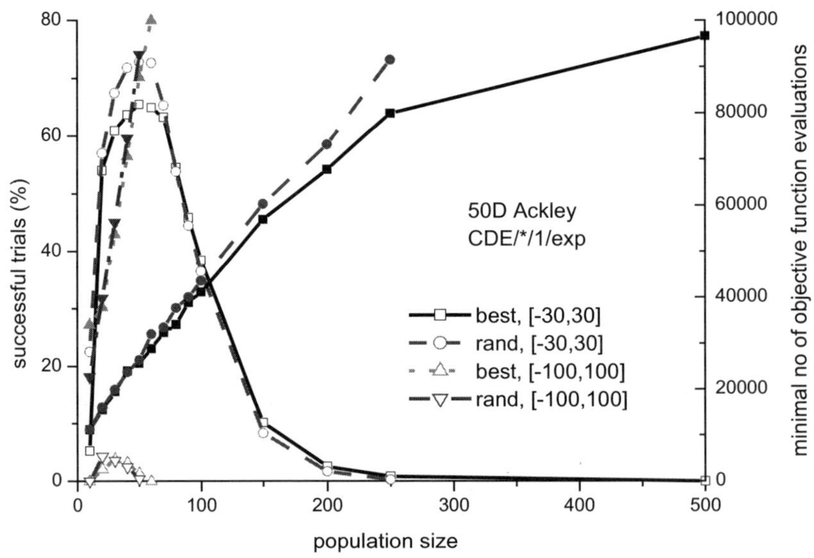

Figure 8.32 Robustness and efficiency of CDE/*/1/exp for the 50-dimensional Ackley function

8.4.6 Schwefel Function 1.2

The optimal population size and the safeguard zone of the optimal population size for the Schwefel function 1.2 are summarized in Table 8.7.

Table 8.2 Optimal population size and safeguard zone of population size for the sphere function

Strategy	Optimal population size				Safeguard zone			
	8D	16D	24D	50D	8D	16D	24D	50D
DDE/best/1/bin	64	56	48	40	[40,80]	[32,80]	[24,80]	[20,150]
	48	48	40	40	[32,80]	[24,80]	[24,80]	[20,100]
CDE/best/1/bin	56	32	32	30	[32,80]	[24,80]	[16,64]	[20,60]
	48	32	24	30	[32,72]	[24,64]	[16,56]	[20,60]
DDE/rand/1/bin	40	32	40	30	[24,64]	[16,64]	[16,64]]	[20,80]
	32	32	32	30	[24,56]	[16,56]	[16,56]	[20,70]
CDE/rand/1/bin	40	32	24	30	[24,64]	[16,64]	[16,64]	[20,80]
	32	32	32	30	[24,48]	[16,56]	[16,56]	[20,70]
DDE/best/1/exp	64	64	56	50	[32,80]	[32,80]	[24,80]	[30,70]
	56	56	48	40	[32,80]	[24,72]	[24,64]	[20,60]
CDE/best/1/exp	64	56	48	50	[32,80]	[24,80]	[24,72]	[20,70]
	56	48	48	40	[32,72]	[24,64]	[24,64]	[20,60]
DDE/rand/1/exp	48	48	48	40	[32,72]	[24,64]	[24,64]	[20,60]
	40	40	40	40	[24,56]	[24,56]	[24,56]	[30,50]
CDE/rand/1/exp	48	40	48	40	[24,72]	[24,64]	[24,64]	[30,60]
	40	40	40	40	[24,56]	[24,56]	[24,56]	[30,50]

Table 8.3 Optimal population size and safeguard zone of population size for step function 2

Strategy	Optimal population size				Safeguard zone			
	8D	16D	24D	50D	8D	16D	24D	50D
DDE/best/1/bin	120	80	64	60	[56,160]	[40,160]	[40,120]	[20,150]
	80	72	48	50	[56,120]	[40,120]	[24,120]	[20,100]
CDE/best/1/bin	72	64	56	60	[40,120]	[32,120]	[32,120]	[20,150]
	72	56	72	60	[32,80]	[32,80]	[24,80]	[20,100]
DDE/rand/1/bin	56	48	48	60	[32,80]	[24,80]	[32,80]	[30,150]
	48	40	40	60	[24,72]	[24,72]	[24,80]	[30,100]
CDE/rand/1/bin	56	48	48	60	[32,80]	[24,80]	[24,80]	[30,120]
	40	40	40	70	[24,72]	[24,64]	[24,80]	[30,100]
DDE/best/1/exp	120	80	80	80	[48,160]	[40,120]	[40,120]	[40,100]
	80	72	64	60	[48,80]	[40,80]	[32,80]	[30,80]
CDE/best/1/exp	80	72	80	70	[40,120]	[40,120]	[40,120]	[40,100]
	80	72	64	60	[40,80]	[40,80]	[40,80]	[30,80]
DDE/rand/1/exp	64	64	56	70	[32,80]	[32,80]	[32,80]	[40,90]
	48	48	56	50	[32,80]	[32,72]	[32,72]	[30,70]
CDE/rand/1/exp	64	64	64	70	[32,80]	[32,80]	[32,80]	[40,90]
	56	48	48	50	[32,80]	[32,72]	[32,72]	[30,70]

8.4.7 Rastrigin Function

The optimal population size and the safeguard zone of the optimal population size for the Rastrigin function are summarized in Table 8.8.

Table 8.4 Optimal population size and safeguard zone of population size for the hyper-ellipsoid function

Strategy	Optimal population size				Safeguard zone			
	8D	16D	24D	50D	8D	16D	24D	50D
DDE/best/1/bin	64	48	40	20	[40,80]	[24,72]	[24,72]	[20,50]
	48	40	32	20	[32,72]	[24,64]	[16,64]	[20,50]
CDE/best/1/bin	56	32	24	20	[32,80]	[16,56]	[16,48]	[20,40]
	48	24	24	20	[24,72]	[24,48]	[16,40]	[20,30]
DDE/rand/1/bin	40	24	24	20	[24,56]	[16,48]	[16,48]	[20,40]
	32	26	24	20	[24,48]	[16,40]	[16,40]	[20,40]
CDE/rand/1/bin	32	24	24	20	[24,56]	[16,48]	[16,48]	[20,40]
	32	24	24	20	[24,48]	[16,40]	[16,40]	[20,40]
DDE/best/1/exp	56	48	40	20	[32,80]	[24,64]	[24,48]	[20,30]
	56	40	32	20	[32,64]	[24,56]	[24,40]	[20,30]
CDE/best/1/exp	64	40	32	20	[32,80]	[24,56]	[24,48]	[20,30]
	48	32	32	20	[24,64]	[24,48]	[24,40]	[20,20]
DDE/rand/1/exp	48	40	32	20	[24,64]	[24,48]	[24,40]	[20,20]
	40	32	32	20	[24,48]	[24,40]	[24,32]	[20,20]
CDE/rand/1/exp	40	32	32	20	[24,56]	[24,48]	[24,40]	[20,20]
	40	32	32	20	[24,48]	[24,40]	[24,32]	[20,20]

Table 8.5 Optimal population size and safeguard zone of population size for the Qing function

Strategy	Optimal population size				Safeguard zone			
	8D	16D	24D	50D	8D	16D	24D	50D
DDE/best/1/bin	40	32	32	30	[32,64]	[24,64]	[24,56]	[20,50]
	40	32	32	30	[24,56]	[16,64]	[16,72]	[20,100]
CDE/best/1/bin	40	32	24	30	[24,56]	[24,48]	[24,40]	[20,40]
	32	24	24	20	[24,56]	[16,40]	[16,40]	[20,40]
DDE/rand/1/bin	16	16	16	20	[16,24]	[16,16]	[16,16]	[10,30]
	16	16	16	20	[16,24]	[16,16]	[16,16]	[10,40]
CDE/rand/1/bin	16	16	16	20	[16,24]	[16,16]	[16,16]	[10,30]
	16	16	16	20	[16,24]	[16,16]	[16,16]	[10,40]
DDE/best/1/exp	48	40	32	30	[24,64]	[24,56]	[24,48]	[20,40]
	40	32	32	30	[24,56]	[24,48]	[24,40]	[20,30]
CDE/best/1/exp	40	32	32	80	[24,64]	[24,48]	[16,48]	[20,40]
	40	32	32	30	[24,56]	[16,48]	[16,40]	[20,30]
DDE/rand/1/exp	24	24	24	20	[16,40]	[16,32]	[16,32]	[20,30]
	24	24	24	20	[16,32]	[16,32]	[16,32]	[20,20]
CDE/rand/1/exp	24	24	24	20	[16,32]	[16,32]	[16,32]	[20,30]
	24	24	24	20	[16,32]	[16,32]	[16,32]	[20,30]

Table 8.6 Optimal population size and safeguard zone of population size for the Schwefel function 2.22

Strategy	Optimal population size				Safeguard zone			
	8D	16D	24D	50D	8D	16D	24D	50D
DDE/best/1/bin	48	40	40	40	[32,72]	[24,64]	[24,64]	[30,70]
	40	32	40	40	[24,64]	[24,56]	[24,64]	[30,70]
CDE/best/1/bin	40	32	32	30	[24,64]	[24,56]	[24,56]	[20,60]
	32	32	32	30	[24,56]	[24,56]	[24,56]	[30,60]
DDE/rand/1/bin	24	24	24	30	[24,48]	[16,48]	[16,56]	[20,70]
	24	24	24	30	[16,40]	[24,48]	[16,48]	[20,60]
CDE/rand/1/bin	24	24	24	20	[16,56]	[24,48]	[24,56]	[20,70]
	24	24	24	30	[16,40]	[16,48]	[16,48]	[20,70]
DDE/best/1/exp	56	40	40	40	[32,72]	[24,64]	[24,56]	[30,50]
	48	40	40	40	[24,64]	[32,48]	[24,48]	[30,40]
CDE/best/1/exp	56	48	40	40	[24,64]	[24,56]	[24,56]	[30,50]
	40	40	40	30	[24,56]	[24,48]	[24,48]	[30,40]
DDE/rand/1/exp	40	32	40	40	[24,56]	[24,48]	[24,48]	[30,40]
	32	32	32	30	[24,48]	[24,48]	[24,40]	[30,40]
CDE/rand/1/exp	40	40	32	40	[24,48]	[24,48]	[24,48]	[30,40]
	32	32	32	30	[24,40]	[24,40]	[24,40]	[30,40]

Table 8.7 Optimal population size and safeguard zone of optimal population size for the Schwefel function 1.2

Strategy	Optimal population size				Safeguard zone			
	8D	16D	24D	50D	8D	16D	24D	50D
DDE/best/1/bin	24	32	56	200	[16,120]	[16,400]	[16,400]	[40,500]
	24	24	32	150	[16,80]	[16,400]	[16,400]	[30,500]
CDE/best/1/bin	24	16	24	—	[16,72]	[16,80]	[16,120]	—
	16	16	16	—	[16,48]	[16,80]	[16,160]	—
DDE/rand/1/bin	16	16	—	—	[8,24]	[16,16]	—	—
	16	16	—	—	[16,24]	[8,48]	—	—
CDE/rand/1/bin	16	16	—	—	[8,24]	[16,16]	—	—
	16	16	—	—	[16,24]	[8,48]	—	—
DDE/best/1/exp	16	16	16	250	[16,40]	[16,200]	[8,400]	[30,500]
	24	16	16	200	[16,48]	[16,160]	[16,400]	[10,500]
CDE/best/1/exp	16	16	16	—	[16,32]	[16,56]	[8,200]	—
	16	16	16	20	[16,32]	[8,40]	[8,120]	[10,50]
DDE/rand/1/exp	16	16	16	—	[8,16]	[8,56]	[8,24]	—
	16	8	8	—	[16,16]	[8,24]	[8,48]	—
CDE/rand/1/exp	16	8	16	—	[8,24]	[8,40]	[8,24]	—
	16	8	8	—	[16,16]	[8,24]	[8,48]	—

Table 8.8 Optimal population size and safeguard zone of optimal population size for the Rastrigin function

Strategy	Optimal population zize				Safeguard zone			
	8D	16D	24D	50D	8D	16D	24D	50D
DDE/best/1/bin	48	48	48	40	[32,80]	[16,80]	[24,80]	[40,70]
	48	48	48	40	[32,72]	[24,80]	[24,72]	[30,60]
CDE/best/1/bin	56	40	48	40	[24,80]	[16,80]	[24,80]	[40,70]
	40	48	40	40	[24,72]	[16,80]	[24,72]	[30,60]
DDE/rand/1/bin	32	32	40	40	[24,48]	[24,56]	[24,64]	[30,70]
	32	32	40	40	[24,40]	[24,48]	[16,80]	[30,60]
CDE/rand/1/bin	32	32	40	40	[24,40]	[24,56]	[24,64]	[30,70]
	32	32	32	40	[24,40]	[24,48]	[24,64]	[30,70]
DDE/best/1/exp	64	48	56	50	[40,72]	[40,64]	[40,64]	[30,40]
	56	56	48	50	[40,64]	[48,56]	[40,56]	[40,60]
CDE/best/1/exp	56	40	56	50	[40,72]	[32,64]	[48,64]	[40,60]
	56	56	48	50	[40,64]	[32,56]	[40,56]	[50,60]
DDE/rand/1/exp	32	40	40	40	[32,48]	[32,48]	[32,48]	[40,50]
	32	40	40	40	[24,40]	[32,40]	[32,40]	[40,40]
CDE/rand/1/exp	32	40	40	40	[32,40]	[32,48]	[32,48]	[40,50]
	32	40	40	40	[24,40]	[32,40]	[40,40]	[40,40]

Table 8.9 Optimal population size and safeguard zone of optimal population size for the Ackley function

Strategy	Optimal population xize				Safeguard zone			
	8D	16D	24D	50D	8D	16D	24D	50D
DDE/best/1/bin	80	56	48	40	[48,80]	[32,80]	[24,80]	[20,100]
	24	24	24	30	[16,56]	[16,56]	[16,56]	[20,50]
CDE/best/1/bin	64	56	48	30	[40,80]	[24,80]	[24,80]	[20,100]
	32	24	32	30	[16,56]	[16,56]	[16,56]	[20,40]
DDE/rand/1/bin	40	40	40	50	[24,64]	[24,64]	[24,72]	[20,100]
	16	24	24	20	[16,40]	[16,48]	[16,48]	[20,40]
CDE/rand/1/bin	40	32	40	40	[24,64]	[24,64]	[24,72]	[20,100]
	16	24	24	20	[16,48]	[16,48]	[16,48]	[20,40]
DDE/best/1/exp	72	72	64	50	[48,80]	[40,80]	[32,72]	[30,70]
	24	32	24	30	[24,32]	[16,56]	[16,56]	[20,50]
CDE/best/1/exp	64	72	56	50	[40,80]	[32,80]	[32,72]	[30,70]
	24	32	24	30	[24,32]	[16,56]	[16,56]	[20,50]
DDE/rand/1/exp	48	48	56	50	[32,64]	[24,64]	[32,64]	[30,60]
	16	24	24	20	[16,32]	[16,48]	[16,48]	[20,50]
CDE/rand/1/exp	48	48	48	50	[32,64]	[32,64]	[32,64]	[30,60]
	16	24	24	20	[16,32]	[16,48]	[16,48]	[20,50]

8.4.8 Ackley Function

The optimal population size and the safeguard zone of the optimal population size for the Ackley function are summarized in Table 8.9. The optimal population size is much smaller and the safeguard zone is significantly narrower when the alternative search space is applied.

8.5 Optimal Mutation Intensity and Crossover Probability for Sphere Function

8.5.1 8-Dimensional Sphere Function

The optimal mutation intensity and crossover probability of differential evolution for the 8-dimensional sphere function are shown in Figures 8.33–8.36. Little difference is observed.

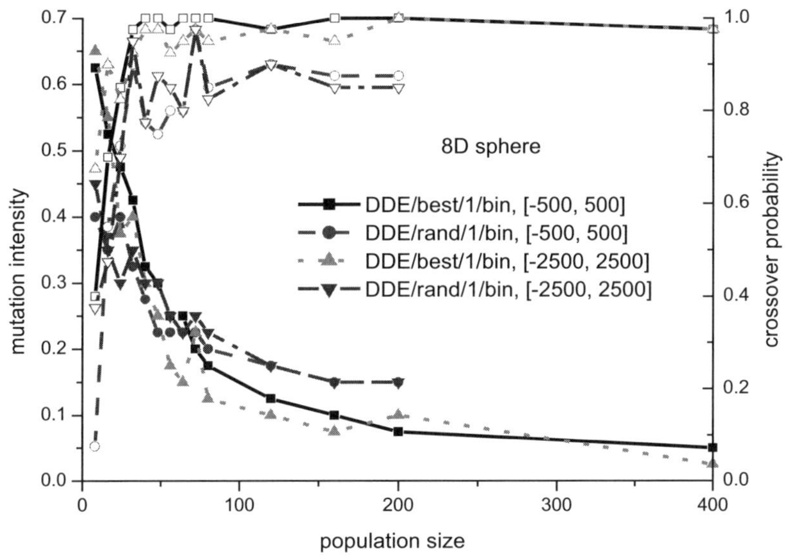

Figure 8.33 Mutation intensity and crossover probability of DDE/*/1/bin for the 8-dimensional sphere function

8.5.2 16-Dimensional Sphere Function

The optimal mutation intensity and crossover probability of differential evolution for the 16-dimensional sphere function are shown in Figures 8.37–8.40.

8.5.3 24-Dimensional Sphere Function

The optimal mutation intensity and crossover probability of differential evolution for the 24-dimensional sphere function are shown in Figures 8.41–8.44.

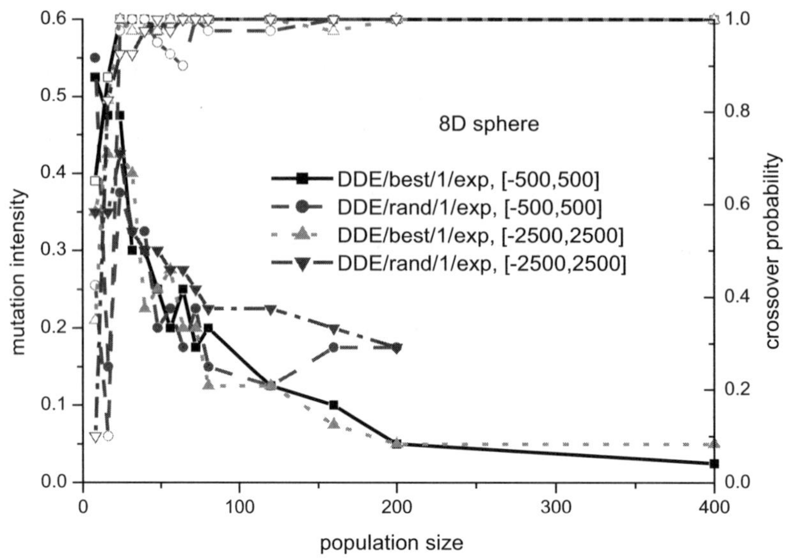

Figure 8.34 Mutation intensity and crossover probability of DDE/*/1/exp for the 8-dimensional sphere function

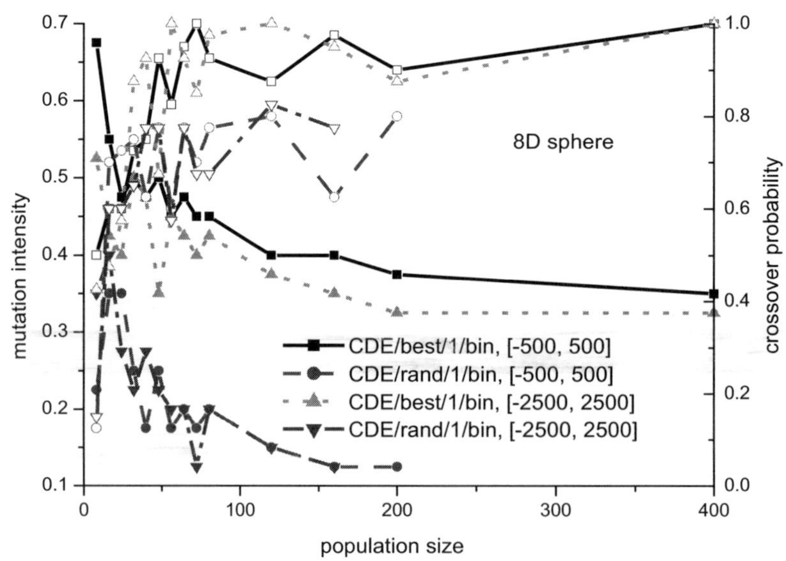

Figure 8.35 Mutation intensity and crossover probability of CDE/*/1/bin for the 8-dimensional sphere function

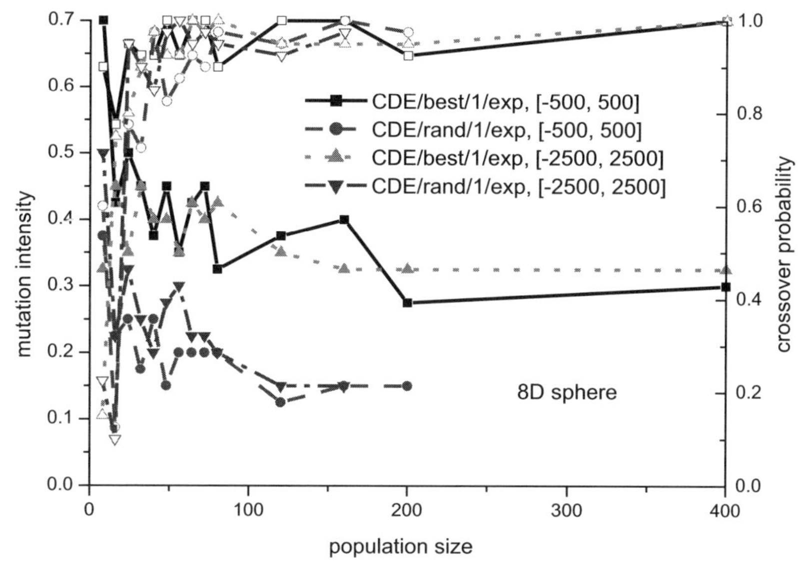

Figure 8.36 Mutation intensity and crossover probability of CDE/*/1/exp for the 8-dimensional sphere function

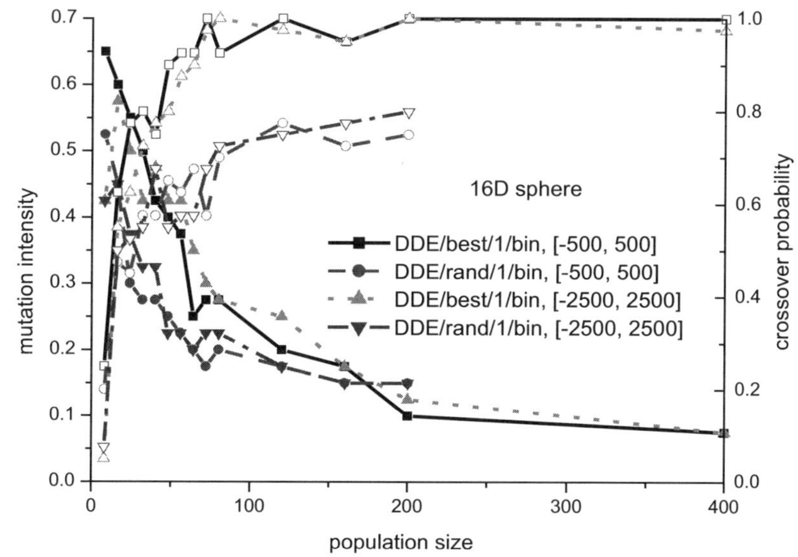

Figure 8.37 Mutation intensity and crossover probability of DDE/*/1/bin for the 16-dimensional sphere function

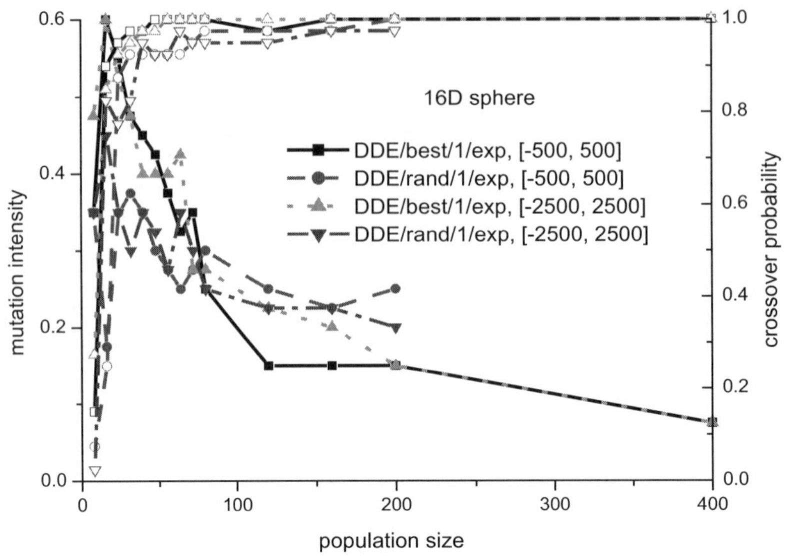

Figure 8.38 Mutation intensity and crossover probability of DDE/*/1/exp for the 16-dimensional sphere function

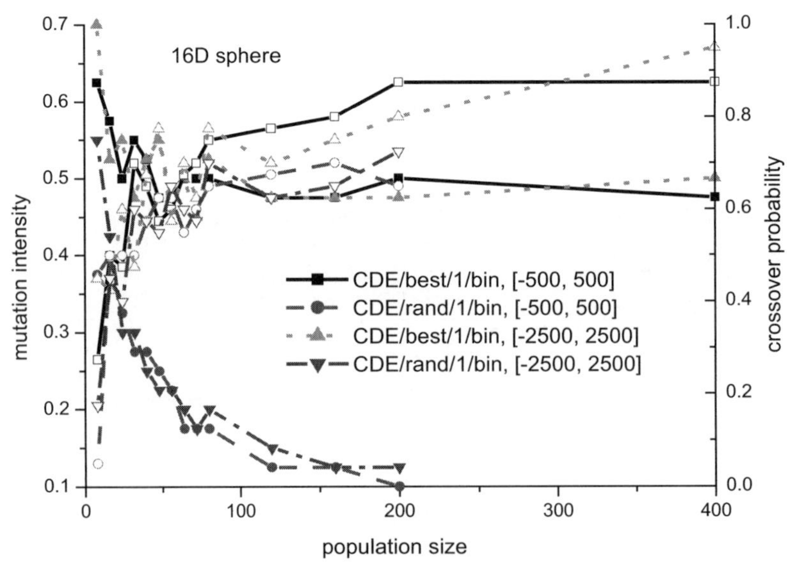

Figure 8.39 Mutation intensity and crossover probability of CDE/*/1/bin for the 16-dimensional sphere function

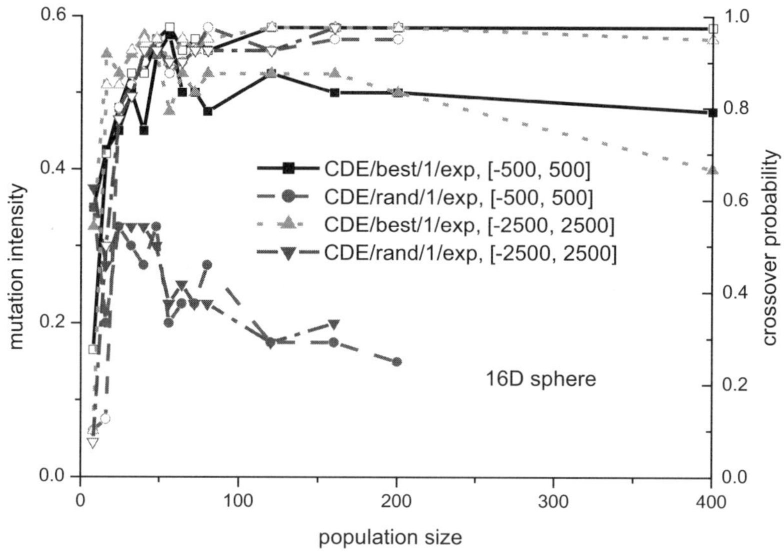

Figure 8.40 Mutation intensity and crossover probability of CDE/*/1/exp for the 16-dimensional sphere function

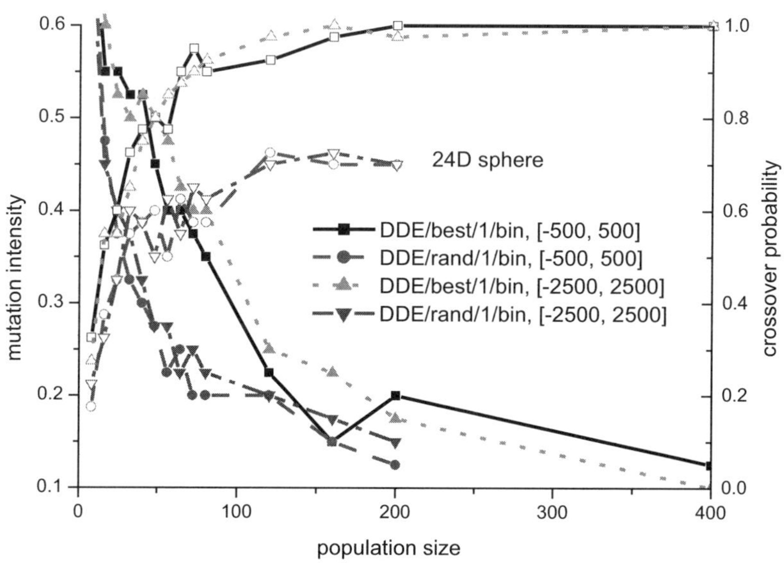

Figure 8.41 Mutation intensity and crossover probability of DDE/*/1/bin for the 24-dimensional sphere function

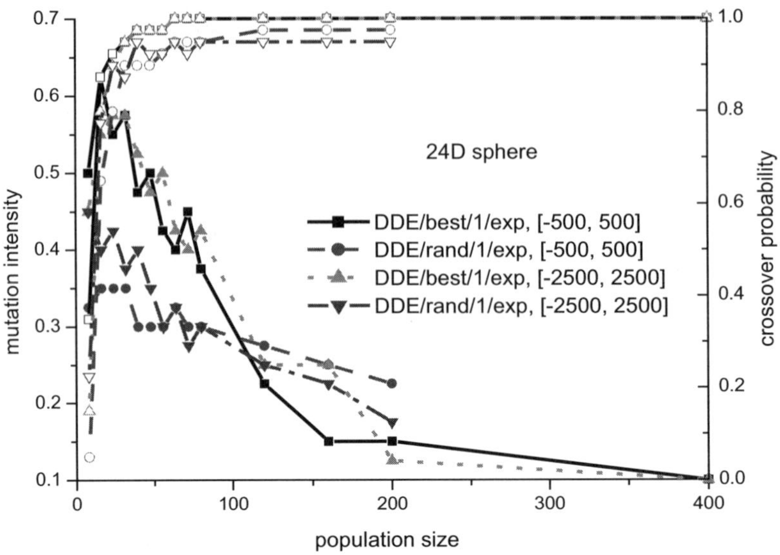

Figure 8.42 Mutation intensity and crossover probability of DDE/*/1/exp for the 24-dimensional sphere function

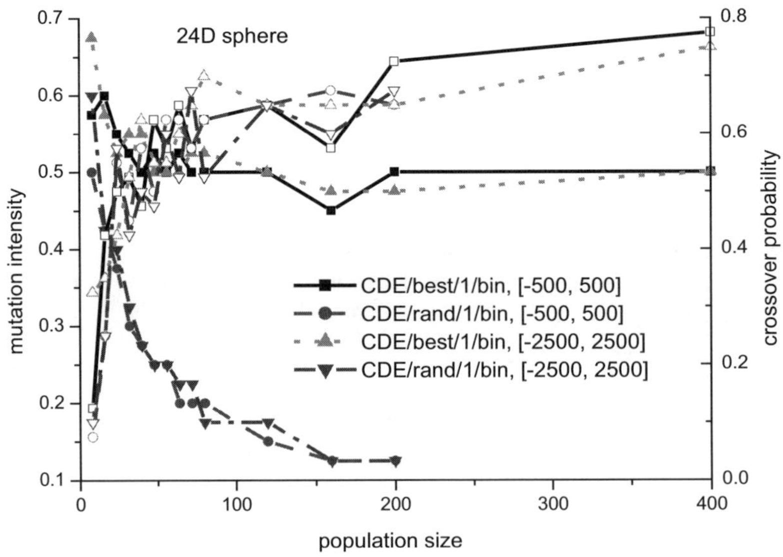

Figure 8.43 Mutation intensity and crossover probability of CDE/*/1/bin for the 24-dimensional sphere function

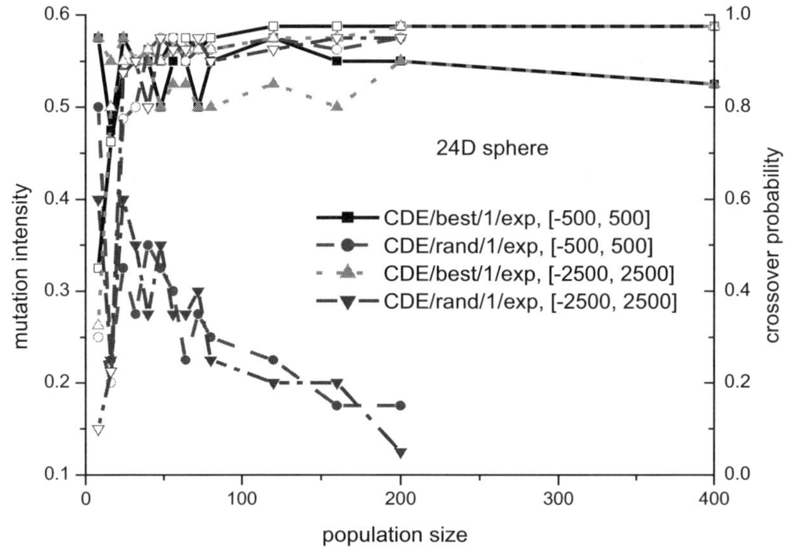

Figure 8.44 Mutation intensity and crossover probability of CDE/*/1/exp for the 24-dimensional sphere function

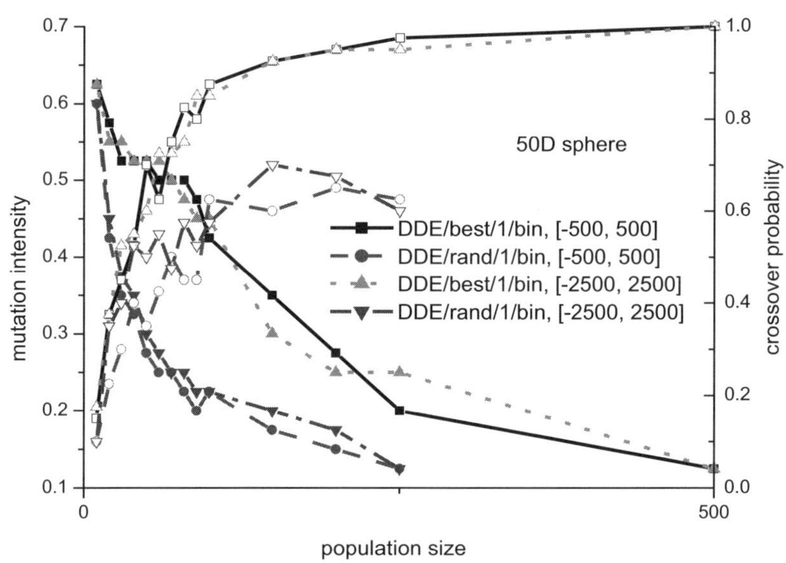

Figure 8.45 Mutation intensity and crossover probability of DDE/*/1/bin for the 50-dimensional sphere function

8.5.4 50-Dimensional Sphere Function

The optimal mutation intensity and crossover probability of differential evolution for the 50-dimensional sphere function are shown in Figures 8.45–8.48.

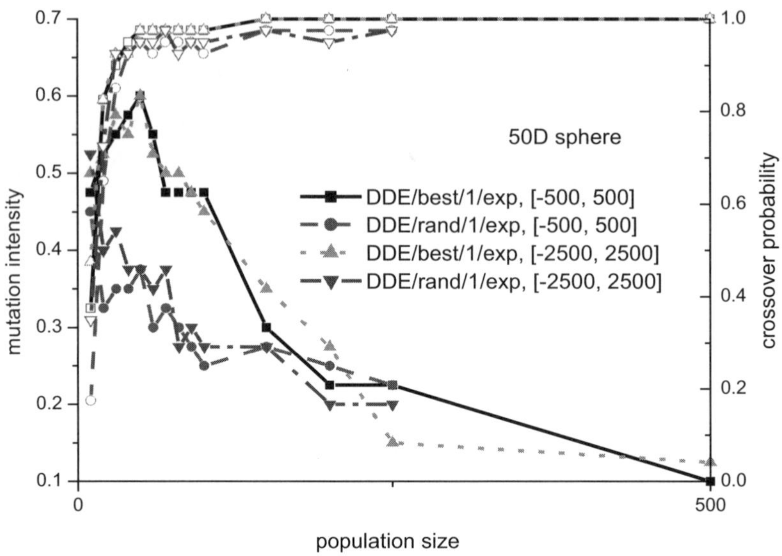

Figure 8.46 Mutation intensity and crossover probability of DDE/*/1/exp for the 50-dimensional sphere function

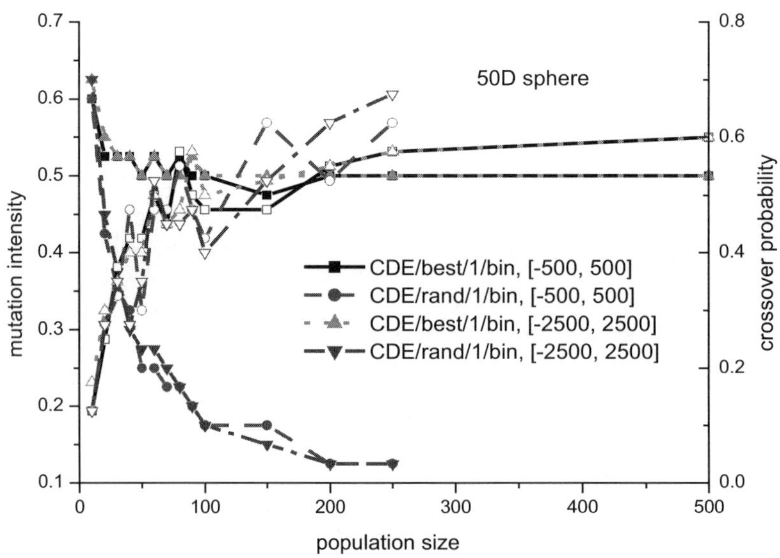

Figure 8.47 Mutation intensity and crossover probability of CDE/*/1/bin for the 50-dimensional sphere function

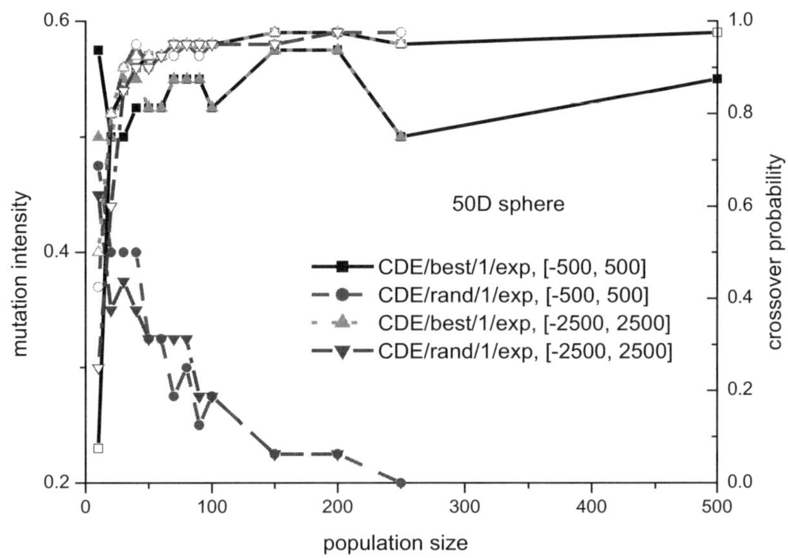

Figure 8.48 Mutation intensity and crossover probability of CDE/*/1/exp for the 50-dimensional sphere function

9

An Introductory Survey on Differential Evolution in Electrical and Electronic Engineering

Advances in electrical and electronic engineering stimulate evolutionary computation. Conversely, electrical and electronic engineering benefits greatly from differential evolution. This chapter presents an introductory survey on applications of differential evolution in electrical and electronic engineering. Later chapters will showcase applications of differential evolution to specific problems in electrical and electronic engineering. Researchers from diverse fields have been invited to make these contributions.

9.1 Communication

9.1.1 Communication Systems

9.1.1.1 Multi-Input Multi-Output System

In recent years high speed data transmission systems have come under increasing pressure due to the rapidly growing demand for wireless communication. However, the channel capacity of the conventional single-input single-output communication system is subject to the Shannon limit. The multi-input multi-output communication system has emerged as a promising alternative. It offers some important advantages such as higher bit-rate and increased capacity. Differential evolution has been applied to determine the optimal antenna number ratio under low signal-to-noise ratio (SNR) conditions [1].

9.1.1.2 Power Allocation

Power control is implemented to improve spectral efficiency. Differential evolution has been applied to optimize power allocation among users in the interleave-division multiple-access (IDMA) system [2–4] and the code-division multiple-access (CDMA) system [5–9].

9.1.1.3 Radio Network Design

The radio network design problem aims to achieve maximum area coverage using a minimal number of antennas without sacrificing service quality by optimizing antenna locations. It is a very important network configuration problem in cellular wireless communication. Differential evolution has been successfully applied to solve this problem [10–13].

9.1.1.4 Rain Attenuation

Rain causes pronounced attenuation in communication systems operating at frequencies above 10 GHz. Therefore, knowledge of rain attenuation is crucial for the design of a reliable communication system. A rain attenuation prediction model based on differential evolution is proposed by Develi [14]. The predicted rain attenuation agrees perfectly well with measured rain attenuation.

9.1.1.5 Telecommunication Flow Modeling

In 1999, differential evolution was applied to telecommunication flow modeling [15–17].

9.1.1.6 Ultrawide-Band Radio System

The ultrawide-band radio system provides short-range high-speed radio services. Differential evolution has been implemented to optimize source pulses and detection template in ultrawide-band radio systems [18].

9.1.2 Communication Codes

We are now in the era of digital communication. In digital communication, information is digitized into codes. As early as 1999, differential evolution was applied to design and optimize erasure codes [19,20].

The explosive growth in information technology has produced a corresponding increase in commercial interest in developing highly efficient data transmission codes as such codes impact everything from signal quality to battery life. The low density parity-check (LDPC) codes, or Gallager codes, are one of the most promising candidates.

The concept of LDPC codes was developed by Robert G. Gallager in his doctoral dissertation at MIR in 1960. It was the first code to allow data transmission rates close to the theoretical maximum, the Shannon limit, and remains, in theory, the most effective developed to date. LDPC codes are in a position to become a standard in the developing market for highly efficient data transmission methods. In 2003, an LDPC code was chosen as the error correcting code in the new DVB-S2 standard for the satellite transmission of digital television.

Differential evolution has been intensively applied to design and optimize LDPC codes, as summarized in Table 9.1. Due to the large number of publications cited in this and subsequent tables, it is impossible to list them at the end of this chapter. Interested readers should refer to the companion website for details of the publications cited.

Table 9.1 Design and optimization of LDPC codes using differential evolution

Application	References
LDPC codes on Rayleigh fading channel	(Hou, Siegel, and Milstein, 2001)
Capacity-approaching irregular LDPC codes	(Richardson, Shokrollahi, and Urbanke, 2001)
LDPC codes for orthogonal frequency division multiplexing (OFDM) transmission	(Mannoni, Declereq, and Gelle, 2002; Mannoni, Declereq, and Gelle, 2004)
LDPC codes for Gaussian broadcast channels	(Berlin and Tuninetti, 2004; Berlin and Tuninetti, 2005)
Irregular LDPC codes of moderate length and high rate	(Yang, Ryan, and Li, 2004)
LDPC coded OFDM for mobile communication via LEO satellites	(Du, Zhang, and Liu, 2005)
Differential space-time modulation based on LDPC codes	(Rui and Xu, 2005)
Generalized LDPC codes on the binary erasure channel	(Paolini, Fossorier, and Chiani, 2006; Cai, Gong, and Huang, 2007)
LDPC code for cooperative relay system in half-duplex mode	(Li, Khojastepour, Yue, Wang, and Madihian, 2007)
LDPC codes for OFDM and spread OFDM in correlated channels	(Serener, Natarajan, and Gruenbacher, 2007; Serener, Natarajan, and Gruenbacher, 2008)
Parallel concatenated LDPC codes	(Wang, Cheng, Xu, and Wang, 2007)
LDPC codes over Rice channel	(Xu and Xu, 2007)

9.2 Computer Engineering

9.2.1 Computer Network and Internet

In 2005, differential evolution was applied to solve dynamic routing in computer networks [21]. Later, it was also applied for Internet routing [22,23].

9.2.2 Cryptography and Security

Cryptography and computer security have been a regular topic at the IEEE Congress on Evolutionary Computation. Differential evolution has been introduced to handle integer factorization problems in cryptography [24] and to detect intrusions in a network [25].

9.2.3 Grid Computing

Grid computing has emerged as a cheap and powerful alternative for multi-processor supercomputers to solve massive computational problems. Differential evolution has been implemented to solve the mapping problem in a grid computing environment aiming at reducing the degree of use of the grid resources [26].

9.2.4 *Parallel Computing*

Nowadays, researchers have been facing more and more computationally expensive problems. Parallel computing often seems to be the only feasible approach for such problems. Multiprocessor scheduling plays a critical role in parallel programs. Differential evolution has been successfully applied to solve multi-processor scheduling problem [27–29].

9.3 Control Theory and Engineering

Control problems occur in almost all engineering fields such as aeronautics, agriculture, chemical engineering, civil engineering, environmental engineering, food engineering, manufacturing industry, mass rapid transit system, material engineering, and power engineering. Differential evolution has been successfully applied to solve numerous control-related application problems.

9.3.1 *System Modeling*

System modeling is fundamental in the simulation and controller design for complex dynamic systems. Different systems have been modeled by using differential evolution [30–33].

9.3.2 *Controller Design*

Differential evolution plays a significant role in controller design. Controllers designed by using differential evolution are summarized in Table 9.2.

Table 9.2 Controllers designed by using differential evolution

Controller	References
Fuzzy logic controller	(Cheong and Lai, 1999; Sastry, Behera, and Nagrath, 1999; Cheong and Lai, 2002; Cheong and Lai, 2005; Xue, Sanderson, Bonissone, and Graves, 2005; Cheong and Lai, 2007a; Cheong and Lai, 2007b; Cheong, Lai, Kim, Hwang, Oh, and Kim, 2007)
Fuzzy phase-plane controller	(Lii, Chiang, Su, and Hwung, 2003)
Model-free learning adaptive controller	(Coelho, Rodrigues Coelho, and Sumar, 2006)
Proportional-integral derivative controller	(Cheng and Hwang, 1998; Hwang and Hsiao, 2002; Kau, Mukherjee, Loo, and Kwek, 2002; Leu, Tsay, and Hwang, 2002; Bingul, 2004; Chang and Hwang, 2004; Pishkenari, Mahboobi, and Meghdari, 2004; de Moura Oliveira, 2005; Gao and Tong, 2006; Tan and Dou, 2007; Zeng and Tan, 2007)

Due to advances in microelectronics, digital controllers have gained popularity in continuous-time control systems. An interesting approach, the digital redesign of analog controllers, aims to replace an existing well-designed analog controller by a digital one along with a sampler and a hold. Differential evolution has been successfully applied to solve this problem [34].

9.3.3 Robotics

The history of implementing differential evolution in robotics is only slightly shorter than that of differential evolution itself. Problems in robotics solved by using differential evolution are summarized in Table 9.3.

Table 9.3 Applications of differential evolution in robotics

Application	References
Multi-sensor fusion	(Joshi and Sanderson, 1996; Joshi and Sanderson, 1997a; Joshi and Sanderson, 1997b; Joshi and Sanderson, 1997c; Joshi and Sanderson, 1999)
Motion (path and trajectory) planning	(Wang and Chiou, 1997b; Aydin and Temeltas, 2002; Aydin and Temeltas, 2003; Aydin and Temeltas, 2004; Aydin and Temeltas, 2005; Oliveira, Saramago, and Oliveira, 2006; Oplatkova and Zelinka, 2007)
Design	(Shiakolas, Koladiya, and Kebrle, 2002a; Shiakolas, Koladiya, and Kebrle, 2002b; Shiakolas, Koladiya, and Kebrle, 2002c; Fung and Yau, 2003; Koladiya, Kebrle, and Shiakolas, 2003; Koladiya, Shiakolas, and Kebrle, 2003; Fung and Yau, 2005; Shiakolas, Koladiya, and Kebrle, 2005)
Coordination and synchronization of locomotion	(Teo and Abbass, 2002; Teo and Abbass, 2003; Teo, 2005a)
Trajectory tracking	Pishkenari, Mahboobi, and Meghdari, 2004
Coordination of robot ants	(De, Ray, Konar, and Chatterjee, 2005a)
Disturbance suppression	(Kwek, Kang, Loo, and Wong, 2005)
Wireless sensor network routing	(Xue, Sanderson, and Graves, 2006)
Navigation	Moreno, Garrido, Martin, and Munoz, 2007; Vahdat, Naser, and Saeed, 2007

9.4 Electrical Engineering

Applications of differential evolution in electrical engineering focus on motor modeling and design, as summarized in Table 9.4.

9.5 Electromagnetics

9.5.1 Antennas and Antenna Arrays

The antenna is of great practical importance in electrical and electronic engineering. Accordingly, it is one of the foci of differential evolution applications in electrical and electronic engineering.

Differential evolution has been applied to synthesize antennas such as the pyramidal horn [35] and microstrip antenna [36–39]. Increasing accuracy has been observed as a result.

Table 9.4 Motor modeling and design by using differential evolution

Application	References
Parameter identification of induction motors	(Ursem and Vadstrup, 2003)
Brushless permanent magnet motor design	(Sykulski, 2004b)
Interior permanent magnet motor design	(Žarko, Ban, and Klarić, 2005; Zarko, Ban, and Lipo, 2005)
Estimation of equivalent circuit parameters of induction machines	(Repo and Arkkio, 2006)
Permanent magnet AC servo motor design	(Zarko, Ban, and Goricki, 2006)
Induction motor design	(Padma, Bhuvaneswari, and Subramanian, 2007)
Universal electric motor design	(Tušar, Korošec, Papa, Filipič, and Šilc, 2007)

Antenna array synthesis is essential. Generally, there are two approaches to synthesizing antenna arrays. In the first approach, mutual coupling between array elements is neglected. Synthesized ideal antenna arrays using differential evolution are summarized in Table 9.5.

In practice, mutual coupling between array elements is inevitable and may degrade array performance significantly. In recent years, this issue has been brought to the attention of

Table 9.5 Synthesized ideal antenna arrays using differential evolution

Ideal Antenna Array	References
Reconfigurable and conformal antenna arrays	(Michalski and Deng, 2001)
Linear array	(Güney, 2002; Yang, Qing, and Gan, 2003; Fan, Jin, Geng, and Liu, 2004; Yang, Gan, and Qing, 2004; Guney, Akdagli, and Babayigit, 2006; Chen, Yang, and Nie, 2007; Guney and Onay, 2007a; Guney and Onay, 2007b)
Time-modulated linear array	(Yang, Gan, and Qing, 2002; Yang, Gan, and Qing, 2003a; Yang, Gan, and Tan, 2003; Fondevila-Gomez, Bre gains, Franceschetti, and Ares, 2005; Yang, Gan, Qing, and Tan, 2005; Yang and Nie, 2007)
Unequally spaced reflectarray of microstrip patch elements	(Kurup, Himdi, and Rydberg, 2003a)
Unequally spaced linear array	(Kurup, Himdi, and Rydberg, 2003b)
Linear array with moving phase center and static excitation amplitude distribution	(Yang, Gan, and Qing, 2003b)
Monopulse antenna array	(Caorsi, Massa, Pastorino, and Randazzo, 2005; Massa, Pastorino, and Randazzo, 2006)
Time-modulated planar array with square lattices and circular boundaries	(Yang and Nie, 2005a)
Steerable linear array	(Rocha-Alicano, Covarrubias-Rosales, and Brizuela-Rodríguez, 2006)
Planar array	(Rocha-Alicano, Covarrubias-Rosales, Brizuela-Rodriguez, and Panduro Mendoza, 2007)

antenna array synthesizers [40]. Differential evolution has been applied to compensate the mutual coupling of time-modulated linear arrays [41] and printed dipole linear arrays [42,43].

9.5.2 Computational Electromagnetics

Most application problems in computational electromagnetics are computationally extremely expensive. Differential evolution is used to trace the optimal higher-order Whitney element to speed up convergence of the finite-element method for three-dimensional microwave and antenna simulations [44].

9.5.3 Electromagnetic Composite Materials

Composite materials have been playing an increasingly important role. Accurate modeling of composite materials has been a great challenge. Differential evolution has been used to model the effective wavenumber [45,46] and to retrieve the effective permittivity tensor [47] of electromagnetic composite materials. Besides more accurate results, effective anisotropy in electromagnetic composite materials with aligned spheroidal inclusion has been observed for the first time.

9.5.4 Electromagnetic Inverse Problems

Electromagnetic inverse problems are of great interest to both scientific researchers and application engineers due to its non-invasive feature. In nature, they are optimization problems. Therefore, differential evolution is directly applicable.

Michalski has pioneered the introduction of differential evolution to solve electromagnetic inverse problems. He applied differential evolution to reconstruct circular-cylindrical conductors and tunnels [48], elliptical-cylindrical conductors and tunnels [49], and homogeneous-dielectric elliptic cylinders [50].

Qing has been very active in using differential evolution to solve electromagnetic inverse scattering of multiple perfectly conducting cylinders in free space [51–56].

9.5.4.1 One-Dimensional Electromagnetic Inverse Problems

Although one-dimensional electromagnetic inverse problems are far from real situations, they are scientifically important. They are also a good starting point for the study of electromagnetic inverse problems. Applications of differential evolution to one-dimensional electromagnetic inverse problems are summarized in Table 9.6.

Table 9.6 Applications of differential evolution to one-dimensional electromagnetic inverse problems

Application	References
Inhomogeneous material filling a waveguide	(Baganas, Kehagias, and Charalambopoulos, 2001; Baganas, 2004)
Bi-anisotropic	(Chen, Wu, Kong, and Grzegorczyk, 2005; Chen, Grzegorczyk, and Kong, 2006)

9.5.4.2 Two-Dimensional Electromagnetic Inverse Problems

Two-dimensional electromagnetic inverse problems have a broad appeal. Real application problems are three-dimensional in nature. However, most electromagnetic inverse problems are computationally unaffordable without two-dimensional approximation. Usually, two-dimensional approximation does not degrade the results too much. Two-dimensional electromagnetic inverse problems solved by using differential evolution are summarized in Table 9.7.

Table 9.7 Two-dimensional electromagnetic inverse problems solved by using differential evolution

Application	References
Well-log data inversion	(Goswami, Mydur, and Wu, 2002; Goswami, Mydur, Wu, and Heliot, 2004; Xing and Xue, 2007)
Electrical impedance tomography	(Li, Rao, He, Xu, Wu, Ge, and Yan, 2003; Li, Rao, He, Xu, Guo, Yan, Wang, and Yang, 2004; Li, Xu, Rao, He, and Yan, 2005; Li, Rao, He, Xu, Wu, Yan, Dong, and Yang, 2005; Bachorec and Dedkova, 2006; Li, Xu, Guo, Wang, Yang, Rao, He, and Yan, 2006)
Dielectric objects buried in a half-space	(Caorsi, Donelli, Massa, Pastorino, and Randazzo, 2004; Massa, Pastorino, and Randazzo, 2004)
Electroencephalography	(Li, Li, He, Rao, Wu, Xu, Shen, and Yan, 2004)
Multiple homogeneous dielectric cylinders	(Krishna and Chen, 2007)
A two-layer dielectric cylinder	(Pastorino, 2007)

9.5.4.3 Three-Dimensional Electromagnetic Inverse Problems

Three-dimensional electromagnetic inverse problems come directly from real applications. Therefore, application engineers are very interested in them. Two kinds of electromagnetic inverse problems, as summarized in Table 9.8, have been solved by using differential evolution.

Table 9.8 Three-dimensional electromagnetic inverse problems solved by using differential evolution

Application	References
Unexploded ordnance inversion	(Chen, O'Neill, Barrowes, Grzegorczyk, and Kong, 2004; Shubitidze, O'Neill, Shamatava, Sun, and Paulsen, 2005a; Shubitidze, O'Neill, Shamatava, Sun, and Paulsen, 2005b; Shubitidze, O'Neill, Barrowes, Shamatava, Fernandez, Sun, and Paulsen, 2007)
Classifying underground targets and simultaneously estimating their burial conditions using ultra-wideband ground penetrating radar	(Strifors, Gaunaurd, and Sullivan, 2004; Strifors, Andersson, Axelsson, and Gaunaurd, 2005)

9.5.5 *Frequency Selective Surfaces*

Frequency selective surfaces have attracted a great amount of attention because of their frequency filtering property. Synthesizing frequency selective surfaces with desirable frequency filtering specification is therefore appealing. Planar arrays synthesized include the dipole array [57,58] and double-square-loop array [59,60].

9.5.6 Microwave Devices

Differential evolution has also been applied to design various microwave devices, as summarized in Table 9.9.

Table 9.9 Microwave devices design by using differential evolution

Microwave Device	References
RF low noise amplifier	(Vancorenland, De Ranter, Steyaert, and Gielen, 2000)
RF mixer	(Vancorenland, Van der Plas, Steyaert, Gielen, and Sansen, 2001)
Langmuir probe in plasma processes	(Nolle, Zelinka, Hopgood, and Goodyear, 2005)
High-power millimeter-wave TM01-TE11 mode converters	(Yang and Qing, 2005)
Multilayer homogeneous coupling structure	(Guney, Yildiz, Kaya, and Turkmen, 2007)

9.5.7 Radar

Differential evolution has been applied to estimate target motion parameters [61], a fundamental problem in radar.

9.6 Electronics

9.6.1 Analysis

Analysis is essential and fundamental for better design and control. Applications of differential evolution in this area summarized in Table 9.10.

Table 9.10 Applications of differential evolution in analysis of electronic circuits and systems

Application	References
Determination of DC operating point of nonlinear circuits	(Crutchley and Zwolinski, 2002; Crutchley and Zwolinski, 2003)
Modeling of laser diode nonlinearity in a radio-over-fiber network	(Akdagli and Yuksel, 2006)
Modeling of hard disk drive servo system	(Svecko and Pec, 2006)

9.6.2 Circuit Design

The circuit is the building block of all electronic systems. A circuit has to be well designed to meet its specific function requirement. Strong design capabilities are critical for the electronic industry. A large amount of investment in the electronics industry has been devoted to research and development. Not surprisingly, differential evolution has found numerous applications in circuit design.

The filter is an essential component in many circuits. Intensive efforts have been devoted to filter design using differential evolution, as summarized in Table 9.11. Besides filters, other circuit components have also been designed by using differential evolution, as summarized in Table 9.12.

Table 9.11 Filter design by using differential evolution

Filter	References
Infinite-impulse response filter	(Storn, 1995; Storn, 1996a; Chang, 2005a; Das and Dey, 2005; Das, Konar, and Chakraborty, 2005; Karaboga, 2005; Storn, 2005; Huang, Wang, and Liu, 2006)
Howling removal unit in modern audio communication applications	(Storn, 1996b)
Switched capacitor filter	(Storn, 1999; Dolivka and Hospodka, 2006)
Analog filter	(Vondras and Martinek, 2001; Martinek and Vondraš, 2002)
Finite-impulse response (FIR) filter	(Karaboga and Cetinkaya, 2004; Chang, 2005b; Karaboga and Cetinkaya, 2005; Karaboga and Cetinkaya, 2006; Tirronen, Neri, Karkkainen, Majava, and Rossi, 2007; Zhao and Peng, 2007)
Scaling filter	(Sampo, 2004a; Sampo, 2004b)
Adaptive FIR filter	(Karaboga and Koyuncu, 2005; Yigit and Karaboga, 2007; Yigit, Koyuncu, and Karaboga, 2007)
All-pass filter	(Ziska and Laipert, 2005; Ziska, Vlcek, and Laipert, 2005; Ziska and Laipert, 2006a; Ziska and Laipert, 2006b; Ziska and Vrbata, 2006; Ziska and Laipert, 2007)
Switched-current filter	(Dolivka and Hospodka, 2007)

Table 9.12 Non-filter circuit design by using differential evolution

Circuit	References
Howling removal unit in modern audio communication applications	(Storn, 1996b)
Delta–sigma modulators	(Francken, Vancorenland, and Gielen, 2000; Francken and Gielen, 2003)
Snubber circuits for low–power DC–DC converters realized with MOSFETs	(Tezak, Dolinar, and Milanovic, 2003; Tezak, Dolinar, and Milanovic, 2004; Tezak, Dolinar, and Milanovic, 2005)
Selected harmonic elimination pulse width modulation inverter	(Huang, Hu, and Czarkowski, 2004; Hu, Huang, and Czarkowski, 2005)
Combinational logic circuits	(Moore and Venayagamoorthy, 2006)
Negative feedback amplifier	(Miranda-Varela, Mecura-Montes, and Sarmiento-Reyes, 2007)
Switchgear	(Kitak, Pihler, Ticar, Stermecki, Biro, and Preis, 2006; Kitak, Pihler, Ticar, Stermecki1, Magele, Bíró, and Preis, 2006; Kitak, Pihler, Ticar, Stermecki, Biro, and Preis, 2007)
Integrated circuit	(Olensek, Burmen, Puhan, and Tuma, 2007)
AM radio receiver	(Perales-Graván and Lahoz-Beltra, 2008)

9.6.3 Fabrication

Differential evolution has been applied to develop a decision-making system for the design, supply, and manufacture of an electronic circuit board product [62].

9.6.4 Packaging

Differential evolution has been applied to improve electronic packaging reliability [63–65].

9.6.5 Testing

All electronic applications have to undergo stringent testing before marketing so that they will not violate any regulations. Differential evolution has been applied to reduce the testing time of electronic applications using system on chip (SoC) design [66].

9.7 Magnetics

Besides magnet motor design (Section 9.4), differential evolution has also been applied to solve many other design problems in magnetics, as summarized in Table 9.13.

Table 9.13 Applications of differential evolution in magnetics

Application	References
Permanent magnet machine	(Pahner, Mertens, de Gersem, Belmans, and Hameyer, 1998; Ouyang, Zarko, and Lipo, 2006)
Radial active magnetic bearing	(Stumberger, Dolinar, Pahner, and Hameyer, 1999; Stumberger, Dolinar, Stumberger, Pahner, and Hameyer, 1999; Stumberger, Dolinar, Pahner, and Hameyer, 2000; Polajzer, Stumberger, Dolinar, and Hameyer, 2002)
Resistive magnet	(Sánchez and Garrido, 2000)
Magnetizer	(Rashid, Farina, Ramirez, Sykulski, and Freeman, 2001; Sykulski, 2004a; Sykulski, 2004b)
Solenoidal superconducting magnetic energy storage	(Coelho and Alotto, 2006)
Loney's solenoid design	(Coelho and Mariani, 2006b; Coelho and Mariani, 2006c)

9.8 Power Engineering

Electricity plays a critical role in everyday life. Power engineering is thus one of fields in electrical and electronic engineering in which differential evolution is implemented most intensively.

9.8.1 Generation

Issues related to power generation include generator modeling, system modeling, scheduling, and system expansion. Differential evolution has demonstrated its potential to solve these problems.

9.8.1.1 Economic Dispatch

The economic dispatch problem [67–77] minimizes system cost by properly allocating the real power demand amongst the online generating units.

As a subproblem of economic dispatch, the reactive power dispatch problem has a significant influence on secure and economic operation of power systems. It determines all kinds of controllable variables, such as reactive power outputs of generators and static VAR compensators, tap ratios of transformers, outputs of shunt capacitors/reactors, and so on, and minimizes transmission losses or other appropriate objective functions, while satisfying a given set of physical and operating constraints [78–84].

9.8.1.2 Generator Modeling

Induction motors play an essential role in power generation. Accurate identification of motor parameters [31,85] is very important for both motor and system analysis and control.

9.8.1.3 Scheduling

Short-term scheduling involves the scheduling for a group of power generation units on a system over a given time horizon to achieve minimal cost while satisfying the hourly power demand constraints. Issues involved such as fuel allocation [86], short-term scheduling of hydrothermal power system with cascaded reservoirs [87,88] and the unit commitment problem [89] have been considered using differential evolution.

9.8.1.4 Static Excitation System

Differential evolution has been successfully applied to tune the static excitation system parameters for better dynamic performance of the Tong-Shiao Generation Station of the Taiwan power system [90].

9.8.1.5 System Expansion

Generation expansion planning is concerned with determining what type of generating units should be commissioned and when the generating units should g online, over a long-range planning horizon [85,91–94]. It involves minimizing cost while meeting reliability, security, and other system constraints. In a competitive market, the objective of generation expansion planning also involves profit maximization and risk management issues.

9.8.2 Distribution

Once generated, electricity has to be transmitted to end users through a distribution system. Differential evolution is used to optimize network configurations of distribution systems. Various aspects including system cost, efficiency, and stability, have been scrutinized.

9.8.2.1 Available Transfer Capability

Available transfer capability [85] is a measure of the transfer capability remaining in the physical transmission network for further commercial activity over the already committed uses without compromising system security. It is an essential measure for power system operation and planning.

9.8.2.2 Cable Design

Losses occur when transmitting electricity by cable. Well-designed cables will improve transmission efficiency. Differential evolution has been applied to optimize the structure parameters of the multilayered conductors of high-temperature superconducting cables to reduce AC losses due to non-uniform current distribution among the multilayer conductors [95].

9.8.2.3 Network Configuration

The distribution network is subject to strict requirements with regard to efficiency and reliability. The installation cost should be as low as possible subject to such requirements being met. Differential evolution has been applied to allocate main feeder sectionalizing switches and lateral fuses so as to minimize total costs for different conditions and configurations for optimal distribution system reliability planning [96–99]. It has also been applied to design an underground cable system [100].

9.8.2.4 Reactive Power Planning

The aim of reactive power planning [101–111] is to determine the new reactive power sources in terms of type, size and location in the network that will result in an adequate voltage control capability by achieving a correct balance between security and economic concerns over a given time horizon, typically 1–3 years. It is performed in coordination with transmission capacity studies that have a longer time horizon and a higher priority.

9.8.2.5 Distribution Network with Distributed Generation

Distributed generation is advantageous in terms of system restructuring, expansion and upgrading, and pollution reduction. Therefore, it has been increasingly utilized in distribution networks in recent years. In accordance with this trend, differential evolution has been applied to help operate distribution system with distributed generation [112,113].

9.8.3 System Operation

Keeping power systems stable is of extreme importance. System monitoring, state analysis and estimation, and control are the three key issues involved in running power systems smoothly [114].

9.8.3.1 State Estimation and Forecast

To keep the system stable, a system operator has to know future system status and take precautions to avoid system failures. Correct estimation of future system status is thus of vital importance. Differential evolution has been applied to estimate the system state of a power system [115], to forecast the short-term load [116,117], and to analyze marginal static voltage stability [118].

9.8.3.2 Control

Nowadays, power systems are operated automatically. The controller is an essential ingredient in automatic operation of power system. Differential evolution has played a role in developing power plant controllers [119] and power system stabilizers [120,121].

9.8.3.3 Fault Identification and Removal

Power system failure may be inevitable. However, prompt response, rapid location and identification, and immediate repair are critical in minimizing the resulting damage. Differential evolution has been implemented to identify worst-case voltage and current harmonic distortion in Singapore's MRT power supply system [122,123] and to locate voltage collapse points [12]. It has also been applied to develop a congestion management system [125,126] and to design an SF_6 gas circuit breaker [127].

9.8.3.4 Network Reconfiguration

Network reconfiguration is the process of changing the topology of distribution systems by altering the open/closed status of switches to reduce power loss. Differential evolution has been found to be a good solver for this issue [85,128–131].

9.8.3.5 Optimal Reactive Power Flow

Optimal reactive power flow is a very important planning and operation optimization problem in power systems. Its main purpose is to minimize the total power losses of the network while maintaining the voltage profile of the network in an acceptable range. It has been successfully solved by using differential evolution [132–134].

9.8.4 Trading

A strategic bidding system to maximize profits of power generators has been developed by implementing differential evolution [135].

9.8.5 Environment Assessment

Assessment of environmental effects [136] is necessary in order to make sure that the introduction of a power system will not have disastrous consequences for the surrounding environment and human beings.

9.9 Signal and Information Processing

9.9.1 Data Clustering

Cluster analysis is used in many different fields as a tool for preliminary and descriptive analysis and for unsupervised classification of objects characterized by different features. It aims to identify homogeneous groups by finding similarities between objects with regard to their characterizing attributes. Data in the same group are sufficiently similar, while data in different groups are sufficiently dissimilar. Useful knowledge can then be extracted.

It is straightforward to state this as an optimization problem for which the objective is to maximize the similarities between objects from the same clusters while minimizing the similarities between objects from different clusters. Different evolution has been found a very good tool for partitioning data [137–143].

9.9.2 Image Processing

Differential evolution is used to reduce noise in positron emission tomography imaging [144].

Color map generation is another interesting application of differential evolution [145]. The colors in digital images are typically represented by three 8-bit numbers representing the responses of the red, green, and blue sensors of the camera. Consequently, an image can have up to 2^{24}, that is, more than 16.8 million different colors. However, when displaying images on limited hardware such as mobile devices – but also for tasks such as image compression or image retrieval – it is desired to represent images with a limited number of different colors. Clearly, the choice of these colors is crucial as it determines the closeness of the resulting image to its original and hence the image quality. The process of finding such a palette or map of representative colors is known as color map generation or color quantization.

9.9.3 Image Registration

Image registration [146,147] is also called image matching. It is an important preliminary step in image processing. The purpose of image registration is to geometrically align two or more images taken, for example, at different times, from different sensors, or from different viewpoints, so that pixels representing the same physical structure may be superimposed.

Image registration problems arise in diverse application fields such as medicine [148–154], physics [155], and remote sensing [156,157]. For example, many hospitals own two or more radiological imaging facilities such as X-ray, computed tomography, magnetic resonance imaging, ultrasound, positron emission tomography or single photon emission computed tomography at the same time. A fundamental problem is the integration of information from multiple images of the same or different subjects, acquired using the same or different imaging methods and possibly at different times.

9.9.4 Pattern Recognition

Pattern recognition aims to automatically identify figures, characters, shapes, forms, and patterns without active human participation in the decision process. Numerous approaches to solving pattern recognition problems by using differential evolution have been reported.

9.9.4.1 Feature Extraction

Pattern recognition has been extensively applied to extract interesting information in astronomy, medicine, robotics, remote sensing, security, and so on. Known application problems of pattern recognition solved by using differential evolution include detection of epileptic patterns in electroencephaloghaph waveforms [158], surface roughness estimation of shot blasted steel bars [159], and computer-assisted colonoscopic diagnosis [160,161].

9.9.4.2 Shape Recognition

A shape recognition system automatically identifies or verifies a shape from a digital image or a video frame from a video source. It is typically used in security systems. Differential evolution has been applied to detect ellipses of limited eccentricity in images with high noise levels [162], leaf recognition [163,164], and face recognition [165].

9.9.4.3 Image Clustering

Image clustering is the process of organizing images into meaningful categories of similar image signatures. The image signatures can be pixels, regions, line elements and so on, depending on the problem encountered. Different researchers have applied differential evolution to cluster images such as satellite images and hand-segmented images.

When an expert's domain expertise is available, the clustering process is usually supervised [166,167]. The signature of each category is determined in advance. The computer system recognizes images according to the signature of each image. When there is little or no expertise, the clustering process is unsupervised [168,169]. The computer identifies inherent patterns in the images and uses a clustering algorithm to order them into discrete classes.

9.9.4.4 Object Tracking

Motion tracking involves the construction of the probability distribution of the current state of an evolving system, given the previous observation history. Different motion tracking problems including monocular human motion tracking [170], license plate tracking [171], contour tracking [172], and digital image correlation [173] have been solved by using differential evolution.

9.9.4.5 Speaker Verification and Identification

Speaker verification aims to authenticate a claimed speaker identity from a voice signal based on speaker-specific characteristics reflected in spoken words or sentences. On the other hand, speaker identification uses an individual's speech to search a database for a reference model that matches a sample submitted by an unknown speaker and returns a corresponding identity if found.

Speaker verification and identification are useful in many applications such as security. Differential evolution has been applied to build a text-independent speaker verification system [174,175] and a multimodal speaker detection system [176].

9.9.5 Signal Processing

Signal processing is the analysis, interpretation, and manipulation of signals including sound, images, biological, and radar. Processing of such signals includes storage and reconstruction, separation of information from noise, compression, and feature extraction. Differential evolution has been applied to implement the arc tangent function in fixed-point digital signal processing [177] and to separate a blind source of nonlinear mixtures [178].

References

[1] Develi, I. (2007) Determination of optimum antenna number ratio based on differential evolution for MIMO systems under low SNR conditions. *Wireless Personal Communications*, **43**(4), 1667–1673.

[2] Li, K., Wang, X. and Li, P. (2005) Analysis and optimization of interleave-division multiple-access communication systems. 2005 IEEE Int. Conf. Acoustics Speech Signal Processing, Philadelphia, PA, March 18–23, vol. 3, pp. 917–920.

[3] Li, K., Wang, X.D. and Ping, L. (2007) Analysis and optimization of interleave-division multiple-access communication systems. *IEEE Transactions on Wireless Communications*, **6**(5), 1973–1983.

[4] Weitkemper, P. and Kammeyer, K.D. (2007) Power optimization of IDMA systems with different target BER constraints. 2007 65th IEEE Vehicular Technology Conf., Dublin, Ireland, April 22–25, pp. 2812–2816.

[5] Weitkemper, P., Zielinski, K., Kammeyer, K.D. and Laur, R. (2006) Optimization of interference cancellation in coded CDMA systems by means of differential evolution. 4th Int. Symp. Turbo Codes Related Topics (also 6th Int. ITG-Conf. Source Channel Coding), April 3–7, Munich, Germany.

[6] Zielinski, K. and Laur, R. (2006) Stopping criteria for constrained optimization with particle swarms. 2nd Int. Conf. Bioinspired Optimization Methods Applications, Ljubljana, Slovenia, October 9–10, pp. 45–54.

[7] Zielinski, K. and Laur, R. (2006) Parameter adaptation for differential evolution with design of experiments. IASTED Int. Conf. Computational Intelligence, San Francisco, California, Nov. 20–22.

[8] Zielinski, K., Weitkemper, P., Laur, R. and Kammeyer, K.D. (2006) Examination of stopping criteria for differential evolution based on a power allocation problem. 10th Int. Conf. Optimization Electrical Electronic Equipment, Brasov, Romania, May 18–19, pp. 149–156.

[9] Zielinski, K., Weitkemper, P., Laur, R. and Kammeyer, K.D. (2006) Parameter study for differential evolution using a power allocation problem including interference cancellation. 2006 IEEE Congress Evolutionary Computation, Vancouver, Canada, July 16–21, pp. 1857–1864.

[10] Priem Mendes, S., Gómez Pulido, J.A., Vega Rodríguez, M.A. *et al.* (2006) A differential evolution based algorithm to optimize the radio network design problem. 2nd IEEE Int. Conf. e-Science Grid Computing, Amsterdam, The Netherlands, Dec., pp. 119–125.

[11] Priem Mendes, S., Gómez Pulido, J.A., Vega Rodríguez, M.A. *et al.* (2007) Fast wide area network design optimisation using differential evolution. 2007 Int. Conf. Advanced Engineering Computing Applications Sciences, Papeete, France, Nov. 4–9, pp. 3–10.

[12] Vega-Rodriguez, M.A., Gomez-Pulido, J.A., Alba, E. *et al.* (2007) Using omnidirectional BTS and different evolutionary approaches to solve the RND problem, in *Computer Aided Systems Theory – EUROCAST 2007* (eds. R. Moreno-Diaz, F. Pichler and A. Quesada Arencibia), Lecture Notes in Computer Science, **4739**, Springer, Berlin, pp., 853–860.

[13] Vega-Rodriguez, M.A., Gomez-Pulido, J.A., Alba, E. *et al.* (2007) Evaluation of different metaheuristics solving the RND problem, in *Applications of Evolutionary Computing* (eds. M. Giacobini *et al.*), Lecture Notes in Computer Science, **4448**, Springer, Berlin, pp. 101–110.

[14] Develi, I. (2007) Differential evolution based prediction of rain attenuation over a LOS terrestrial links situated in the southern United Kingdom. *Radio Science*, **42**(3), Art. No. RS3011.

[15] Fischer, M.M., Hlavackova-Schindler, K. and Reismann, M. (1999) An evolutionary mutation-based algorithm for weight training in neural networks for telecommunication flow modelling. Computational Intelligence Modelling, Control Automation. Evolutionary Computation and Fuzzy Logic for Intelligent Control, Knowledge Acquisition and Information Retrieval, Vienna, Austria, February 17–19, Concurrent Systems Engineering Series, vol. 55, pp. 54–59.

[16] Fischer, M.M., Hlavackova-Schindler, K. and Reismann, M. (1999) A global search procedure for parameter estimation in neural spatial interaction modelling. *Papers Regional Science*, **78**(2), 119–134.

[17] Fischer, M.M., Reismann, M. and Hlavackova-Schindler, K. (1999) Parameter estimation in neural spatial interaction modelling by a derivative free global optimization method. 4th Int. Conf. GeoComputation, Fredericksburg, VA, July 25–28.

[18] Shan, D.M., Chen, Z.N. and Wu, X.H. (2005) Signal optimization for UWB radio systems. *IEEE Transactions on Antennas and Propagation*, **53**(7), 2178–2184.

[19] Shokrollahi, A. and Storn, R. (1999) Design of efficient erasure codes with differential evolution. Genetic Evolutionary Computation Conf., Orlando, FL, July 13–17.

[20] Shokrollahi, A. and Storn, R. (2000) Design of efficient erasure codes with differential evolution. 2000 IEEE Int. Symp. Information Theory, Sorrento, Italy, June 25–30.

[21] De, D., Ray, S., Konar, A. and Chatterjee, A. (2005) A fuzzy logic controller based dynamic routing algorithm with SPDE based differential evolution approach. 2005 Genetic Evolutionary Computation Conf., Washington, D.C., June 25–29.

[22] Wang, J., Wang, X., Liu, P. and Huang, M. (2006) A differential evolution based flexible QoS multicast routing algorithm in NGI. 2006 7th Int. Conf. Parallel Distributed Computing Applications Technologies, Dec., pp. 250–253.

[23] Rocha, M., Sousa, P., Cortez, P. and Rio, M. (2007) Evolutionary computation for quality of service internet routing optimization, in *Applications of Evolutionary Computing* (eds. M. Giacobini *et al.*), Lecture Notes in Computer Science, **4448**, Springer, Berlin, pp. 71–80.

[24] Meletiou, G.C., Laskari, E.C. and Vrahatis, M.N. (2005) Problems of cryptography as discrete optimization tasks. *Nonlinear Analysis*, **63**(5–7), e831–e837.

[25] Chen, Y.H., Zhang, L. and Abraham, A. (2006) Comparison study of evolutionary flexible neural networks for intrusion detection system. *WSEAS Transactions on Information Science and Applications*, **3**(7), 1332–1340.

[26] De Falco, I., Scafuri, U., Tarantino, E. and Della Cioppa, A. (2007) A distributed differential evolution approach for mapping in a grid environment. 15th EUROMICRO Int. Conf. Parallel Distributed Network-based Processing, Napoli, Italy, Feb. 7–9, pp. 442–449.

[27] Rae, A. and Parameswaran, S. (1998) Application-specific heterogeneous multiprocessor synthesis using differential-evolution. 11th Int. Symp. System Synthesis, Hsinchu, Taiwan, December 2–4, pp. 83–88.

[28] Rzadca, K. and Seredynski, F. (2005) Heterogeneous multiprocessor scheduling with differential evolution. 2005 IEEE Congress Evolutionary Computation, Edinburgh, UK, Sept. 2–5, 2005, vol. 3, pp. 2840–2847.

[29] Kong, X., Xu, W. and Liu, J. (2006) A permutation-based differential evolution algorithm incorporating simulated annealing for multiprocessor scheduling with communication delays, in *Computational Science – ICCS 2006* (ed. V.N. Alexandrov), Lecture Notes Computer Science, **3991**, Springer, Berlin, pp. 514–521.

[30] Cheng, S.L. and Hwang, C. (2001) Optimal approximation of linear systems by a differential evolution algorithm. *IEEE Transactions on Systems, Man, and Cybernetics*, **31**(6), 698–707.

[31] Ursem, R.K. and Vadstrup, P. (2003) Parameter identification of induction motors using differential evolution. 2003 IEEE Congress Evolutionary Computation, Canberra, Australia, December 8–12, vol. 2, pp. 790–796.

[32] Gong, M.G., Du, H.F. and Jiao, L.C. (2006) Optimal approximation of linear systems by artificial immune response. *Science China F – Information Sciences*, **49**(1), 63–79.

[33] Svecko, R. and Pec, M. (2006) Robust modelling of hard-disk drives servo system. 25th IASTED Int. Conf. Modelling Identification Control, Lanzarote, Canary Island, Spain, Feb. 6–8, pp. 225–230.

[34] Hwang, C.Y., Chang, F.P. and Hwang, J.H. (2003) Digital redesign of analog controllers based on minimizing integral of squared errors. *Optimal Control Applications Methods*, **24**(1), 1–22.

[35] Guney, K. and Karaboga, D. (2004) New narrow aperture dimension expressions obtained by using a differential evolution algorithm for optimum gain pyramidal horns. *Journal of Electromagnetic Waves and Applications*, **18**(3), 321–339.

[36] Akdagli, A. (2007) A closed-form expression for the resonant frequency of rectangular microstrip antennas. *Microwave Optical Technology Letters*, **49**(8), 1848–1852.

[37] Akdagli, A. (2007) CAD formulas for patch dimensions of rectangular micorstrip antennas with various substrate thickness. *Microwave Optical Technology Letters*, **49**(9), 2197–2201.

[38] Akdagli, A. (2007) A novel expression for effective radius in calculating the resonant frequency of circular microstrip patch antennas. *Microwave Optical Technology Letters*, **49**(10), 2395–2398.

[39] Guney, K. and Sarikaya, N. (2007) Resonant frequency calculation for circular microstrip antennas with a dielectric cover using adaptive network-based fuzzy inference system optimized by various algorithms. *Progress Electromagnetics Research*, **72**, 279–306.

[40] Razavi, F., Aghdam, K.M.P. and Kamarei, M. (2005) Synthesis of antenna arrays using gradient based differential evolution algorithm in presence of mutual coupling. 2005 IEEE Wireless Microwave Technology Conf., Florida, April 7–8.

[41] Yang, S. and Nie, Z. (2005) Mutual coupling compensation in time modulated linear antenna arrays. *IEEE Transactions on Antennas and Propagation*, **53**(12), 4182–4185.

[42] Yang, S., Nie, Z. and Yang, F. (2005) Mutual coupling compensation in small antenna arrays by the differential evolution algorithm. 2005 Asia-Pacific Microwave Conf, Suzhou, China, December 4–7, vol. 5.

[43] Yang, S.W., Nie, Z.P. and Wu, Y.J. (2007) A practical array pattern synthesis approach including mutual coupling effects. *Electromagnetics*, **27**(1), 53–63.

[44] Rekanos, I.T. and Yioultsis, T.V. (2006) Convergence enhancement for the vector finite element modeling of microwaves and antennas via differential evolution. *AEU – International Journal of Electronics and Communications*, **60**(6), 428–434.
[45] Qing, A., Xu, X. and Gan, Y.B. (2003) Effective wave number of composite materials with oriented randomly distributed inclusions. 2003 IEEE AP-S Int. Symp., June 22–27, Columbus, Ohio, vol. 4, pp. 655–658.
[46] Qing, A., Xu, X. and Gan, Y.B. (2005) Anisotropy of composite materials with inclusion with orientation preference. *IEEE Transactions on Antennas and Propagation*, **53**(2), 737–744.
[47] Qing, A., Xu, X. and Gan, Y.B. (2004) Effective permittivity tensor of composite material with aligned spheroidal inclusion. *Journal of Electromagnetic Waves and Applications*, **18**(7), 899–910.
[48] Michalski, K.A. (2000) Electromagnetic imaging of circular-cylindrical conductors and tunnels using a differential evolution algorithm. *Microwave and Optical Technology Letters*, **27**(5), 330–334.
[49] Michalski, K.A. (2001) Electromagnetic imaging of elliptical-cylindrical conductors and tunnels using a differential evolution algorithm. *Microwave and Optical Technology Letters*, **28**(3), 164–169.
[50] Michalski, K.A. (2001) Electromagnetic imaging of homogeneous-dielectric elliptic cylinders using a differential evolution algorithm combined with a single boundary integral equation method. URSI Int. Symp. Electromagnetic Theory, Victoria, Canada, May 13–17.
[51] Qing, A. (2003) Electromagnetic inverse scattering of multiple two-dimensional perfectly conducting objects by the differential evolution strategy. *IEEE Transactions on Antennas and Propagation*, **51**(6), 1251–1262.
[52] Qing, A. (2003) Electromagnetic inverse scattering of multiple perfectly conducting cylinders by differential evolution strategy with individuals in groups (GDES). 2003 IEEE AP-S Int. Symp., June 22–27, Columbus, Ohio, vol. 1, pp. 519–522.
[53] Qing, A. (2004) Electromagnetic inverse scattering of multiple perfectly conducting cylinders by differential evolution strategy with individuals in groups (GDES). *IEEE Transactions on Antennas and Propagation*, **52**(5), 1223–1229.
[54] Qing, A. and Gan, Y.B. (2005) Electromagnetic inverse problems (invited), in *Encyclopedia of RF and Microwave Engineering*, vol. **2** (ed. K. Chang), John Wiley, pp. New York, 1200–1216.
[55] Qing, A. (2006) Dynamic differential evolution strategy and applications in electromagnetic inverse scattering problems. *IEEE Transactions on Geoscience and Remote Sensing*, **44**(1), 116–125.
[56] Qing, A. (2007) A parametric study on differential evolution based on benchmark electromagnetic inverse scattering problem. 2007 IEEE Congress Evolutionary Computation, Singapore, Sept. 25–28, pp. 1904–1909.
[57] Luo, X.F., Lee, C.K. and Qing, A. (2004) Design of frequency selective surfaces (FSS) using differential evolution strategy (DES). 27th ESA Antenna Technology Workshop Innovative Periodic Antennas Electromagnetic Bandgap Left Handed Materials Fractal Frequency Selective Surfaces, Santiago de Compostela, Spain, Mar. 9–11, pp. 201–207.
[58] Luo, X.F., Qing, A. and Lee, C.K. (2005) Application of the differential-evolution strategy to the design of frequency-selective surfaces. *International Journal of RF and Microwave Computer-Aided Engineering*, **15**(2), 173–180.
[59] Luo, X.F., Teo, P.T., Qing, A. and Lee, C.K. (2004) Design of double-square-loop frequency selective surfaces using differential evolution strategy coupled with equivalent circuit model. The 4th Int. Conf. Microwave Millimeter Wave Technology, Beijing, China, Aug. 18–21, pp. 94–97.
[60] Luo, X.F., Teo, P.T., Qing, A. and Lee, C.K. (2005) Design of double-square-loop frequency selective surfaces using differential evolution strategy coupled with equivalent circuit model. *Microwave Optical Technology Letters*, **44**(2), 159–162.
[61] Wei, G., Wu, S. and Mao, E. (2003) Differential evolution for target motion parameter estimation. 2003 IEEE Int. Conf. Neural Networks Signal Processing, Nanjing, China, Dec. 14–17, vol. 1, pp. 563–566.
[62] Xue, F., Sanderson, A.C. and Graves, R.J. (2005) Multi-objective evolutionary decision support for design-supplier-manufacturing planning. 2005 IEEE Conf. Automation Science Engineering, Edmonton, Alta., Canada, August 1–2, pp. 197–202.
[63] Han, Z.X., Xu, L., Wei, R. *et al.* (2004) Reliability-based design optimization for land grid array solder joints under thermo-mechanical load. Thermal Mechanical Simulation Experiments Microelectronics Microsystems, Brussels, Belgium, May 10–12, pp. 219–224.
[64] Xu, L., Han, Z.X., Ren, W. *et al.* (2004) Design optimization approaches for the thermo-mechanical reliability of land grid array solder joints. 2004 Int. Conf. Business Electronic Product Reliability Liability, Shanghai, China, April 27–30, pp. 107–112.
[65] Xu, L., Reinikainen, T., Wei, R. *et al.* (2004) A simulation-based multi-objective design optimization of electronic packages under thermal cycling and bending. *Microelectronics Reliability*, **44**(12), 1977–1983.

[66] Zhan, J.Y. and Xiong, G.Z. (2006) Optimal bus assignment scheme for system on chip test scheduling. *Jisuanji Jicheng Zhizao Xitong/Computer Integrated, Manufacturing Systems*, **12**(10), 1693–1697.

[67] Pérez-Guerrero, R.E. and Cedeño-Maldonado, J.R. (2005) Economic Power Dispatch with Non-Smooth Cost Functions Using Differential Evolution. Proceedings of the 37th Annual North American Power Symposium, pp. 183–190.

[68] Pérez-Guerrero, R.E. and Cedeño-Maldonado, J.R. (2005) Differential Evolution Based Economic Environmental Power Dispatch. Proceedings of the 37th Annual North American Power Symposium, pp. 191–197.

[69] Coelho, L.S. and Mariani, V.C. (2006) Combining of chaotic differential evolution and quadratic programming for economic dispatch optimization with valve-point effect. *IEEE Transactions on Power Systems*, **21**(2), 989–996.

[70] Coelho, L.S. and Mariani, V.C. (2006) Particle swarm optimization with quasi-Newton local search for solving economic dispatch problem. 2006 IEEE Int. Conf. Systems Man Cybernetics, Taipei, Taiwan, Oct. 8–11, vol. 1, pp. 3109–3113.

[71] Thitithamrongchai, C. and Eua-Arporn, B. (2006) Economic load dispatch for piecewise quadratic cost function using hybrid self-adaptive differential evolution with augmented Lagrange multiplier method. 2006 Int. Conf. Power System Technology, Chongqing, China, October 22–26, pp. 1–8.

[72] Thitithamrongchai, C. and Eua-Arporn, B. (2006) Hybrid self-adaptive differential evolution method with augmented Lagrange multiplier for power economic dispatch of units with valve-point effects and multiple fuels. 2006 IEEE PES Power Systems Conf., Atlanta, Georgia, October 29–November 1, pp. 908–914.

[73] Wang, S.K., Liu, C.W. and Chiou, J.P. (2006) Ant direction hybrid differential evolution for solving economic dispatch of power system. 2006 IEEE Int. Conf. Systems Man Cybernetics, Taipei, Taiwan, Oct. 8–11, pp. 1154–1159.

[74] Chiou, J.P. (2007) Variable scaling hybrid differential evolution for large-scale economic dispatch problems. *Electric Power Systems Research*, **77**(3–4), 212–218.

[75] Coelho, L.D.S. and Mariani, V.C. (2007) Improved differential evolution algorithms for handling economic dispatch optimization with generator constraints. *Energy Conversion Management*, **48**(5), 1631–1639.

[76] Sinha, N., Purkayastha, B.S. and Purkayastha, B. (2007) Optimal combined non-convex economic and emission load dispatch using NSDE. Int. Conf. Computational Intelligence Multimedia Applications, Sivakasi, Tamil Nadu, India, Dec. 13–15, vol. 1, pp. 473–480.

[77] Wang, S.K., Chiou, J.P. and Liu, C.W. (2007) Non-smooth/non-convex economic dispatch by a novel hybrid differential evolution algorithm. *IET Generation Transmission Distribution*, **1**(5), 793–803.

[78] Liang, C.H., Duan, X.Z., Zhong, Z.Y. and Huang, J.B. (2006) Parallel reactive power optimization based on differential evolution and PC-cluster. *Automation Electric Power Systems*, **30**(1), 34–39.

[79] Abbasy, A., Tabatabaii, I. and Hosseini, S.H. (2007) Optimal reactive power dispatch in electricity markets using a multiagent-based differential evolution algorithm. Int. Conf. Power Engineering Engergy Electrical Drives, Setubal, Portugal, April 12–14, pp. 249–254.

[80] Bakare, G.A., Krost, G., Venayagamoorthy, G.K. and Aliyu, U.O. (2007) Differential evolution approach for reactive power optimization of Nigerian grid system. IEEE Power Engineering Society General Meeting, Tampa, FL, June 24–28, pp. 3348–3353.

[81] Bakare, G.A., Krost, G., Venayagamoorthy, G.K. and Aliyu, U.O. (2007) Comparative application of differential evolution and particle swarm techniques to reactive power and voltage control. Int. Conf. Intelligent Systems Applications Power Systems, Taiwan, China, Nov. 5–8, pp. 393–398.

[82] Bhattacharyya, B. and GoswAmi, S.K. (2007) Reactive power optimization through evolutionary techniques: a comparative study of the Ga, De and PSO. *Intelligent Automation Soft Computing*, **13**(4), 461–469.

[83] Bhattacharyya, B. and GoswAmi, S.K. (2007) A decomposition approach for combined heuristic and differential evolution method for the reactive power problem, in *Pattern Recognition and Machine Intelligence* (ed. A. Ghosh *et al.*), Lecture Notes in Computer Science, **4815**, Springer, Berlin, pp. 261–268.

[84] Zhang, F.T., Song, J.H., Li, J. and Cheng, X.L. (2007) A hybrid differential evolution method for optimal reactive power optimization. *Power System Technology*, **31**(9), 33–37.

[85] Wong, K.P. and Dong, Z.Y. (2005) Differential evolution, an alternative approach to evolutionary algorithm. 13th Int. Conf. Intelligent Systems Application Power Systems, Arlington, VA, Nov. 6–10, pp. 73–83.

[86] Babu, B.V. and Angira, R. (2002) Optimization of non-linear chemical processes using evolutionary algorithm. Int. Symp. 55th Annual Session of IIChE, OU, Hyderabad, December 19–22.

[87] Kremer, B. and Dolinar, D. (2005) Optimizacija proizvodnje elektricne energije na dereguliranem trgu (Optimal short-term hydro-thermal scheduling). *Elektrotehniski Vestnik/Electrotechnical Review*, **72**(2–3), 97–106.

[88] Lakshminarasimman, L. and Subramanian, S. (2006) Short-term scheduling of hydrothermal power system with cascaded reservoirs by using modified differential evolution. *IEE Proceedings. C, Generation, Transmission, and Distribution*, **153**(6), 693–700.

[89] Keles, A. (2007) Binary differential evolution for the unit commitment problem. 2007 Genetic Evolutionary Computation Conf., London, July 7–11, pp. 2765–2768.

[90] Wu, C.J., Chuang, Y.S., Huang, C.P. and Chang, Y.P. (2007) Tuning of static excitation systems to improve dynamic performance for generators serving electric arc furnaces loads. *IEEE Transactions on Energy Conversion*, **22**(2), 350–357.

[91] Kannan, S., Slochanal, S.M.R., Subbaraj, P. and Padhy, N.P. (2004) Application of particle swarm optimization technique and its variants to generation expansion planning problem. *Electric Power Systems Research*, **70**(3), 203–210.

[92] Kannan, S., Slochanal, S.M.R. and Padhy, N.P. (2005) Application and comparison of metaheuristic techniques to generation expansion planning problem. *IEEE Transactions on Power Systems*, **20**(1), 466–475.

[93] Subramanian, K., Mary Raja Slochanal, S., Subramanian, B. and Padhy, N.P. (2006) Application and comparison of metaheuristic techniques to generation expansion planning incorporating both bilateral and multilateral transactions. *International Journal of Emerging Electric Power Systems*, **6**(1), 1–14.

[94] Kanna, S., Slochanal, S.M.R., Baskar, S. and Murugan, P. (2007) Application and comparison of metaheuristic techniques to generation expansion planning in the partially deregulated environment. *IET Generation Transmission Distribution*, **1**(1), 111–118.

[95] Liu, X., Wang, S. and Qiu, J. (2007) Improved particle swarm optimization with its applications to parameter optimization of high temperature superconducting cables. *Journal of Xi'an Jiaotong University*, **41**(2), 219–222 & 240.

[96] Chang, H.C. and Tsao, T.F. (2005) Optimal distribution system reliability planning: a differential evolution based approach. *International Journal of Power and Energy Systems*, **25**(3), 151–158.

[97] Tsao, T.F., Chang, Y.P. and Tseng, W.K. (2005) Reliability and costs optimization for distribution system placement problem. 2005 IEEE PES Transmission Distribution Conf. Asia Pacific, Dalian, China, August 15–18.

[98] Niu, W., Liu, Z., Zhang, J. and Yan, J. (2007) Substation locating and sizing in power system based on GIS and differential evolution algorithm. *Automation Electric Power Systems*, **31**(18), 82–86.

[99] Liu, J., Liu, Z., Huang, W. *et al.* (2007) An intelligent distribution network planning method based on geographical differential evolution. *Automation Electric Power Systems*, **31**(2), pp. 32–35.

[100] Kovač, N., Sarajčev, I. and Poijak, D. (2006) A numerical-stochastic technique for underground cable system design. *IEE Proceedings C, Generation, Transmission, and Distribution*, **153**(2), 181–186.

[101] Chang, T.T. and Chang, H.C. (1998) Application of differential evolution to passive shunt harmonic filter planning. 8th Int. Conf. Harmonics Quality Power, Athens, Greece, Oct. 14–16, 1998, vol. 1, pp. 149–153.

[102] Chang, T.T. and Chang, H.C. (2000) An efficient approach for reducing harmonic voltage distortion in distribution systems with active power line conditioners. *IEEE Transactions on Power Delivery*, **15**(3), 990–995.

[103] Chang, H.C. and Chang, T.T. (2001) Optimal installation of three-phase active power line conditioners in unbalanced distribution systems. *Electric Power Systems Research*, **57**(3), 163–171.

[104] Su, C.T. and Lee, C.S. (2002) Modified differential evolution method for capacitor placement of distribution systems. IEEE PES Transmission Distribution Conf., Yokohama, Japan, Oct. 6–10, vol., pp 208–213.

[105] Chang, Y.P. and Wu, C.J. (2004) Design of harmonic filters using combined feasible direction method and differential evolution. 2004 Int. Conf. Power System Technology, Singapore, Nov. 21–24, pp. 812–817.

[106] Chiou, J.P., Chang, C.F. and Su, C.T. (2004) Ant direction hybrid differential evolution for solving large capacitor placement problems. *IEEE Transactions on Power Systems*, **19**(4), 1794–1800.

[107] Chang, Y.P. and Wu, C.J. (2005) Optimal multiobjective planning of large-scale passive harmonic filters using hybrid differential evolution method considering parameter and loading uncertainty. *IEEE Transactions on Power Delivery*, **20**(1), 408–416.

[108] Chang, Y.P. and Wu, C.J. (2005) Hybrid differential evolution and probability method applied to optimal planning of large scale harmonics filters. *International Journal of Electrical Engineering*, **12**(1), 63–73.

[109] Chiou, J.P., Chang, C.F. and Su, C.T. (2006) Capacitor placement in large-scale distribution systems using variable scaling hybrid differential evolution. *International Journal of Electrical Power and Energy Systems*, **28**(10), 739–745.

[110] Cuello-Reyna, A.A. and Cedeño-Maldonado, J.R. (2006) A differential evolution approach to optimal reactive power planning. 2006 IEEE PES Transmission and Distribution Conference and Exposition Latin America, Caracas, Aug., pp. 1–7.

[111] Chang, C.F., Wong, J.J., Chiou, J.P. and Su, C.T. (2007) Robust searching hybrid differential evolution method for optimal reactive power planning in large-scale distribution systems. *Electric Power Systems Research*, **77**(5–6), 430–437.

[112] Niknam, T., Ranjbar, A.M., Shirani, A.R. *et al.* (2005) Optimal operation of distribution system with regard to distributed generation: a comparison of evolutionary methods. 2005 IEEE Industry Applications Conf., Hong Kong, Oct. 2–6, vol. 4, pp. 2690–2697.

[113] Olamaie, J. and Niknam, T. (2006) Daily volt/var control in distribution networks with regard to DGs: a comparison of evolutionary methods. 2006 IEEE Power India Conf., New Delhi, India, April 10–12.

[114] Lee, K.Y. (2005) Tutorial on intelligent optimization and control for power systems: an introduction. 13th Int. Conf. Intelligent Systems Application Power Systems, Arlington, VA, Nov. 6–10, pp. 2–5.

[115] Figueroa, N.G. and Cedeno, J.R. (2004) A differential evolution solution approach for power system state estimation. IASTED Int. Conf. Power Energy Systems, Clearwater Beach, FL, Nov. 28–Dec. 1, pp. 413–418.

[116] Niu, D.X., Wang, H.Q. and Gu, Z.H. (2005) Short-term load forecasting using general regression neural network. 2005 Int. Conf. Machine Learning Cybernetics, Guangzhou, China, August 18–21, vol. 7, pp. 4076–4082.

[117] Gu, Z.H., Niu, D.X. and Wang, H.Q. (2006) Research on application of general regression neural network in short-term load forecasting. *Electric Power*, **39**(4), 11–14.

[118] Li, C. and Chen, Y.P. (2006) Application of the improved differential evolution calculation method based on continuous power flow to the analysis of marginal static voltage stability. *Power Engineering*, **26**(5), 756–760.

[119] Van Sickel, J.H., Lee, K.Y. and Heo, J.S. (2007) Differential evolution and its applications to power plant control. Int. Conf. Intelligent Systems Applications Power Systems, Taiwan, Nov. 5–8, pp. 535–540.

[120] Wu, C.J. and Chuang, Y.S. (2005) Design of decentralized output feedback power system stabilizers using hybrid differential evolution. 2005 IEEE Power Engineering Society General Meeting, San Francisco, CA, June 12–16, vol. 2, pp. 1404–1411.

[121] Chuang, Y.S. and Wu, C.J. (2006) A damping constant limitation design of power system stabilizer using hybrid differential evolution. *Journal of Marine Science and Technology*, **14**(2), 84–92.

[122] Chang, C.S., Lu, L.R. and Wang, F. (2003) Application of differential evolution for harmonic worst-case identification of mass rapid transit power supply system. 2003 IEEE Congress Evolutionary Computation, Canberra, Australia, Dec. 08–12, pp. 593–599.

[123] Chang, C.S., Lu, L.R. and Wang, F. (2004) Application of differential evolution in harmonic worst-case identification of mass rapid transit power supply system. *International Journal of Systems Science*, **35**(13–14), 731–739.

[124] Goh, S.H., Dong, Z.Y. and Saha, T.K. (2007) Locating voltage collapse points using evolutionary computation techniques. 2007 IEEE Congress Evolutionary Computation, Singapore, Sept. 25–28, pp. 2923–2930.

[125] Cuello-Reyna, A. and Cedeno-Maldonado, J.R. (2005) OPF framework for congestion management in deregulated environments using differential evolution. 37th Annual North American Power Symp., Ames, IA, Oct. 23–25, pp. 127–133.

[126] Cuello-Reyna, A.A. and Cedeno-Maldonado, J.R. (2005) OPF framework for congestion management in deregulated environments using differential evolution. 8th IASTED Int. Conf. Power Energy Systems, Ames, IA, Oct. 24–26, pp. 177–184.

[127] Kim, H.K., Chong, J.K., Park, K.Y. and Lowther, D.A. (2007) Differential evolution strategy for constrained global optimization and application to practical engineering problems. *IEEE Transactions on Magnetics*, **43**(4), 1565–1568.

[128] Su, C.T. and Lee, C.S. (2003) Network reconfiguration of distribution systems using improved mixed-integer hybrid differential evolution. *IEEE Transactions on Power Delivery*, **18**(3), 1022–1027.

[129] Chiou, J.P., Chang, C.F. and Su, C.T. (2005) Variable scaling hybrid differential evolution for solving network reconfiguration of distribution systems. *IEEE Transactions on Power Systems*, **20**(2), 668–674.

[130] Su, C.T., Chang, C.F. and Lee, C.S. (2005) Distribution network reconfiguration for loss reduction by hybrid differential evolution. *Electric Power Components Systems*, **33**(12), 1297–1312.

[131] Chang, C.F. and Chiou, J.P. (2007) Optimal feeder reconfiguration for loss reduction by robust searching hybrid differential evolution. *Cybernetics Systems*, **38**(4), 429–450.

[132] Su, S., Chung, C.Y., Wong, K.P. *et al.* (2006) Fault tolerant differential evolution based optimal reactive power flow. Int. Conf. Machine Learning Cybernetics, Dalian, China, August 13–16.

[133] Liang, C.H., Chung, C.Y., Wong, K.P. and Duan, X.Z. (2007) Parallel optimal reactive power flow based on cooperative co-evolutionary differential evolution and power system decomposition. *IEEE Transactions on Power Systems*, **22**(1), 249–257.

[134] Liang, C.H., Chung, C.Y., Wong, K.P. *et al.* (2007) Study of differential evolution for optimal reactive power flow. *IET Generation Transmission Distribution*, **1**(2), 253–260.
[135] Yin, X., Zhao, J., Saha, T.K. and Dong, Z.Y. (2007) Developing GENCO's strategic bidding in an electricity market with incomplete information. IEEE Power Engineering Society General Meeting, Tampa, FL, June 24–28, pp. 1686–1692.
[136] Kovac, N., Poljak, D. and Bozic, N. (2005) Assessment of maximal magnetic flux density value from a power substation based on the discrete measured data. 18th Int. Conf. Applied Electromagnetics Communications/Joint COST 284 Workshop, Dubrovnik, Croatia, Oct. 12–14.
[137] Abbass, H.A. (2003) Pareto neuro-evolution: constructing ensemble of neural networks using multi-objective optimisation. 2003 IEEE Congress Evolutionary Computation, Canberra, ACT, Australia, Dec. 8–12, vol. 3, pp. 2074–2080.
[138] Paterlini, S. and Krink, T. (2004) High performance clustering with differential evolution. 2004 IEEE Congress Evolutionary Computation, Portland, OR, June 19–23, vol. 2, pp. 2004–2011.
[139] Kwedlo, W. (2006) Parallelizing evolutionary algorithms for clustering data, in *Parallel Processing and Applied Mathematics* (ed. R. Wyrzykowski), Lecture Notes Computer Science, **3911**, Springer, Berlin, pp. 430–438.
[140] Paterlini, S. and Krink, T. (2006) Differential evolution and particle swarm optimisation in partitional clustering. *Computational Statistics Data Analysis*, **50**(5), 1220–1247.
[141] Zaharie, D. (2006) Distributed clustering based on representatives evolved by crowding differential evolution. 12th Int. Mendel Conf. Soft Computing, Brno, Czech Republic, May 31–June 2, pp. 51–56.
[142] Zhang, L.B., Ma, M., Liu, X.H. *et al.* (2006) Differential evolution fuzzy clustering algorithm based on kernel methods, in *Rough Sets and Knowledge Technology* (ed. G. Wang *et al.*), Lecture Notes Computer Science, **4062**, Springer, Berlin, pp. 430–435.
[143] Das, S., Abraham, A. and Konar, A. (2008) Automatic clustering using an improved differential evolution. *IEEE Transactions on Systems, Man, and Cybernetics*, **38**(1), 218–237.
[144] Gupta, V., Chan, C.C. and Sian, P.T. (2007) A differential evolution approach to PET image de-noising. Annual Int. Conf. IEEE Engineering Medicine Biology Society, Aug. 22–26, pp. 4173–4176.
[145] Schaefer, G. and Nolle, L. (2005) A hybrid differential evolution approach to colour map generation. 20th European Conf. Modeling Simulation, Sankt Augustin, Germany, May 28–31, pp. 434–436.
[146] Thomas, P. and Vernon, D. (1997) Image registration by differential evolution. 1st Irish Machine Vision Image Processing Conf., Magee College, University of Ulster, pp. 221–225.
[147] Phan, H.V., Lech, M. and Nguyen, T.D. (2004) Registration of 3D range images using particle swarm optimisation, in *Advances in Computer Science – ASIAN 2004: Higher Level Decision Making* (ed. M.J. Maher), *Lecture Notes Computer Science*, **3321**, Springer, Berlin, pp. 223–235.
[148] Salomon, M. (2001) Parallelizing differential evolution for 3D medical image registration. *Technique et Science Informatiques*, **20**(5), 605–627.
[149] Salomon, M., Perrin, G.R. and Heitz, F. (2001) Differential evolution for medical image registration. Int. Conf. Artificial Intelligence, vol. 2, Las Vegas, June 25–28, pp. 123–129.
[150] He, R. and Narayana, P.A. (2002) Improvement to global optimization of mutual information on retrospective registration of magnetic resonance image. Annual Int. Conf. IEEE Engineering Medicine Biology, Houston, TX, Oct. 23–26, vol. 2, pp. 1009–1010.
[151] Dony, R. and Xu, X. (2004) Differential evolution with powell's direction set method in medical image registration. 2004 IEEE Int. Symp. Biomedical Imaging, Arlington, VA, April 15–18, pp. 732–735.
[152] Talbi, H. and Batouche, M. (2004) Hybrid particle swarm with differential evolution for multimodal image registration. 2004 IEEE International Conf. Industrial Technology, Tunisia, December 8–10, vol. 3, pp. 1567–1572.
[153] Xu, X. and Dony, R.D. (2004) Differential evolution with Powell's direction set method in medical image registration. 2nd Int. Symp. Biomedical Imaging: Macro Nano, Arlington, VA, April 15–18, vol. 1, 732–735.
[154] Xia, Z., Huang, X., Zhou, X. *et al.* (2008) Registration of 3-D CT and 2-D flat images of mouse via affine transformation. *IEEE Transactions on Information Technology in Biomedicine*, **12**(5), 569–578.
[155] Telenczuk, B., Ledesma-Carbato, M.J., Velazquez-Muriel, J.A. *et al.* (2006) Molecular image registration using mutual information and differential evolution optimisation. 2006 3rd IEEE Int. Symp. Biomedical Imaging: Macro Nano, Arlington, VA, April 6–9, pp. 844–847.
[156] De Falco, I., Cioppa, A.D., Maisto, D. *et al.* (2007) Satellite image registration by distributed differential evolution, in *Applications of Evolutionary Computing* (eds. M. Giacobini *et al.*), Lecture Notes in Computer Science, **4448**, Springer, Berlin, pp. 251–260.

[157] De Falco, I., Maisto, D., Scafuri, U. *et al.* (2007) Distributed differential evolution for the registration of remotely sensed images. 15th EUROMICRO Int. Conf. Parallel Distributed Network-based Processing, Napoli, Italy, Feb. 7–9, pp. 358–362.

[158] Neelaveni, R., Gurusamy, G. and Hemavathy, L. (2001) Adaptive genetic algorithm and differential evolution based backpropagation neural network for epileptic pattern recognition. National Conf. Technology Information Communication Entertainment, Cochin, India, February 23–24, pp. 26–30.

[159] Lyden, S., Kalviainen, H. and Nykanen, J. (2004) Surface roughness estimation of shot blasted steel bars using machine vision. SPIE, Philadelphia, PA, Oct. 25–27, vol. 5608, pp. 278–289.

[160] Magoulas, G.D., Plagianakos, V.P. and Vrahatis, M.N. (2001) Hybrid methods using evolutionary algorithms for on-line training. IEEE Int. Joint Conf. Neural Networks, Washington, DC, July 15–19, vol. 3, pp. 2218–2223.

[161] Magoulas, G.D., Plagianakos, V.P. and Vrahatis, M.N. (2004) Neural network-based colonoscopic diagnosis using on-line learning and differential evolution. *Applied Soft Computing*, **4**(4), 369–379.

[162] Kasemir, K.U. and Betzler, K. (2003) Detecting ellipses of limited eccentricity in images with high noise levels. *Image Vision Computing*, **21**(2), 221–227.

[163] Du, J.X., Huang, D.S., Wang, X.F. and Gu, X. (2005) Neural network-based shape recognition using generalized differential evolution training algorithm. 2005 IEEE Int. Joint Conf. Neural Networks, July 31–August 4, vol. 4, pp. 2012–2017.

[164] Du, J.X., Huang, D.S., Wang, X.F. and Gu, X. (2007) Shape recognition based on neural networks trained by differential evolution algorithm. *Neurocomputing*, **70**(4–6), 896–903.

[165] Chen, Y.H. and Zhao, Y. (2006) Face recognition using DCT and hierarchical RBF model, in *Tools and Algorithms for the Construction and Analysis of Systems* (ed. O. Grumberg and M. Huth), Lecture Notes Computer Science, **4224**, Springer, Berlin, pp. 355–362.

[166] Zhu, Q.Y., Qin, A.K., Suganthan, P.N. and Huang, G.B. (2005) Evolutionary extreme learning machine. *Pattern Recognition*, **38**(10), 1759–1763.

[167] De Falco, I., Della Cioppa, A. and Tarantino, E. (2006) Automatic classification of handsegmented image parts with differential evolution, in *Applications of Evolutionary Computing* (ed. F. Rothlauf *et al.*), Lecture Notes Computer Science, **3907**, 403–414.

[168] Omran, M.G.H., Salman, A. and Engelbrecht, A.P. (2005) Self-adaptive differential evolution, in *Computational Intelligence and Security* (ed. Y. Hao *et al.*), Lecture Notes Computer Science, **3801**, Springer, Berlin, pp. 192–199.

[169] Omran, M.G.H., Engelbrecht, A.P. and Salman, A. (2006) Using the ring neighborhood topology with self-adaptive differential evolution in *Advances in Natural Comnputation* (eds. L. Jiao *et al.*), Lecture Notes in Computer Science, **4221**, Springer, Berlin, pp. 976–979.

[170] Du, M. and Guan, L. (2006) Monocular human motion tracking with the DE-MC particle filter. 2006 IEEE Int. Conf. Acoustics Speech Signal Processing, Toulouse, France, May 14–19, vol. 2, pp. 205–208.

[171] Yalın, I.K. and Gökmen, M. (2006) Integrating differential evolution and condensation algorithm for license plate tracking. 18th Int. Conf. Pattern Recognition, Hong Kong, September 20–24, pp. 658–661.

[172] Zhang, H., Huang, Z.H. and Yu, D.Y. (2006) Contour tracking using particle filter and differential evolution. *Guangdian Gongcheng/Opto-Electronic Engineering*, **33**(2), 29–32.

[173] Pan, B. and Xie, H.M. (2007) Digital image correlation method with differential evolution. *Journal of Optoelectronics – Laser*, **18**(1), 100–103.

[174] Ganchev, T., Fakotakis, N., Tasoulis, D.K. and Vrahatis, M.N. (2004) Generalized locally recurrent probabilistic neural networks for text-independent speaker verification. IEEE Int. Conf. Acoustics Speech Signal Processing, Montreal, Que, Canada, May 17–21, vol. 1, pp. 141–144.

[175] Ganchev, T.D., Tasoulis, D.K., Vrahatis, M.N. and Fakotakis, N.D. (2007) Generalized locally recurrent probabilistic neural networks with application to text-independent speaker verification. *Neurocomputing*, **70**(7–9), 1424–1438.

[176] Besson, P., Popovici, V., Vesin, J.M. *et al.* (2008) Extraction of audio features specific to speech production for multimodal speaker detection. *IEEE Transactions on Multimedia*, **10**(1), 63–73.

[177] Wu, W., Liu, X.H., Li, Z.R. and Wang, F.X. (2005) Implementation of arc tangent function using assembly in fixed-point DSP based on differential evolution. *Systems Engineering Electronics*, **27**(5), 926–928.

[178] Gao, Y. and Xie, S.L. (2006) Blind source separation of nonlinear mixtures based on functional link artificial neural networks and differential evolution algorithm. *Journal of Electronics Information Technology*, **28**(1), 50–54.

10

Flexible QoS Multicast Routing in Next-Generation Internet

J. Wang, X. Wang, M. Huang

School of Information Science and Technology, Northeastern University, Shenyang, PR China

10.1 Introduction

The next generation internet (NGI) has received much attention in recent years due to rapidly increasing demand for real-time multimedia services that combine audio, video and data streams. Consequently, demand for network resources has considerably increased. In addition, such network resources have to be distributed simultaneously to a great number of groups of users [1]. This imposes stringent communication requirements on bandwidth, delay, delay jitter, cost and other quality of service (QoS) metrics [2].

QoS routing selects the feasible path of high overall resource efficiency. This is the first step toward achieving end-to-end QoS guarantees. At the same time, multicasting employs a tree network structure and allows several users to get the content of paid-for sessions at the same time without packet duplication. Hence, multicasting with QoS is a promising solution for group multimedia applications.

The fundamental issue in QoS multicast routing is to develop simple, effective and robust multicast routing algorithms [3,4]. This is very important in deploying NGI. Several solutions ([5–24]), such as integrated service, differentiated service, multi-protocol label switching, traffic engineering, and QoS multicast routing algorithms, have been proposed.

10.2 Mathematical Model

10.2.1 Problem Model

A network can be modeled as a graph $G(N, E)$, where N is the set of nodes and E the set of edges. Parameters for node $v \in N$ include delay, delay jitter and packet loss rate, while parameters for

Differential Evolution Anyong Qing

edge $e \in E$ are available bandwidth, delay and error rate. For simplicity, the parameters of a node are merged into those of its downstream edge. Thus, an edge e has the following parameters: available bandwidth $b(e) \in [b_{\min}, b_{\max}]$, delay $d(e) \in [d_{\min}, d_{\max}]$, delay jitter $j(e) \in [j_{\min}, j_{\max}]$ and error rate $l(e) \in [l_{\min}, l_{\max}]$.

Users' requirements are defined as $R(s, M, \mathbf{V}_B, \mathbf{V}_D, \mathbf{V}_J, \mathbf{V}_L)$, where $s \in N$ is the source node, $M = \{m_1, m_2, \ldots, m_k\}$ is the set of destination nodes, $\mathbf{V}_B = (\Delta_{B1}, \Delta_{B2}, \ldots, \Delta_{Bk})$ is the vector of users' bandwidth requirements, $\Delta_{Bi} = [\Delta_{Bi}^{\text{low}}, \Delta_{Bi}^{\text{high}}]$ represents the bandwidth constraint interval of the ith user, $\mathbf{V}_D = (\Delta_{D1}, \Delta_{D2}, \ldots, \Delta_{Dk})$ is the vector of users' delay requirements, $\Delta_{Di} = [\Delta_{Di}^{\text{low}}, \Delta_{Di}^{\text{high}}]$ represents delay constraint interval of the ith user, $\mathbf{V}_J = (\Delta_{J1}, \Delta_{J2}, \ldots, \Delta_{Jk})$ is the vector of users' delay jitter requirements, $\Delta_{Ji} = [\Delta_{Ji}^{\text{low}}, \Delta_{Ji}^{\text{high}}]$ represents delay jitter constraint interval of the ith user, $\mathbf{V}_L = (\Delta_{L1}, \Delta_{L2}, \ldots, \Delta_{Lk})$ is the vector of users' error rate requirements, and $\Delta_{Li} = [\Delta_{Li}^{\text{low}}, \Delta_{Li}^{\text{high}}]$ represents error rate constraint interval of the ith user. A QoS multicast routing tree $T(s, M)$ supporting communication of s and M should be found.

10.2.2 Management of Inaccurate Parameters

Multicast routing under inaccurate network information has received significant attention in recent years. Generally, there are two methods to deal with inaccurate information, probability theory and fuzzy mathematics. In this chapter, fuzzy mathematics is used.

For a parameter σ with inaccurate interval $[\sigma_L, \sigma_H]$ and a triangular fuzzy number $A(\sigma) = (a_1, c_1)$ where a_1 is the center and c_1 is the breadth of the interval, the corresponding membership function is defined as [25]

$$\mu_A(x) = \begin{cases} 1 - \dfrac{x - a_1}{c} & a_1 - c_1 < x < a_1 + c_1 \\ \varepsilon & x = a_1 \pm c_1 \\ 0 & \text{otherwise} \end{cases} \tag{10.1}$$

where $a_1 = (\sigma_L + \sigma_H)/2$, $c_1 = (\sigma_H - \sigma_L)/2$, and ε is a positive decimal and much smaller than 1.

Assume that the QoS constraint interval Δ is related to a fuzzy set $F(\Delta) = (a_2, c_2)$ and its membership function is

$$\mu_F(x) = \begin{cases} 1 & a_2 - c_2 \le x \le a_2 + c_2 \\ 0 & \text{otherwise.} \end{cases} \tag{10.2}$$

Here, $a_2 = (\Delta^{\text{low}} + \Delta^{\text{high}})/2$, $c_2 = (\Delta^{\text{high}} - \Delta^{\text{low}})/2$.

Given two fuzzy sets A and F, the nearness degree is defined as follows:

$$T(A, F) = \{\bigvee_{x \ge 0} [\mu_A(x) \wedge \mu_F(x)]\} \wedge \{\bigwedge_{x \ge 0} [\mu_A(x) \vee \mu_F(x)]\}. \tag{10.3}$$

The credibility degree that A is less than F is

$$R(A, F) = \begin{cases} 0.5 & a_1 = a_2 \\ 1 - 0.5R(A, F) & a_1 < a_2 \\ 0.5R(A, F) & a_1 > a_2. \end{cases} \tag{10.4}$$

Let $P(s, m_i)$ be the communication route from s to $m_i \in M$, $1 \le i \le k$, on $T(s, m)$. The bandwidth credibility degree is defined as

$$R_B(m_i) = \min_{e \in P(s,m_i)} R\{F(\Delta_{Bi}), A[b(e)]\}. \tag{10.5}$$

Similarly, the delay credibility degree, the delay jitter credibility degree, and the error rate credibility degree are respectively defined as

$$R_D(m_i) = R\{A[d(m_i)], F(\Delta_{Di})\}, \tag{10.6}$$

$$R_J(m_i) = R\{A[j(m_i)], F(\Delta_{Ji})\}, \tag{10.7}$$

$$R_L(m_i) = R\{A[l(m_i)], F(\Delta_{Li})\}, \tag{10.8}$$

where

$$A[d(m_i)] = ([d(m_i)_{\min} + d(m_i)_{\max}]/2, [d(m_i)_{\max} - d(m_i)_{\max}]/2) \tag{10.9}$$

$$A[j(m_i)] = ([j(m_i)_{\min} + j(m_i)_{\max}]/2, [j(m_i)_{\max} - j(m_i)_{\max}]/2) \tag{10.10}$$

$$A[l(m_i)] = ([l(m_i)_{\min} + l(m_i)_{\max}]/2, [l(m_i)_{\max} - l(m_i)_{\max}]/2) \tag{10.11}$$

are fuzzy numbers for delay, delay jitter and error rate respectively, in which

$$d(m_i)_{\min} = \sum_{e \in P(s,m_i)} d_{\min} \quad \text{and} \quad d(m_i)_{\max} = \sum_{e \in P(s,m_i)} d_{\max},$$

$$j(m_i)_{\min} = \sum_{e \in P(s,m_i)} j_{\min} \quad \text{and} \quad j(m_i)_{\max} = \sum_{e \in P(s,m_i)} j_{\max},$$

and

$$l(m_i)_{\min} = 1 - \prod_{e \in P(s,m_i)} (1 - l_{\min}) \quad \text{and} \quad l(m_i)_{\max} = 1 - \prod_{e \in P(s,m_i)} (1 - l_{\max}).$$

The credibility degree of $P(s, m_i)$ is accordingly defined as

$$R(m_i) = \min_{I \in B,D,J,L} R_I(m_i). \tag{10.12}$$

Finally, the credibility degree of $T(s, m)$ is defined as

$$R[T(s, M)] = \min_{m_i \in M} R(m_i). \tag{10.13}$$

10.2.3 User's QoS Satisfaction Degree

The user's QoS satisfaction degree reflects the user's satisfaction degree of a multicast tree. It is also a fuzzy notion. A fuzzy membership function and fuzzy rules are constructed in order to express it.

10.2.3.1 Membership Function

For the *i*th user bandwidth constraint, let $I_{BL} = (3\Delta_{Bi}^{\mathrm{low}} + \Delta_{Bi}^{\mathrm{high}})/4$ and $I_{BH} = (\Delta_{Bi}^{\mathrm{low}} + 3\Delta_{Bi}^{\mathrm{high}})/4$. Three fuzzy sets $F_{\mathrm{low}}(\sigma)$, $F_{\mathrm{middle}}(\sigma)$ and $F_{\mathrm{high}}(\sigma)$, related to user's satisfaction degree low *L*, middle *M* and high *H* respectively, are designed for each QoS parameter σ based on the trisection method of fuzzy mathematics [26].

The membership function of a path whose bandwidth satisfaction degree is low corresponding to fuzzy set $F_{\mathrm{low}}(B)$ is defined as

$$\mu_{BL}(x) = 1 - \Phi[(x - I_{BL})/\beta_B] \tag{10.14}$$

where β_B is a positive constant and

$$\Phi(x) = \int_{-\infty}^{x} \frac{1}{2\pi} e^{-t^2/2} dt.$$

The membership function of a path whose bandwidth satisfaction degree is middle corresponding to fuzzy set $F_{\mathrm{middle}}(\mathrm{B})$ is defined as

$$\mu_{BM}(x) = \Phi[(x - I_{BL})/\beta_B] - \Phi[(x - I_{BH})/\beta_B]. \tag{10.15}$$

The membership function of a path whose bandwidth satisfaction degree is high corresponding to fuzzy set $F_{\mathrm{high}}(\mathrm{B})$ is defined as

$$\mu_{BH}(x) = \Phi[(x - I_{BH})/\beta_B] \tag{10.16}$$

The membership degrees of satisfaction degree corresponding to bandwidth satisfaction degree for the *i*th user are $T[\mathrm{F}(\Delta_{Bi}), F_{\mathrm{low}}(\mathrm{B})]$, $T[\mathrm{F}(\Delta_{Bi}), F_{\mathrm{middle}}(\mathrm{B})]$, and $T[\mathrm{F}(\Delta_{Bi}), F_{\mathrm{high}}(\mathrm{B})]$.

Similarly, for the *i*th user delay $(X = D)$, delay jitter $(X = J)$ and error rate $(X = L)$ constraints, let $I_{XL} = (3\Delta_{Xi}^{\mathrm{low}} + \Delta_{Xi}^{\mathrm{high}})/4$ and $I_{XH} = (\Delta_{Xi}^{\mathrm{low}} + 3\Delta_{Xi}^{\mathrm{high}})/4$. The membership functions of a path whose delay, delay jitter, or error rate satisfaction degrees are low are defined as

$$\mu_{XL}(x) = \Phi[(x - I_{XH})/\beta_X]. \tag{10.17}$$

Bandwidth is a concave parameter while delay, delay jitter and error rate are additive or multiplicative parameters. For a route, the minimum bandwidth needs to be superior to bandwidth requirement while the maximum delay, delay jitter and error rate need to be inferior to delay, delay jitter and error rate requirements. Thus, the definition here is different from that in (10.14). Similarly, formula (10.19) is different from that in (10.16).

The membership functions of a path whose delay, delay jitter, or error rate satisfaction degrees are middle are defined as

$$\mu_{XM}(x) = \Phi[(x - I_{XL})/\beta_X] - \Phi[(x - I_{XH})/\beta_X]. \tag{10.18}$$

The membership functions of a path whose delay, delay jitter, or error rate satisfaction degrees are high are defined as

$$\mu_{XH}(x) = 1 - \Phi[(x - I_{XL})/\beta_X]. \tag{10.19}$$

The membership degrees of satisfaction degree corresponding to these QoS satisfaction degree can be computed similarly.

10.2.3.2 Fuzzy Rules

The corresponding fuzzy rules are defined in Table 10.1. Here the P_i, $i = 1, 2, 3, 4$, denote bandwidth, delay, delay jitter or error rate.

Table 10.1 Fuzzy rules

Rule	P_1	P_2	P_3	P_4	Fuzzy result
1	*L*	*L*	*L*	*L*	*EL*
2	*L*	*L*	*L*	*M*	*EL*
3	*L*	*L*	*L*	*H*	*VL*
4	*L*	*L*	*M*	*M*	*VL*
5	*L*	*L*	*M*	*H*	*L*
6	*L*	*L*	*H*	*H*	*M*
7	*L*	*M*	*M*	*M*	*L*
8	*L*	*M*	*M*	*H*	*M*
9	*L*	*M*	*H*	*H*	*H*
10	*L*	*H*	*H*	*H*	*VH*
11	*M*	*M*	*M*	*M*	*M*
12	*M*	*M*	*M*	*H*	*H*
13	*M*	*M*	*H*	*H*	*VH*
14	*M*	*H*	*H*	*H*	*EH*
15	*H*	*H*	*H*	*H*	*EH*

10.2.3.3 QoS Satisfaction Degree Computing

The centre of each fuzzy result is defined as: $el = 0$, $vl = 16$, $l = 33$, $m = 50$, $h = 66$, $vh = 83$, $eh = 100$.

The ith user's QoS satisfaction degree is

$$Q(m_i) = \frac{el \cdot EL + el \cdot VL + l \cdot L + m \cdot M + h \cdot H + vh \cdot VH + eh \cdot EH}{100(EL + VL + L + M + H + VH + EH)}. \tag{10.20}$$

The QoS satisfaction degree of *T*(*s*, *M*) is

$$Q[T(s,M)] = \sum_{i=1}^{k} Q(m_i)/k. \tag{10.21}$$

The goal of the proposed algorithm is to find a multicast tree with maximum reliability degree and user's QoS satisfaction degree. That is:

$$\text{maximize}\, R[T(s,M)] \times Q[T(s,M)] \tag{10.22}$$

for all $m_i \in D$ subject to

$$\{b_{\max} | \forall e \in P(s,m_i)\} \geq \Delta_{Bi}^{\text{low}}, \tag{10.23}$$

$$d(m_i)_{\min} \leq \Delta_{Di}^{\text{high}}, \tag{10.24}$$

$$j(m_i)_{\min} \leq \Delta_{Ji}^{\text{high}}, \tag{10.25}$$

$$l(m_i)_{\min} \leq \Delta_{Li}^{\text{high}}. \tag{10.26}$$

To solve the problem, Dijkstra *k*th shortest path algorithm is first used to find candidate routes that satisfy Equations 10.23–10.26 and constructs a set of routes $\{p_m^1, p_m^2 \ldots, p_m^j\}$. Here, p_m^j is the *j*th route from *s* to m_i. Then the multicast tree is generated by choosing a route from each set of routes randomly. The position of the route in the route sets is labeled, and the *k* integers generate the code of *T*(*s*, *M*). Then DE/best/1/bin is applied to solve the problem. In particular, the mutation intensity is linearly adapted.

10.3 Performance Evaluation

Simulations have been done on NS2 (Network Simulator 2) platforms. The proposed algorithm (DEQM), fuzzy QoS multicast routing algorithm based on general GA in [27] (GAQM) and multicast routing algorithm based on the Prim algorithm [28], pp. 168–174] have been performed over CERNET (T1), CERNET2 (T2), NSFNET (T3) and a virtual topology (T4, generated by Waxman2 with average node degree 3) [29]. The topology of the networks concerned is shown in Figure 10.1.

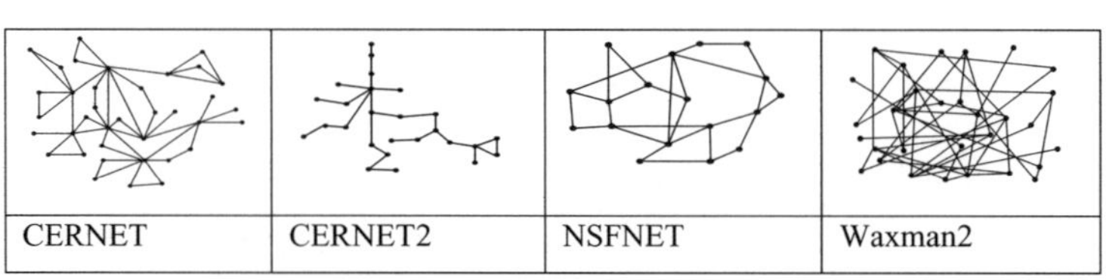

Figure 10.1 Topology of the first network

10.3.1 Request Success Rate

The number of request times that the optimal solution is found as a percentage of the total multicast requests is called the request success rate. Due to the dynamic characteristics of NGI, in actual communication, if the real-time value of QoS parameter of a found multicast tree can meet users' requirements, the multicast tree is successful. The result is shown in Figure 10.2.

Simulation has shown that the request success rate of DEQM is higher than that of the other two algorithms, especially in network T4 whose connectivity is low. The success of DEQM is due to elaborate management of inaccurate network information.

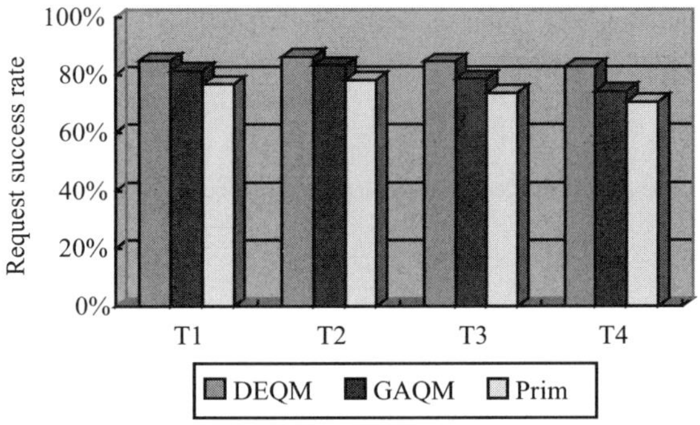

Figure 10.2 Request success rate

10.3.2 Credibility Degree of Meeting User's Requirement and User's QoS Satisfaction Degree

The credibility degrees of meeting users' requirement and users' QoS satisfaction degrees are shown in Figures 10.3 and 10.4. Similarly, both credibility degree of meeting users' requirements and users' QoS satisfaction degree are higher for DEQM than for the other two algorithms. Moreover, with increasing network topology complexity, there is no obvious

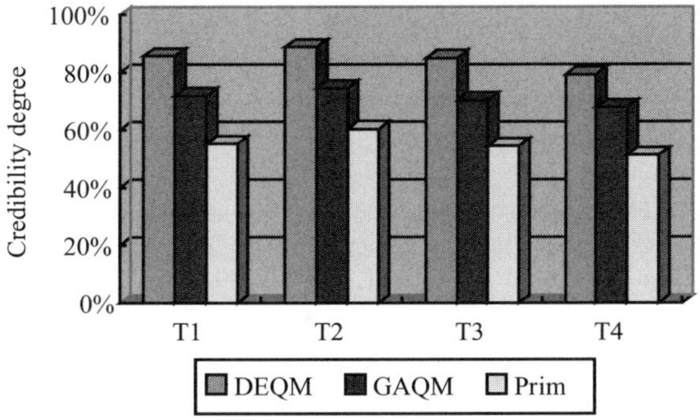

Figure 10.3 Credibility degree of meeting users' requirements

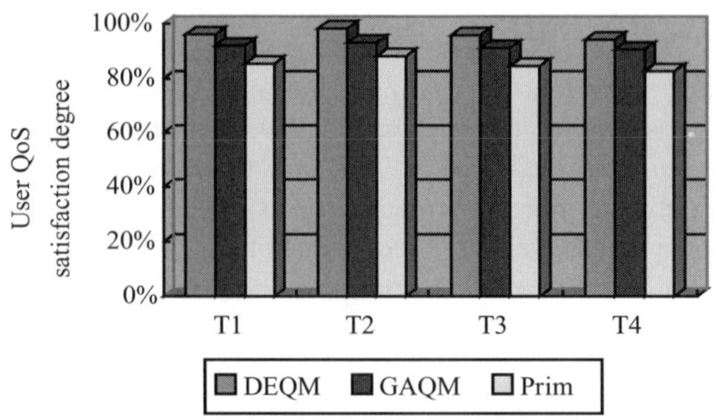

Figure 10.4 Comparison of users' QoS satisfaction degree

fluctuation on the values of the two indicators. Apparently, DEQM has the advantage of scalability.

10.4 Conclusions

In this chapter, a QoS multicast routing algorithm based on differential evolution is proposed. Inaccurate parameters of NGI and flexibility QoS requirements of users are considered by means of the principle of fuzzy mathematics. The multicast routing tree with maximum credibility degree of meeting user QoS requirements and maximum user QoS satisfaction degree within multiple users' flexible QoS constraints is found. Simulations have shown that the proposed algorithm is both feasible and effective.

10.5 Acknowledgement

This work was supported by the National High-Tech Research and Development Plan of China under Grant No. 2006AA01Z214; the National Natural Science Foundation of China under Grant No. 60673159 and No. 70671020; Program for New Century Excellent Talents in University; the Key Project of Chinese Ministry of Education under Grant No. 108040; Specialized Research Fund for the Doctoral Program of Higher Education under Grant No. 20060145012 and No. 20070145017; the Natural Science Foundation of Liaoning Province under Grant No. 20062022.

References

[1] Wang, D., Ergun, F. and Xu, Z. (2005) Unicast and multicast routing with multiple constraints, in *Quality of Service in Multiservice IP Networks* (eds. M. Ajmone Marsan, G. Bianchi, M. Listanti and M. Meo), Lecture Notes in Computer Science, **3375**, Springer, Berlin, pp. 481–494.

[2] Yuan, X. (2002) Heuristic algorithms for multiconstrained quality-of-service routing. *IEEE/ACM Transactions on Networking*, **10**(2), 244–256.

[3] Wang, Z. and Crowcroft, J. (1996) Quality-of-service for supporting multimedia application. *IEEE Journal on Selected Areas in Communications*, **14**(7), 1228–1234.

[4] Ghaboosi, N. and Haghighat, A.T.(Dec. 18–21, 2006) A tabu search based algorithm for multicast routing with QoS constraints. 9th Int. Conf. Information Technology, Bhubaneswar, pp. 33–39.

[5] Kompella, V.P., Pasquale, J.C. and Polyzo, G.C. (1993) Multicast routing for multimedia communication. *IEEE/ACM Transaction Network*, **28**(1), 286–292.
[6] Wang, X.W., Huang, M. and Liu, J.R. (2001) Research on quality of service based destination node joining and leaving algorithms for multimedia group communication. *Chinese Journal of Computers*, **24**(8), 838–844.
[7] Moses, C., Joseph, N. and Baruch, S. (2004) Resource optimization in QoS multicast routing of real-time multimedia. *IEEE/ACM Transactions on Networking*, **12**(2), 340–348.
[8] Shavitt, Y., Winkler, P. and Wool, A. (2004) On the economics of multicasting. *Netnomics*, **6**(1), 1–20.
[9] Wang, L., Li, Z.Z., Song, C.Q. and Yan, T. (2004) A dynamic multicast routing algorithm with inaccurate information satisfying multiple QoS constraints. *Acta Electronica Sinica*, **32**(8), 244–1247.
[10] Chakrabarti, A. and Manimaran, G. (2005) Reliability constrained routing in QoS networks. *IEEE/ACM Transactions on Networking*, **13**(3), 662–675.
[11] Cohen, A., Korach, E., Last, M. and Ohayon, R. (2005) Fuzzy-based path ordering algorithm for QoS routing in non-deterministic communication networks. *Fuzzy Sets and Systems*, **150**(3), 401–417.
[12] Gatani, L., Lo, R.G. and Gaglio, S. (2005) An efficient distributed algorithm for generating multicast distribution trees, 2005. Int. Conf. Parallel Processing Workshops, June 14–17, 2005, pp. 477–484.
[13] Kim, D.K., Kim, K.I., Hwang, I.S. and Kim, S.H. (2005) Hierarchical overlay multicast based on host group model and topology-awareness. 7th Int. Conf. Advanced Communication Technology, Phoenix Park, pp. 335–339.
[14] Moussaoui, O., Ksentini, A., Naïmi, M. and Gueroui, A. (2005) Multicast tree construction with QoS guaranties, in *Management of Multimedia Networks and Services* (eds. J.D. Royo and G. Hasegawa), Lecture Notes in Computer Science, **3754**, Springer, Berlin, pp. 96–108.
[15] Siachalou, S. and Georgiadis, L. (2005) Algorithms for precomputing constrained widest paths and multicast trees. *IEEE/ACM Transactions on Networking*, **13**(5), 1174–1187.
[16] Wang, X.W., Li, J. and Huang, M. (2005) An integrated QoS multicast routing algorithm based on tabu search in IP/DWDM optical internet, in *High Performance Computing and Communications* (eds. L.T. Yang *et al.*), Lecture Notes in Computer Science, **3726**, Springer, Berlin, pp. 111–116.
[17] Lowu, F. and Baryamureeba, V. (2006) On efficient distribution of data in multicast networks: QoS in scalable networks, in *Large-Scale Scientific Computing* (eds. I. Lirkov, S. Margenov and J. Wasniewski), Lecture Notes in Computer Science, **3743**, Springer, Berlin, pp. 518–525.
[18] Wang, X.W., Gao, N., An, G.Y. and Huang, M. (2006) A fuzzy integral and microeconomics based QoS multicast routing scheme in NGI. 6th IEEE Int. Conf. Computer Information Technology, Seoul, pp. 106–111.
[19] Chen, N.S., Li, L.Y. and Ke, Z.W. (2007) QoS multicast routing algorithm based on QGA. *IFIP Int. Conf. Network Parallel Computing Workshops, Liaoning, China*, pp. 683–688.
[20] Galatchi, D. (2007) A QoS multicast routing protocol for mobile Ad-Hoc networks. *8th Int. Conf. Telecommunications Modern Satellite Cable Broadcasting Services, Serbia, Niš*, pp. 27–30.
[21] Gong, B.C., Li, L.Y., Wang, X.L. and Jiang, T. (2007) A novel QoS multicast routing algorithm based on ant algorithm. Int. Conf. Wireless Communications Networking Mobile Computing, Shanghai, China, pp. 2025–2028.
[22] Chen, S. and Nahrstedt, K. (1998) Distributed quality-of-service routing in high-speed networks based on selective probing. 23rd Annual Conf. Local Computer Networks, Lowell, MA, pp. 80–89.
[23] Takahashi, A. and Kinoshita, T. (2006) A design and operation model for agent-based flexible distributed system. Proceedings of the IEEE/WIC/ACM International Conference on Intelligent Agent Technology (IAT '06), pp. 57–65.
[24] Böringer, R., Schmidt, S. and Mitschele-Thiel, A. (2007) A QoS-aware multicast routing protocol for wireless access networks. 3rd Euro NGI Conference on Next Generation Internet Networks, pp. 119–126.
[25] Zhang, P., Li, M.L. and Wang, S. (2006) Multicast QoS routing and partition with fuzzy parameters. *Chinese Journal of Computers*, **29**(2), 279–285.
[26] Yang, G.B. and Gao, Y.Y. (2002) *Principle and Application of Fuzzy Mathematics*. South China University of Technology Press, Guangzhou.
[27] Chen, P. and Dong, T.L. (2004) A fuzzy genetic algorithm for QoS multicast routing. *Computer Communications*, **34**(4), 50–54.
[28] Yan, W.M. and Wu, W.M. (1997) *Data Structure*. Tsinghua University Press, Beijing.
[29] Waxman, B.M. (1988) Routing of multipoint connections. *IEEE Journal on Selected Areas in Communications*, **6**(9), 1617–1622.

11

Multisite Mapping onto Grid Environments

I. De Falco[1], A. Della Cioppa[2], D. Maisto[1], U. Scafuri[1], E. Tarantino[1]

[1] *Institute of High Performance Computing and Networking, National Research Council of Italy (ICAR-CNR), Via P. Castellino 111, 80131 Naples, Italy*

[2] *Natural Computation Lab, DIIIE, University of Salerno, Via Ponte don Melillo 1, 84084 Fisciano (SA), Italy*

11.1 Introduction

The computational power provided by the vast majority of computing systems is seldom fully exploited. In fact, except for brief and rare spans of time, the resources used for the execution of applications are almost always a modest percentage of those actually available. Obviously, when these systems are some hundreds or thousands (potentially even millions) in number, the possible aggregated and coordinated use of their locally unemployed resources would make available a computational power which is unconceivable on any physically assemblable system, even a parallel one.

The recent development of powerful computers and low-cost high-speed network technologies allows the physical construction of computational grids [1], that is, geographically distributed infrastructures which generally aggregate heterogeneous, large-scale, and multiple-institutional resources to provide transparent, secure and coordinated access to various computing resources (supercomputers, clusters, scientific instruments, databases, storage devices, etc.). In recent years many different kinds of computational grids supporting very promising paradigms for high-performance throughput have been built. For example, the Desktop Grids, which bring together thousands of internet users around the world donating the power of their idle home desktop computers ([2], [3], pp. 431–450), [4,5]), have received growing attraction.

In any case grids represent today a platform to exploit the unemployed computational power available on these very large collections of interconnected heterogeneous processing elements. These elements, in general all running on common grid middleware (e.g., Globus Service [6]),

Differential Evolution Anyong Qing

are organized in sites containing different computing systems each constituted by one or more processing nodes.

Hence a grid user, who does not have sufficient local resources to execute his own parallel computationally expensive application, could conveniently benefit from a grid which provides access to computing power far beyond the current availability threshold at a single site. However, in order to develop programs able to exploit this potential, besides classical problems of parallel computing, programmers must deal with grid-specific problems such as advance reservation, co-allocation and multisite launching [7].

Generally a grid-aware application is decomposed into communicating tasks, where each task may have various execution requirements [8,9]. Appropriate decisions are to be made about task mapping onto grid nodes. For each, a node is selected from those assuring an adequate quantity of locally unemployed computational resources so that the application executes in as short a time as possible.

The problem becomes even more complex when, as often happens, one desires to determine mapping solutions which, besides guaranteeing short execution time and efficient use of resources, must also satisfy user-dependent requirements [9], such as performance and quality of service (QoS) [10]. In such circumstances resources at a single site could be insufficient to meet all these needs. Thus, a multisite mapping tool, able to choose among resources spread over multiple sites and obtain high throughput while fulfilling application requirements with the networked grid computing resources, must be designed.

In this chapter we deal with the multisite mapping problem, assuming that the middleware layer level supplies suitable services for co-allocation. This hypothesis is necessary because, in absence of information about the timing of communication, the execution of communicating tasks proceeds only if their simultaneous allocation is guaranteed [7]. With these premises, we deploy the application tasks onto the nodes aiming at minimizing the application execution time and contemporaneously optimizing resource utilization.

The classical approach [11–14] takes into account the grid user's point of view and aims at minimizing the completion time of applications. Here the problem is considered from the grid manager's point of view. The goal is to find the solution which, in addition to fulfilling user's QoS requests, allows to achieve the best possible balance, in terms of time, for the use of the grid nodes involved by the solution. This view leads towards the discovery of solutions which do not use for each task the most powerful available nodes if, due to overlapping, their employment does not reduce the execution time of the whole application. In such a way the manager uses resources suitable to the task requirements and avoids keeping busy the most powerful nodes which could be more profitably exploited for further applications.

Mapping algorithms for traditional parallel and distributed systems on homogeneous and dedicated resources do not work well in these situations ([15], pp. 279–307). Some approaches have been proposed to deal with different aspects of the problem [16,17]. Unfortunately, the mapping problem, already known to be NP-complete on conventional parallel systems, gets an additional twist when the resources, apart from being heterogeneous and geographically distributed, present dynamic characteristics arising from both local loads and network [18,19].

Among different heuristic approaches proposed, there has been particular interest in evolutionary algorithms which find at least a suboptimal solution in a reasonable amount of time. Valuable results have been achieved [11–13,17,20–25].

Here, a multi-objective differential evolution based on the Pareto method [26] is proposed to provide a set of possible mapping solutions, each with a different balance between use of resources and QoS requests. Unlike the other existing evolutionary methods, which simply search for one system per site onto which to map the whole application, we deal with a multisite and multisystem approach. In fact, at a site we can have more systems each of which can have one or more processing elements (i.e., nodes), such as a conventional workstation or a multi-computer. Our approach performs a multisite mapping, spreading the application tasks among the systems which can belong to different sites. In the case of multi-computer systems, among all their processing elements, our approach chooses only those which, on the basis of their features, turn out to be the most suitable for the application to be mapped. By doing so, we consider the nodes making up the systems as the lowest computational units taking their reliabilities and their actual loads into account.

11.2 Working Environment

11.2.1 Grid Framework and Task Scheduling

In this work we refer to a computational grid which, organized in separate sites, is constituted by numerous computing systems. These can be both desktop computers and up-to-date powerful multi-computers (i.e., parallel systems commercialized by manufacturers as IBM, HP, and so on, or else arranged in clusters obtained by interconnecting conventional machines and known as Beowulf systems). Each system is supposed to operate in time-shared mode and has two different queues, one for processes locally submitted and the other for remote tasks presented via the grid to be executed during idle times. The tasks in the local queue are scheduled on the basis of the locally established policy. However, a strategy with priority must be adopted for those in the remote queue. In particular, the same priority is assigned to tasks of the same application. On the other hand, submission time is considered for different applications: tasks of an application submitted earlier are assigned higher priority. The task scheduling of the remote queue does not assume a fair distribution of CPU time. Instead, the time allowed for the execution of the tasks present in the remote queue will be fully dedicated to the processes with higher priorities. In other words, the processor will execute a task with a lower priority if and only if tasks in the ready state with higher priorities do not exist. Consequently, the execution of a process will be interrupted only if, once its time slice has elapsed, another task belonging to the same application exists in the queue, or because the current task is moved to a waiting state owing to a not-ready communication, or else because a suspended process of a different application with higher priority has become ready again (the scheduler must be pre-emptive).

With respect to the previous assumptions, our grid is organized into clients, that is, parallel application submitters, and resource providers (owned by institutions or individuals) which donate their computing resources during idle times. We hypothesize a unique mapping server which, given a submitted application and known local workloads of grid resources (hence, known average idle times), performs the co-allocation of tasks on the grid resources. Otherwise stated, the user submits his application subdivided into tasks to the unique mapper/scheduler on the whole grid which, having established the mapping, co-allocates the tasks to the selected processing elements. This implies that the generic task of a grid application shares the computing resources of the assigned node with local and remote processes on the basis of the priorities.

Assuming knowledge of the time instant at which the task starts to 'run' and of the amount of computation to perform, to calculate the execution time our mapping server operates as if all the remaining resources of the node are assigned to that task and not to others. Clearly, the idle time of a processing node corresponding to a not-ready communication could be conveniently exploited to carry out the execution of other tasks (of the same or other applications) allocated.

When a task of a different application is assigned to a node, it is evident that if the tasks previously assigned are not pending and the load remains constant, the time needed to execute this new task is the sum of the time calculated as if the first ones were never allocated and the time required for the running of the first ones.

11.2.2 Grid Mapping

To focus on the mapping problem we need information on the number and status of both accessible and required resources. We assume that our application is subdivided into P tasks (required resources) to be mapped to n ($n \leq P$) nodes, that is, accessible resources. Each node is identified by an integer value in the range [1, N], where N is the total number of available grid nodes.

We need to know a priori the number of instructions α_i computed per time unit on each node i and the communication bandwidth β_{ij} between any couple of nodes i and j. Note that β_{ij} is the generic element of an $N \times N$ symmetric matrix $\boldsymbol{\beta}$ with very high values on the main diagonal, that is, β_{ii} is the bandwidth between two tasks on the same node. This information is supposed to be contained in tables based either on statistical estimates in a particular time span or gathered by tracking periodically and by forecasting dynamically resource conditions [27,28]. For example, in the Globus Toolkit [10], which is standard grid middleware, similar information is gathered by the Grid Index Information Service [28].

Since grids address non-dedicated resources, their own local workloads must be considered to evaluate the computation time of the tasks. There exist several prediction methods to face the challenge of non-dedicated resources [29,30]. For example, we assume that we know the average load $l_i(\Delta t)$ of node i at a given time span Δt with $l_i(\Delta t) \in [0.0, 1.0]$, where 0.0 means a node completely discharged and 1.0 a node locally loaded at 100%. Hence $(1 - l_i(\Delta t))^* \alpha_i$ represents the power fraction of node i available for the execution of grid tasks.

As to the resources requested by the application task, we assume that we know, for each task k, the number of instructions γ_k and the amount of communications ψ_{km} between the kth and the mth task $\forall m \neq k$ to be executed. Obviously, ψ_{km} is the generic element of a $P \times P$ symmetric matrix $\boldsymbol{\psi}$ with all null elements on the main diagonal. All this information can be obtained either by a static program analysis, or by using smart compilers or by other tools which automatically generate them. For instance, the Globus Toolkit includes an XML standard format to define application requirements [28].

Finally, information about the degree of reliability of any component of the grid must be provided. This is expressed in terms of fraction of actual operativity π_z for the processor z and λ_w for the link connecting to internet the site w to which z belongs. These values can be gathered by means of a historical and statistical analysis in [0.0, 1.0].

11.3 Differential Evolution for Grid Mapping

On the basis of the assumptions about the application and the available resources made in Section 11.2.2, it is possible to define the encoding and the fitness functions for our mapping tool based on a differential evolution technique.

11.3.1 Encoding

In general, any mapping solution should be represented by a vector $\boldsymbol{\mu}$ of P integers ranging in the interval $[1, N]$. To obtain $\boldsymbol{\mu}$, the real values provided by differential evolution in the interval $[1,N + 1[$ are truncated before evaluation. The truncated value $\lfloor \mu_i \rfloor$ denotes the node onto which task i is mapped.

Given the nature of the problem, a number of points need to be borne in mind. As long as the mapping is considered by characterizing the tasks in terms of their computational needs γ_k only, the mapping problem is a classical NP-complete optimization problem, in which the allocation of a task does not affect that of the other ones, unless, of course, one does attempt to load more tasks on the same node. However, if communications ψ_{km} are also taken into account, it becomes more complicated. The communication bandwidths existing within any cluster are higher than those between clusters, therefore, allocating a task on a given node in a given cluster may require allocating other tasks on the same node or in the same cluster so as to reduce communication and execution times.

Such a problem is a typical example of epistasis, that is, a situation in which the value taken by a variable influences those of other variables. This situation is also deceptive, since one solution can be transformed into another with better fitness only by passing through intermediate solutions worse than both current and best ones. Let us consider, as an example, the circumstance in which two or more tasks are allocated onto nodes belonging to a cluster with a given communication bandwidth, while a suitable number of at least equally fast nodes yet with higher bandwidth is available on another cluster. It is extremely improbable that all those tasks will migrate at once from the slow cluster to the fast one by using classical differential evolution. In fact, what very likely happens is that a newly generated solution proposes to keep some of those tasks in the former cluster and to move some others to the latter. This allocation leads to a worse fitness value, so this solution will be discarded, although it should be saved since it might help to reach the optimal solution by further evolution.

To overcome this problem we have introduced a new operator of cluster mutation into classical differential evolution. It is applied with a probability p_m any time a new individual must be produced. When this mutation is carried out in the current solution, a gene is randomly chosen and the node value contained in it, related to a cluster C_i, is equi-probabilistically modified into another one which is related to another cluster, say C_j. Then, any other task assigned to C_i in the current solution randomly migrates to another node belonging to C_j by inserting into the related gene a random value within the bounds for C_j.

11.3.2 Fitness

Resource consumers who submit various applications, and resource providers who share their resources, usually have different motivations when they join the grid. These incentives should be included in objective functions for mapping. Currently, most objective functions in grid computing are inherited from traditional parallel and distributed systems. Grid users and resource providers may have different demands to satisfy. For example, users may be interested in the total cost of running their application, while providers may pay more attention to the throughput of their resources in a particular time interval. Thus objective functions have to meet different goals.

To evaluate fitness we make use of the information on the number and status of both the available and requested resources contained within the data structures defined in Section 11.2.2.

Furthermore, given the two goals described in Section 11.1, we have two fitness functions, one for the time of use of resources and the other for their reliability.

11.3.2.1 Use of Resources

The total time needed to execute task *i* on node *j* is evaluated on the basis of the computation power and of the bandwidth available after deducting the local workload:

$$\tau_{ij} = \tau_{ij}^{\text{comp}} + \tau_{ij}^{\text{comm}}, \tag{11.1}$$

where τ_{ij}^{comp} and τ_{ij}^{comm} are the computation and communication times requested.

Let τ_j^s be the summation on all the tasks assigned to the *j*th node by the current mapping. This value is the time spent by node *j* in executing computations and communications of all the tasks assigned to it. Clearly, τ_j^s is equal to zero for all the nodes not included in the vector $\boldsymbol{\mu}$.

Considering that all the tasks are co-scheduled, the minimum time required to complete the application execution is given by the maximum value among all the τ_j^s. Thus, the fitness function is

$$\Phi_1(\boldsymbol{\mu}) = \max_{1 \le j \le N} \{\tau_j^s\}. \tag{11.2}$$

The goal of the evolutionary algorithm is to search for the smallest fitness value, that is, to find a mapping which uses the grid resources for the shortest time.

11.3.2.2 Reliability

The reliability is evaluated as:

$$\Phi_2(\boldsymbol{\mu}) = \prod_{i=1}^{P} \pi_{\lfloor \mu_i \rfloor} \lambda_w, \tag{11.3}$$

where w is the site to which node $\lfloor \mu_i \rfloor$ belongs.

It should be noted that the first fitness function should be minimized, while the second should be maximized. It is evident that we are dealing with a two-objective problem which can be profitably faced by designing and implementing a multi-objective differential evolution algorithm based on the concept of the so-called Pareto optimal set.

11.4 Experiments in Predetermined Conditions

11.4.1 Experimental Setup

Before conducting any kind of experiment the structure of the available resources and the features of the machines belonging to each site must be known. Generally a grid environment is composed of sites, each containing many heterogeneous computation devices (parallel machines, clusters, supercomputers, dedicated systems, etc.).

Since a generally accepted set of heterogeneous computing benchmarks does not exist and considering that the determination of a representative set of such benchmarks remains a current and unresolved challenge, the grid architecture is conceived so that an immediate check of the mapping solution proposed can be done 'by hand'. In particular, by choosing conveniently the computing capabilities and the communication bandwidths of the grid nodes, and their load conditions, it is in some cases possible, by suitable arrangement of experiments, to promptly detect the optimal solutions. In these situations, the goodness of the achieved suboptimal solutions can be rapidly verified for comparison with the optimal ones.

Our simulations refer to a grid which aggregates 184 nodes subdivided into six sites denoted by *A*, *B*, *C*, *D*, *E* and *F*, with 32, 24, 48, 20, 24 and 36 nodes, respectively. Moreover each site of *B*, *C*, *E* and *F* constitutes two clusters with different numbers of nodes and performance. This grid structure is outlined in Figure 11.1.

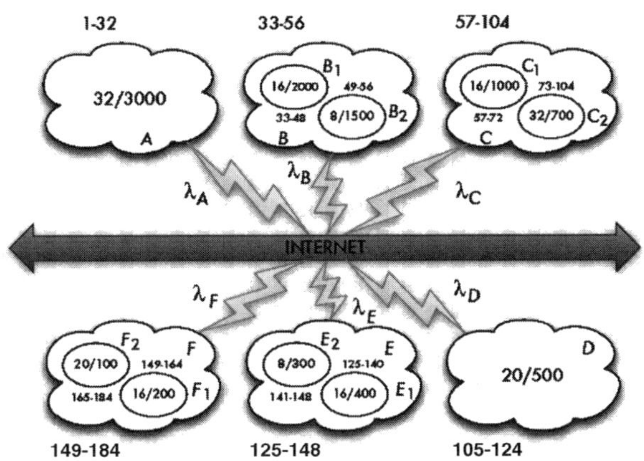

Figure 11.1 The grid architecture

Henceforth we shall denote the nodes by means of the numbers shown outside each cloud in Figure 11.1. For instance, 35 is the third node in cluster B_1, while 109 is the fifth node in cluster *D*. Each node is characterized by a set of static and dynamic resource properties. Without loss of generality, we suppose that all nodes belonging to the same cluster have the same nominal power α expressed in terms of millions of instructions per second (MIPS). The numbers inside the clouds in Figure 11.1 denote the cluster size and the computational capacity. For example, site B is composed of subsites B_1, a cluster made up of 16 nodes with $\alpha = 2000$, and B_2, a cluster made up of 8 nodes with $\alpha = 1500$. Note that the power of clusters decreases from 3000 to 100 from *A* to F_2.

For simplicity we have hypothesized four communication typologies for each node. The first is the bandwidth β_{ii} available when tasks are mapped onto the same node (*intranode communication*), the second is the bandwidth β_{ij} between nodes *i* and *j* belonging to the same cluster (*intracluster communication*), to different clusters of the same site (*intrasite communication*) or to different sites (*intersite communication*). The intranode bandwidths β_{ii}, usually higher than β_{ij}, have all been fixed to 10 Gbit/s. For each link, the input and output bandwidths are assumed equal. In our case, the intersite, intrasite and intracluster bandwidths are reported in Table 11.1.

Table 11.1 Intersite, intrasite, and intracluster bandwidths (Mbit/s)

	A	B_1	B_2	C_1	C_2	D	E_1	E_2	F_1	F_2
A	100									
B_1	2	200								
B_2	2	75	200							
C_1	4	4	4	400						
C_2	4	4	4	35	400					
D	8	8	8	8	8	800				
E_1	16	16	16	16	16	16	1000			
E_2	16	16	16	16	16	16	75	1000		
F_1	32	32	32	32	32	32	32	32	2000	
F_2	32	32	32	32	32	32	32	32	100	2000

Observe that the intracluster bandwidth increases from 100 to 2000 from A to F while the bandwidth between clusters belonging to the same site is assumed lower. For example, each node of B_1 communicates with a node of B_2 with a bandwidth of 75 Mbit/s. Moreover, we assume that we know the average load of available grid resources for the time span of interest and also the reliability of the nodes and the internet links.

As to control parameters of differential evolution, population size is 100. The maximum number of generations is 4000 and $p_m = 0.3$. For the first half of generations, $F = 1.4$, $p_c = 0.1$. However, F and p_c vary randomly in the ranges [0.8, 2] and [0.1, 0.4] in the second half.

The differential evolution algorithm has been implemented in C language and all the experiments have been executed on a 1.5 GHz Pentium 4 computer. For each test problem, 20 executions have been carried outs to reduce randomness. At the end of the experiments the best results are extracted and presented. However, the average findings confirm the robustness of our approach. The time for each execution depends on the number of tasks. For example, the time for 20 and 30 tasks is 60 and 200 seconds, respectively.

Having defined the evolutionary parameters and the grid characteristics, different scenarios must be devised to evaluate the effectiveness of our multi-objective differential evolution. Henceforth we denote by $\boldsymbol{\mu}_{\Phi_1}$ and $\boldsymbol{\mu}_{\Phi_2}$ the best solutions found in terms of lowest maximal resource utilization time and of highest reliability.

11.4.2 Experiment 1

The first experiment has been carried out with applications composed of different number of tasks each with $\gamma_k = 90\,000$ Mega-Instructions (MIs), $\psi_{km} = 0$, $\forall k,m \in [1, \ldots, P]$, $l_i(\Delta t) = 0$, $\lambda_w = 1.0$ and $\pi_z = 1.0$ for all nodes. The best time values achieved for Φ_1 are shown in Table 11.2.

Table 11.2 Use of resources for applications in the absence of communication and load

Number of tasks	Time achieved (s)	Optimal time expected (s)
30	30	30
32	30	30
45	45	45
70	90	60
80	90	60

It is observed from Table 11.2 that optimal solutions are obtained for the first three cases, that is, the solution which involves the most powerful nodes. For example, the optimal solution related to the mapping of 45 tasks is $\boldsymbol{\mu}_{\Phi_1} = \{38, 39, 40, 41, 42, 43, 44, 45, 1, 2, 3, 4, 5, 6, 7, 8, 9, 10, 11, 12, 13, 14, 15, 16, 17, 18, 19, 20, 21, 22, 23, 24, 25, 26, 27, 28, 29, 30, 31, 32, 33, 34, 35, 36, 37\}$. It is evident that tasks have been distributed on the most powerful nodes of A and B.

Suboptimal solutions are obtained for the other two cases. This means that some tasks are not mapped onto the most powerful nodes when the number of tasks increases.

11.4.3 Experiment 2

Here, $\gamma_k = 0$ MIs $\psi_{km} = 4000$ Mbit, $\forall k,m \in [1, \ldots, P]$. Load and reliability settings are unchanged. As expected, the solutions found implicate nodes with the highest bandwidths. For example, the solution for mapping 20 communicating tasks is $\boldsymbol{\mu}_{\Phi_1} = \{171, 172, 173, 174, 175, 176, 177, 178, 179, 180, 181, 182, 183, 184, 165, 166, 167, 168, 169, 170\}$, while that for mapping 30 communicating tasks is $\boldsymbol{\mu}_{\Phi_1} = \{174, 175, 176, 177, 178, 179, 180, 181, 182, 183, 184, 174, 175, 176, 177, 178, 179, 180, 181, 182, 183, 165, 166, 167, 168, 169, 170, 171, 172, 173\}$. All tasks have been allocated on F_2 whose nodes have the highest communication bandwidths.

In the latter case the 30 tasks have been mapped onto the 20 nodes of cluster F_2. Some of the nodes are used more than once despite the existence of cluster F_1 with equal communication bandwidths. The reason is that in this solution the placement of all tasks on the same cluster avoids intrasite communications which, as shown in Table 11.1, are slower than the intracluster ones.

11.4.4 Experiment 3

Load and reliability for the third experiment are unchanged. It consists of 20 tasks, subdivided in two blocks of 10 processes. Task i $(1 \leq i \leq 10)$ executes $\gamma_i = 90$ K MIs and sends $\psi_{im} = 320$ Mbit to task $m = (i + 10)$ of the second block. Task j $(11 \leq j \leq 20)$ performs $\gamma_j = 500$ MIs and sends 24 000 Mbit to each process of the second block. The best result found is $\boldsymbol{\mu}_{\Phi_1} = \{49, 50, 51, 52, 53, 54, 55, 1, 1, 3, 155, 156, 157, 158, 159, 160, 161, 162, 163, 164\}$. The resource use time of this solution is 110.5 s. This time is used for communication of 24 000 Mbit among tasks of the second block by grid nodes with the highest bandwidths ($\psi = 2000$ Mbit) and computation of 500 MIs by the most powerful grid nodes among them. In fact, for the placement of tasks of the second block, the aforementioned solution prefers nodes of F_1 because they have the same bandwidth but a power double compared to that of the nodes of F_2. Instead, the tasks of the first block have been mapped in a way not to exceed the above optimal time.

By setting $l_i(\Delta t) = 0.9$ for nodes in the range $[33, \ldots, 164]$, the allocation is $\boldsymbol{\mu}_{\Phi_1} = \{7, 2, 11, 11, 8, 8, 3, 3, 4, 4, 166, 167, 168, 169, 170, 171, 172, 173, 174, 175\}$. It is worth observing that in this event the tasks have been allocated on the discharged nodes.

11.4.5 Experiment 4

The fourth test consists of 30 tasks subdivided into three blocks of cardinality 10. Each task i $(1 \leq i \leq 10)$ of the first block executes $\gamma_i = 90$ K MIs and sends $\psi_{im} = 320$ Mbit to task $m = (i + 20)$ of the third block. Each task j $(11 \leq j \leq 20)$ of the second block performs

$\gamma_j = 14$ K MIs and sends 2000 Mbit to each process of the same block and $\psi_{ja} = 320$ Mbit to task $a = (j + 10)$. Finally, task l $(21 \leq l \leq 30)$ of the third block carries out $\gamma_l = 500$ MIs and sends 14 000 Mbit to processes of the same block. The best mapping obtained is $\boldsymbol{\mu}_{\Phi_1} = \{8, 11, 5, 6, 21, 12, 23, 16, 10, 17, 63, 64, 65, 66, 67, 68, 69, 70, 71, 72, 175, 176, 177, 178, 179, 180, 181, 182, 183, 184\}$. As expected, the first ten tasks are allocated on the most powerful grid nodes of A, tasks of the second block are mapped onto nodes of C_1 which have a good balance between power and bandwidth, while the last ten are placed on F_2 which presents the highest values of intracluster bandwidths.

11.5 More Realistic Experiments

Having established that the system works properly in predetermined conditions, we can test it over a broad range of more realistic scenarios. This is done by carrying out more complex experiments which take account of node loads, computational powers and bandwidths, and fixing specific node and internet link reliabilities. Findings are in line with those expected.

11.5.1 Experiment 5

Here, applications are composed of tasks with $\gamma_k = 90$ K MIs $\psi_{km} = 0, \forall k, m \in [1, \ldots, P]$, and $l_i(\Delta t) = 0.9$ for all nodes except those discharged ones shown in Table 11.3.

Table 11.3 Power and indices of discharged nodes per site and cluster

Site	α	Node indices
A	3000	2,4,6,8,10,12,13–18
B_1	2000	33–38
B_2	1500	51–54
C_1	1000	61–68
C_2	700	73–80

Thus we have 12 nodes with $\alpha = 3000$, 6 with $\alpha = 2000$, 4 with $\alpha = 1500$, 8 with $\alpha = 1000$ and 8 with $\alpha = 700$ for a total of 38 nodes completely discharged. The best time values achieved for Φ_1 are shown in Table 11.4 as a function of number of tasks, together with the expected optima.

Table 11.4 Use of resources for applications in presence of computation and load

Number of tasks	Time achieved (s)	Optimal time expected (s)
10	30	30
12	45	30
16	60	45
26	90	60
35	300	90

Most of the solutions involve the most powerful nodes and correctly discard the loaded nodes. As an example, for the ten tasks in the first row of Table 11.4, the solution is $\boldsymbol{\mu}_{\Phi_1} = \{6, 4, 8, 14, 15, 2, 10, 17, 13, 12\}$. It selects 10 of the 12 discharged nodes in A.

Note that, according to the fixed loads, the lowest available power on the least powerful nodes of F_2 is equal to 10 MIPS. Consequently, the solutions are much closer to the pertaining optima reported in Table 11.4 than to the worst case of 9000 s obtained by dividing $\gamma_k = 90$ K MIs with the worst residual power of 10 MIPS. For example, the solution related to the placement of 16 tasks is $\boldsymbol{\mu}_{\Phi_1} = \{12, 15, 10, 12, 13, 36, 10, 6, 38, 8, 14, 17, 2, 4, 6, 4\}$. It is suboptimal because it does not use 16 of the 18 most powerful nodes available. Instead, it assigns to some of the unloaded nodes of A two tasks and excludes some nodes of B.

11.5.2 Experiment 6

From now on, the conditions of the reliability values are also varied. Therefore, the second objective is also of concern.

An application of 30 tasks is considered. For each task, $\gamma_k = 90$ K MIs, $\psi_{km} = 0$, $l_i(\Delta t) = 0$, $\lambda_w = 1.0$. Node π_z values are given in Table 11.5.

Table 11.5 Reliability for the nodes

Sites	A	B_1	B_2	C_1	C_2	D	E_1	E_2	F_1	F_2
Nodes	1–32	33–48	49–56	57–72	73–104	105–124	125–140	141–148	149–164	165–184
π_z	0.9	0.9	0.9	0.955	0.975	0.985	0.995	0.999	0.99	0.9

Any execution of the differential evolution mapper finds several solutions, all being non-dominated in the final Pareto front. The allocations at the extremes of the front are $\boldsymbol{\mu}_{\Phi_1} = \{19, 20, 21, 22, 23, 24, 25, 26, 27, 28, 29, 30, 31, 32, 3, 4, 5, 6, 7, 8, 9, 10, 11, 12, 13, 14, 15, 16, 17, 18\}$ with $\Phi_1 = 30.0$ s and $\Phi_2 = 0.042$, and $\boldsymbol{\mu}_{\Phi_2} = \{158, 159, 160, 125, 126, 127, 128, 129, 130, 131, 132, 133, 134, 135, 136, 137, 138, 139, 140, 128, 129, 149, 150, 151, 152, 153, 154, 155, 156, 157\}$ with $\Phi_1 = 450.0$ s and $\Phi_2 = 0.81$. $\boldsymbol{\mu}_{\Phi_2}$ utilizes nodes of A, the most powerful ones while $\boldsymbol{\mu}_{\Phi_2}$ uses the most reliable nodes contained in the sites E_1, E_2 and F_1.

Furthermore, the system provides some other non-dominated solutions better balanced in terms of the two goals, such as $\boldsymbol{\mu} = \{63, 13, 86, 28, 35, 32, 125, 89, 47, 127, 22, 16, 15, 113, 20, 9, 6, 14, 61, 30, 84, 104, 57, 114, 128, 133, 5, 12, 92, 75\}$ with $\Phi_1 = 225.0$ s and $\Phi_2 = 0.15$. It is up to the user to choose, among the proposed solutions, the one which best suits his or her needs.

11.5.3 Experiment 7

This experiment is slightly different from the previous one. It adds node loads as shown in Table 11.6.

Table 11.6 Load for the nodes

Nodes	1–16	17–32	33–48	49–56	57–72	73–104	105–124	125–140	141–148	149–164	165–184
$l_i(\Delta t)$	0.9	0.0	0.0	0.0	0.0	0.0	0.0	0.9	0.0	0.9	0.9

The mappings at the extremes of the front obtained are $\boldsymbol{\mu}_{\Phi_1} = \{24, 25, 26, 27, 28, 29, 30, 31, 32, 33, 34, 35, 36, 37, 38, 39, 40, 41, 42, 43, 44, 45, 46, 47, 18, 19, 20, 21, 22, 23\}$ with $\Phi_1 = 45.0$ s and $\Phi_2 = 0.042$, and $\boldsymbol{\mu}_{\Phi_2} = \{125, 126, 127, 128, 129, 130, 131, 132, 133, 134, 135, 136, 137, 138, 139, 124, 125, 126, 127, 128, 129, 130, 131, 132, 133, 134, 135, 136, 137, 124\}$ with $\Phi_1 = 4500.0$ s and $\Phi_2 = 0.86$. $\boldsymbol{\mu}_{\Phi_2}$ uses the most reliable nodes contained in C_2.

The load presence has resulted in an increase in the execution time. In fact, $\boldsymbol{\mu}_{\Phi_1}$ excludes the first 16 nodes of A because they are loaded and uses 15 nodes of B_1 which are now the most powerful ones. In contrast, $\boldsymbol{\mu}_{\Phi_2}$ still uses the most reliable nodes which are loaded and causes a significant increase in the resource utilization time.

11.5.4 Experiment 8

This example consists of 30 tasks subdivided into three blocks of cardinality 10. Each task i $(1 \leq i \leq 10)$ of the first block executes $\gamma_i = 90$ K MIs and sends $\psi_{im} = 320$ Mbit to the task $m = (i + 20)$ of the third block. Each task j $(11 \leq j \leq 20)$ of the second block performs $\gamma_j = 10$ K MIs and sends 1000 Mbit to each process of the same block and $\psi_{ja} = 320$ Mbit to task $a = (j + 10)$. Finally, task l $(21 \leq l \leq 30)$ of the third block carries out $\gamma_l = 500$ MIs and sends 6000 Mbit to processes of the same block. The load and the reliabilities are the same as for the previous experiment. The allocations at the extremes of the front obtained are $\boldsymbol{\mu}_{\Phi_1} = \{22, 23, 24, 25, 26, 27, 28, 29, 30, 31, 105, 106, 107, 108, 109, 110, 111, 112, 113, 114, 115, 116, 117, 118, 119, 120, 121, 122, 123, 124\}$ with $\Phi_1 = 68.5$ s and $\Phi_2 = 0.258$, and $\boldsymbol{\mu}_{\Phi_2} = \{133, 134, 135, 136, 137, 138, 139, 150, 151, 152, 153, 154, 155, 156, 157, 158, 159, 160, 161, 162, 163, 164, 125, 126, 127, 128, 129, 130, 131, 132\}$ with $\Phi_1 = 4500.0$ s and $\Phi_2 = 0.802$.

11.5.5 Experiment 9

Here, the reliability of the first 20 nodes of site A has been set to 0.999 instead of 0.9. The mappings at the extremes of the front obtained are $\boldsymbol{\mu}_{\Phi_1} = \{13, 14, 15, 16, 17, 18, 19, 20, 21, 22, 105, 106, 107, 108, 109, 110, 111, 112, 113, 114, 115, 116, 117, 118, 119, 120, 121, 122, 123, 124\}$ with $\Phi_1 = 68.5$ s and $\Phi_2 = 0.594$, and $\boldsymbol{\mu}_{\Phi_1} = \{14, 4, 136, 12, 8, 7, 13, 2, 6, 9, 12, 15, 98, 9, 19, 10, 2, 2, 2, 8, 12, 3, 8, 1, 19, 14, 4, 17, 11, 5\}$ with $\Phi_1 = 6970.0$ s and $\Phi_2 = 0.943$.

Comparing the solutions with the previous test, it can be noticed that in $\boldsymbol{\mu}_{\Phi_1}$, which still uses 10 nodes of A and 20 nodes of D, 9 of the 10 nodes of A with the highest reliability have been selected. This choice has left the execution time unchanged but has notably improved the pertaining reliability. As to the second objective, the increased number of nodes with high reliability has improved Φ_2 from 0.802 to 0.943. However, it has a higher resource utilization time.

11.6 Conclusions

This chapter deals with the mapping problem in a grid environment by means of a differential evolution algorithm aiming at minimizing the use of grid resources while fulfilling user QoS requests and environment specifications at the same time.

To provide the user with a set of possible mapping solutions, each with a different balance of use of resources and QoS values, a multi-objective differential evolution based on the Pareto optimality criterion is proposed. In particular, the reliability of links and nodes is considered.

The evolutionary approach proposed can be used in a variety of heterogeneous frameworks because it does not rely on any specific communication subsystem model. Experimental results are outlined and discussed on several allocation scenarios differing for applications, node loads, communication and computation task requirements, bandwidths and reliability. The promising results achieved show that a multi-objective differential evolution is a viable approach to face the grid multisite mapping problem.

References

[1] Buyya, R., Abramson, D., Giddy, J. and Stockinger, H. (2002) Economic models for resource management and scheduling in grid computing. *Concurrency and Computation: Practice and Experience*, **14** (13–15), 1507–1542.

[2] Anderson, D.P. (2004) BOINC: a system for public-resource computing and storage. 5th IEEE/ACM Int. Workshop Grid Computing, Pittsburgh, pp. 4–10.

[3] Nabrzyski, J., Schopf, J.M. and Weglarz, J. (eds) (2004) *Grid Resource Management: State of the Art and Future Trends*, Kluwer Academic, Norwell, MA.

[4] Knodo, D., Taufer, M., Karanicolas, J. *et al.* (2004) Characterizing and evaluating desktop grids: an empirical study. Int. Parallel Distributed Processing Symp., Santa Fe, NM, pp. 26–35.

[5] Choi, S.J., Baik, M.S., Gil, J.M. *et al.* (2006) Adaptive group scheduling mechanism using mobile agents in peer-to-peer grid computing environment. *Applied Intelligence*, **25**(2), 199–221.

[6] Foster, I. and Kesselmann, C. (eds) (2003) *The Grid 2: Blueprint for A New Computing Architecture*, Morgan Kaufmann, San Francisco.

[7] Mateescu, G. (2003) Quality of service on the grid via metascheduling with resource co-scheduling an co-reservation. *International Journal of High Performance Computing Applications*, **17**(3), 209–218.

[8] Laforenza, D. (2002) Grid programming: some indications where we are headed. *Parallel Computing*, **28**(12), 1733–1752.

[9] Doğan, A. and Özgüner, F. (2006) Scheduling of a meta-task with QoS requirements in heterogeneous computing systems. *Journal of Parallel and Distributed Computing*, **66**(2), 181–196.

[10] Foster, I. (2005) Globus Toolkit version 4: software for service-oriented systems, in *Network and Parallel Computing* (eds. H. Jin, D. Reed and W. Jiang), Lecture Notes in Computer Science, **3779**, Springer, Berlin, pp. 2–13.

[11] Singh, H. and Youssel, A. (1996) Mapping and scheduling heterogeneous task graphs using genetic algorithms. Heterogeneous Computing Workshop (HCW96), Honolulu, HI, pp. 86–97.

[12] Shroff, P., Waston, D.W., Flann, N.S. and Freund, R.F. (1996) Genetic simulated annealing for scheduling data-dependent tasks in heterogeneous environments. Heterogeneous Computing Workshop, Honolulu, HI, pp. 98–104.

[13] Wang, L., Siegel, J.S., Roychowdhury, V.P. and Maciejewski, A.A. (1997) Task matching and scheduling in heterogeneous computing environments using a genetic-algorithm-based approach. *Journal of Parallel and Distributed Computing*, **47**(1), 8–22.

[14] Blythe, J., Jain, S., Deelman, E. *et al.* (2005) Task scheduling strategies for workflow-based applications in grids. IEEE Int Symp Cluster Computing Grid, Cardiff, Wales, pp. 759–767.

[15] Foster, I. and Kesselman, C. (eds) (1998) *The Grid: Blueprint for a Future Computing Infrastructure*, Morgan Kaufmann, San Francisco.

[16] Eshaghian, M.M. and Shaaban, M.E. (1994) Cluster-M programming paradigm. *International Journal of High Speed Computing*, **6**(2), 287–309.

[17] Braun, T.D., Siegel, H.J., Beck, N. *et al.* (2001) A comparison of eleven static heuristics for mapping a class of independent tasks onto heterogeneous distributed computing systems. *Journal of Parallel and Distributed Computing*, **61**(6), 810–837.

[18] Ibarra, O.H. and Kim, C.E. (1977) Heuristic algorithms for scheduling independent tasks on non-identical processors. *Journal of Association Computing Machinery*, **24**(2), 280–289.

[19] Fernandez-Baca, D. (1989) Allocating modules to processors in a distributed system. *IEEE Transactions on Software Engineering*, **15**(11), 1427–1436.

[20] Kwok, Y.K. and Ahmad, I. (1997) Efficient scheduling of arbitrary task graphs to multiprocessors using a parallel genetic algorithm. *Journal of Parallel and Distributed Computing*, **47**(1), 58–77.

[21] Abraham, A., Buyya, R. and Nath, B. (2000) Nature's heuristics for scheduling jobs on computational grids. 8th Int. Conf. Advanced Computing Communication, Cochin, India, pp. 45–52.

[22] Kim, S. and Weissman, J.B. (2004) A genetic algorithm based approach for scheduling decomposable data grid applications. Int. Conf. Parallel Processing, Montreal, Quebec, Canada, pp. 406–413.

[23] Wu, A.S., Yu, H., Jin, S. *et al.* (2004) An incremental genetic algorithm approach to multiprocessor scheduling. *IEEE Transactions on Parallel Distributed Systems*, **15**(9), 824–834.

[24] Song, S., Kwok, Y.K. and Hwang, K. (2005) Security-driven heuristics and a fast genetic algorithm for trusted grid job scheduling. 19th IEEE International Parallel Distributed Processing Symp, Denver, CO, pp. 65–74.

[25] Aggarwal, M., Kent, R.D. and Ngom, A. (2005) Genetic algorithm based scheduler for computational grids. 19th Int. Symp. High Performance Computing Systems Applications, Guelph, Ontario, Canada, pp. 209–215.

[26] Fonseca, C.M. and Fleming, P.J. (1995) An overview of evolutionary algorithms in multiobjective optimization. *Evolutionary Computation*, **3**(1), 1–16.

[27] Fitzgerald, S., Foster, I., Kesselman, C. *et al.* (1997) A directory service for configuring high-performance distributed computations. 6th Symp. High Performance Distributed Computing, Portland, OR, pp. 365–375.

[28] Czajkowski, K., Fitzgerald, S., Foster, I. and Kesselman, C. (2001) Grid information services for distributed resource sharing. 10th Symp. High Performance Distributed Computing, San Francisco, pp. 181–194.

[29] Wolski, R., Spring, N. and Hayes, J. (1999) The network weather service: a distributed resource performance forecasting service for metacomputing. *Future Generation Computer Systems*, **15**(5–6), 757–768.

[30] Gong, L., Sun, X.H. and Watson, E. (2002) Performance modeling and prediction of non-dedicated network computing. *IEEE Transactions on Computers*, **51**(9), 1041–1055.

12

Synthesis of Time-Modulated Antenna Arrays

Shiwen Yang, Yikai Chen

Department of Microwave Engineering, School of Electronic Engineering, University of Electronic Science and Technology of China, Chengdu, PR China

12.1 Introduction

An array is a sampled aperture excited only at points where radiators or receivers called elements are located. Antenna array synthesis seeks a set of optimal complex weights and/or element locations to generate a required beam pattern. It can be cast into an optimization problem, where complex excitations and layout of elements are the optimization parameters, and the goals may include maximizing the directivity, minimizing the *sidelobe levels* (SLLs), placing null constraints, and so on.

Far-field radiation patterns with low/ultra-low SLLs can be realized by tapering the excitation amplitude distributions in antenna arrays. However, in conventional antenna arrays, *dynamic range ratios* (DRR) of the excitations are rather high for low/ultra-low sidelobe antenna arrays, which result in stringent error tolerance requirements in practical design.

Time-modulated antenna arrays were proposed as an alternative and promising means of realizing low/ultra-low SLL patterns [1–17], in which only very low or even uniform amplitude excitation DRRs are necessary. As compared to the conventional antenna arrays, time-modulated arrays consist of simple on–off switching of antenna elements in predetermined sequence, such that, the SLL can be reduced once the output of the array has been filtered. Since the additional design freedom 'time' can be easily, rapidly, and accurately adjusted for different time sequences, stringent requirements on various error tolerances can be relaxed. Thus hardware implementation is much easier.

Synthesis of time-modulated antenna arrays is non-convex, and local optimization techniques are easily trapped in local optima. Hence, population-based stochastic optimization methods are necessary. The authors have been active in applying differential evolution to

Differential Evolution Anyong Qing

synthesis of time-modulated antenna arrays [3,4,7,10–13,16,17]. Reported applications include suppressing sideband radiations as well as controlling SLLs, shaped power pattern synthesis, and mutual power compensation. In this chapter, the latest advances in synthesis of time-modulated antenna arrays using differential evolution are reported.

12.2 Antenna Arrays

12.2.1 Principle of Pattern Multiplication

The far-field radiation pattern for an N-element linear array with isotropic elements as shown in Figure 12.1 is given by

$$AF(\theta) = \sum_{i=1}^{N} w_i e^{j2\pi x_i \sin\theta/\lambda}, \tag{12.1}$$

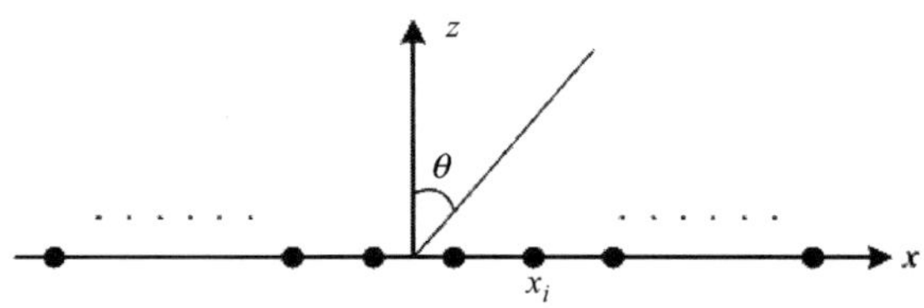

Figure 12.1 Configuration of a linear array

where $w_i = A_i \exp(j\varphi_i)$ is the complex weight for the ith element located at x_i, λ is the wavelength at the operating frequency, and θ is the angle measured from the broadside of the linear array.

In particular, the progressive phase caused by the space distribution of each element is compensated by the complex weights for a specific direction θ_m, that is, $\varphi_i = -2\pi x_i \sin(\theta_m)/\lambda$. Thus, the exponential terms in (12.1) are added in phase, and the main beam is steered in the direction of θ_m. From this point of view, if the progressive phase in each element is compensated for various θ_m, by electronically controlling the phase of the complex weights φ_i, the beam is scanned.

Proceed now to a more general situation, where the linear array with isotropic elements can be readily extended to a linear array with identical non-isotropic elements. By taking into account the effect of the non-isotropic antenna element, the far-field radiation pattern $FF(\theta, \varphi)$ of a linear array can be calculated from

$$FF(\theta, \varphi) = EP(\theta, \varphi) \times AF(\theta), \tag{12.2}$$

where $EP(\theta, \varphi)$ is the radiation pattern of the single non-isotropic antenna element.

The far-field radiation pattern expressed in (12.2) illustrates the principle of pattern multiplication stated as follows: The field pattern of an array with identical non-isotropic antenna elements is the product of the pattern of the individual non-isotropic antenna element and the pattern of an array of isotropic elements having the same locations and complex weights.

12.2.2 Planar Antenna Arrays

For a rectangular planar array consisting of $M \times N$ isotropic elements in the (x, y) plane, the far-field pattern in the direction of (θ, φ) calculated by a standard *Woodward–Lawson* method [18,19] is given by

$$AF(u, v) = \sum_{p=-P}^{P} \sum_{q=-Q}^{Q} a_{pq} \Phi_u(p) \Phi_v(q), \tag{12.3}$$

where a_{pq} is the sampling field of a desired array factor $f_d(u, v)$ at the Woodward–Lawson sampling points $u_p = p\lambda/Md_x$ and $v_q = q\lambda/Nd_y$, and $\Phi_u(p)$ and $\Phi_v(q)$ are given by

$$\Phi_u(p) = \sin[M(u - u_p)\beta d_x/2]/\{M\sin[(u - u_p)\beta d_x/2]\}, \tag{12.4}$$

$$\Phi_v(q) = \sin[N(v - v_q)\beta d_y/2]/N\sin[(v - v_q)\beta d_y/2], \tag{12.5}$$

where β is the free space propagation constant.

Equation 12.3 indicates that the far-field representation of the synthesized pattern is a summation of the weighted $\Phi_u(p)\Phi_v(q)$ component, which is a sampling function of the desired pattern. Notice that the sampling function is an array factor of a uniform planar array with an SLL of -13.5 dB, which goes against the aim of minimizing the SLL. Apparently, the sampling function for $f_d(u, v)$ should not be limited to the $\Phi_u(p)\Phi_v(q)$ function, as suggested by Sikina [20]. In fact, any type of aperture distribution that can produce low-SLL patterns can be selected as sampling function.

A Taylor aperture distribution is employed here to serve as the local basis function for the representation of the desired far field. Consequently, the SLL of the synthesized footprint pattern can be well controlled by only adjusting the design parameters in the Taylor synthesis method.

The new sampling function with Taylor distribution can be represented by

$$\Phi_{u\_T}(p) = \sum_{m=1}^{M} i_m e^{j\beta d_x (m-1)(u-u_p)}, \tag{12.6}$$

$$\Phi_{v\_T}(q) = \sum_{n=1}^{N} i_n e^{j\beta d_y (n-1)(v-v_q)}, \tag{12.7}$$

where i_m and i_n are referred to as the excitations of a linear antenna array with Taylor distribution. By substituting (12.6) and (12.7) into (12.3), the element excitation I_{mn} can be given by

$$I_{mn} = \sum_{p=-P}^{P} \sum_{q=-Q}^{Q} a_{pq} i_m i_n e^{-j\beta[(m-1)d_x u_p + (n-1)d_y v_q]}. \tag{12.8}$$

12.2.3 Time-Modulated Linear Antenna Arrays

For an N-element linear array of isotropic elements with a half-wavelength uniform spacing, each element is controlled by a high speed RF switch. An associated switch on–off time function $U_k(t)$ ($k = 1, 2, \ldots, N$) is defined to present the on and off time in a modulation period T_p. The far-field pattern of the time-modulated linear array with static excitation $w_k = A_k \exp(j\varphi_k)$, $k = 1, 2, \ldots, N$, is given by

$$E(\theta, t) = e^{j2\pi f_0 t} \sum_{n=1}^{N} w_n U_n(t) e^{j(n-1)\pi \sin\theta}, \tag{12.9}$$

where f_0 is the center frequency.

It is assumed that the switch on–off time function $U_k(t)$ for the kth element is periodic with modulation period T_p, and is switched on for τ_k ($0 \le \tau_k \le T_p$) in each period T_p. If the time-modulated array works in a continuous state, then the transmitted pulse width T is also T_p, and the pulse repetition frequency is $f_p = 1/T_p$.

By decomposing (12.9) into Fourier series with different frequency components separated by f_p, the mth-order ($m = 0, \pm 1, \ldots, \pm\infty$) Fourier component of (12.9) is given by

$$E_m(\theta, t) = e^{j2\pi f_0 t} \sum_{n=1}^{N} a_{mn} e^{j(n-1)\pi \sin\theta} e^{j2\pi m f_p t}, \tag{12.10}$$

where a_{mk} is excitation for the mth harmonic and is given by

$$a_{mn} = w_n \tau_n \sin(\pi m f_p \tau_n) e^{-j\pi m f_p \tau_n} / (T_p \pi m f_p \tau_n). \tag{12.11}$$

As can be seen from (12.11), the new design freedom τ_k introduced in the time-modulated antenna array can be used to alleviate the problem of high DRR in low/ultra-low sidelobe array.

At center frequency f_0 ($m = 0$), a_{0k} can be used to synthesize low/ultra-low sidelobe patterns, by controlling both static excitation and switch-on time intervals τ_k. The sideband radiations ($m \neq 0$) can either be suppressed to enhance the radiation efficiency or utilized for special applications.

12.2.4 Time-Modulated Planar Array

The far field of a $M \times N$ time-modulated planar array with elements located at the intersections of a rectangular grid with spacing of d_x and d_y in the (x, y) plane can be described as follows:

$$AF(u, v, t) = e^{j2\pi f_0 t} \sum_{m=1}^{M} \sum_{n=1}^{N} A_{mn} U_{mn}(t) e^{j\beta[(m-1)d_x u + (n-1)d_y v]}, \tag{12.12}$$

where A_{mn} is the static complex excitation, $U_{mn}(t)$ is the switch on–off time function, and each element is switched on for τ_{mn} ($0 \le \tau_{mn} \le T_p$) in one modulation period T_p by a high speed RF switch. Here, the modulation period T_p is chosen as the transmitted pulse width T, $u = \sin\theta\cos\varphi$, $v = \sin\theta\cos\varphi$.

By decomposing (12.12) into Fourier series with different frequency components, the kth-order ($k = 0, \pm1, \ldots, \pm\infty$) Fourier component of (12.12) is given by

$$E_k(u, v) = e^{j2\pi f_0 t} \sum_{m=1}^{M} \sum_{n=1}^{N} a_{mn}^k e^{j\beta[(m-1)d_x u + (n-1)d_y v]} e^{j2\pi k f_p t}, \tag{12.13}$$

where a_{mn}^k is the excitation for the kth harmonic and is given by

$$a_{mn}^k = A_{mn}\tau_{mn} \sin(\pi k f_p \tau_{mn}) e^{-j\pi k f_p \tau_{mn}} / (T_p \pi k f_p \tau_{mn}). \tag{12.14}$$

From (12.14) we know that a time-modulated antenna array with very low DRR in static excitation implies that the high-speed RF switches should have the ability to control the on–off state with high precision, which offers a new challenge in time-modulated arrays.

If I_{mn} are regarded as the excitations at f_0, we have

$$a_{mn}^0 = A_{mn}\tau_{mn}/T_p = |A_{mn}|\tau_{mn} e^{j\arg A_{mn}}/T_p = |I_{mn}| e^{j\arg I_{mn}}. \tag{12.15}$$

12.3 Synthesis of Multiple Patterns from Time-Modulated Arrays

12.3.1 Motivations

Multiple patterns generated from only one antenna array are often required in many terrestrial or vehicular applications, due to their limited space for antenna installation on vehicles. In order to simplify the feed network or maximize the total radiated power of the solid-state drivers of an antenna array, multiple patterns are usually switched from one to another by only turning the phase excitations [21], and a pre-assigned common amplitude distribution is usually preferred for multiple pattern synthesis while switching among different phase distributions for different radiation patterns [22,23]. However, multiple patterns synthesized by conventional antenna arrays usually require a pre-assigned amplitude excitation with relatively higher DRRs. Although uniform amplitude excitation can be used in the multiple pattern synthesis, the SLLs of the synthesized multiple patterns are usually high, making them undesirable for some applications. Moreover, the number of multiple patterns synthesized by such a phase-only synthesis approach is rather limited. Thus it is expected that by exploiting the advantages of the time-modulation technique and the powerful search ability of differential evolution, optimized common amplitude exactions and different phase and time sequences can be obtained for ease of multiple pattern implementations.

12.3.2 Synthesis Examples

A time-modulated linear array of 20 isotropic elements with $\lambda/2$ spacing is considered for the synthesis of multiple patterns, including a sum pattern, a flat-topped pattern, a cosec-squared pattern and a difference pattern. The excitation amplitude distribution of the time-modulated linear array is pre-assigned as a -30 dB SLL discrete Taylor ($\bar{n} = 5$) pattern [24], where the DRR is about 3.95 and is lower than that of the Gaussian distribution in Ref. [25]. The multiple patterns optimized using differential evolution at center frequency are presented in Figure 12.2.

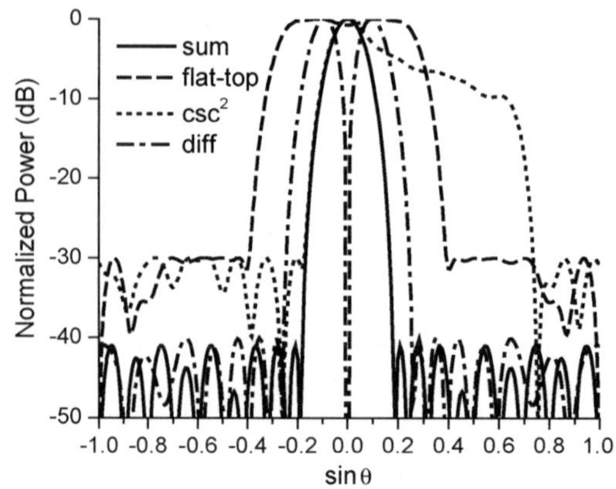

Figure 12.2 Multiple patterns at f_0, synthesized from a 20-element time-modulated linear array with discrete Taylor $\bar{n}$ amplitude distribution

It is observed that the SLLs are suppressed to -30 dB for both of the flat-topped pattern and the cosec-squared pattern, and even -40 dB for both of the sum pattern and the difference pattern, an improvement of 10 dB and 20 dB, respectively, as compared to those of the results in Ref. [25]. Again, the non-uniform amplitude linear array in Ref. [25] cannot be used to synthesize a difference pattern, since its amplitude distribution was pre-assigned as a Gaussian distribution. Other beam characteristics such as the beam widths, transition widths and the ripple levels are similar to those of [25].

Figure 12.3 shows the corresponding first sideband patterns normalized to the maxima of their respective central frequency patterns in Figure 12.2. The maximum sideband levels for all the four patterns are suppressed to less than -15 dB and are relatively small.

Figure 12.4(a) plots the pre-assigned common excitation amplitude distribution, and Figures 12.4(b) and (c) show the corresponding differential evolution optimized phase excitations and switch-on time intervals for all four patterns, respectively.

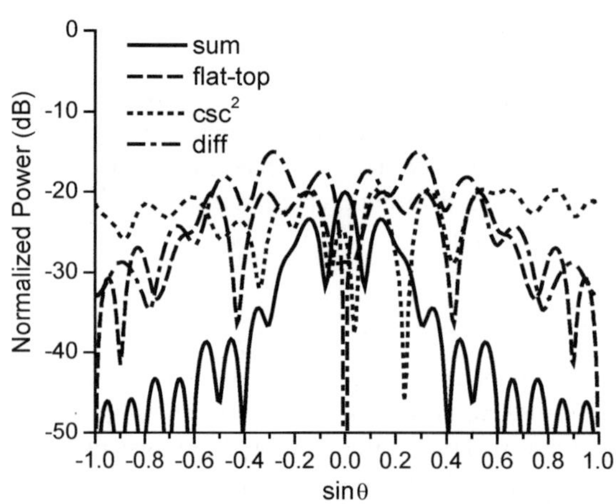

Figure 12.3 First sideband patterns at f_0 + prf of Figure 12.2

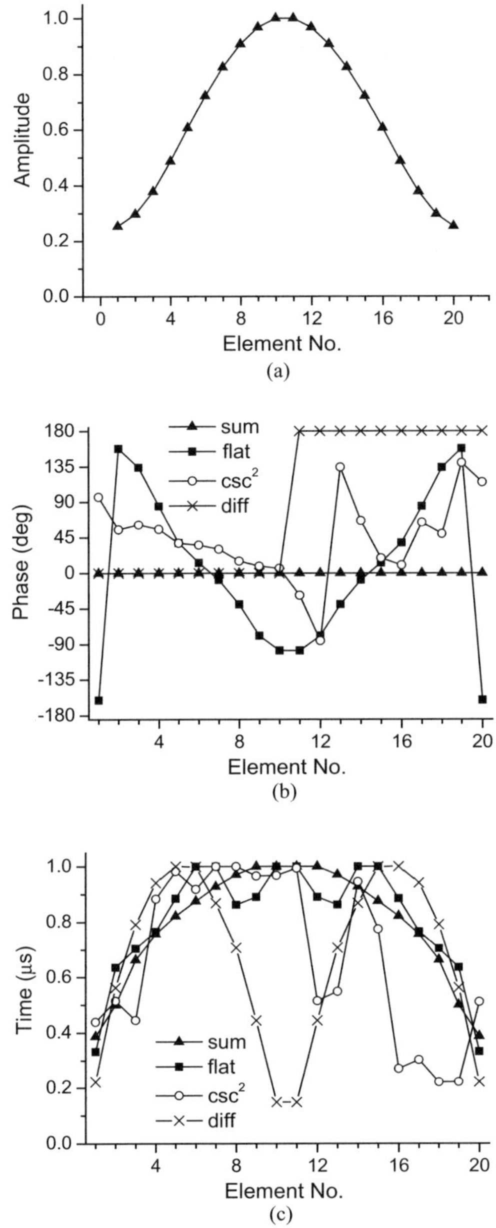

Figure 12.4 Optimized excitations of Figure 12.2: (a) common amplitude; (b) phases; (c) switch-on time intervals

The calculated directivities of the time-modulated linear array are about 11.49 dBi for the sum pattern, 5.45 dBi for the flat-top pattern, 8.17 dBi for the cosec-squared pattern, and 8.70 dBi for the difference pattern, respectively. As compared to the directivities of the conventional linear array with Gaussian amplitude distribution, the time-modulated linear array has a reduction in directivity of only about 0.44 dBi for the sum pattern, 0.13 dBi for the flat-top pattern and 1.03 dBi for the cosec-squared pattern, respectively.

12.4 Pattern Synthesis of Time-Modulated Planar Arrays

12.4.1 Introduction

The time-modulation technique has brought great convenience to the synthesis of linear antenna arrays. We believe that it also brings similar convenience – such as uniform or low DRRs of amplitude distribution – to planar array designs.

12.4.2 Monopulse Antennas

The most time-consuming part of the iterative procedure of differential evolution is the conventional element-by-element field calculation of the synthesized pattern planar array synthesis problems; thus a fast computation method is employed to speed up the evaluations of array patterns. The fast computation method is based on the two-dimensional fast Fourier transform technique [17,26,27], and the fitness value of each group of design parameters can be derived from the calculated array patterns.

Since the number of elements within a given array aperture can be reduced in planar arrays with triangular grid and hexagonal shape [17,28–30], planar arrays with triangular grid and hexagonal shape are usually preferred as compared to those with rectangular grid. A standard hexagonal planar array with isosceles triangular grids is shown in Figure 12.5. It is illustrated in Figure 12.5 that the hexagonal planar array consists of Q concentric hexagonal-ring arrays, and an additional single element is located at the center.

The sum pattern, difference pattern, and double-difference pattern at f_0 are plotted in Figure 12.6. Note that low SLLs of −25.2 dB, −23.7 dB and −22.1 dB have been obtained for

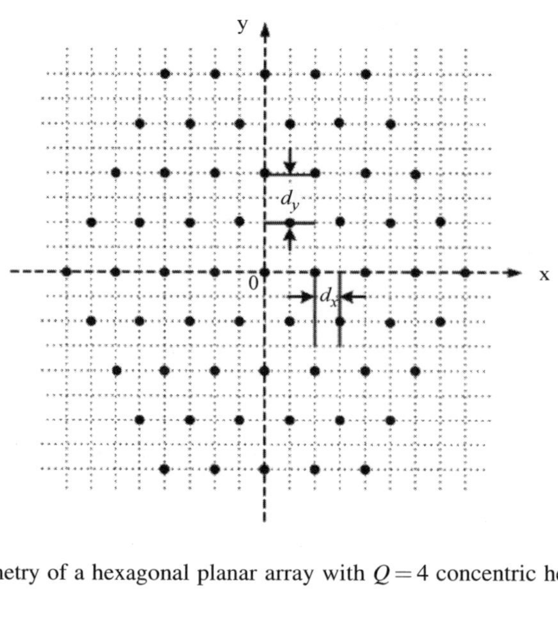

Figure 12.5 Geometry of a hexagonal planar array with $Q = 4$ concentric hexagonal-ring arrays

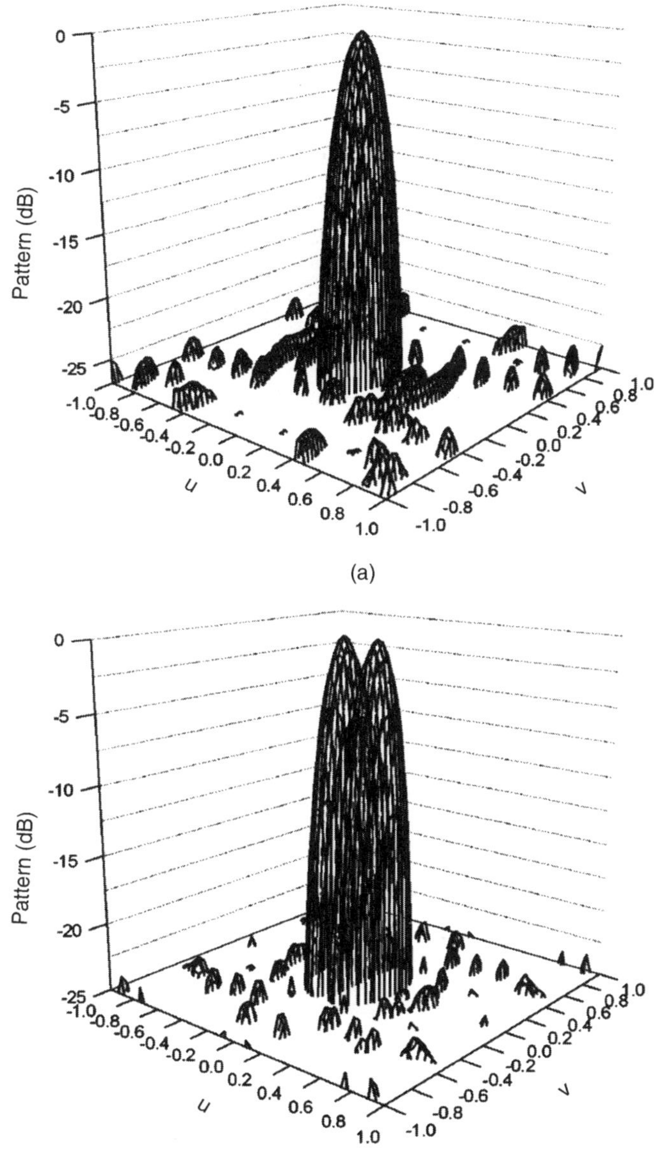

Figure 12.6 Radiation patterns for the time-modulated hexagonal planar array at the center frequency f_0: (a) sum pattern; (b) difference pattern; (c) double-difference pattern

the sum, difference, and difference pattern, respectively. In addition, the maximum sideband levels have also been suppressed to sufficiently low levels of −25.5 dB, −25.4 dB, and −25.2 for the sum, difference, and difference pattern, respectively.

The directivity reduction is investigated for the cases of sum, difference, and double-difference patterns in this $Q = 8$ hexagonal planar array. The directivity for the time-modulated

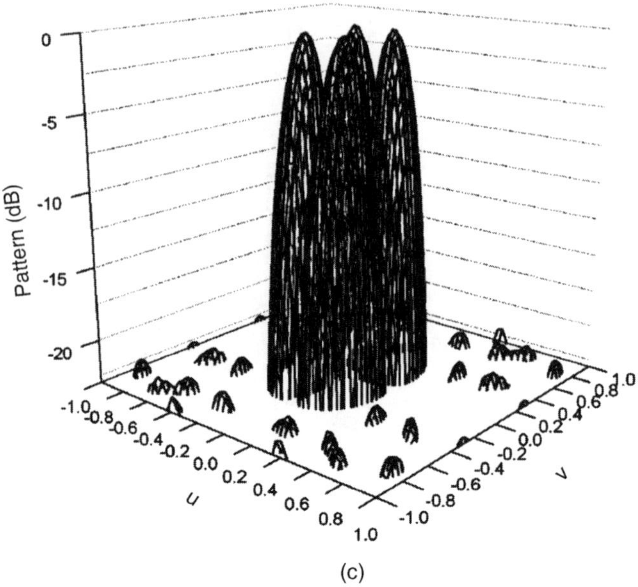

Figure 12.6 (*Contined*)

and reference static array is presented in Table 12.1. It can be seen that the directivity reduction is 0.37 dB, 0.43 dB, and 0.74 dB for the cases of sum, difference, and double-difference patterns, respectively. Thus the differential evolution method is again demonstrated to be an effective tool for the suppression of sideband radiations.

Table 12.1 Directivity comparison between the time-modulated array and the corresponding static array

Directivity (dB)	Reference static array	Time-modulated array
Sum	22.11	21.75
Difference	19.00	18.56
Double- difference	16.26	15.52

12.5 Adaptive Nulling with Time-Modulated Antenna Arrays

12.5.1 Introduction

Adaptive arrays used in polluted electromagnetic environmenta are aimed at separating a desired signal from interfering signals. Common adaptive algorithms, such as the Applebaum adaptive algorithm, least mean-square and recursive least-square algorithm have been proposed to improve the signal-to-interference-plus-noise ratio in the presence of interference sources and background noise. However, these methods require analog amplitude and phase weights on each element; although variable amplitude weights are very desirable from a theoretical point of view, the implementation of amplitude control is extremely costly and is rarely used in practical systems, thus the inherent drawbacks in such methods prohibit their widespread use.

Another class of phase-only adaptive nulling algorithm that adjusts the least significant bits of digital phase shifters to reduce the total output power from array has been studied extensively in recent years [31–36]. We observe from these studies that adaptive nulling with variable amplitude and phase is the most efficient method, and the phase-only adaptive nulling methods are capable of steering the array nulls precisely in the direction of interferences in most cases. However, previous studies on phase-only adaptive nulling methods have also indicated that amplitude control in adaptive nulling is not normally considered as a viable option both for their high cost and difficulties in hardware implementation, thus phase-only adaptive nulling method deserves much interest. Moreover, it also implied that if the difficulties involved in amplitude control were eliminated, then amplitude–phase adaptive nulling methods would be preferred. On the other hand, the time-modulated antenna arrays, which have regained their popularity for array designs with critical amplitude requirements, are generally considered as an alternative to overcome the difficult amplitude control problems in conventional antenna arrays [6,8,9,13,16,17,37]. Therefore, time-modulated antenna arrays are applied to adaptive nulling problems and are expected to avoid the difficulties exists in conventional adaptive nulling antenna arrays.

12.5.2 Formulation

A *hybrid differential evolution* (HDE) is applied to adjust some least significant bits of the digital phase shifters and small values of perturbations $\Delta_{\tau n}$ imposed on the on–off time weights τ_n for the minimization of the total output power and perturbations on the main beam. The search space for the digital phase shifter settings is discrete, and that for the perturbations $\Delta_{\tau n}$ is continuous, thus HDE is employed to deal with optimization problems with a hybrid of both continuous and discrete optimization parameters [37,38].

Assuming the P least significant bits are taken as the adjustable nulling phase in a digital phase shifter with B bits ($P < B$), then the nulling phase is represented by

$$\delta_n = 2\pi \sum_{i=-B}^{P-(B+1)} b_i 2^i, \tag{12.16}$$

where the b_i are the binary bits defined for the representation of δ_n. From (12.16) we know that the nulling phases can only take digitalized values for the digital phase shifters, and these digitalized values are varied with a minimum value of $2\pi/2^B$. The perturbations $\Delta_{\tau n}$ imposed on the on–off time weights τ_n are assigned with a small perturbation range of $[-\Delta_\tau, \Delta_\tau]$. For convenience, a vector $\mathbf{v} = [\Delta_{\tau n}, \delta_n]$ is defined to denote all the optimization parameters involved in the adaptive nulling method.

In conventional antenna arrays, the total output power term is taken as the cost function immediately; however, in the time-modulated antenna arrays considered here, nulls are only required to be placed in the array pattern at the center frequency f_0 ($m = 0$), thus only the output power at the center frequency serves as the term in the cost function for the purpose of array pattern nulling. On the other hand, considering the inherent drawback that there are sideband radiations ($m \neq 0$) in time-modulated arrays [3], part of the radiated power is shifted to the sidebands, and in order to enhance the radiation efficiency, the sideband radiations should be suppressed by adding the corresponding term into the cost function, as in array synthesis problems [3,8,9]. Therefore, in order to place nulls adaptively and suppress the sideband radiations in the adaptive nulling approach, the cost function

$$
\begin{aligned}
f^n(\mathbf{v}) = {} & 20w_1\log_{10}\left[\sum_{i=1}^{N_I+1} s_i \sin\phi_i \left|\sum_{n=1}^{2N} w_n^{(0)} e^{j(n-N-0.5)\beta d\cos\phi_i}\right|\right] \\
& + w_2 \frac{SBL_{\max}(\mathbf{v})}{SBL_0} U(SBL_{\max}(\mathbf{v}) - SBL_0)\big|_{m\neq 0}
\end{aligned} \qquad (12.17)
$$

is constructed, where $w_n^{(0)} = e^{j(\delta_n+\Delta_n)}(\tau_n + \Delta_{\tau n})/T_p$ is the adjusted weight at the nth element, and the first term in (12.17) represents the total noncoherent output power at the center frequency; w_1 and w_2 are the weighting factors of each term to emphasize the different contributions to the cost function; s_i represents the signal strength of source i; ϕ_i is the incident angle of source i; $N_I + 1$ is the number of signal sources, including one desired signal and N_I interfering sources; $2N$ is total number of elements in the time-modulated linear antenna array; U is a Heaviside step function, and the second term in (12.17) represents the relative error of the sideband level $SBL_{\max}(\mathbf{v})$ with respect to the desired level SBL_0.

12.5.3 Synthesis Examples

Figure 12.7 shows the diagram of a time-modulated adaptive array controlled by HDE. The HDE starts by initializing the population randomly with a uniform probability distribution.

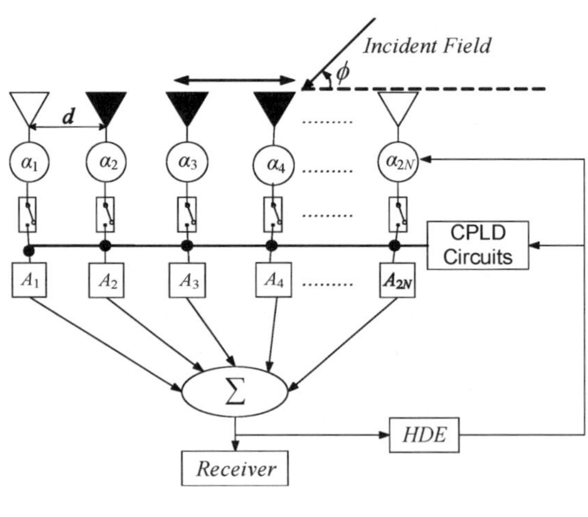

Figure 12.7 Diagram of an adaptive time-modulated antenna array controlled by HDE

New nulling chromosomes are generated from old ones, and chromosomes with lower cost function values are retained while those large values are discarded, thus the output power at the center frequency and the peak levels of the radiation pattern at the first sideband $f_0 + f_p$ (which are the maximum sideband levels [17]) should be measured for each chromosome. However, both classic differential evolution and modified differential evolution [37] have a large population size (usually 3 to 5 times the number of optimization parameters) to ensure that the optimum can be found successfully; the large population size will slow the algorithm immediately and make the algorithm unsuitable for real-time applications such as adaptive nulling. On the other hand, a crucial requirement for adaptive nulling is that the algorithm should be fast, but a global minimum is not necessary since a -50 dB null can reject an interference as well as a -90 dB null does. From this view of point, a relatively small population is used in the HDE to accelerate the nulling speed, but the small population selected should also be large enough to keep the search out of local minima.

In order to show the capabilities of the adaptive nulling method, a 40-element time-modulated liner array with -30 dB Chebyshev pattern at the center frequency is modeled. The equivalent amplitude taper at the center frequency is formed from a tapered on–off time sequence, and uniform static amplitude excitations are employed to avoid the difficult static amplitude tapering. Elements are spaced $\lambda_0/2$ apart at the center frequency f_0, phase shifters have six-bit accuracy and only the two least significant bits are taken as the adaptive nulling phase; the search range of the perturbations to be imposed on the on–off time weights τ_n is set to be $[-0.23T_p, 0.23T_p]$. Two interfering sources are assumed to appear at $u = 0.62$ and 0.72, each 60 dB stronger than the desired signal power, where $u = \cos\phi$ is the cosine space coordinate. This adaptive nulling problem has been studied in conventional antenna arrays in [34], however, we will discuss it again with regard to time-modulated antenna arrays. The quiescent and adapted array patterns at f_0 and the first two sidebands are plotted in Figures 12.8 and 12.9, respectively.

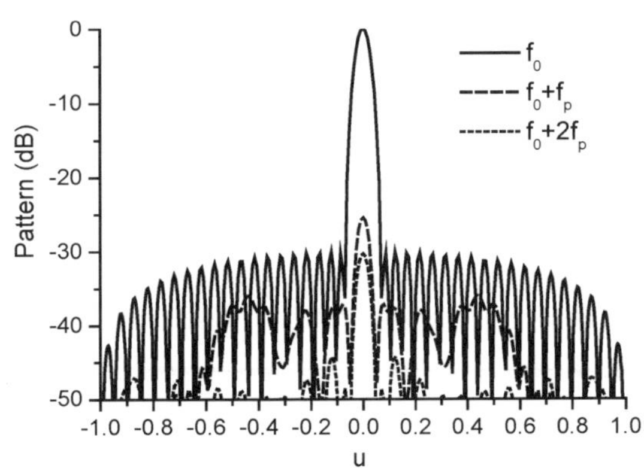

Figure 12.8 The quiescent time-modulated antenna array patterns, with a -30 dB Chebyshev pattern at the center frequency

As can be seen from Figure 12.9, nulls deeper than -50 dB are created in the pattern at the center frequency, and the distortion brought to the main beam is rather small, with an increased SLL of -27 dB, this improved SLL (over that obtained in Ref. [34]) benefiting from the additional freedom of degree in time-modulated arrays. The maximum sideband level was suppressed to -16.7 dB.

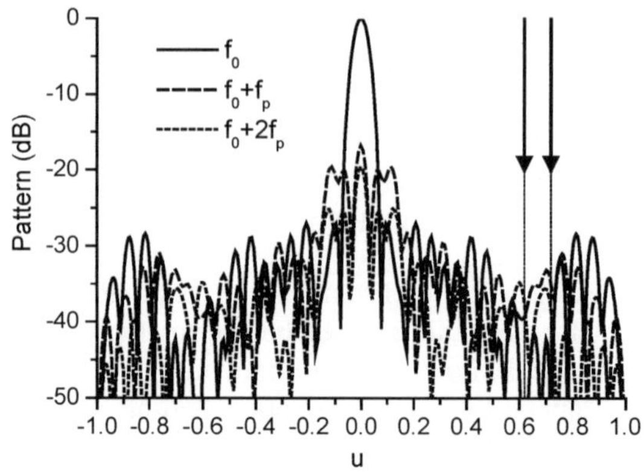

Figure 12.9 The adapted time-modulated antenna array patterns, with suppressed sideband radiations

Convergence of the algorithm is shown in Figure 12.10; it can be observed that about 250 power measurements are required to form the deep null in the two interference directions. Since the number of optimization parameters has been doubled as compared to those phase-only nulling methods, and another term has been included in the cost function for the sideband level suppression, the convergence speed of the optimization algorithm is lowered immediately, leading to a larger number of power measurements than those in phase-only methods [34].

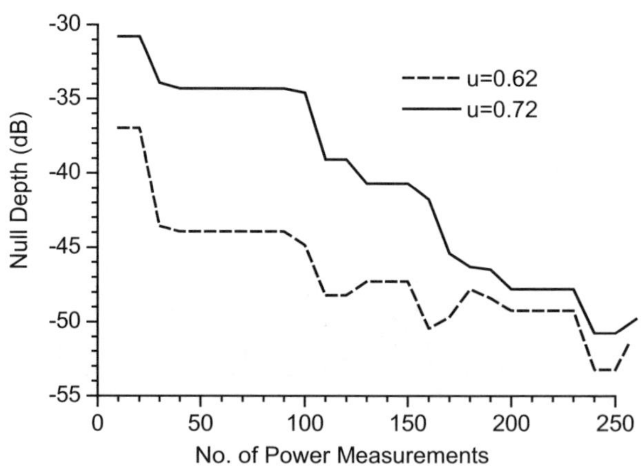

Figure 12.10 Null depth as a function of the number of power measurements

Finally, a proof based on fair comparison is given to to illustrate the adaptive nulling speed. The same time-modulated antenna array has been considered, whereas the term for the sideband level suppression is not included in the cost function. The resulting adapted pattern and convergence rate are shown in Figures 12.11 and 12.12, respectively. It is noted that the maximum sideband level goes up to about -12 dB, due to the exclusion of the sideband level

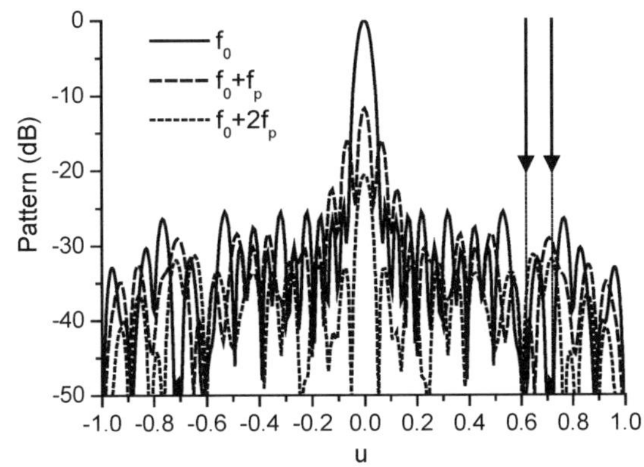

Figure 12.11 The adapted time-modulated antenna array patterns (sideband radiations have not been suppressed)

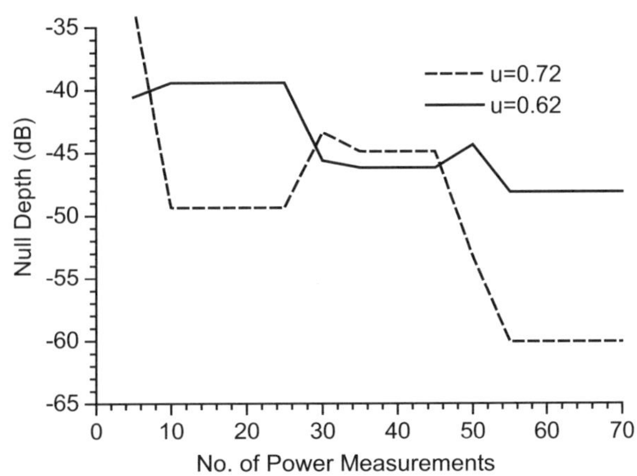

Figure 12.12 Null depth as a function of the number of power measurements

suppression term in the cost function. It can be also observed that only 70 power measurements are needed for null forming, which illustrates that the efficiency of the adaptive nulling method proposed herein is the same as the phase-only adaptive nulling method based on a genetic algorithm [34], although the number of optimization parameters is doubled due to the new degree of freedom introduced. Moreover, thanks to the new degree of freedom in time-modulated antenna arrays, the resulting SLL of the pattern at f_0 (about -25 dB) is lower than that obtained in [34]. To this end, we can observe that the pattern distortion can be reduced through the introduction of the time-modulation technique, and the convergence rate of adaptive nulling method can be improved greatly if the sideband radiations are not considered. Even so, it is strongly recommended that the sideband radiations should be suppressed if no further utilization has been taken from them, partly for the reason that the convergence rate in the sideband suppression case is still in the acceptable range.

References

[1] Kummer, W., Villeneuve, A., Fong, T. and Terrio, F. (1963) Ultra-low sidelobes from time-modulated arrays. *IEEE Transactions on Antennas and Propagation*, **11**(15), 633–639.

[2] Lewis, B. and Evins, J. (1983) A new technique for reducing radar response to signals entering antenna sidelobes. *IEEE Transactions on Antennas and Propagation*, **31**(6), 993–996.

[3] Yang, S., Gan, Y. and Qing, A. (2002) Sideband suppression in time-modulated linear arrays by the differential evolution algorithm. *IEEE Antennas and Wireless Propagation Letters*, **1**, 173–175.

[4] Yang, S., Gan, Y. and Qing, A. (Jun. 22–27, 2003) Low sidelobe phased array antennas with time modulation. IEEE AP-S Int. Symp., Columbus, OH, USA, vol. 4, pp. 200–203.

[5] Yang, S., Gan, Y. and Qing, A. (2003) Moving phase center antenna arrays with optimized static excitations. *Microwave Optical Technology Letters*, **38**(1), 83–85.

[6] Yang, S., Gan, Y. and Tan, P. (2003) A new technique for power-pattern synthesis in time-modulated linear arrays. *IEEE Antennas and Wireless Propagation Letters*, **2**, 285–287.

[7] Yang, S., Gan, Y. and Tan, P. (2004) Comparative study of low sidelobe time modulated linear arrays with different time schemes. *Journal of Electromagnetic Waves and Applications*, **18**(11), 1443–1458.

[8] Fondevila, J., Brégains, J.C., Ares, F. and Moreno, E. (2004) Optimizing uniformly excited linear arrays through time modulation. *IEEE Antennas and Wireless Propagation Letters*, **3**(1), 298–301.

[9] Yang, S., Gan, Y.B., Qing, A. and Tan, P.K. (2005) Design of a uniform amplitude time modulated linear array with optimized time sequences. *IEEE Transactions on Antennas and Propagation*, **53**(7), 2337–2339.

[10] Yang, S., Nie, Z. and Yang, F. (2005) Mutual coupling compensation in small antenna arrays by the differential evolution algorithm. Asia-Pacific Microwave Conf., Suzhou, China.

[11] Yang, S. and Nie, Z. (2005) Time modulated planar arrays with square lattices and circular boundaries. *International Journal of Numerical Modelling – Electronic Networks Devices and Fields*, **18**(6), 469–480.

[12] Yang, S. and Nie, Z. (2005) Mutual coupling compensation in time modulated linear antenna arrays. *IEEE Transactions on Antennas and Propagation*, **53**(12), 4182–4185.

[13] Yang, S. and Nie, Z. (2007) Millimeter-wave low sidelobe time modulated linear arrays with uniform amplitude excitations. *International Journal of Infrared Millimeter Waves*, **28**(7), 531–540.

[14] Tennant, A. and Chambers, B. (2007) A two-element time-modulated array with direction-finding properties. *IEEE Antennas and Wireless Propagation Letters*, **6**, 64–65.

[15] Brégains, J.C., Fondevila, J., Franceschetti, G. and Ares, F. (2008) Signal radiation and power losses of time-modulated arrays. *IEEE Transactions on Antennas and Propagation*, **56**(6), 1799–1804.

[16] Yang, S. and Nie, Z. (2006) The four dimensional linear antenna arrays. Fourth Asia-Pacific Conf. Environ. Electromagn., Dalian, China, pp. 692–695.

[17] Chen, Y., Yang, S. and Nie, Z. (2008) Synthesis of optimal sum and difference patterns from time modulated hexagonal planar arrays. *International Journal of Infrared Millimeter Waves*, **29**, 933–945.

[18] Woodward, P. (1946) A method of calculating the field over a plane aperture required to produce a given polar diagram. *Proceedings of the IEEE*, **93**, 1554–1558.

[19] Woodward, P.M. and Lawson, J.D. (1948) The theoretical precision with which an arbitrary radiation-pattern may be obtained from a source of finite extent. *Proceedings of the IEEE*, **95**, 63–370.

[20] Sikina, T. (1986) A method of antenna pattern synthesis. IEEE AP-S Int. Symp., Philadelphia, PA, USA, vol. 24, pp. 323–326.

[21] Deford, J.F. and Gandhi, O.P. (1988) Phase-only synthesis of minimum peak sidelobe patterns for linear and planar arrays. *IEEE Transactions on Antennas and Propagation*, **36**(2), 191–201.

[22] Bucci, M., Mazzarella, G. and Panariello, G. (1991) Reconfigurable arrays by phase-only control. *IEEE Transactions on Antennas and Propagation*, **39**(7), 919–925.

[23] Diaz, X., Rodriguez, J.A., Ares, F. and Moreno, E. (2000) Design of phase-differentiated multiple-pattern antenna arrays. *Microwave Optical Technology Letters*, **26**(1), 53–54.

[24] Villeneuve, A. (1984) Taylor patterns for discrete arrays. *IEEE Transactions on Antennas and Propagation*, **32**(10), 1089–1093.

[25] Dürr, M., Trastoy, A. and Ares, F. (2000) Multiple-pattern linear antenna arrays with single prefixed amplitude distributions: modified Woodward-Lawson synthesis. *Electronics Letters*, **36**(16), 1345–1346.

[26] Wang, L., Fang, D. and Sheng, W. (2003) Combination of genetic algorithm (GA) and fast Fourier transform (FFT) for synthesis of arrays. *Microwave Optical Technology Letters*, **37**(1), 56–59.

[27] Jiang, M., Chen, R., Zhou, L. *et al.* (2005) Synthesis of arrays with genetic algorithm (GA) and nonuniform fast fourier transform (NFFT). Asia-Pacific Microwave Conf., Suzhou, China.

[28] Sharp, E.D. (1961) A triangular arrangement of planar-array elements that reduces the number needed. *IRE Transactions on Antennas and Propagation*, **9**(2), 126–129.

[29] Hsiao, J.K. (1972) Properties of a nonisosceles triangular grid planar phased array. *IEEE Transactions on Antennas and Propagation*, **20**(4), 415–421.

[30] Botha, E. and McNamara, D.A. (1993) The synthesis of difference patterns for planar arrays with rectangular and hexagonal lattices. IEEE AP-S Int Symposium, vol. 3, pp. 1274–1277.

[31] Haupt, R.L. and Shore, R.A. (1984) Experimental partially adaptive nulling in a low sidelobe phased array. IEEE AP-S Int. Symp., Boston, vol. 2, pp. 823–826.

[32] Steyskal, H., Shore, R.A. and Haupt, R.L. (1986) Methods for null control and their effects on the radiation pattern. *IEEE Transactions on Antennas and Propagation*, **34**(3), 404–409.

[33] Haupt, R.L. (1988) Adaptive nulling in monopulse antennas. *IEEE Transactions on Antennas and Propagation*, **36**, **2**, 202–208.

[34] Haupt, R.L. (1997) Phase-only adaptive nulling with a genetic algorithm. *IEEE Transactions on Antennas and Propagation*, **45**(6), 1009–1015.

[35] Chung, Y.C. and Haupt, R.L. (1999) Adaptive nulling with spherical arrays using a genetic algorithm. IEEE AP-S Int Symp., Orlando, FL, USA, vol. 3, pp. 2000–2003.

[36] Haupt, R.L. (2006) Adaptive antenna arrays using a genetic algorithm. IEEE Mt. Workshop Adapt. Learn. Syst., Logan, UT, USA, pp. 249–254.

[37] Chen, Y., Yang, S. and Nie, Z. (2008) The application of a modified differential evolution strategy to some array pattern synthesis problems. *IEEE Transactions on Antennas and Propagation*, **56**(7), 1919–1927.

[38] Caorsi, S., Massa, A., Pastorino, M. and Randazzo, A. (2005) Optimization of the difference patterns for monopulse antennas by a hybrid real/integer-coded differential evolution method. *IEEE Transactions on Antennas and Propagation*, **53**(1), 372–376.

13

Automated Analog Electronic Circuits Sizing

Jernej Olenšek, Árpád Bűrmen, Janez Puhan, Tadej Tuma

University of Ljubljana, Faculty of Electrical Engineering, Tržaška 25, 1000 Ljubljana, Slovenija

13.1 Introduction

The main focus of this chapter is device sizing in analog integrated circuit (IC) design. Most modern electronic systems consist of both digital signal processing components and analog interfaces to the outside world. There are many tools for the efficient design of the digital portion of the system. However, the analog design often represents a bottleneck in the design flow. Analog design consists of two main stages. First, the circuit topology must be selected. Then the appropriate device parameters (transistor dimensions, resistances, capacitances, bias voltages and currents, etc.) must be determined so that the final circuit satisfies the specified design goals (gain, bandwidth, speed, offset, noise, area, etc.). Topology selection depends greatly on the designer's knowledge and experience, but it is not the main focus of this chapter, and we assume that the topology is fixed. We are concerned with device sizing. Device parameters can be determined by hand, based on experience. However, this is not the optimal approach. The parameter space is often very large and the design specifications usually consist of many conflicting design goals. This makes manual device sizing extremely hard, since many trials are required to satisfy all the design goals. To evaluate the performance of every trial circuit, a circuit simulator such as SPICE can be used. Since numerical simulations can be very time-consuming, the entire device sizing process can take days or even weeks to complete. In practice, the design time is usually limited, which raises the need for computer software that can automatically solve the problem faster and with greater accuracy than the designer can do by hand. The basis for all work presented in this chapter is SPICE OPUS [1], a version of the original SPICE circuit simulator from the University of California at Berkeley. Unlike other similar simulators, SPICE OPUS is also capable of automatically solving the problem of device sizing with the use of various optimization methods [2].

Differential Evolution Anyong Qing

13.2 Cost Function

The objective of device sizing is to balance several conflicting design goals, so the problem is multi-objective by nature. However, multi-objective optimization methods have only been recently used for practical applications and are usually applied to problems with only a few objectives. We concentrate on single-objective methods, since they have been the subject of studies for a much longer time and are more common in practice. To be able to use such methods for device sizing, all specified design goals must first be incorporated into a single real-valued function referred to as the cost function (CF), whose value reflects the quality of the circuit.

The CF can be defined as a sum of so-called penalty functions [3]. For every design goal a separate penalty function is constructed. Its value increases as the corresponding performance measure gets further away from the specified design goal.

Figure 13.1 shows an example of a penalty function $f(m)$ for performance measure m, which is desired to be as small as possible (e.g., circuit area, noise). The design goal is denoted by g. The penalty factor p and tradeoff factor t, combined with norm n, allow the designer to specify the two sections of the function. The penalty region ($f(m) > 0$) determines the contribution of this specific performance measure to the overall CF value, when the corresponding design goal is yet not achieved. The tradeoff region ($f(m) <= 0$) is designed to reward trial circuits that satisfy or surpass the specific design goal.

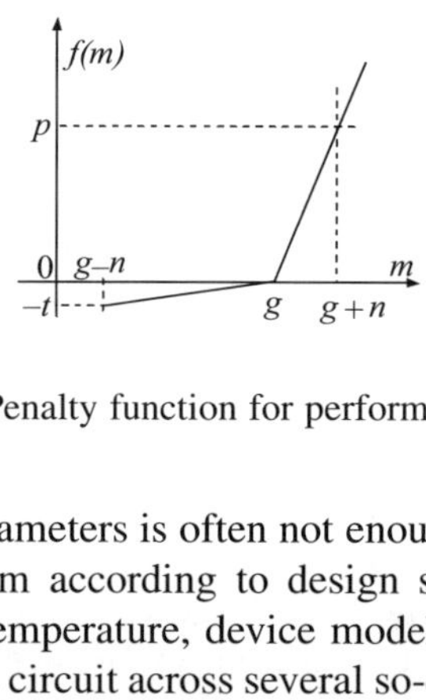

Figure 13.1 Penalty function for performance measure m

Finding the optimal device parameters is often not enough. The final circuits must also be robust, that is, they must perform according to design specifications even under varying environmental conditions (e.g., temperature, device model parameters, power supply). This can be achieved by simulating the circuit across several so-called corner points or corners [3]. Every corner represents a different combination of values of the environmental parameters and requires a separate circuit simulation. The penalty functions are constructed for every corner separately. The worst corner of every performance measure (the one with the largest penalty function value) determines the contribution of this particular performance measure to the overall CF value:

$$\mathrm{CF} = \sum_{i=1}^{N_m} \max_{1 \le j \le N_c} f_{ij}(m_{ij}), \tag{13.1}$$

where m_{ij} denotes the value of ith performance measure in the jth corner and $f_{ij}(m_{ij})$ is the corresponding penalty function value. N_m and N_c denote the number of design goals and corners, respectively.

The problem of device sizing is now a problem of finding the minimum of the CF. If all tradeoff factors are zero, CF = 0 indicates that all design goals are achieved. With positive tradeoff factors this is not necessarily true since some design goals may be surpassed (negative contribution to the CF value) while others may not even be satisfied (positive contribution to the CF value). This is usually resolved by setting the tradeoff factors to much smaller values than the penalty factors so that tradeoffs cancel out only a negligible part of the penalties. In the experiments presented later all tradeoff factors were set to zero so the CF is always nonnegative.

13.3 Hybrid Differential Evolution

13.3.1 Motivations

Device sizing is noisy, non-differentiable, nonlinear, and multimodal optimazation problem. Gradient optimization methods are therefore inappropriate. Instead, direct search methods [4–6] are applicable. Here we use DESA, a hybrid algorithm based on differential evolution (DE) and simulated annealing (SA).

DE [7] is a simple yet very powerful and popular optimization method. The greedy selection scheme prevents the population from escaping from such regions so the method can get trapped in local suboptimal solutions. SA [8] is another very popular optimization method. The main feature of SA is the probabilistic Metropolis criterion that controls the transitions from the current point to the trial point with probability $P = \exp\{-[f(\mathbf{x}^n) - f(\mathbf{x})]/T\} \leq 1$, where $\mathbf{x}^n$ denotes a trial point obtained by random perturbation of the current point $\mathbf{x}$, and T is annealing temperature. Transitions to solutions with lower CF value are always accepted, while transitions to solutions with higher CF value are not always rejected.

DESA is a hybrid method that utilizes features from DE and SA in a way that ties both methods closely together.

13.3.2 Algorithm Description

The flow chart of the method is depicted by Figure 13.2. Like DE, DESA uses a population of M points to guide the search process. Every point is held by a separate sampler. The sampling strategy is the original self-adaptive sampling borrowed from DE, augmented with random sampling from SA. The greedy selection of DE is replaced by the Metropolis criterion. To control trial point generation, every sampler is assigned a different fixed crossover probability P_c^i, $i = 1, 2, \ldots, M$, and a so-called random sampling radius R^i, that controls the length of random steps. To avoid the difficulties of selecting the initial temperature and the appropriate cooling schedule, DESA uses an entirely different approach. Instead of decreasing the temperature with time, a constant temperature T^i is assigned to every sampler. At every iteration, a single sampler is activated. To balance the global and local search, DESA tries to send bad points to samplers with low T and R. This temperature transition mechanism is probabilistic and will occasionally allow the transition of bad points to samplers with large R and T. Samplers with bad points have a higher probability of being activated. This means that small R and T will be used more often but occasional activation of a sampler with large R and T will allow generation of more diverse trial points to prevent stagnation in local minima.

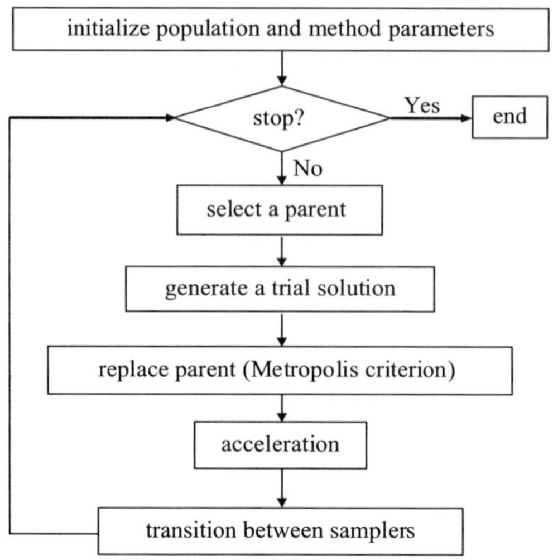

Figure 13.2 Flow chart for DESA

DESA requires some input parameters from the user. They are the number of samplers $M \geq 4$ (population size), minimal temperature $T^M > 0$ (temperature for the last sampler), minimal random sampling radius $R^M > 0$ (radius parameter for the last sampler), and stopping distance $D_{\text{stop}} > 0$. Default values for these parameters are $M = 20$, $T^M = 10^{-6}$, $R^M = 10^{-6}$, and $D_{\text{stop}} = 10^{-4}$.

13.3.2.1 Initialization

The initial population can be generated randomly, but DESA uses the Latin hypercube approach which allows a more thorough exploration of the search space. Every optimization variable interval is first divided into M equal subintervals. Then M points are randomly generated so that every subinterval for every optimization variable is included in the initial population. This is very important in algorithms that use crossover operators. The values of parameters inside subintervals are chosen randomly.

Before the actual optimization some additional method parameters must be set. These parameters are the temperature, the sampling radius, and the crossover probability for every sampler. They are all initialized in the same way. The values of the parameters for the first and the last sampler are fixed. T^1 is set to the CF difference between the worst and the best point in the initial population, $R^1 = 1$, $P_c^1 = 0.1$, $P_c^M = 0.5$.

The remaining samplers are then calculated as

$$\zeta^i = \zeta^1 \exp\left(-\frac{(i-1)\log(\zeta^1/\zeta^M)}{M-1}\right), \tag{13.2}$$

where ζ is T, R, or P_c.

13.3.2.2 Trial Point Generation

At every iteration, a single sampler is activated. It is denoted by superscript *it*. DESA selects the sampler that holds the worst point in the current population, but any sampler that does not hold the best point can be selected here. The point held by the active sampler is the target point DESA is trying to improve. A trial point is generated using a combination of an operator similar to the original DE operator and a random move. The procedure is given by Figure 13.3 where $\mathbf{x}^m$ and $\mathbf{x}^g$ denote the mutated point and the generated trial point. $U[0,1]$ denotes a random number uniformly distributed in [0,1].

The main difference in trial point generation between DE and DESA is the addition of a random step r. In many random search methods the normal distribution is used to generate random steps; however, DESA uses the Cauchy distribution. It has heavier tails which increases the probability of longer jumps through the search space which in turn increases the explorative ability of the method.

activate a sampler (denoted by *it*)
select randomly $ic1, ic2, ic3$, $1 \le ic1 \le M$, $1 \le ic2 \le M$, $1 \le ic3 \le M$, $ic1 \ne ic2 \ne ic3 \ne it$
set $w = U[0,1] \cdot 2$
$\mathbf{x}^m = \mathbf{x}^{ic1} + (\mathbf{x}^{ic2} - \mathbf{x}^{ic3}) \cdot w$
for $i = 1, 2, \ldots N$ do
  $n = n + 1$
  $r = R^{it} \cdot \tan\{\pi \cdot (U[0,1] - 0.5)\}$
  if $U[0,1] < P_c^{it}$
    $x^g(i) = x^m(i) + r$
  else
    $x^g(i) = x^m(i) + r$
  end
  if $x^g(i) >$ upper bound
    $x^g(i) = x^{it}(i) + [\text{upper bound} - x^{it}(i)] \cdot U[0,1]$
  end
  if $x^g(i) <$ lower bound
    $x^g(i) = x^{it}(i) + [\text{lower bound} - x^{it}(i)] \cdot U[0,1]$
  end
end do

Figure 13.3 Trial point generation

13.3.2.3 Replacement

In this phase of the algorithm the generated trial point $\mathbf{x}^g$ is submitted to the Metropolis criterion with temperature T^{it}. If the criterion is satisfied, $\mathbf{x}^g$ replaces $\mathbf{x}^{it}$ in the next iteration. Better points are always accepted. If the trial point $\mathbf{x}^g$ is worse than the current target point, the transition depends on the sampler that holds the target point. If this sampler has large T, $\mathbf{x}^g$ will be accepted with high probability which increases the chances of the algorithm escaping from a local minimum.

13.3.2.4 Acceleration

There are many different mechanisms to speed up the convergence of optimization methods. DESA uses a very simple procedure. Every time a new best point is found, acceleration occurs.

This works by constructing a quadratic model function based on three collinear points in the search space. The first point il is a randomly selected point from the current population using roulette wheel selection. The second point is the centroid of the population points $\mathbf{c}$. These two points define a search direction $\mathbf{d} = \mathbf{c} - \mathbf{x}^{il}$. The third point $\mathbf{p}^3$ is obtained by making a random move from $\mathbf{x}^{il}$ in the direction $\mathbf{d}$. If the quadratic model function obtained is not convex, the best of these three points is returned. For the convex case the minimum of the model function is returned. The point returned replaces $\mathbf{x}^{il}$ if it has lower CF value. The procedure is shown in Figure 13.4.

```
select x^il and get c
w = 1 + U[0,1]
p^3 = x^il + (c - x^il) · w
if p^3 is out of bound
   p^3 = (c + x^il) / 2
end
construct quadratic model function through ,x^il c, and p^3
if model function convex
   find minimum of the model function p^m
   if p^3 is out of bound
      p^m = (p^m + x^il) / 2
   end
else
   set p^m as the best of x^il, c, and p^3
end
if(f(p^m) < f(x^il)) replace x^il with p^m
```

Figure 13.4 Acceleration mechanism

13.3.2.5 Transition between Samplers

One of the main problems of the original SA algorithm is the selection of the appropriate cooling schedule. This means the selection of the initial temperature, the number of steps at every temperature stage and the temperature reduction mechanism. If the cooling is too fast the algorithm can get trapped in a local minimum, and if the cooling is too slow the optimization becomes prohibitively slow. In DESA the cooling schedule is not needed because temperature changes are achieved by simply exchanging points between samplers which operate at different but fixed temperatures. After every trial point generation, replacement, and acceleration phase we randomly select a sampler g^{is} from the population. Then samplers g^{it} and g^{is} exchange their points in the search space with probability $P = \min(1, \exp(-(f(\mathbf{x}^{is}) - f(\mathbf{x}^{it})) * (1/T^{is} - 1/T^{it})))$. This mechanism is quite different from the original idea of SA. Here the idea is to send bad solutions to samplers with low T and R, where trial points are generated with the efficient DE operator with small random component, and most of the trial points that do not improve the CF are rejected. Occasionally the transition of bad points to samplers with high T and R is also allowed. When such samplers are activated, random steps are large and uphill transitions are accepted with higher probability which prevents stagnation and allows the method to escape from a local minimum. If the algorithm cannot find an acceptable solution, this target point will eventually end up at samplers with low T and R and the process starts over. If a good solution is found, the next target point is likely to be at a sampler with higher T and R so DESA performs (at least for a while) a more global search until the target point ends up at samplers with low T and R. With this mechanism DESA also performs a kind of re-annealing and further improves the chances of escaping from a local minimum.

13.3.2.6 Termination Criteria

Several criteria can be used to terminate optimization methods. In practice the time available for the optimization is always limited so the maximal number of CF evaluations (CFE) is a logical choice for termination. The maximal distance between points in the population and the current best point is also used in the termination condition. When this distance falls below a user-defined stopping distance D_{stop} the algorithm is terminated. The third termination criterion is the CF value difference between the best and the worst point in the current population. When this difference becomes smaller than the user-defined minimal temperature (T^M) the algorithm is terminated. In the experiments described later the tradeoff factors were set to zero for all cases so CF $= 0$ was also used as a termination condition.

13.4 Device Sizing

DESA was implemented in C language and integrated into SPICE OPUS. The following paragraphs present several real-world ICs in different configurations that are used to examine its performance.

13.4.1 Test Cases

A detailed description of the optimization problem is given only for the DAMP1 case as shown in Figure 13.5, which is an amplifier circuit with 27 optimization variables including 3 optimization variables from 3 resistors, 2 from 2 capacitors, 2 (width and length) from identical transistors NM0 and NM1, 2 (width and length) from identical transistors NM3, NM5, NM7, and NM8, 2 (width and length) from identical transistors PM0 and PM1, 2 (width and

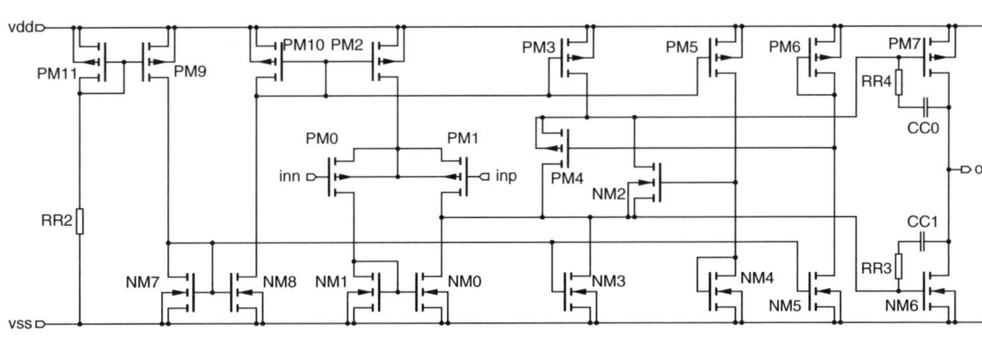

Figure 13.5 Topology for DAMP1 and DAMP1-5c

length) from identical transistors PM2, PM3, PM5, and PM10, 2 optimization variables (width and length) from identical transistors PM9 and PM11, and 12 optimization variables (widths and lengths) from transistors NM2, NM4, NM6, PM4, PM6, and PM7.

We do not optimize transistor multiplier factors. The case requires an OP, a DC, an AC, and a transient analysis to measure 15 circuit properties specified as design goals. The performance measures and their target values are listed in Table 13.1.

Table 13.1 Performance measures for DAMP1

Performance measure	Goal	Performance measure	Goal
Circuit area (m^2)	$<10^{-8}$	Current consumption (mA)	<1
AC gain (dB)	>70	Unity gain bandwidth (MHz)	>5
Bandwidth (Hz)	>500	Phase margin (°)	>60
Gain margin (°)	>10	Maximum derivative of gain magnitude	<0
Output voltage swing (V)	>1.6	DC gain (dB)	>60
Settling time (ms)	<300	Overshoot (%)	<1
Slew rate (V/s)	$>5\times10^6$	Rise time (ns)	<200
Fall time (ns)	<200		

In this first configuration we are not interested in the varying environmental conditions so a single (nominal) corner with typical device model parameters, $T=27\,°C$, and $V_{dd}=1.8$ V, is considered.

The DAMP1-5c case considers the same topology and design goals as DAMP1, but requires a more robust final circuit. For this purpose the case includes five corners to account for varying temperature, supply voltage, and device model parameters: typical NMOS, PMOS, $T=27\,°C$, $V_{dd}=1.8$ V; fast NMOS, fast PMOS, $T=50\,°C$, $V_{dd}=2.0$ V; slow NMOS, slow PMOS, $T=0\,°C$, $V_{dd}=1.6$ V; fast NMOS, fast PMOS, $T=0\,°C$, $V_{dd}=1.6$ V; slow NMOS, slow PMOS, $T=50\,°C$, $V_{dd}=2.0$ V. Although only five corners are considered, they represent the most extreme combinations of environmental parameters. If the circuit satisfies the design goals in these corners, it is very likely that the performance will be the same or even better at other combinations of the environmental parameters.

The remaining cases are described more briefly. The DAMP2 circuit is shown in Figure 13.6. It is another amplifier circuit with 15 optimization parameters. The case considers 14 different corners and requires an OP, a DC, two AC, a transient, and a noise analysis to measure 13 circuit properties.

The LFBUFF and LFBUFF-5c cases shown in Figure 13.7 deal with a buffer circuit with 36 optimization variables. To satisfy 13 specified design goals an OP, an AC, a DC, and a transient analysis are required. The LFBUFF case considers only the nominal corner, while LFBUFF-5c case deals with five different corners.

The NAND gate circuit in Figure 13.8 is fairly simple with only three optimization parameters. Three different corners are considered while trying to satisfy nine design goals. The case requires an OP analysis and two transient analyses to measure these nine circuit properties.

The DELAY case in Figure 13.9 considers 12 optimization variables and a single nominal corner. To achieve the six specified design goals an OP and a transient analysis are required.

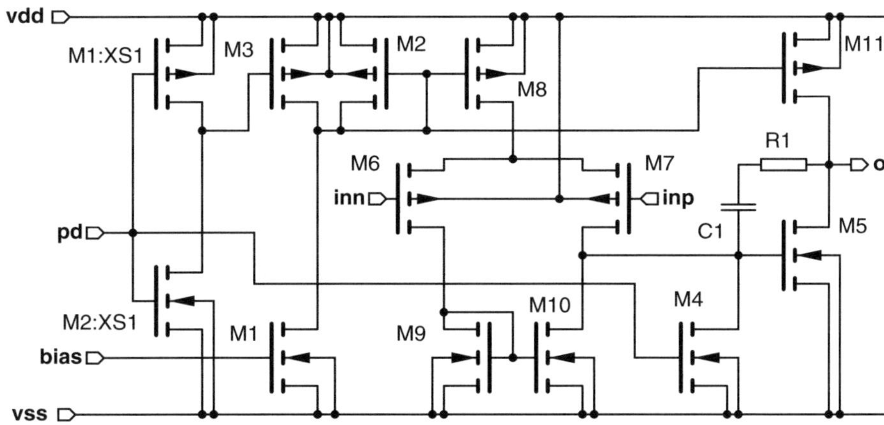

Figure 13.6 Topology for DAMP2

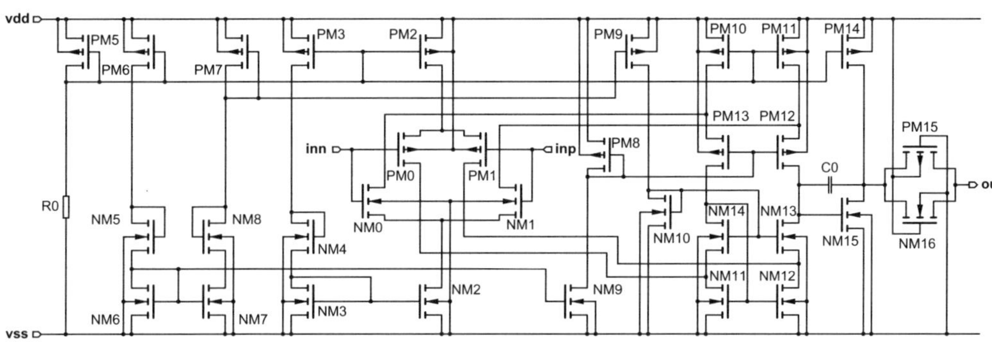

Figure 13.7 Topology for LFBUFF and LFBUFF-5c

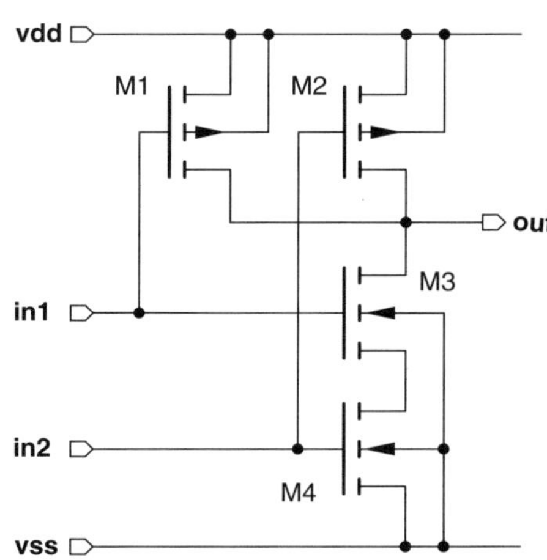

Figure 13.8 Topology for NAND

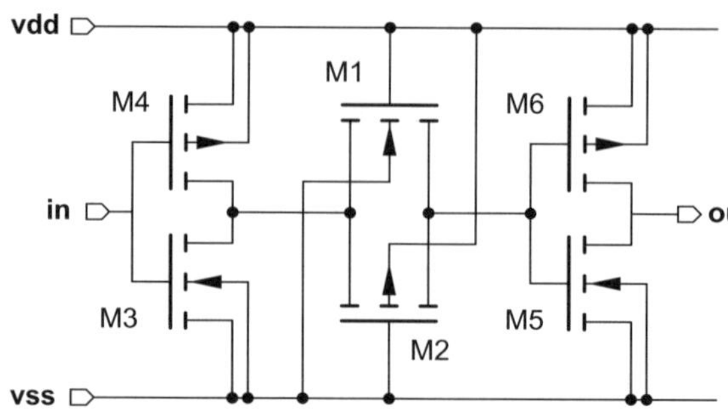

Figure 13.9 Topology for DELAY

To examine the performance of DESA for IC optimization, it is compared with a method already implemented in SPICE OPUS – a multistart version of the constrained simplex method (multistart COMPLEX). The method produced good results for IC device sizing. The original constrained simplex method [9] dates back to 1965 and proved to be very successful in practical applications. In SPICE OPUS it is extended with the probabilistic multistart concept which makes the global search possible and allows the detection of several local minima [10]. With this modification it is also possible to show that the method finds the true global minimum of the CF with probability 1 if the number of restarts is large enough.

13.4.2 Optimization Results

Optimization results for all seven test cases are presented in Table 13.2. Since every optimization run takes a considerable amount of time, every circuit is optimized only once. For every test case the number of design variables (vars), the number of design goals (goals),

Table 13.2 IC optimization results

Case		DESA	Multistart COMPLEX
DAMP1	CFE for CF < 1	5843	5208 (2)
Vars = 27	CFE for CF < 0.5	7818	/
Goals = 15	Final CFE	0	0.087
Corners = 1	CFE for minimal CF	63 863	68 471 (15)
	Final CFE	63 863	100 000 (24)
DAMP1-5c	CFE for CF < 10	2490	1356 (1)
Vars = 27	CFE for CF < 5	3598	21 281 (5)
Goals = 15	Final CF	1.806	3.425
Corners = 5	CFE for minimal CF	250 021	231 312
	Final CFE	300 000	300 000
DAMP2	CFE for CF < 20	989	420 (1)
Vars = 15	CFE for CF < 10	11 282	3926 (3)
Goals = 13	Final CF	5.926	7.487
Corners = 14	CFE for minimal CF	251 244	326 011 (231)
	Final CFE	365 343	500 000 (352)
LFBUFF	CFE for CF < 10	1781	414 (1)

Table 13.2 (*Continued*)

Case		DESA	Multistart COMPLEX
Vars = 36	CFE for CF < 1	9523	1339 (1)
Goals = 13	Final CF	0	0.512
Corners = 1	CFE for minimal CF	79 377	3310 (1)
	Final CFE	79 377	100 000
LFBUFF-5c	CFE for CF < 10	1941	989 (1)
Vars = 36	CFE for CF < 5	5852	13 523 (4)
Goals = 13	Final CF	2.330	4.313
Corners = 5	CFE for minimal CF	229 658	157 612 (41)
	Final CFE	300 000	300 000 (78)
NAND	CFE for CF < 500	282	40 (1)
Vars = 3	CFE for CF < 200	507	94 (1)
Goals = 9	Final CF	166.641	166.686
Corners = 3	CFE for minimal CF	2208	5818 (47)
	Final CFE	2986	10 000 (81)
DELAY	CFE for CF < 2×10^4	38 698	660 (1)
Vars = 12	CFE for CF < 1×10^4	84 302	21 101 (28)
Goals = 6	Final CF	0	6183.500
Corners = 3	CFE for minimal CF	183 687	114 794 (143)
	Final CFE	183 687	200 000 (250)

and the number of corner points (corners) is given. The table shows the number of CFE needed to reach a solution with the given CF value, the CF value of the final solution, the number of CFE needed to find the final solution, and the number of CFE when the methods are terminated. For the multistart COMPLEX method the number of restarts needed to reach a particular solution is given in brackets. Table 13.3 shows the detailed results for the DAMP1 case. The methods are compared according to number of CFE needed to reach a solution of the given quality.

Table 13.3 Results for the DAMP1 case

Measurement	goal	DESA	Multistart COMPLEX
circuit area [m2]	$< 10^{-8}$	$8 \cdot 10^{-9}$	$5.95 \cdot 10^{-9}$
current consumption [A]	$< 10^{-3}$	$437 \cdot 10^{-6}$	$530 \cdot 10^{-6}$
AC gain [dB]	> 70	70.4	70
unity gain bandwidth [Hz]	$> 5 \cdot 10^{6}$	$17.2 \cdot 10^{6}$	$15.7 \cdot 10^{6}$
bandwidth [Hz]	> 500	$1.38 \cdot 10^{3}$	$2.23 \cdot 10^{3}$
phase margin [o]	> 60	65.9	90.3
gain margin [o]	> 10	33.4	15.6
max.derivative of gain magnitude	< 0	$-104 \cdot 10^{-9}$	$-39.9 \cdot 10^{-9}$
output voltage swing [V]	> 1.6	1.6	1.57
DC gain [dB]	> 60	69.3	66.7
settling time [ms]	< 300	174	168
overshoot [%]	< 1	$696 \cdot 10^{-3}$	$885 \cdot 10^{-3}$
slew rate [V/s]	$> 5 \cdot 10^{6}$	$7.44 \cdot 10^{6}$	$7.48 \cdot 10^{6}$
rise time [ns]	< 200	64.4	64.1
fall time [ns]	< 200	57.5	81.1

The results confirm that DESA is a promising approach for automated IC device sizing. It was able to find the true global minimum for cases DAMP1, LFBUFF, and DELAY (tradeoff factors were set to zero, so CF = 0 indicates the global minimum). For these cases all specified design goals were satisfied. The multistart COMPLEX method came close to the minimum for the DAMP1 and LFBUFF cases. For the DELAY case the quality of the final solution was poor. For all the cases where the final CF > 0 we can make no claims about the true global minimum, since there is no way of knowing whether CF = 0 is achievable at all with the given topology, parameter bounds, and design goals. The fact that we know almost nothing about the CF landscape, the location of the global minimum, and the minimal CF value is a major problem in many practical applications and makes the selection of the appropriate optimization method and the optimization itself very difficult. But even though both methods failed to completely satisfy all design goals for these cases, we can still claim that DESA is capable of finding considerably better solutions with fewer CFE than the multistart COMPLEX method. One may also notice that even though cases DAMP1 and DAMP1-5c (similarly for LFBUFF and LFBUFF-5c) deal with the same circuit topology and design goals, designing a robust circuit by considering several corners is much more challenging than nominal design.

13.4.3 Analysis

It is clear that the initial progress is slower for DESA than for the multistart COMPLEX method. Fast initial progress often results in premature convergence to a local minimum for many optimization methods. Several restarts in the multistart COMPLEX method avoid this problem to some degree; however, the method wastes too much time zooming in on low-quality local minima. DESA is designed to perform a more thorough global search and is capable of finding the global minimum with higher probability. Random sampling and the Metropolis criterion increase population diversity and allow the method to escape from low-quality local minima. The population of points and the efficient self-adaptive sampling from DE assure the cooperation between individuals needed to guide the search toward promising regions of the search space. DESA also does not spend too much time on local search. While the acceleration mechanism speeds up the convergence, it is a simple procedure and requires only three CFE. The real fine-tuning only occurs when several individuals come close together, which indicates a region of high-quality solutions.

With these features DESA balances between exploration (searching through unexplored regions of the parameter space) and exploitation (using the knowledge of the entire population to zoom in on high-quality solutions). It is also possible to show that, under some mild assumptions, DESA finds the true global minimum of the CF with probability 1.

13.4.4 CF Profiles

To better understand the difficulties of IC optimization Figures 13.10 and 13.11 show the CF profile for a small subset of the optimization parameters for the DAMP1 test case. Every curve represents a sweep through one of the 27 optimization parameters while others are kept constant. The curves intersect at a point with x-axis value zero representing the center of the profile. The figures show the profiles of the CF at the initial and final points of the optimization. Even though the profiles do not include all 27 optimization parameters, they clearly show why IC optimization is so difficult. The CF is highly nonlinear and noisy. The sensitivity of the CF to

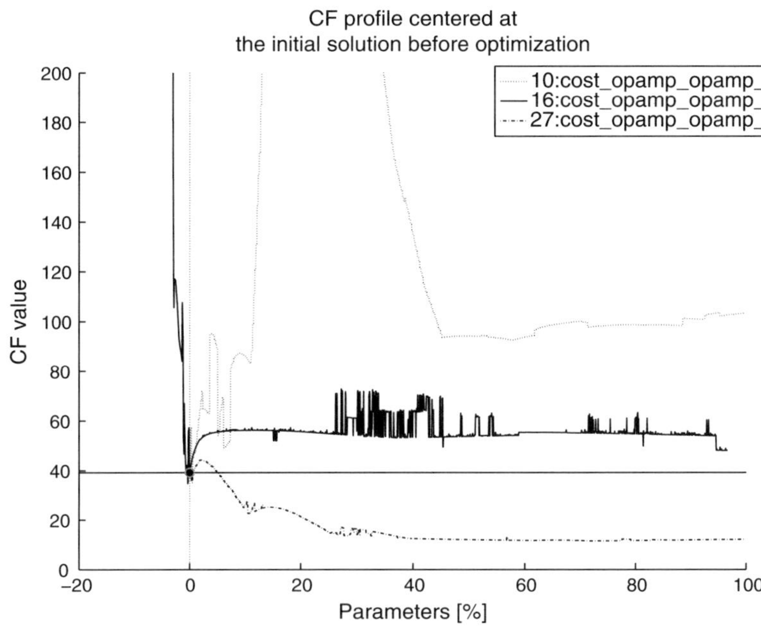

Figure 13.10 Cost profile of DAMP1 at the initial point

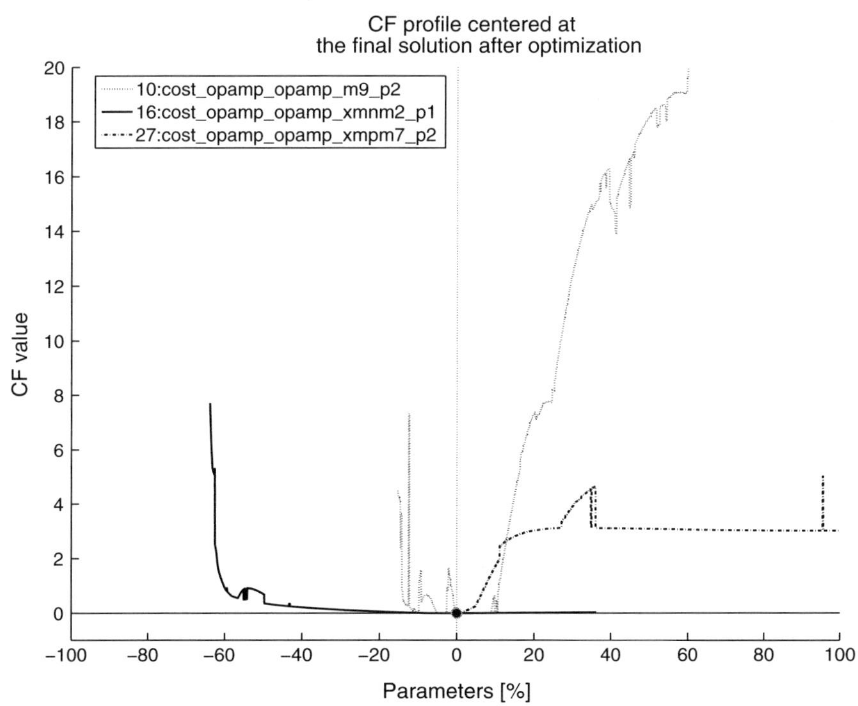

Figure 13.11 Cost profile of DAMP1 at the final point found by DESA

different parameters varies considerably and even when only a single optimization variable is swept, local minima are still clearly visible.

We should mention that the material presented in this chapter does not cover all aspects of analog IC design. There are some issues we did not include in our experiments, such as the worst case and mismatch issues, yield estimation, and layout issues [11,12]. They can be included in the existing framework as separate analyses and measurements. Simulation-based design is also not the only way to design analog ICs. In [13–17], for example, the authors use symbolic analysis to obtain reduced-order symbolic models for the linear transfer functions of the circuit which can reduce the design time. In [18,19] the authors used two different modules to design analog circuits. The first compiles the SPICE netlist describing the circuit and design specifications into a C-style cost function, and the second then uses simulated annealing to find its minimum. Other researchers developed specialized simulators to speedup the evaluation of trial circuits; see http://www.fe.uni-lj.si/spice/ [20]. In [18,21] interested readers can find more extensive surveys of analog design strategies and automated device sizing methods, and in [22,23] the authors present an overview of more recent design methodologies for the design of large systems on chip. The methods involve the extraction of nonlinear macro models of the circuits which are then used to evaluate the circuit performance.

13.5 Conclusions

Many real-world problems can be represented as optimization problems of some kind. The area of numerical optimization is very wide and many researches are developing various methods to solve these problems. Differential evolution has proved to be simple but very powerful and has become one of the most popular optimization methods. Simulated annealing is also a very popular method based on an entirely different concept. Since there is no ultimate optimization method that outperforms all others on all possible optimization problems, hybridization of different approaches can often produce very good results. For the purpose of automated IC device sizing with SPICE OPUS we combined the efficient self-adaptive sampling mechanism from differential evolution with random sampling and the Metropolis selection criterion from simulated annealing. The result was a hybrid global population-based optimization method (DESA) with provable global convergence and very promising performance. Comparison with the multistart COMPLEX method, which has produced the best results in SPICE OPUS so far, confirms that DESA can efficiently solve complex analog IC device sizing problems.

References

[1] *SPICE OPUS circuit simulator homepage*. Faculty of Electrical Engineering, Electronic Design Automation Laboratory, 2005.

[2] Gielen, G.G.E. (2005) Cad tools for embedded analogue circuits in mixed signal integrated systems on chip. *Proceedings of the IEEE: Computers & Digital Techniques*, **152**(3), 317–332.

[3] Bűrmen, A., Strle, D., Bratkovič, F. *et al.* (2003) Automated robust design and optimization of integrated circuits by means of penalty functions. *AEU International Journal of Electronics Communications*, **57**(1), 47–56.

[4] Wright, H.H. (1995) Direct search methods: once scorned, now respectable, 1995. Dundee Biennial Conf. Numerical Analysis, University of Dundee, pp. 191–208.

[5] Powell, M.J. (1998) Direct search algorithms for optimization calculations. *Acta Numerica*, **7**, 287–336.

[6] Lewis, R.M., Torczon, V.J. and Trosset, M.W. (2000) Direct search methods: then and now. *Journal of Computational and Applied Mathematics*, **124**(1–2), 191–207.

[7] Storn, R. and Price, K.V. (1997) Differential evolution – a simple and efficient heuristic for global optimization over continuous spaces. *Journal of Global Optimization*, **11**(4), 341–359.

[8] Kirkpatrick, S., Gelatt, C.D. and Vecchi, M.P. (1983) Optimization by simulated annealing. *Science*, **220**(4598), 671–680.

[9] Box, M.J. (1965) A new method of constrained optimization and a comparison with other methods. *Computer Journal*, **8**, 42–52.

[10] Puhan, J., Bűrmen, A. and Tuma, T. (2003) Analogue integrated circuit sizing with several optimization runs using heuristics for setting initial points. *Canadian Journal of Electrical Computer Engineering*, **28**(3), 105–111.

[11] Antreich, K. J., Graeb, H.E. and Wieser, C.U. (1994). Circuit analysis and optimization driven by worst-case distances. *IEEE Trans. Computer Aided Design Integrated Circuits Systems*, **13**(1), 703–717.

[12] Schwencker, R., Schenkel, F., Pronath, M. and Graeb, H. (2002) Analog circuit sizing using adaptive worst-case parameter sets. Design Automation Test Europe Conf. Exhibition, Paris, pp. 581–585.

[13] Gielen, F., Wambacq, P. and Sansen, W. (1990) Symbolic analysis methods and applications for analog circuits: a tutorial overview. *Proc. IEEE*, **82**, 287–304.

[14] Wambacq, P., Vanhienen, J., Gielen, G. and Sansen, W. (1991) A design tool for weakly nonlinear analog integrated circuits with multiple inputs (mixers, multipliers). *IEEE CICC*, San Diego, CA, 5.1/1–5.1/4.

[15] Yu, Q. and Sechen, C. (1996) A unified approach to the approximate symbolic analysis of large analog integrated circuits. *IEEE Trans. Circuits Systems I*, **43**(8), 656–669.

[16] Shi, C.J., and Tan, X. (1997) Symbolic analysis of large analog circuits with determinant decision diagrams. *IEEE/ACM Int. Conf. Computer Aided Design*, San Jose, CA, Nov. 9-13, 366–373.

[17] Daems, W., Gielen, G. and Sansen, G. (1999) Circuit complexity reduction for symbolic analysis of analog integrated circuits. *36th Design Automation Conf.*, New Orleans, LA, June 21-25, 958–963.

[18] Ochotta, E.S., Rutenbar, R.A. and Carley, L.R. (1996) Synthesis of high performance analog circuits and astrx/oblx. *IEEE Trans. Computer Aided Design Integrated Circuits Systems*, **15**(3), 273–294.

[19] Ochotta, E., Mukherjee, R., Rutenbar, and Carley, L. (1998) *Practical Synthesis of High-performance Analog Circuits*. Norwell, MA: Kluwer Academic.

[20] Alpaydın, G., Balkır, S. and Dündar, G. (2003) An evolutionary approach to automatic synthesis of high-performance analog integrated circuits. *IEEE Transactions on Evolutionary Computation*, **7**(3), 240–252.

[21] Gielen, G.G.E. and Rutenbar, R.A. (2000) Computer-aided design of analog and mixed-signal integrated circuits. *Proceedings of the IEEE*, **88**(12), 1825–1852.

[22] Gielen, G.G.E. (2005) Cad tools for embedded analogue circuits in mixed signal integrated systems on chip. *Proceedings of the IEEE: Computers & Digital Techniques*, **152**(3), 317–332.

[23] Rutenbar, R.A., Gielen, G.G.E. and Roychowdhury, J. (2007) Hierarchical modeling, optimization, and synthesis for system-level analog and rf designs. *Proceedings of the IEEE*, **95**(3), 640–669.

14

Strategic Bidding in a Competitive Electricity Market

Xia Yin[1], Junhua Zhao[2], Zhaoyang Dong[3], Kit Po Wong[3]

[1] *Stanwell Corporation, Brisbane, Australia*
[2] *School of Economics, University of Queensland, Australia*
[3] *Department of Electrical Engineering, Hong Kong Polytechnic University, Hong Kong*

14.1 Electrical Energy Market

14.1.1 Electricity Market Deregulation

The power system was long considered as a 'natural monopoly' and had long been dominated by vertically integrated utilities before deregulation [1]. From the early 1990s, a trend toward electricity market deregulation began throughout the world. Currently, several semi-deregulated markets are operating in a number of countries, including the USA, Australia and some European countries.

Generally speaking, deregulation aims to achieve the following benefits: decreasing costs and more accurate pricing.

14.1.1.1 Decreasing Costs

Competition provides much stronger cost-minimizing incentives than typical regulation and drives suppliers to propose cost-saving innovations more quickly. The innovations include labor saving techniques, more efficient repairs, cheaper plant construction costs and proper investment strategies.

14.1.1.2 Holding Price Down to Marginal Cost

While holding down prices, competition also provides incentives for more accurate pricing. Because it imposes the real-time wholesale spot price on the retailer's marginal purchases,

Differential Evolution Anyong Qing

wholesale competition should encourage real-time pricing for retail customers. A competitive retailer should have an added incentive to provide the option of real-time retail pricing because that would reflect its costs.

14.1.2 Major Market Elements

Several major market elements should be considered to construct a concise but expressive model well describing the market mechanism and bidding procedure. Figure 14.1 depicts a restructured electricity market operation [1].

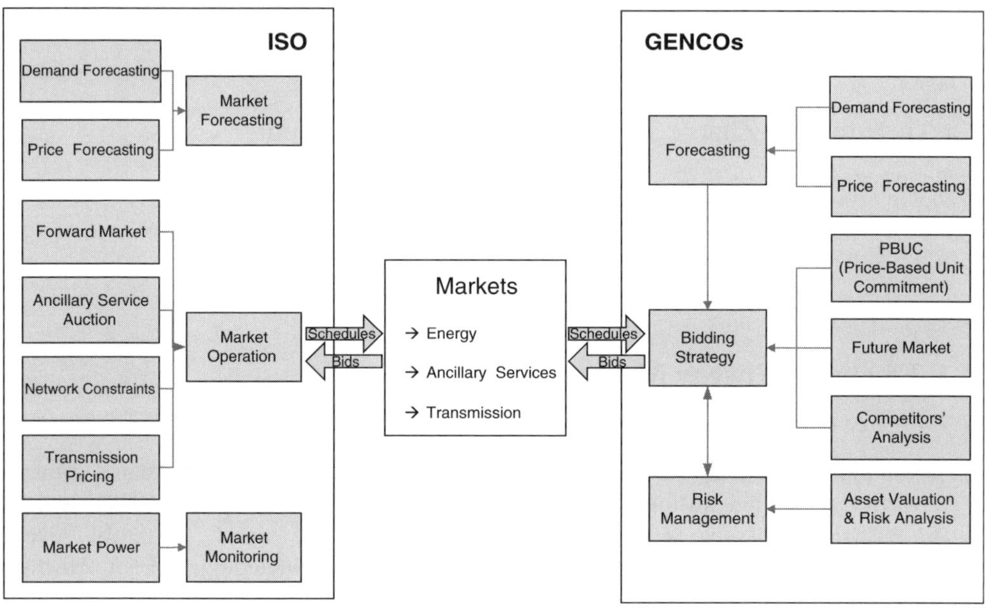

Figure 14.1 Deregulated electricity market operation

14.1.2.1 Models of Market Mechanism

There are three basic models of the deregulated electricity market [1–2]:

14.1.2.1.1 PoolCo Model A PoolCo is a centralized marketplace that clears the market for suppliers and customers. An independent system operator (ISO) in a PoolCo implements the economic dispatch based on maximum social welfare and produces a market clearing price (MCP). In this market, a seller who bids too high may not be able to sell and a buyer who bids too low may not be able to buy. Winning bidders are paid the uniform MCP which is equal to the highest bid of the winners.

14.1.2.1.2 Bilateral Contracts Model Bilateral contracts are negotiable agreements on trading power between suppliers and customers. The bilateral model is very flexible as trading parties specify their contract terms. Bilateral contracts are often used by traders to alleviate risks.

14.1.2.1.3 Hybrid Model The hybrid model incorporates various features of the previous two models. In this model, any customer would be allowed to negotiate bilateral contracts with sellers or choose to purchase power at the spot market price. In our research, we assume the market applies the hybrid model.

14.1.2.2 Market Structure and Operation

The deregulation of the electricity market has greatly changed the structure of the market and therefore the role of traditional entities. In the current market, they can function independently and can be categorized into market operators and market participants.

14.1.2.2.1 Independent System Operators The ISO is the leading entity and its functions determine the market rules. As an independent controller of the grid, the ISO administers transmission tariffs, maintains the system security, coordinates maintenance scheduling, and has a role in coordinating long-term planning.

14.1.2.2.2 Generation Companies A generation company (GENCO) operates and maintains existing generating plants. In the deregulated electricity market, generators employ individual trading profit maximization, rather than cost minimization, as their major objective. Therefore, building optimal bidding strategies consisting of sets of price–production pairs is essential for achieving the maximum profit and has become a major concern for GENCOs.

14.1.2.2.3 Distribution Companies A distribution company (DISCO) distributes electricity through its facilities to customers in a certain geographical region.

14.1.2.2.4 Transmission Companies A transmission company (TRANSCO) transmits electricity using a high-voltage, bulk-transport system from GENCOs to DISCOs for delivery to customers. The transmission system is the most crucial element in electricity markets. A TRANSCO has the role of building, owning, maintaining, and operating the transmission system in a certain geographical region to provide services for maintaining the overall reliability of the electrical system.

14.1.2.2.5 Customer A customer is the end-user of electricity with certain facilities connected to the distribution system or transmission system, according to the customer's size.

14.1.2.2.6 Other Market Entities There are some other market entities in electricity market, including RETAILCOs, aggregators, brokers, marketers, etc. Their introduction is omitted here and can be found in [1].

14.1.2.3 Market Types

Based on trading, the market types include the energy market, ancillary services market, and transmission market.

14.1.2.3.1 Energy Market The energy market is where the competitive trading of electricity occurs. The energy market is a centralized mechanism that facilitates energy trading between buyers and sellers.

14.1.2.3.2 Ancillary Services Market Ancillary services are needed for the power system to operate reliably. In the restructured industry, ancillary services are mandated to be unbundled from energy and are procured through the market competitively.

14.1.2.3.3 Transmission Market In a restructured power system, the transmission network is where competition occurs among suppliers in meeting the demands of large users and DISCOs. The commodity traded in the transmission market is transmission rights.

14.1.2.4 Market Processes

In an electricity market, a vast amount of information that must be collected and passed among market participants. Some examples are bidding, real-time dispatch and metering information. An overview of the market data flows and main systems is shown in Figure 14.2.

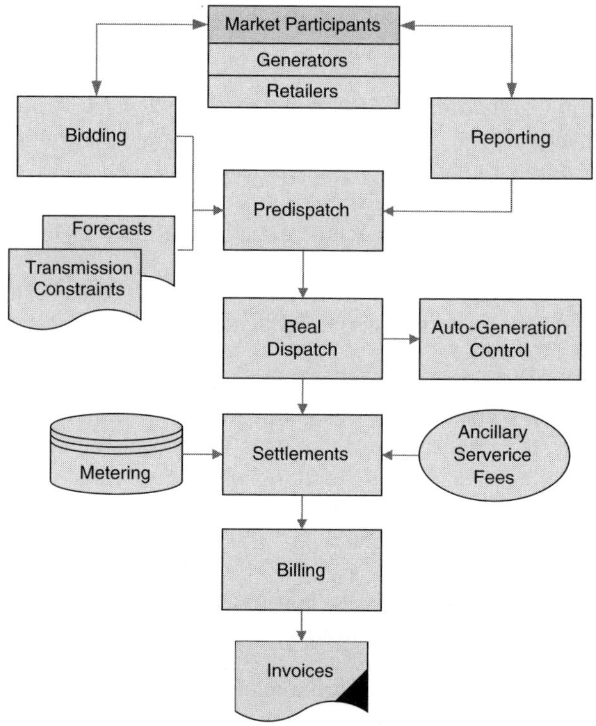

Figure 14.2 A market information flow

14.1.2.5 Unique Characteristics of the Electricity Market

Electricity is a unique commodity and it is unlike any other product that is bought and sold. So the electricity market cannot be examined and discussed by using exactly the same principles that are traditionally used by economists in discussions about market places and competitions.

Electricity, by its nature, has five features that distinguish it from other products and make the market for electricity different. The first feature is that electricity is unidentifiable. This means that once power is injected into the transmission system by a generator, there is no way of knowing its origin. Electrons do not carry any special markings that indicate where they are originated. Thus, end-customers have no way of knowing which generator is supplying them. Secondly, electricity is homogeneous. The product is standardized to certain specifications within areas and within countries. For example, the frequency in Australia is 50 Hz. Thirdly, electricity is not storable. Excess power cannot be saved, which means that the supply must equal demand at all times and distinguishes electricity from all other products that are bought and sold. Another important distinguishing feature of electricity is transportation. The transmission system for electricity is different from the transportation system used for other products. All power flow must occur along power lines that are in place and have limits. Electricity must follow the laws of physics. If a line is constrained, no more transportation of power can occur on that line at that time. This feature is highlighted in ([2], pp. 335–341). Moreover, one generator's behavior can impact the supply of other competitors in the market. If a generator uses the transmission system to its advantage, it can prevent another generator from dispatching.

The above unique characteristics of electricity mean that the electricity market is different from other markets and should be treated differently. Also, the potential market power may arise where it is not expected. The effect of market power leads, directly or indirectly, to higher prices for consumers.

14.2 Bidding Strategies in an Electricity Market

14.2.1 Introduction

The deregulation of the power industry across the world has greatly increased market competition by reforming the traditionally integrated power utility into a competitive electricity market, which essentially consists of the day-ahead energy market, real-time energy market and ancillary services market. Therefore, in a deregulated environment, GENCOs are faced with the problem of optimally allocating their generation capacities to different markets for profit maximization purposes. Moreover, the GENCOs have greater risks than before because of the significant price volatility in the spot energy market introduced by deregulation. Bidding strategies are essential for maximizing profit and have been extensively studied [3–6]. Usually, developing optimal bidding strategies is based on the GENCO's own costs, anticipation of other participants' bidding behaviors and power systems' operation constraints. The PoolCo model is a widely employed electricity market model. In this model, GENCOs develop optimal bidding strategies, which consist of sets of price–production pairs. The ISO implements the market clearing procedure and sets the MCP [7]. Theoretically, GENCOs should bid at their marginal cost to achieve profit maximization if they are in a perfectly competitive market. However, the electricity market is more akin to an oligopoly market and GENCOs may achieve benefits by bidding at a price higher than their marginal cost. Therefore, developing an optimal bidding strategy is essential for achieving the maximum profit and has become a major concern for GENCOs.

Identifying the potential for the abuse of market power is another main objective in investigating bidding strategies. There is a widespread belief among regulators and policy analysts that deregulation of the electricity generating industry will yield economies in the cost of power supply by introducing competition. However, because the electricity industry has a relatively small number of firms, the benefits that would accrue from lower electricity prices may be offset. In particular, in the normal operation of markets, the price can be well above the short-run marginal cost of production as a result of pricing strategies adopted by rational firms. In economics terms, a supplier has market power when it can raise its price above the level dictated by competition [8]. Thus it is important to have as much information and clarity as possible about these market power effects, so that they can be mitigated before they manifest themselves to the detriment of consumers.

14.2.2 Auction and Bidding Protocols

An auction is a market institution with an explicit set of rules determining resource allocation and prices on the basis of bids from the market participants. The auction mechanism has been a preferred choice for setting prices for electricity markets. It is an economically efficient mechanism to allocate demand to suppliers, and the structure of the electricity market in many countries is based on auctions. For example, as part of the Australian National Electricity Market (NEM), scheduled generators are required to submit offers to the market indicating the volume of electricity they are prepared to produce for a specified price. Figure 14.3 shows a GENCO's typical bid curve. Then the bids are aggregated to determine which GENCOs will

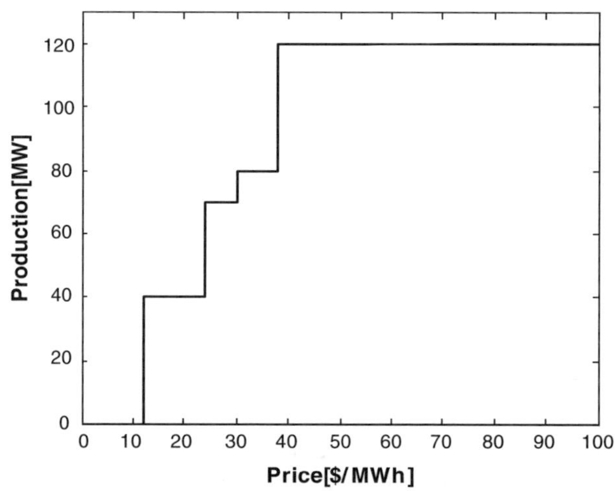

Figure 14.3 Typical GENCO bidding curve

dispatch into the market, at what time and at what volume. This process of balancing the supply and demand in the market is called scheduling and it also prioritizes dispatch based on cost-efficiency of supply. Energy offers from GENCOs are stacked in increasing price order until demand is met. As energy demand increases, more expensive generators are dispatched. The scheduling of generators, however, may be constrained by the capacity of the interconnectors between the regions. When this occurs, more expensive generators will be dispatched to meet the demand within the region and this is also one reason for the difference in the electricity spot price between regions in the NEM. As shown in Figure 14.4, a marginal clearing price is set at the intersection point between the aggregated demand and supply curves

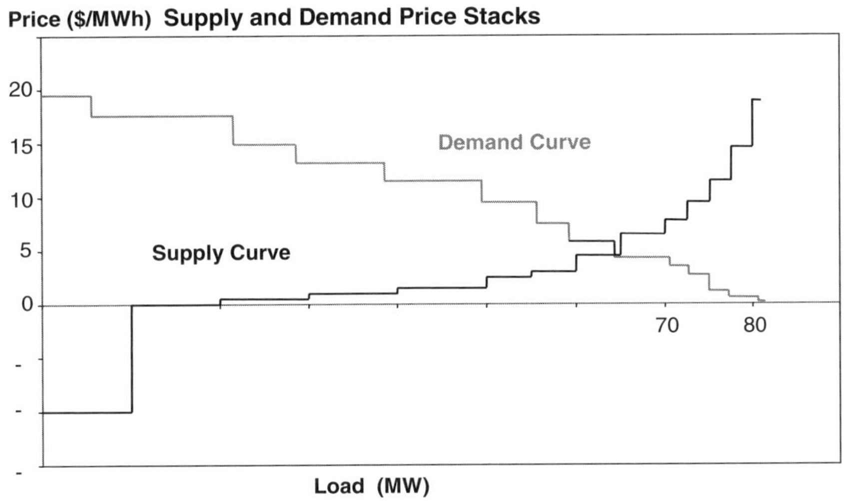

Figure 14.4 Market clearing price obtained by the intersection of demand and supply curves

for each dispatching period. The spot price for a half-hour trading period, which consists of six dispatching periods, would be the average of the prices of the six dispatching periods. All GENCOs winning the auction are paid at the uniform clearing price.

Bidding has a strong relationship with the auction. Bidding strategies should be developed according to market models and activity rules. The auction rules and bidding protocols are the most important of these rules.

14.2.2.1 Auction Methods

Auction methods can be categorized as static or dynamic. In static auctions, bidders submit sealed bids simultaneously. In dynamic auctions, bidders can observe other competitors' bids and may revise their own sequentially. In terms of discriminating pricing or non-discriminating pricing, bidders in static auctions are paid their offered prices or a uniform price.

Auctions can also be classified as open or sealed-bid. Open auctions may be classified as English (descending) or Dutch (ascending). Sealed-bid auctions are non-discriminating auctions. Almost all operating electricity markets employ the sealed-bid auction with uniform market price.

14.2.2.2 Bidding Protocols

The bidding protocols can be classified as multipart bid or single-part bid according the price components included in bids.

14.2.2.2.1 Multipart Bid A multipart bid, also called a complex bid, consists of separate prices for ramps, start-up costs, shut-down costs, no-load operation, and energy. In another words, both cost structure and technical constraints are included in this kind of bid. By employing multipart bid protocols, bid prices, technical constraints and related economic information can be taken into account. A well-known example of the multipart bid is the electricity market in England and Wales.

14.2.2.2.2 Single-Part Bid In single-part bid, generators bid only independent prices for each hour. Based on the intersection of supply and demand bid curves, the market clearing process is conducted to decide the win bids, MCP and schedules for each dispatch period. This scheme is decentralized. Generators need to internalize all involved costs and technique constraints in developing their bids to make their own unit commitments, while in multipart bid these will be done by the market operator (MO). The single-part bid has been implemented in several electricity markets, such as Australia, California, Norway, and Sweden.

There are many publications on building bidding strategies for this type of market [3,5,7].

14.2.3 Other Factors Relevant to Bidding

In order to deal with the uncertainties in an electricity market, different constraints have been considered in developing optimal bidding strategies. One of the main constraints is risk. Methods for building optimal bidding strategies for generators according to their degree of risk aversion are discussed in [9–11].

14.2.3.1 Contracts

In a deregulated market, GENCOs compete through both the spot market and bilateral market. Consequently, GENCOs have to consider their various categories of contracts when they construct their bids in the spot market [6,12].

14.2.3.2 Transmission and Technical Constraints

The influence of congestion has been considered in [13] and other technical limitations of generating units are comprised in the optimization problem in [10,14].

14.2.3.3 Coordination with Ancillary Services Market

The development of optimally coordinated bidding strategies in energy and spinning reserve markets is considered in [15,16]. Each generator bids a linear energy supply function and a linear spinning reserve supply function to the energy and spinning reserve market. The two markets are dispatched separately to minimize customer payments. To obtain maximum profit, each generator chooses the coefficient in the linear supply functions, subject to expectations about how his or her rivals will bid. A genetic algorithm is applied to solve it.

14.2.3.4 Rivals' Information

Sometimes it is assumed that rival GENCOs' cost information, bid information or benefit function is public and available. For example, in [17], it is assumed that a GENCO knows all the other competitors' cost information. In [5], each GENCO anticipates a value for the bid from each of the other market rivals. In [18], the parameters in rivals' bids are assumed to be available as discrete distributions. In [19], it is assumed that each market participant can estimate its competitors' benefit functions and their minimum and maximum values. However, these assumptions are often impractical because most of the cost, benefit and bid information of rivals are confidential in an electricity market.

Given this background, the profit of each generator will be subject to the information it has. Therefore, in order to design realistic optimal bidding strategies with incomplete information, the Asymmetrical behaviors of suppliers should be modeled correctly. Unfortunately, there is a general lack of research in this direction so far.

14.2.3.5 Single-Sided and Double-Sided Auctions

Bidding strategies are also subject to the market mechanism. In [20], a framework for testing and modifying bidding strategies by using a genetic algorithm was proposed. In this chapter, DISCOs and GENCOs buy and sell electricity via double auctions implemented in a regional commodity exchange. The efficiency and competitiveness of double-sided and single-sided (supply-only) auctions are compared in [21].

14.2.3.6 Different Types of Generators

Empirical analysis of different bidding behaviors of different types of generators in NEM can be found on the National Electricity Market Management Company (NEMMCO) web site, http://www.nemmco.com.au.

In a deregulated electricity market, GENCOs have the freedom to design bidding strategies by taking several factors into account [22]. The main factors influencing bidding strategies include technical limitations of generating units and various categories of contracts. Usually, generators' limitation on energy generation will greatly influence the amount of contracted electricity. Thermal contracted generators enter contracts covering the entire trading day subject to their own technique constraints. They usually bid low prices in order to be scheduled. If they cannot be dispatched and the spot price is higher than the contract spike price, the generator will have to buy the electricity from the pool market to meet the contract. This will result in revenue losses. Some hydro power stations with limited water resources and gas power stations with high fuel costs are peak generating units. Most of their costs are recovered by bidding strategies on the spot market. Gas power stations always wait for the high-price periods to bid on the market and recover their costs. It is difficult for them to enter long-term contracts. At the same time, they are usually contracted by the system operator to provide ancillary services or as system reserve units during peak hours.

14.3 Application of Differential Evolution to Strategic Bidding Systems

14.3.1 Introduction

In this section, the problem of bidding decision-making is studied from the viewpoint of a GENCO. A nonlinear relationship between generators' bidding productions and MCP is estimated. A new method is proposed for dealing with the incomplete information to construct optimal bidding strategies. This method is based on the support vector machine (SVM) [23] with a differential evolution bidding strategy optimizer and Monte Carlo simulation. The SVM is an advanced technique that has attracted extensive research in data mining and has been successfully applied to electricity price forecasting [24]. It has been proven to be effective in estimating the nonlinear functional relationship, hence is employed in the proposed method to estimate the relationship between MCP and GENCOs' bid productions.

We first assume that the bidding production of the GENCO in each dispatch interval follows a normal distribution. The parameters of these normal distributions can be estimated from historical bidding data. Having obtained the probability density function (PDF) and its parameters (mean and standard deviation), the SVM is employed to estimate the nonlinear relationship between bidding productions and MCP from publicly available historical market data. To deal with the problem with inherent stochastic structures, Monte Carlo simulation is used to acquire approximate solutions by performing statistical sampling experiments. In each iteration of the Monte Carlo simulation, the SVM outcomes are used in a bidding strategy optimizer (differential evolution) to maximize the GENCO's profit during its self-schedule and then provide the optimal strategies comprising price–production pairs. By estimating the price distribution from historical price data, the proposed method is able to construct a series of price scenarios according to generators' different attitudes to the risk. The proposed method is therefore useful for both the risk-averse and the risk seeker.

The main contributions of this methodology are as follows:

14.3.1.1 Prediction of Rivals' Behaviors

By employing an advanced data mining technique, rivals' behaviors can be accurately predicted based on publicly available market data. This method therefore effectively solves the incomplete information problem commonly found in the electricity market.

14.3.1.2 Uncertainty Handling

The Monte Carlo simulation and statistical estimation techniques are used to reliably handle the uncertainties involved in designing the optimal bids.

14.3.2 Problem Formulation

14.3.2.1 Market Mechanism

The mechanism of a day-ahead electricity market is introduced as follows. Firstly, each GENCO uses a self-schedule algorithm to determine its optimal self-schedule. Secondly, each GENCO applies a bidding strategy to optimize its self-schedule in order to get the maximum profit. Thirdly, market participants submit their bids in each trading interval to the MO. For example, in NEM, these bids are composed of the power prices and power offered in different intervals. Finally, the MO determines the generation and load dispatch, as well as the MCP, via its economic dispatch algorithm, which selects the cheapest available generating resource. The supplier that sets up the MCP is the most expensive one, and other suppliers are all paid at this price which is above their bidding prices.

The GENCO's behavior depends on several factors such as the daily demand and forecasted price, the maximum system capacity, the generation reserve predicted, the transfer between different regions, bilateral contracts and its degree of risk aversion under price uncertainty.

14.3.2.2 Proposed Bidding Strategy Model

In the proposed bidding strategy model, suppose that there are n independent GNECOs and a group of loads. Each GENCO submits its bid of a price–production pair in each trading interval to the MO. A nonlinear relationship is assumed between the GENCOs' bidding productions and the MCP. We also assume that in each trading interval, the production of each generator obeys the normal distribution. The mean value μ_{ti} and standard deviation σ_{ti} can be estimated from historical bidding data. For each GENCO, there are technical constraints of output $P_{i,\min}$ and $P_{i,\max}$, $1 \le i \le n$. The estimated MCP λ_t at trading interval t in \$/MWh can be calculated by:

$$\lambda_t = f[p_{t1}, p_{t2}, \ldots, p_{tn}], \quad 1 \le t \le T, \tag{14.1}$$

where $p_{ti} \sim N(\mu_{ti}, \sigma_{ti}^2)$ is the bidding productions of GENCO i at trading interval t in MW, and T is the number of trading intervals in a day. In NEM, there are 288 trading intervals in a business day. Here, therefore, we set $T = 288$. λ_t is estimated by SVM, as is discussed in more detail in the next section.

The profit maximization problem can be formulated as follows:

$$\begin{aligned}
&\text{maximize} \sum_{t=1}^{T} \alpha[\lambda_t p_{ti} - C_{ti}(p_{ti})] \\
&\text{subject to } \lambda_t = \mu_\lambda(t) + z_{1-\alpha}\sigma_\lambda(t), \\
&\quad \mu_\lambda(t) = f(p_{t1}, p_{t2}, \cdots, p_{tn}), \\
&\quad P_{i\min} v(t) \le p_{it} \le P_{i\max}[v(t) - z(t+1)] + z(t+1)SD \\
&\quad -RDv(t-1) - SDz(t) \le p_t - p_{t-1} \le RUv(t-1) + SUy(t), \\
&\quad y(t) - z(t) = v(t) - v(t-1), \\
&\quad y(t) + z(t) \le 1, \\
&\quad z(t) \in \{0, 1\},
\end{aligned} \tag{14.2}$$

where α is the confidence level chosen by the GENCO to adjust the MCP to take account of its attitude to risk, $C_{ti}(p_{ti})$ is the cost to the ith GENCO of generating p_{ti} amount of power at time t, $\mu_\lambda(t)$ is the mean of MCP at time t, $\sigma_\lambda(t)$ is the variance of historical price data, $z_{1-\alpha}$ is the one-sided $1-\alpha$ critical value of the standard normal distribution, $P_{i,\min}$ and $P_{i,\max}$ are the minimum and maximum power output of GENCO i in MW, $y(t)$, $z(t)$, $v(t)$ are the running, start-up and shut-down status changes, and *SD*, *SU*, *SD* and *RD* are shut-down ramp, start-up ramp, ramp-down, and ramp-up rate limits in MW/h.

Based on the assumption that MCP is normally distributed, we can guarantee that the real MCP will be greater than λ_t with probability α. Therefore, bidding at λ_t will have a probability of α of being dispatched. GENCOs can determine their risks by selecting different α. A probability $\alpha > 0.5$ represents a risk-averse GENCO, otherwise the GENCO is a risk seeker.

For simplicity, in the optimization, the variable production cost will be used as GENCO's operation cost, while neglecting fixed, shut-down, and start-up costs in our function. This objective function will be used to optimize the GENCO's bids with the MCP obtained by SVM as the λ_t variable and then differential evolution will be used as the optimizer to solve the self-scheduling problem and get the optimal production of electricity.

14.3.3 Proposed Methodology

14.3.3.1 Outline of the Proposed Method

A detailed description of the proposed method is now given. As discussed in the preceding section, it is essential to estimate the function $f(\cdot)$ and parameters μ_{ti} and σ_{ti} before the optimization can be performed. These estimates can be obtained by SVM and standard statistical methods. Then the Monte Carlo simulation is conducted. In each iteration of the Monte Carlo simulation, rival GENCOs' productions p_{ti} at time t are randomly generated with the estimated density distributions. These productions can then be employed by the SVM to give the predicted MCP at time t. Differential evolution is used to solve the optimization problem and obtain the optimal production at time t.

Given training production data p_{ti}^d which is the production of GENCO i at time t, day d, the production distribution parameters of GENCO i can then be estimated as

$$\mu_{ti} = \sum_{d=1}^{D} p_{ti}^d / D, \tag{14.3}$$

$$\sigma_{ti} = \sqrt{\sum_{d=1}^{D} (\mu_{ti} - p_{ti}^d)^2 / (D-1)}. \tag{14.4}$$

The second step of the proposed method is to estimate $f(\cdot)$ using the SVM. For each time point t, we assume that the real MCP λ_t and the productions p_{ti} of all GENCOs are both available. In SVM training, the p_{ti} are predictor variables (inputs), and λ_t is selected as the response variable (output). The relationship between λ_t and p_{ti} can be accurately estimated because the SVM can approximate any nonlinear functional relationship.

Having estimated $f(\cdot)$,μ_{ti}, and σ_{ti}, the Monte Carlo simulation and differential evolution can be employed to solve the problem of designing the optimal bidding strategies. Monte Carlo simulation is applied to obtain the optimal bids by performing random experiments. In each iteration of the Monte Carlo simulation, the outcomes of SVM are used in a differential evolution based bidding strategy optimizer to maximize the GENCO's profit during its self-schedule and then provide the optimal strategies. By estimating the price distribution from historical price data, the proposed method is able to construct a series of price scenarios according to GENCOs' different attitudes to risks. The proposed method is therefore useful for both the risk-averse and the risk seeker. More details about the Monte Carlo method and differential evolution are given in the following sections.

The flowchart of the proposed bidding strategy method is shown in Figure 14.5.

14.3.3.2 Support Vector Machine

The support vector machine will be used in the experiments as the estimating model. We give a brief introduction to SVM for completeness.

SVM is a machine learning method proposed by Vladimir Vapnik *et al.* at Bell Laboratories [23]. This method has received increasing attention in recent years because of its excellent

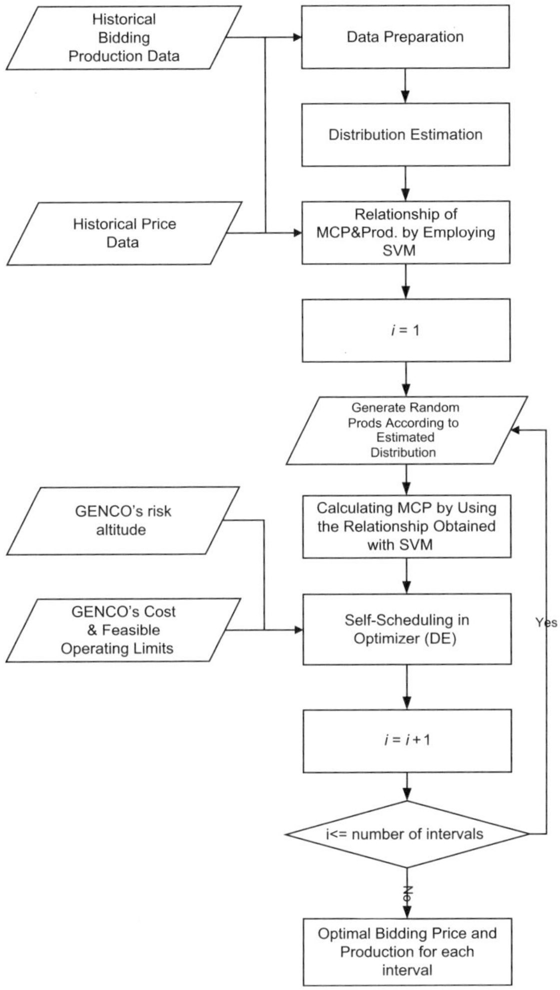

Figure 14.5 Flow chart of the proposed bidding strategy method

performance in both classification and regression. Generally, SVM method is based on Vapnik's work on statistical learning theory [23]. Unlike previous regression methods such as regression trees or neural networks which minimize the empirical risk for the learning machine so as to achieve high forecasting accuracy, the SVM tries to achieve a balance between the empirical risk and the learning capacity of the learning machine. This idea leads to the principle of structural risk minimization, which is the basis of SVM.

The simplest form of SVM is the *maximal margin classifier*. It is used to solve the simplest problem, regression with linear training data. Consider the training data $\{(X_1, y_1), \ldots, (X_l, y_l)\} \subset R^n \times R$, which we assume to be linear. This means that there exists a hyperplane $\langle W, X\rangle + b = 0$ on which $y_l(\langle W, X_i\rangle + b) > 0$, where $\langle W, X\rangle$ is the dot product of W and X.

The margin is defined as the distance from the hyperplane to the nearest data point. The aim of the maximal margin classifier is to find the hyperplane with largest margin, called the maximal hyperplane. Without loss of generality, we assume that the two points with label $+1$ and -1 are nearest to the hyperplane, X_1 and X_2. Note that the rescaling of W and b will not really change the hyperplane, so we have:

$$\begin{aligned} \langle W, X_1 \rangle + b &= 1, \\ \langle W, X_2 \rangle + b &= -1. \end{aligned} \tag{14.5}$$

The maximal margin can be calculated as

$$r = (\langle W, X_1 \rangle + b - \langle W, X_2 \rangle - b)/||W|| = 2/||W||. \tag{14.6}$$

Therefore, the maximal margin classifier problem can be written in the following form:

$$\begin{aligned} &\text{minimize } ||W||^2/2 \\ &\text{subject } to\ y_i(\langle W, X_i \rangle + b) \geq 1, \quad 1 \leq i \leq l. \end{aligned} \tag{14.7}$$

The Lagrange multiplier method can be used to solve this optimization problem.

In most real-world problems, the training data are not linearly separable. There are two ways to modify the linear SVM to suit the nonlinear case. The first is to introduce some slack variables to tolerate some training errors to decrease the influence of the noise in the training data. Another method is to use a map function $\Phi(X)$: $R^n \rightarrow H$ to map the training data from the input space into some high-dimensional feature space, so that they will become linearly separable in the feature space. Then the SVM can be applied in the feature space. Note that the training data used in the SVM are only in dot product form, therefore, after the mapping the SVM algorithm will only depend on the dot product of $\Phi(X)$. If we can find a function that can be written in the kernel function form $K(X_1, X_2) = \langle \Phi(X_1), \Phi(X_2) \rangle$, the mapping function $\Phi(X)$ will not need to be explicitly calculated in the algorithm. Here the radial basis kernel [17] is used:

$$K(X, Y) = e^{-||X-Y||^2/(2\sigma^2)}. \tag{14.8}$$

14.3.3.3 Monte Carlo Simulation to Obtain the Optimal Bidding Strategies

Monte Carlo simulation solves the stochastic optimization problem by performing statistical sampling experiments [24]. Before the Monte Carlo simulation, rival GENCOs' bidding productions are assumed to follow normal distributions. At each iteration of the Monte Carlo simulation, the bidding productions of rivals are randomly generated based on the estimated distribution. Considering the random productions as fixed numbers, building the optimal bidding strategy for the ith generator becomes a one-parameter search problem for which the bids from the other suppliers are fixed through the random sampling procedure. Differential evolution can then be used to solve the deterministic optimization problem and obtain a bidding strategy. After a number of iterations, the mean values of bidding production and price, which are obtained in each iteration, will be selected as the optimal bid.

14.4 Case Study

14.4.1 Case Study Problem

A detailed case study is carried out based on a power system consisting of 11 thermal units with quadratic production costs and generation technical limits. This case study has three major objectives. Firstly, we are interested in whether the proposed method can properly estimate rival generators' productions. Secondly, in the experiments, the MCPs are obtained with SVM based on the random samples of estimated bidding production distribution. The profits obtained using the proposed method are compared with the real profits of each GENCO obtained from real market data, to demonstrate the effectiveness of the proposed method. Thirdly, the bidding results according to different risk levels α are presented, and the influence of the risk level, which depends on the risk attitude of generators, on the bidding strategies is studied.

The case study is conducted based on real market data from NEM. Historical bidding data are publicly available after market clearing and are quite extensive, including all bids submitted, the unit that submitted the bid and the economic entity that controls the unit.

NEM has been operating reliably and efficiently since December 1998. NEMMCO administers and operates a competitive wholesale electricity market where around 165 000 GWh of electricity is traded annually. This makes NEM a good choice for case studies.

Eleven major generators located in NEM are selected in the case study. Their capacities and technical constraints can be found in [25]. The case study is based on real MCP and bidding production data of NEM from September 17 to October 16, 2006. It is assumed in the case study that one of the 11 generators designs its bidding strategy with the proposed method, while other generators' bidding strategies are identical to the real bidding data.

14.4.2 Analysis Result

Figures 14.6–14.8 show the GENCOs' real bidding productions and the estimated bidding productions obtained with the proposed method. It is evident that the proposed method can accurately estimate rival GENCOs' behaviors.

To analyze the profit obtained with the proposed bidding strategy technique, an empirical case study is conducted to compare the difference between the real profits and the profits

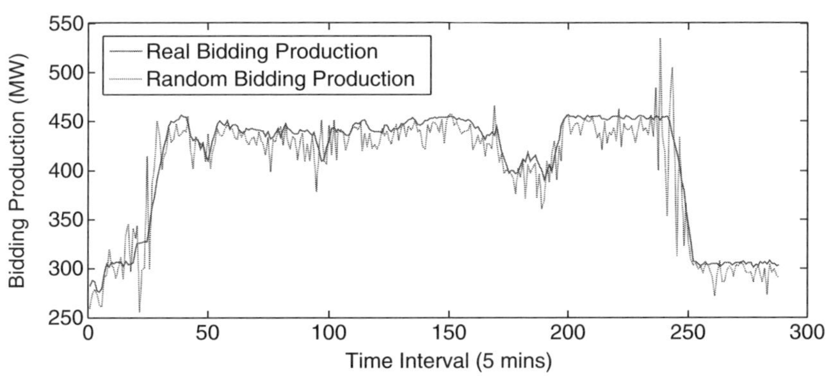

Figure 14.6 The real bidding production versus random bidding production of generator 1

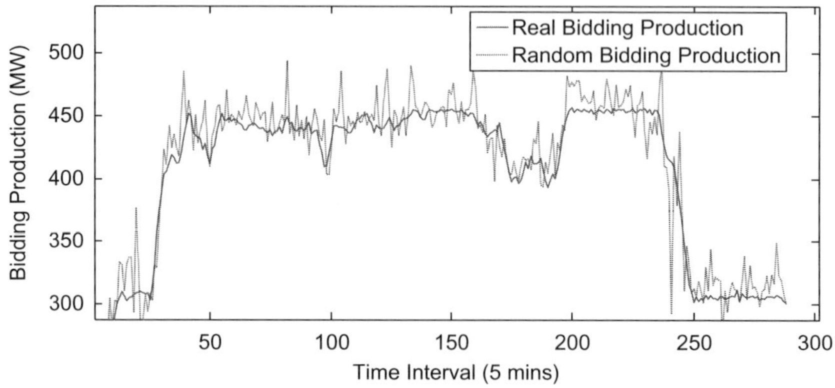

Figure 14.7 The real bidding production versus random bidding production of generator 2

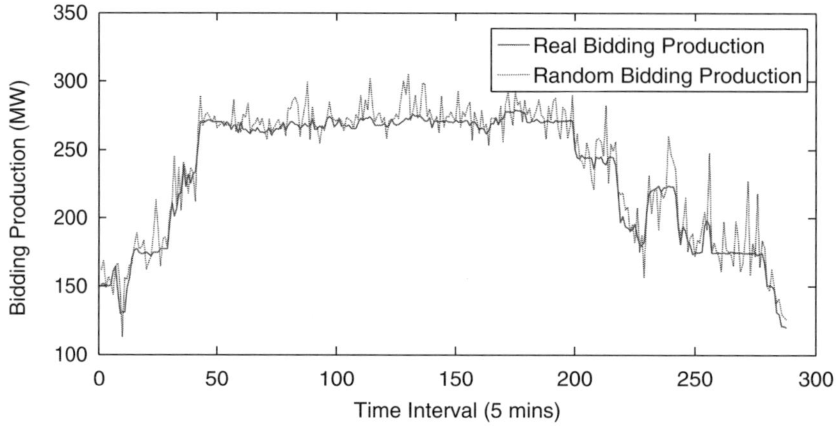

Figure 14.8 The real bidding production versus random bidding production of generator 3

obtained with the proposed method. The results of this study can be used to demonstrate how the bidding strategies optimize GENCO profits.

The case study consists of 11 rounds. In each round, GENCO i, $1 \leq i \leq 11$, is assumed to design its bidding strategies with the proposed method, while other generators follow their actual bidding strategies. For each generator using the proposed method, its profit achieved in the experiment is compared with its real profit calculated from historical data. The results are shown in Figure 14.9.

As clearly illustrated in Figure 14.9, the profits obtained with our method are similar to the real profits for generators G1, G2 and G5. However, the proposed method results in significantly greater profits than the actual profits for all the other eight generators. Moreover, the proposed method can increase the profit by at least 10 % for these eight generators, and raised the profit of G6 by 120 %. The results prove that the proposed method is highly effective on most occasions.

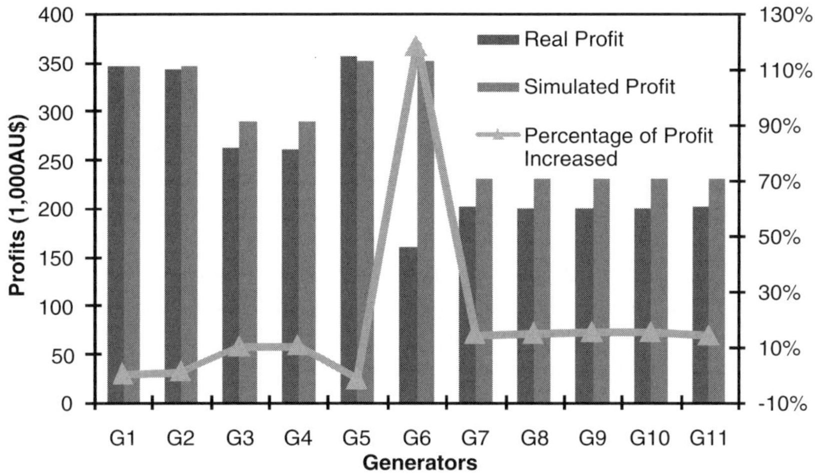

Figure 14.9 Real profits versus simulated profits of 11 generators

To further investigate the performance of the proposed method, we check the historical productions of all 11 generators. According to the historical productions, G1, G2 and G5 are basically base-load generators. Their productions are very close to their maximum capacities at most occasions. The proposed method cannot further increase their productions and therefore cannot significantly improve their profits. On the other hand, theother eight GENCOs' productions are usually far from their maximum capacities. The proposed method can thus perform well.

As discussed in Section 14.2, the proposed method is applicable to both risk-averse and risk-seeking GENCOs, by selecting suitable confidence level α. To study the influence of α on profit, the profits of generator G1 as obtained with the proposed method with respect to different α are plotted in Figure 14.10. Clearly, the profit is increased when α is decreased from 90 % to 70 %. A large α indicates that the generator tends to bid a low price to make sure it will be dispatched.

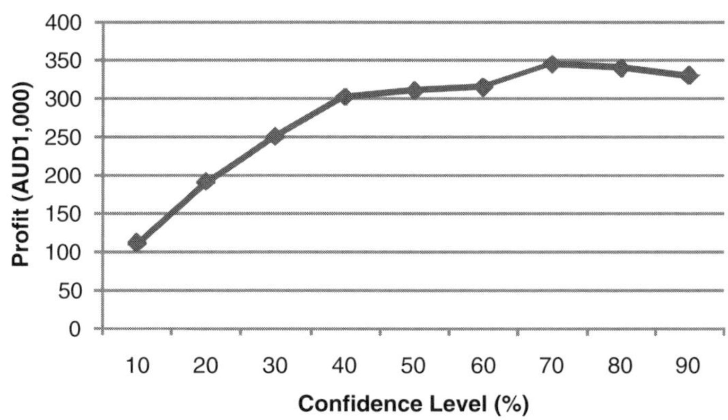

Figure 14.10 Profits of generator G1 by setting different confidence levels

However, bidding a low price may decrease the MCP and thus decrease the generator's own profit, especially for the large GENCO with market power. In contrast, when α decreases from 70 % to 10 %, profit decreases significantly.

This phenomenon implies that, although a small α can increase the bidding price, the risk of not being dispatched is also significantly increased. The profit therefore drops dramatically. In summary, a neutral risk level should usually be set in order to obtain the optimal profit. Choosing an α which is either too large or too small can degrade the performance of the proposed method.

The bidding prices and productions for generator G1 in four time intervals are listed in Tables 14.1. Notice that the thermal unit in the case study has technical constraints. In order to make profits from high price intervals, they will lose some profit in the intervals before the high price, especially when the price is fluctuating.

Table 14.1 Scenarios for a risk seeker in four intervals

Time →	6 a.m.		10 a.m.		4 p.m.		8 p.m.	
Confidence (%)↓	Price ($/MWh)	Production (MW)	Price ($/MWh)	Production (MW)	Price ($/MWh)	Production (MW)	Price ($/MWh)	Production (MW)
10	25.33	263	39.67	420	36.68	420	22.57	420
20	24.45	186	39.14	420	36.35	420	19.97	420
30	23.81	186	38.76	340	36.12	420	18.1	391
40	23.27	186	38.43	340	35.92	314	16.5	196
60	22.25	186	37.82	340	35.54	186	13.5	186
70	21.71	186	37.5	340	35.34	289	11.9	0
80	21.07	186	37.11	237	35.1	186	10.03	0
90	20.19	186	36.58	186	34.78	186	7.43	186

Another result found in the study is that the decision whether to take risk is sometimes determined by the market share of the GENCO. In a fully competitive market, participants intend to bid at their marginal costs. Unfortunately, the electricity market is more like an oligopoly market in which only a few GENCOs compete. Some participants will act as price-setters in market competition, while other smaller participants will act as price takers. For the price-setters, therefore, it will be more feasible to take the risk of making more profits since they could bid a high price and still be dispatched. In contrast, price-takers should choose to avoid the risk, because if they bid a high price which is higher than the real MCP, they will probably lose the opportunity to be dispatched and lose their profits.

14.5 Conclusions

Designing the optimal bidding strategy is a challenging task for generators in the deregulated electricity market. Existing methods usually assume that the rivals' information, such as cost information, bidding parameters and benefit functions, is known. However, in reality, most of the above information is confidential. In this chapter the optimal bidding problem with incomplete information is studied and a novel approach is proposed to solve it. Statistical methods are used to estimate rivals' bidding productions. Furthermore, the support vector machine is applied to approximate the relationship between bidding productions and MCP. Based on the estimated rivals' bidding productions and MCP, the optimal bidding problem is transformed into a stochastic optimization problem, which is solved by the Monte Carlo simulation and differential evolution.

Experiments based on real market data demonstrate that the proposed method can well estimate rivals' behavior. Moreover, significantly improved profits are obtained with the proposed method in the experiment. The proposed method is therefore effective in designing optimal bidding strategies without knowledge of rivals' confidential information. This research provides a highly useful tool for generating companies in the deregulated market environment.

References

[1] Shahidehpour, M., Yamin, H. and Li, Z. (2002) *Market Operations in Electric Power Systems: Forecasting, Scheduling, and Risk Management*, IEEE, Wiley-Interscience, New York.

[2] Ilic, M., Galiana, F. and Fink, L. (eds) (1998) *Power Systems Restructuring: Engineering and Economics*, Kluwer Academic, Norwell, MA.

[3] Li, C.A., Svoboda, A.J., Guan, X. and Singh, H. (1999) Revenue adequate bidding strategies in competitive electricity markets. *IEEE Transactions on Power Systems*, **14**(2), 492–497.

[4] David, A.K. and Fushuan, W. (2000) Strategic bidding in competitive electricity markets: a literature survey. IEEE Power Engineering Society Summer Meeting, Seattle, WA, July 16–20, 2000, vol. 4, pp. 2168–2173.

[5] Gountis, V.P. and Bakirtzis, A.G. (2004) Bidding strategies for electricity producers in a competitive electricity marketplace. *IEEE Transactions on Power Systems*, **19**(1), 356–365.

[6] Niu, H., Baldick, R. and Zhu, G. (2005) Supply function equilibrium bidding strategies with fixed forward contracts. *IEEE Transactions on Power Systems*, **20**(4), 1859–1867.

[7] He, Y. and Song, Y.H. (2002) Integrated bidding strategies by optimal response to probabilistic locational marginal prices. *IEE Proceedings C: Generation, Transmission and Distribution*, **149**(6), 633–639.

[8] Mas-Colell, A., Whinston, M.D. and Green, J.R. (1995) *Microeconomic Theory*, Oxford University Press, New York.

[9] Conejo, A.J., Nogales, F.J. and Arroyo, J.M. (2002) Price-taker bidding strategy under price uncertainty. *IEEE Transactions on Power Systems*, **17**(4), 1081–1088.

[10] Rodriguez, C.P. and Anders, G.J. (2004) Bidding strategy design for different types of electric power market participants. *IEEE Transactions on Power Systems*, **19**(2), 964–971.

[11] Yin, X., Zhao, J.H. and Dong, Z.Y.(Oct. 30–Nov. 2, 2006) Optimal GENCO's Bidding Strategies under Price Uncertainty in Poolco Electricity Market. 7th IET Int. Conf. Advances Power System Control Operation Management, Hong Kong.

[12] Chen, X., He, Y., Song, Y.H. *et al.* (2004) Study of impacts of physical contracts and financial contracts on bidding strategies of GENCOs. *International Journal of Electrical Power & Energy Systems*, **26**(9), 715–723.

[13] Tengshun, P. and Tomsovic, K. (2003) Congestion influence on bidding strategies in an electricity market. *IEEE Transactions on Power Systems*, **18**(3), 1054–1061.

[14] David, A.K. (1993) Competitive bidding in electricity supply. *IEE Proceedings C: Generation, Transmission and Distribution*, **140**(5), 421–426.

[15] Wen, F.S. and David, A.K. (2002a) Coordination of bidding strategies in day-ahead energy and spinning reserve markets. *International Journal of Electrical Power & Energy Systems*, **24**(4), 251–261.

[16] Wen, F.S. and David, A.K. (2002) Optimally co-ordinated bidding strategies in energy and ancillary service markets. *IEE Proceedings C: Generation, Transmission and Distribution*, **149**(3), 331–338.

[17] Li, T. and Shahidehpour, M. (2005) Strategic bidding of transmission-constrained GENCOs with incomplete information. *IEEE Transactions on Power Systems*, **20**(1), 437–447.

[18] Zhang, D.Y., Wang, Y.J. and Luh, P.B. (2000) Optimization based bidding strategies in the deregulated market. *IEEE Transactions on Power Systems*, **15**(3), 981–986.

[19] Kian, A.R. and Cruz, J.B. Jr. (2002) Nash strategies for load serving entities in dynamic energy multi-markets. 35th Annual Hawaii Int. Conf. System Sciences, Big Island, HI, Jan. 7–10, pp. 730–738.

[20] Richter, C.W. Jr. and Sheble, G.B. (1998) Genetic algorithm evolution of utility bidding strategies for the competitive marketplace. *IEEE Transactions on Power Systems*, **13**(1), 256–261.

[21] Kian, A.R., Cruz, J.B. Jr. and Thomas, R.J. (2005) Bidding strategies in oligopolistic dynamic electricity double-sided auctions. *IEEE Transactions on Power Systems*, **20**(1), 50–58.

[22] Mielczarski, W., Michalik, G. and Widjaja, M. (1999) Bidding strategies in electricity markets. IEEE Int. Conf. Power Industry Computer Applications, Santa Clara, CA, May 16–21, 1999, pp. 71–76.

[23] Vapnik, V.N. (2000) *The Nature of Statistical Learning Theory*, 2nd edn. Springer, New York.

[24] Xu, Z., Dong, Z.Y. and Liu, W.Q. (2003) Short-term electricity price forecasting using wavelet and SVM techniques. 3rd Int. DCDIS Conf. Engineering Applications Computational Algorithms, Guelph, Ontario, Canada.

[25] Fishman, G.S. (1996) *Monte Carlo: Concepts, Algorithms, and Applications*. Springer, New York.

15

3D Tracking of License Plates in Video Sequences

İlhan Kubilay Yalçın[1], Muhittin Gökmen[2]

[1] *TUBITAK MRC, Information Technologies Institute, Kocaeli, Turkey*

[2] *Computer Engineering Department, Istanbul Technical University, Istanbul, Turkey*

15.1 Introduction

In this chapter, we present a 3D visual tracking application utilizing differential evolution enhanced particle filtering. The aim of visual tracking is to improve the performance of an automated vehicle identification (AVI) system. AVI is an important research issue which continues to attract attention in the machine vision community. Among its potential applications include automatic barrier systems, automatic payment of parking or highway toll fees, automatic detection of stolen vehicles, and automatic calculation of traffic density.

The information carried on license plates enables us to identify a vehicle and its owner. A suitable and promising solution to vehicle identification is camera-based visual license plate recognition (VLPR). This approach is applicable because it does not require vehicles to carry additional equipment such as special RF transmitters.

VLPR systems can be separated into three major parts: license plate detection (segmentation), license plate tracking, and license plate recognition. License plate detection involves a search over the given image for that part which has license plate pattern characteristics. However, locating the license plate region in each sequential video frame is an expensive process. Tracking the license plate in consecutive video frames is a more affordable alternate.

The process of tracking an object in a video sequence involves identifying its location while the object or the camera itself changes position. There are many approaches depending on the tracked object, degrees of freedom and the setup utilized. Two-dimensional tracking aims to locate the image regions belonging to an object on the 2D image plane. The 2D tracking depends on 2D transformations which model the desired motion. Generally an adaptive model is needed. 2D tracking can provide an object's position by its centroid and scale or by a 2D affine

Differential Evolution Anyong Qing

transformation [1–3]. Considering deformations, more sophisticated models can be utilized such as splines [4], deformable templates [5] or articulated models [6]. Although these models try to localize the object in 2D, they never discover the real 3D position of the object. On the other hand, 3D tracking tries to identify all six degrees of freedom defining the camera coordinate system relative to the real world and the 3D location and orientation of the object.

In this extensive work, we will focus on online model-based 3D tracking using a single camera. Our tracking application is restricted to a single camera and rigid objects.

The primary objective of this work is tracking license plate in consecutive video frames utilizing a 3D rigid shape model, 3D dynamic motion model and stochastic tracking algorithms. Locating a license plate in 3D enables us to predict the license plate location more precisely, because the drift of the license plate in video frames is the result of a motion in 3D space. Utilizing stochastic tracking algorithms can yield better estimation of a multimodal probability distribution, thus adapts the system to the multimodal nature of the image segmentation problem.

The 3D tracking problem has been the subject of research over a long period of time. It is an optimization problem and can be treated in a probabilistic framework for the state estimation of a dynamic system. Sequential Monte Carlo methods such as particle filtering [7] were introduced to solve non-linear dynamic systems. Prominent particle filtering techniques include the condensation algorithm [7–10], genetic condensation algorithm [11], and hybrid sampling [12].

Inspired by the optimization mechanism in differential evolution (DE) and condensation algorithm, we propose a new method, the differential evolution Markov Chain (DEMC) particle filter, to tackle the limitations of the condensation algorithm. DEMC is more effective than standard re-sampling techniques and avoids sample degeneracy and impoverishment without violating the multimodal nature of condensation algorithm.

15.2 3D License Plate Tracking Acquisition Setup

The license plate tracking video acquisition setup is shown in Figure 15.1. The acquisition system consists of a monochrome camera, mounted on a tripod, facing the road ahead at a slight inclination. The camera has fixed pre-mounted optics with focal length of 28–280 mm. Images are digitized at a resolution of 704 × 288 pixels.

We utilize a pinhole camera model [13–15], also referred to as perspective projection. This model connects a point in the 3D space to its perspective projection on the camera plane.

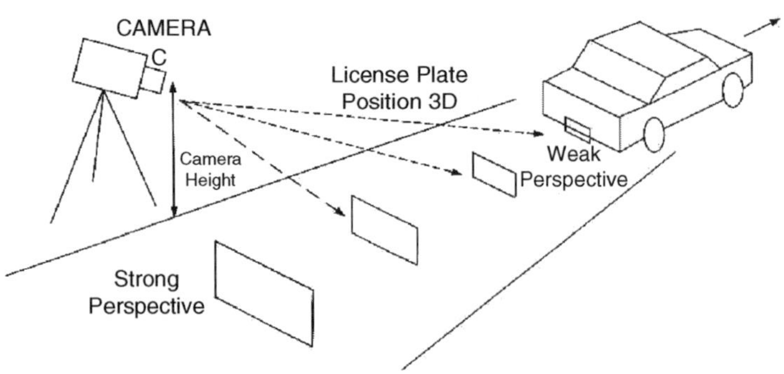

Figure 15.1 VLPR outdoor system

Using homogeneous coordinates in the projective space we get

$$\mathbf{c}^i = \begin{bmatrix} x \\ y \\ s \end{bmatrix} = \begin{bmatrix} \alpha_u & 0 & u_0 & 0 \\ 0 & \alpha_v & v_0 & 0 \\ 0 & 0 & 1 & 0 \end{bmatrix} \cdot \begin{bmatrix} \mathbf{R} & \mathbf{T} \\ \mathbf{0} & \mathbf{1} \end{bmatrix} \cdot \mathbf{c}^s, \tag{15.1}$$

where $\mathbf{c}^i$ represents the homogeneous coordinates of the projection of point P on the image, $\mathbf{c}^s = [X, Y, Z, 1]$ are the homogeneous coordinates of point P in the scene reference, $\alpha_u = f/k_u$, $\alpha_v = f/k_v$, f is the focal length, k_u and k_v are the pixel width and height, u_0 and v_0 are the image coordinates of the projection center, $\mathbf{R}$ is a 3×3 rotation matrix, and $\mathbf{T}$ is a 3×1 translation vector.

A rotation matrix $\mathbf{R}$ can always be written as the product of three matrices representing rotations around the X, Y, and Z axes. Taking Euler angles α, β, γ as rotation angles around X, Y, and Z axis yields

$$\mathbf{R} = \mathbf{R}_\alpha \cdot \mathbf{R}_\beta \cdot \mathbf{R}_\gamma = \begin{bmatrix} 1 & 0 & 0 \\ 0 & \cos\alpha & -\sin\alpha \\ 0 & \sin\alpha & \cos\alpha \end{bmatrix} \cdot \begin{bmatrix} \cos\beta & 0 & \sin\beta \\ 0 & 1 & 0 \\ -\sin\beta & 0 & \cos\beta \end{bmatrix} \cdot \begin{bmatrix} \cos\gamma & -\sin\gamma & 0 \\ \sin\gamma & \cos\gamma & 0 \\ 0 & 0 & 1 \end{bmatrix}, \tag{15.2}$$

$$\mathbf{T} = [\Delta x \quad \Delta y \quad \Delta z]^T, \tag{15.3}$$

where Δx, Δy, and Δz are the spatial drift in X, Y and Z cartesian coordinates.

In most 3D tracking methods, the internal parameters or intrinsic camera parameters, f, k_u, k_v, u_0 and v_0 are assumed to be fixed and known through camera calibration [16]. In our camera setup, the camera is mounted on a tripod, facing the road ahead at a slight inclination. The camera coordinate system is assumed to be fixed and known in advance. Therefore, estimation of external camera parameters such as rotation matrix and translation vector is unnecessary. However, in the presence of camera ego-motion, external camera parameters have to be estimated continuously.

15.3 Statistical Bayesian Estimation and Particle Filtering

The 3D license plate tracking problem is to find the matrices $\mathbf{R}$ and $\mathbf{T}$ so that the image features of an input image correspond to the perspective projection of the rotated and shifted 3D object model. The rotation and translation are performed in the world coordinate system.

The state (3D pose) of a license plate can be represented by a n_s-dimensional state vector $\mathbf{s}_k$. The state vector at time step k can be represented by $[X, Y, Z, \alpha, \beta, \gamma]^T$. The system is in one of the states $\mathbf{s}_k^n$ with probability $p(\mathbf{s}_k^n)$.

The state evolution with respect to time can be modeled by a discrete-time nonlinear dynamic system

$$\mathbf{s}_k = f(\mathbf{s}_{k-1}, \mathbf{w}_{k-1}), \tag{15.4}$$

where $\mathbf{w}_k$ is an n_w-dimensional independent and identically distributed process noise sequence.

The objective of tracking is to recursively estimate $\mathbf{s}_k$ from n_b-dimensional measurements

$$\mathbf{b}_k = \mathbf{h}_k(\mathbf{s}_k, \mathbf{n}_k), \tag{15.5}$$

where $\mathbf{n}_k$ is an n_n-dimensional independent and identically distributed measurement noise sequence. In particular, we seek filtered estimates of $\mathbf{s}_k$ based on the set of all available measurements $\mathbf{b}_{lk} = \{\mathbf{b}_i,\ 1 \le i \le k\}$ up to time k.

From a Bayesian perspective, the tracking problem recursively calculates some degree of belief in the state $\mathbf{s}_k$ at time k, taking different values, given the data $\mathbf{b}_{lk}$. Thus, it is required to construct the probability density function (pdf) $p(\mathbf{s}_k|\mathbf{b}_{lk})$. It is assumed that the initial pdf, $p(\mathbf{s}_0)$, of the state vector, also known as the prior, is available. Then, in principle, the pdf may be obtained recursively in two stages: prediction and update.

Suppose that the required pdf at time $k-1$ is available. The prediction stage involves using the system model to obtain the prior pdf of the state at time k via the Chapman–Kolmogorov equation

$$p(\mathbf{s}_k|\mathbf{b}_{lk}) = \int p(\mathbf{s}_k|\mathbf{s}_{k-1})p(\mathbf{s}_{k-1}|\mathbf{b}_{l(k-1)})d\mathbf{s}_{k-1}. \tag{15.6}$$

At time step k, a measurement $\mathbf{b}_k$ becomes available, and this may be used to update the prior (update stage) via Bayes' rule,

$$p(\mathbf{s}_k|\mathbf{b}_{lk}) = p(\mathbf{b}_k|\mathbf{s}_k)p(\mathbf{s}_k|\mathbf{b}_{l(k-1)})/p(\mathbf{b}_k|\mathbf{b}_{l(k-1)}). \tag{15.7}$$

These recurrence relations form the basis for the optimal Bayesian solution. This recursive propagation of the posterior density is only a conceptual solution. In general, it cannot be determined analytically. Instead, the target probability density $p(\mathbf{s})$ is approximated by the importance density $q(\mathbf{s})$.

The optimum importance density $q(\mathbf{s})$ is given in the terms of target density $p(\mathbf{s})$ as

$$q(\mathbf{s}_k|\mathbf{s}^i_{k-1}, \mathbf{b}_k)_{\text{optimum}} = p(\mathbf{s}_k|\mathbf{s}^i_{k-1}, \mathbf{b}_k), \tag{15.8}$$

which can be simplified as

$$p(\mathbf{s}_k|\mathbf{s}^i_{k-1}, \mathbf{b}_k) = p(\mathbf{b}_k|\mathbf{s}_k)p(\mathbf{s}_k|\mathbf{s}^i_{k-1})/p(\mathbf{b}_k|\mathbf{s}^i_{k-1}). \tag{15.9}$$

The importance density $q(\mathbf{s})$ is obtained by the DEMC particle filter [17,18] as shown in Figure 15.2.

Given $\mathbf{s}^i_{k-1}$, we know how many times it has been resampled in the previous time step, and we can perform DEMC sampling in order to generate the required particles. We will use $p(\mathbf{s}_k|\mathbf{s}^i_{k-1})$ for generating initial values to the DEMC chains. This initial samples are generated by $\mathbf{s}^n_k = f(\mathbf{s}^n_{k-1}) + N(0,\sigma)$. The importance density is reduced to $p(\mathbf{b}_k|\mathbf{s}_k)p(\mathbf{s}_k|\mathbf{s}^i_{k-1})$ because the denominator is the same for all proposals generated. The proposals are generated and rejected or accepted according to the DEMC algorithm. After the DEMC sampling step, the generated samples will have higher likelihood values, significantly reducing the sample degeneracy.

Initialization : Generate samples $\mathbf{s}_0^n$ and π_0^n

Iterate for $k = 1, 2, \ldots$ and at time step k, propagate the probability density as follows:

NSS_n is the number of samples that needs to be sampled from sample $\mathbf{s}_{k-1}^n$ assigned by the previous re-sampling step. Draw samples $\mathbf{s}_k^{mn}$ given $\mathbf{s}_{k-1}^n$ according to $p(\mathbf{s}_k|\mathbf{s}_{k-1})$ where $m = 1,\ldots,3\times NSS_n$ (which guaranties that the minimum number of propagated samples from each $\mathbf{s}_{k-1}^n$ is greater or equal to 3). $\mathbf{s}_k^{mn}$ are the samples that are propagated from $\mathbf{s}_{k-1}^n$ according to $\mathbf{s}_k^{mn} = f\left(\mathbf{s}_{k-1}^n\right) + \mathbf{v}_m$ where $\mathbf{v}_m : N(0, \sigma)$

Iterate for $i = 1, \ldots, I$ where I is the number of DEMC iterations. For each m, n,

Calculate new weights, $\pi_t^{mn} = p\left(\mathbf{b}_k \middle| \mathbf{s}_k^{mn}\right) p\left(\mathbf{s}_k^{mn} \middle| \mathbf{s}_{k-1}^n\right)$, where $p\left(\mathbf{b}_k \middle| \mathbf{s}_k^{mn}\right)$ is the likelihood value of the measurement $\mathbf{b}_k$ given sample $\mathbf{s}_k^{mn}$

Generate a perturbed sample $\mathbf{s}_k^{pn} = \mathbf{s}_k^{mn} + a\left(\mathbf{s}_k^{Rbest,n} - \mathbf{s}_k^{mn}\right) + c\left(_k^{R1,n} - \mathbf{s}_k^{R2,n}\right) + U(-b,b)$ where $\mathbf{s}_k^{Rbest,n}$ is the sample with greatest likelihood in set $\mathbf{s}_k^{mn}$.

Assign a weighted temperature $T = TempCoef \cdot p\left(\mathbf{s}_k^{pn} \middle| \mathbf{s}_{k-1}^n\right)$ to reduce the probability of generating samples that are less probable according to the time propagation

If $log(\pi_k^{mn}) > T \times log[U(0,1)]$, $\mathbf{s}_k^{mn}$ is a new generated sample from $p\left(\mathbf{s}_k \middle| \mathbf{s}_{k-1}^n\right)$ and $\mathbf{s}_k^{mn} = \mathbf{s}_k^{pn}$.

For each sample $\mathbf{s}_{k-1}^n$, calculate likelihood value $\pi_k^n = p\left(\mathbf{b}_k \middle| \mathbf{s}_{k-1}^n\right) = \sum_{m=1}^{M} p\left(\mathbf{b}_k \middle| \mathbf{s}_k^{mn}\right) p\left(\mathbf{s}_k^{mn} \middle| \mathbf{s}_{k-1}^n\right)$.

All samples $\mathbf{s}_k^{mn}$ initially drawn from $p\left(\mathbf{s}_k^{mn} \middle| \mathbf{s}_{k-1}^n\right)$ have the same likelihood π_k^n

Calculate the estimation of $\mathbf{s}_k$, by $\hat{\mathbf{s}}_k = p\left(\mathbf{b}_k \middle| \mathbf{s}_{k-1}^n\right) = \frac{1}{NSS_{best}} \sum_{m=1}^{M} \mathbf{s}_k^{m,best}$. $\mathbf{s}_k^{m,best}$ are the propagated samples from $\mathbf{s}_{k-1}^n$ with the highest π_k^n.

Combine all $\mathbf{s}_k^{mn}$ into $\mathbf{s}_k^n$ and store the new weights as π_k^n and normalize so that $\sum_n \pi_k^n = 1$.

Resample N samples from $\mathbf{s}_k^n$ according to π_k^n and assign the calculated re-sampling as NSS_n for each sample

Figure 15.2 DEMC particle filter

The mechanism of DEMC sampling also increases the variance of the samples generated, which diminishes the sample impoverishment. We have slightly modified the proposal generation of the DEMC algorithm. We have also utilized the best sample in the chain in order to generate new proposals by using an analogy to the DE/rand-to-best/1/exp version of DE algorithm. We aim to reduce the number of iterations for DEMC sampling by shifting the proposals to the best sample.

15.4 3D License Plate Tracking Using DEMC Particle Filter

The likelihood values $\pi_k^n = p(\mathbf{b}_k^n|\mathbf{s}_k^n)$ are obtained by template matching. The matching is performed against the template that is extracted from the initial frame. A sample video frame is shown in Figure 15.3. The template extracted from this frame is shown in Figure 15.4.

There is no update on this template during the tracking. The template is normalized in order to obtain an image with zero mean and unit variance. Template matching is then performed as follows:

$$\text{likelihood} = \sum_{n=0}^{T_x}\sum_{m=0}^{T_y}[\hat{I}_{xy}(n,m)-T(n,m)]^2 \tag{15.10}$$

where T_x and T_y are the template dimensions, $T(n, m)$ is the zero-mean, unit-variance template and $\hat{I}_{xy}(n, m)$ is the extracted image region around (x, y), which is processed to have zero mean and unit variance. Because of the 3D nature of the system state, the extracted image region $\hat{I}_{xy}(n, m)$ also needs to be reverse warped according to the sample's Euler angles.

In Figure 15.5, likelihood values are calculated for a region around the true license plate coordinates (in a 100×50 pixels window on the video frame as shown in Figure 15.3). For the sake of simplicity, in this example only the first two values of the state vector (X, Y) are changed, thus no warping and scaling are applied.

The warping, shifting and scaling need to be performed according to the sample state values, namely the 3D coordinates X, Y, Z and the Euler angles α, β, γ. 3D license plate corners are first projected to the image plane by the camera model and each 2D corner coordinate is fed to the warping function. The perspective transform is applied to the given 2D corner coordinates. The result of the warping operation is shown in Figure 15.6.

Figure 15.3 Video frame sample

Figure 15.4 Template

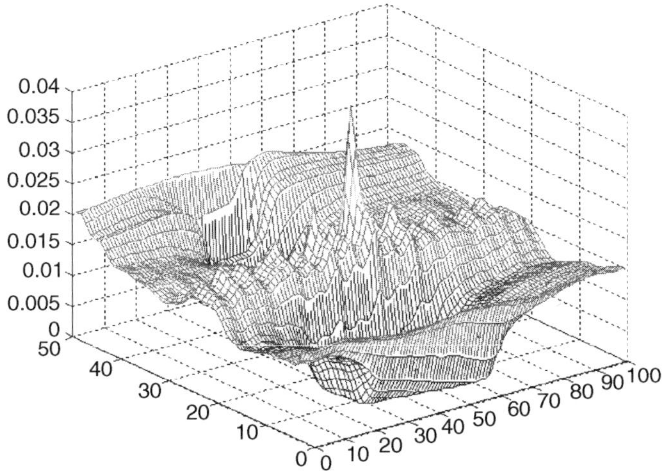

Figure 15.5 Likelihood function plot

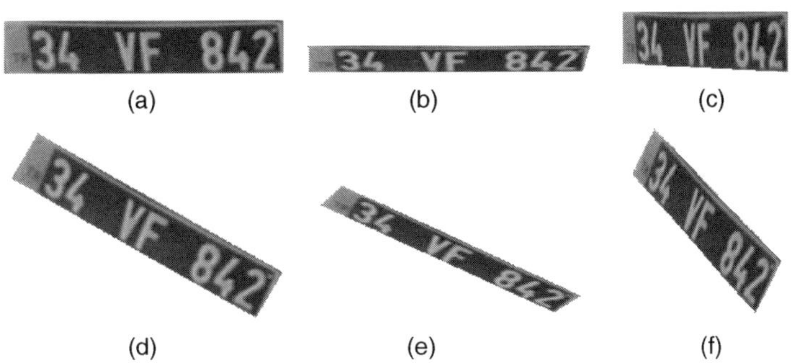

Figure 15.6 3D projection of the license plate at various Euler angles (a) $\alpha = 0°$, $\beta = 0°$, $\gamma = 0°$; (b) $\alpha = 60°$, $\beta = 0°$, $\gamma = 0°$; (c) $\alpha = 0°$, $\beta = 60°$, $\gamma = 0°$; (d) $\alpha = 0°$, $\beta = 0°$, $\gamma = 30°$; (e) $\alpha = 60°$, $\beta = 0°$, $\gamma = 30°$; (f) $\alpha = 0°$, $\beta = 60°$, $\gamma = 30°$

15.5 Comparison

We compare the license plate tracking performance of four algorithms: the auxiliary particle filter, condensation algorithm, genetic condensation algorithm and DEMC particle filter. These algorithms are evaluated by tracking license plates on 12 different test video sequences. The tests are performed with a calibrated static camera. The camera calibration method in [16] is applied which requires a couple of calibration images, in which the calibration pattern is clearly seen. An 8×4 chessboard pattern is used as the calibration pattern is in our application.

The ground truth for each frame is determined manually. A simple GUI program helps select the license plate corners from video sequences. The 12 test videos include 108 vehicles. Each vehicle is tracked a maximum of 25 frames after the initialization frame. In total, 2205 frames of tracking are used in the evaluation. The filters are utilized with the same number of particles and with the same number of likelihood calculations. Considering the number of particles N, the condensation algorithm has complexity $O(N)$, the auxiliary particle filter has $O(2N)$, the DEMC particle filter has $O(9N)$ (in the case of two DEMC iterations), and the genetic condensation has $O(4N)$.

The initial speed for all the filters at the initialization step is given as 20 cm per frame, which is 18 km/h (at 25 fps). The speeds of vehicles in the test sequences are generally between 10 and 50 km/h. These speeds are very common in urban areas. If the tracking is performed in highways, the configuration for initial speed, process noise variance and so on should be tuned

Figure 15.7 Test video sequence 1: (a) calibration image; (b) frame 12; (c) frame 44; (d) frame 116; (e) frame 156; (f) frame 252; (g) frame 376; (h) frame 476

Table 15.1 Average error per corner for the same computation load

Algorithm	Number of particles	Average error per corner
Condensation	360	4.39
Auxiliary particle filter	180	17.42
Genetic condensation	90	4.49
DEMC particle filter	40	3.92

Table 15.2 Average error per corner for the same number of particles

Algorithm	Number of particles	Average error per corner
Condensation	360	4.39
Auxiliary particle filter	360	8.34
genetic condensation	360	2.55
DEMC particle filter	360	2.69

accordingly. The videos are captured by a Sony Handy cam, which can capture 704×576 at the frame rate of 25 fps. This video is interleaved; therefore, we needed to discard even rows of the image, resulting in a 704×288 resolution. The tracking is performed on grayscale images. Color information is discarded in tracking, which makes the approach also applicable to IR cameras. Sample frames from the first video are presented in Figure 15.7.

Total corner displacement error is given as pixels per corner. The results obtained for the 12 test videos are summarized in Tables 15.1 and 15.2.

For the same computation load, the DEMC particle filter and genetic condensation algorithm provide a significant improvement over standard condensation algorithm. Besides, the DEMC particle filter yields less corner displacement error than the genetic condensation algorithm especially when the number of particles is decreased. It is remarkable that the DEMC particle filter provides this improvement without spoiling the multimodal nature of the particle filtering. On the other hand, the genetic condensation algorithm's behavior can decrease the multimodal nature because of its selective evolution step over the whole sample space. The DEMC particle filter divides the sample space into subspaces which are determined by the resampling stage. These subspaces are resampled individually, thus the multimodal nature is protected. In the case of an excessive number of particles and for the same number of particles, the DEMC particle filter performs slightly worse than the genetic condensation algorithm. This loss of performance is compensated by the robust nature of the DEMC particle filter which hardly ever fails to track the right license plate coordinates.

15.6 Conclusions

A new algorithm, the DEMC particle filter, is presented in order to overcome the drawbacks of the standard condensation algorithm. To the best of our knowledge, DEMC particle filtering is the first attempt to embed DEMC sampling into the resampling step of the particle filtering, in order to improve the final sample set, while preserving the multimodal nature of the 3D tracking problem, namely estimating the spatial position and 3D orientation of the license plate from sequential video frames. It is utilized for 3D license plate tracking in video sequences and is compared with known statistical filters, such as the auxiliary particle filter, condensation algorithm and genetic condensation algorithm. Better and robust tracking performance is achieved at higher computational complexity.

References

[1] Wren, C.R., Azarbayejani, A., Darrell, T. and Pentland, A.P. (1997) Pfinder: real-time tracking of the human body. *IEEE Transactions on Pattern Analysis and Machine Intelligence*, **19**(7), 780–785.

[2] Comaniciu, D., Ramesh, V. and Meer, P. (2003) Kernel-based object tracking. *IEEE Transactions on Pattern Analysis and Machine Intelligence*, **25**(5), 564–577.

[3] Jepson, A., Fleet, D.J. and El-Maraghi, T. (2003) Robust on-line appearance models for vision tracking. *IEEE Transactions on Pattern Analysis and Machine Intelligence*, **25**(10), 1296–1311.

[4] Blake, A. and Isard, M. (1998) *Active Contours*. Springer-Verlag, New York.

[5] Yuille, A., Hallinan, P. and Cohen, D. (1992) Feature extraction from faces using deformable templates. *International Journal of Computer Vision*, **8**(2), 99–111.

[6] Cham, T.J. and Rehg, J. (1999) A multiple hypothesis approach to figure tracking. IEEE Computer Society Conf. Computer Vision Pattern Recognition, Fort Collins, CO, June 23–25, 1999, vol. 2, pp. 239–245.

[7] Isard, M. and Blake, A. (1998) A mixed-state condensation tracker with automatic model-switching. 6th Int. Conf. Computer Vision, Bombay, India, Jan. 4–7, 1998, pp. 107–112.

[8] MacCormick, J. and Blake, A. (1998) A probabilistic contour discriminant for object localisation. 6th Int. Conf. Computer Vision, Bombay, India, Jan. 4–7, 1998, pp. 390–395.

[9] Li, B. and Chellappa, R. (2000) Simultaneous tracking and verification via sequential posterior estimation. IEEE Computer Society Conf. Computer Vision Pattern Recognition, Hilton Head Island, SC, June 13–15, 2000, vol. 2, pp. 110–117.

[10] Philomin, V., Duraiswami, R. and Davis, L. (2000) Pedestrian tracking from a moving vehicle. IEEE Intelligent Vehicles Symp., Dearborn, MI, Oct. 3–5, 2000, pp. 350–355.

[11] Ye, Z. and Liu, Z.Q. (2005) Genetic condensation for motion tracking. Int. Conf. Machine Learning Cybernetics, vol. 9, pp. 5542–5547.

[12] Le, L., Dai, X.T. and Hage, G. (2004) A particle filter without dynamics for robust 3D face tracking. Computer Vision Pattern Recognition Workshop, May 27–June 2, 2004, pp. 70–73.

[13] Gavrila, D. (1999) The visual analysis of human movement: a survey. *Computer Vision and Image Understanding*, **73**(1), 82–98.

[14] Chia, K., Cheok, A. and Prince, S. (2002) Online 6 DOF augmented reality registration from natural features. Int. Symp. Mixed Augmented Reality, pp. 305–313.

[15] Genc, Y., Riedel, S., Souvannavong, F. and Navab, N. (2002) Marker-less tracking for augmented reality: a learningbased approach. Int. Symp. Mixed Augmented Reality, pp. 295–304.

[16] Zhang, Z. (2000) A flexible new technique for camera calibration. *IEEE Transactions on Pattern Analysis and Machine Intelligence*, **22**(11), 1330–1334.

[17] Ter Braak, C.J.F. (2006) A Markov chain Monte Carlo version of the genetic algorithm differential evolution: easy Bayesian computing for real parameter spaces. *Statistical Computing*, **16**(3), 239–249.

[18] Strens, M., Bernhardt, M. and Everett, N. (July 8–12, 2002) Markov chain Monte Carlo sampling using direct search optimization. 19th Int. Conf. Machine Learning, Sydey, Australia, pp. 602–609.

16

Color Quantization

Gerald Schaefer[1], Lars Nolle[2]

[1] *School of Engineering and Applied Science, Aston University, Birmingham, United Kingdom*
[2] *School of Science and Technology, Nottingham Trent University, Nottingham, United Kingdom*

16.1 Introduction

The colors in digital images are typically represented by three 8-bit numbers representing the responses of the red, green, and blue sensors of the camera. Consequently, an image can have up to 2^{24}, that is, more than 16 million different colors. However, when displaying images on limited hardware such as mobile devices, but also for tasks such as image compression or image retrieval [1], it is desired to represent images with a limited number of different colors. Clearly, the choice of these colors is crucial as it determines the closeness of the resulting image to its original and hence the image quality. The process of finding such a palette or map of representative colors is known as color map generation or color quantization. It is known to constitute an NP-hard problem [2]. Many different algorithms have been introduced that aim to find a palette that allows for good image quality of the quantized image [2,3].

In this chapter we apply differential evolution (DE) to the color quantization problem. As DE is a black-box optimization algorithm, it does not require any domain-specific knowledge yet is usually able to provide a near-optimal solution. We evaluate the effectiveness of our approach by comparing it with several purpose-built color quantization algorithms [2,3]. The results obtained show that even without any domain-specific knowledge, DE is able to provide similar image quality as standard quantization algorithms. We also combine DE with a standard clustering algorithm, k-means, which is guaranteed to find a local minimum. The resulting hybrid algorithm is shown to further improve the effectiveness of the search and to outperform other color quantization techniques in terms of image quality of the quantized images.

Differential Evolution Anyong Qing

16.2 Differential Evolution Based Color Map Generation

The objective of color quantization is to minimize the total error introduced through the application of a color map in order to produce the image with the best possible quality. In our application, the color map C for an image I, a codebook of k color vectors, should be chosen so as to minimize the error function

$$E(C,I) = \frac{1}{\sum_{j=1}^{k} l_j} \sum_{i=1}^{k} \sum_{j=1}^{l_i} ||C_i - I_j|| + p(C,I), \tag{16.1}$$

where

$$p(C,I) = \sum_{i=1}^{k} \delta a_i, \quad a_i = \begin{cases} 1 & \text{if } l_i = 0, \\ 0 & \text{otherwise}, \end{cases} \tag{16.2}$$

where l_i is the number of pixels I_j represented by color C_i of the color map, $||\cdot||$ is the Euclidean distance in RGB space, and δ is a constant ($\delta = 10$ in our experiments).

The objective function $E(C, I)$ used is hence a combination of the mean Euclidean distance and a penalty function. The penalty function $p(C, I)$ was integrated in order to avoid unused palette colors by adding a constant penalty value to the error for each entry in the map that is not used in the resulting picture.

16.3 Hybrid Differential Evolution for Color Map Generation

Typically, there is some variation in error values within a population. This indicates that although DE is able to find fairly good solutions, that is, solutions from within the region around the global optimum, it rarely exploits that region completely. Therefore, in a second step, we combine DE with a standard k-means clustering algorithm [4] to provide a stacked hybrid optimization method. k-means clustering is guaranteed to converge towards the local clustering minimum by iteratively carrying out the following two steps:

- Each input vector should be mapped to its closest codeword by a nearest-neighbor search.
- The input vectors assigned in each class (i.e., for each codeword) are best represented by the centroid of the vectors in that class.

Hence, in this hybridized algorithm, DE is responsible for identifying the region in the search space that will contain the global optimum, while the k-means component will then descend into the minimum present in that region.

16.4 Experimental Results

In our experiments we have taken a set of three standard images commonly used in the color quantization literature (*Sailboat, Airplane, and Pool*) and applied both our pure DE-based and the hybrid DE color map generation algorithm with a target palette size of 16 colors.

We have also implemented three popular color quantization algorithms so that we can generate comparative results with our new algorithm. The algorithms we have tested are:

- the popularity algorithm (PA) [2], where, following a uniform quantization to 5 bits per channel, the *n* colors that are represented most often form the color palette;
- median cut (MC) quantization [2], an iterative algorithm that repeatedly splits (by a plane through the median point) color cells into sub-cells;
- Neuquant [3], where a one-dimensional self-organizing Kohonen neural network is applied to generate the color map.

For all algorithms, pixels in the quantized images were assigned to their nearest neighbors in the color palette to provide the best possible image quality. The results are shown in Figures 16.1–16.3 which give, for each image, the original, unquantized image together with the images color quantized by all five algorithms.

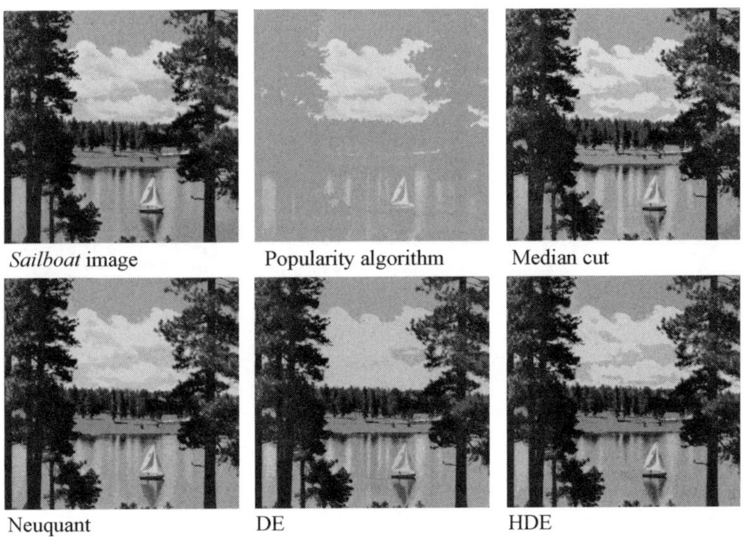

Figure 16.1 Original *Sailboat* image and quantized images

Take the pool image in Figure 16.2 as an example. It is clear that the popularity algorithm performs very poorly on this image as it assigns virtually all of the colors in the palette to green and achromatic colors. Median cut is better but still provides fairly poor color reproduction as most of the colors in the quantized image are fairly different from the original. The same holds true for the images produced by Neuquant. Here the most obvious artifact is the absence of an appropriate red color in the color palette. A far better result is achieved by DE, although the red is not very accurate either and the color of the cue is greenish instead of brown. Clearly the best image quality is achieved by applying hybrid DE. Although the color palette has only 16 entries, all colors of the original image are accurately presented, including the red ball and the color of the billiard cue.

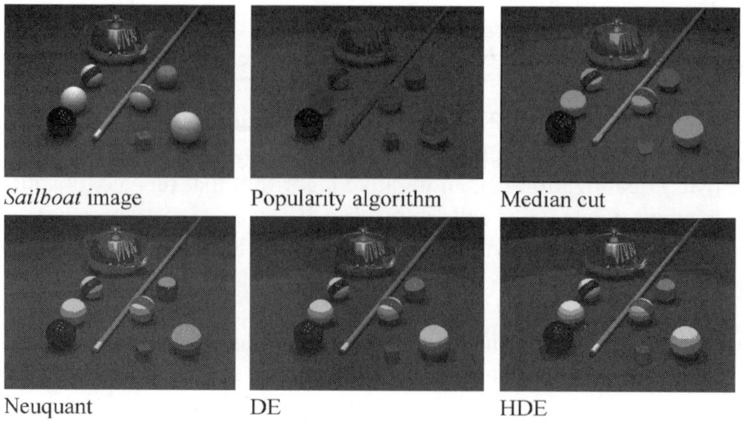

Figure 16.2 Original *Pool* image and quantized images

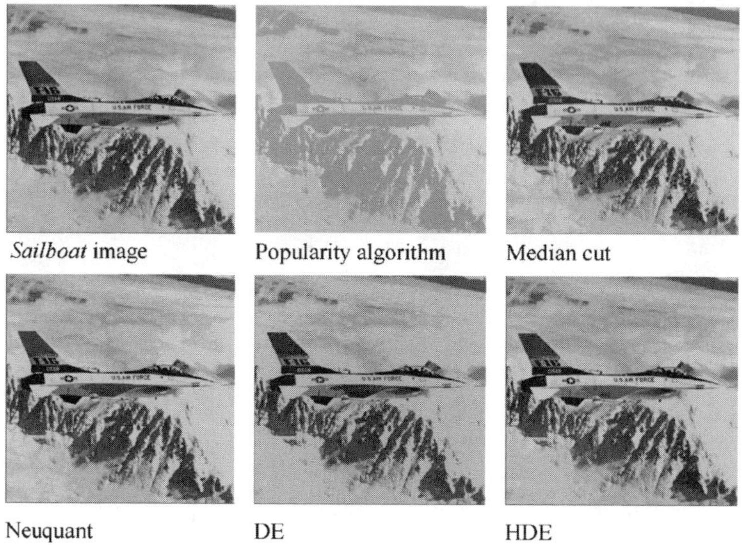

Figure 16.3 Original *Airplane* image and quantized images

To better visualize the difference between various algorithms, we also generate for each quantized image an error image that represents the difference between the original and the palletized image. To do so, we calculate the absolute error at each pixel, invert the resulting image and apply a gamma function to enhance the contrast. The results are shown in Figures 16.4–16.6 which again clearly demonstrate the superiority of the hybrid DE quantization algorithm.

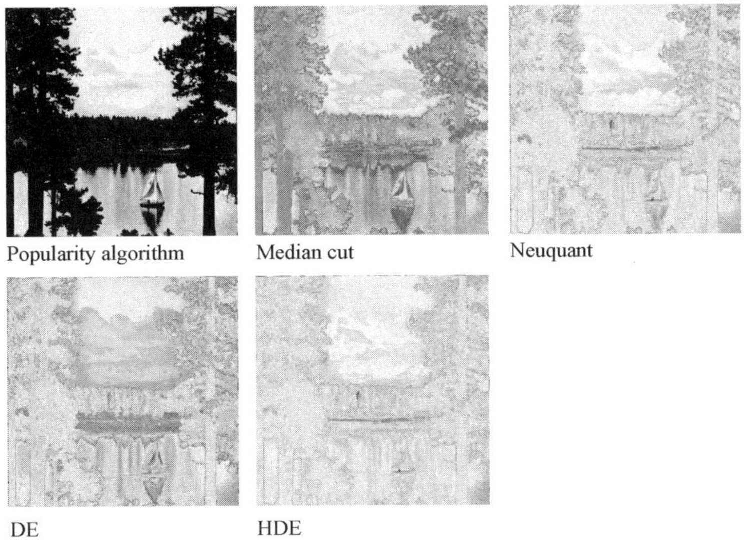

Figure 16.4 Error images of quantized *Sailboat* image

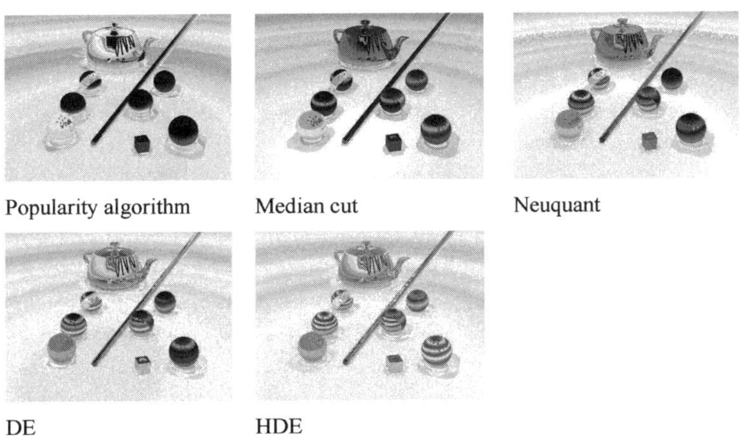

Figure 16.5 Error images of quantized *Pool* image

Apart from a visual comparison, we are of course also interested in an objective evaluation which for color quantization is typically expressed in terms of mean-squared error (MSE) and the peak-signal-to-noise-ratio (PSNR) which are defined as

$$\mathrm{MSE}(I_1, I_2) = \frac{1}{nm}\sum_{i=1}^{n}\sum_{j=1}^{m}\{[R_1(i,j) - R_2(i,j)]^2 + [G_1(i,j) - G_2(i,j)]^2 + [B_1(i,j) - B_2(i,j)]^2\} \tag{16.3}$$

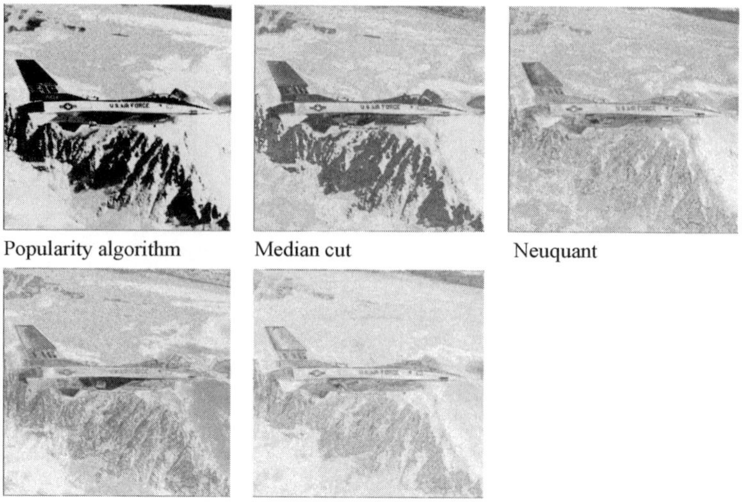

Popularity algorithm — Median cut — Neuquant — DE — HDE

Figure 16.6 Error images of quantized *Airplane* image

and

$$\mathrm{PSNR}(I_1, I_2) = 10\log_{10}\left[\frac{255^2}{MSE(I_1, I_2)}\right], \tag{16.4}$$

where $R(i, j)$, $G(i, j)$, and $B(i, j)$ are the red, green, and blue pixel values at location (i, j), and n and m are the dimensions of the images.

The results are listed in Table 16.1. Differential evolution clearly outperforms the popularity algorithm and median cut algorithm while doing slightly worse than Neuquant. In general, it can therefore be judged as providing image quality similar to specialized color quantization results.

On the other hand, the hybrid differential evolution was clearly able to further improve the performance of differential evolution alone and provides the best image quality for all images with a mean PSNR of 30.35 dB, an improvement of about 3 dB over Neuquant, the next best performing algorithm.

Table 16.1 Quantization results

	MSE				PSNR (dB)			
	Sailboat	Pool	Airplane	All	Sailboat	Pool	Airplane	All
PA	8707.41	669.87	1668.06	3681.78	8.73	19.87	15.91	14.84
MC	409.35	226.85	240.74	292.31	22.01	24.57	24.32	23.63
Neuquant	135.42	127.24	97.59	120.08	26.81	27.08	28.24	27.38
DE	105.13	44.74	45.57	65.15	24.64	28.61	27.46	26.9
HDE	105.13	44.74	45.57	65.15	27.91	31.62	31.52	30.35

16.5 Conclusions

In this chapter we have shown that a hybrid differential evolution algorithm can be successfully applied to the color map generation problem. A standard DE approach was combined with a k-means clustering technique where the DE part identifies the region of a good minimum and k-means descends to the local minimum. Experimental results obtained on a set of common test images demonstrate not only that this approach can be effectively employed but also that it clearly outperforms dedicated color quantization algorithms.

References

[1] Schaefer, G., Qiu, G. and Finlayson, G. (2000) Retrieval of palettized color images. *Proceedings of SPIE*, **3972**, 483–493.

[2] Heckbert, P.S. (1982) Color image quantization for frame buffer display. *ACM SIGGRAPH Computer Graphics*, **16** (3), 297–307.

[3] Dekker, A.H. (1994) Kohonen neural networks for optimal color quantization. *Network Computation in Neural Systems*, **5**(3), 351–367.

[4] Linde, Y., Buzo, A. and Gray, R.M. (1980) An algorithm for vector quantizer design. *IEEE Transactions on Communications*, **28**(1), 84–95.

Index